AF362059

Plant-Derived Natural Products and Their Biomedical Properties

Plant-Derived Natural Products and Their Biomedical Properties

Editors

Efstathia Papada
Charalampia Amerikanou
Marisa Colone

Basel • Beijing • Wuhan • Barcelona • Belgrade • Novi Sad • Cluj • Manchester

Editors

Efstathia Papada
UCL Division of Medicine
University College London
London
United Kingdom

Charalampia Amerikanou
Department of Dietetics and
Nutritional Science, School of
Health Science and Education
Harokopio University
Athens
Greece

Marisa Colone
National Center for Drug
Research and Evaluation
Istituto Superiore di Sanità
(Italian National Institute of
Health)
Rome
Italy

Editorial Office
MDPI
St. Alban-Anlage 66
4052 Basel, Switzerland

This is a reprint of articles from the Special Issue published online in the open access journal *Life* (ISSN 2075-1729) (available at: www.mdpi.com/journal/life/special_issues/Plant_Derived_Natural_Products).

For citation purposes, cite each article independently as indicated on the article page online and as indicated below:

Lastname, A.A.; Lastname, B.B. Article Title. *Journal Name* **Year**, *Volume Number*, Page Range.

ISBN 978-3-0365-9647-1 (Hbk)
ISBN 978-3-0365-9646-4 (PDF)
doi.org/10.3390/books978-3-0365-9646-4

Contents

About the Editors

Efstathia Papada

Dr Efstathia Papada is a Lecturer in Nutrition in the Division of Medicine, University College London, UK. She is also the Programme Lead for the MSc Dietetics (pre-registration) course in the same division. She is a Registered Dietitian both in Greece and in the UK. She graduated from the Department of Dietetics and Nutritional Science, Harokopio University, Athens. She then continued her studies with an MSc in Clinical and Public Health Nutrition at UCL. Following her passion for research, she began her PhD degree at Harokopio University, investigating the effects of a nutritional supplement on the clinical course of patients with inflammatory bowel disease.

After completing her doctoral studies, Dr Papada continued her postdoctoral research, and she was seconded to a renowned pharmaceutical company in Frankfurt to study the effects of a nutritional supplement on a mouse model of steatohepatitis using molecular biology protocols as an Experienced Researcher under the Horizon 2020-funded project, MAST4HEALTH. In the context of the same project, she was then seconded to a bioinformatics SME spin-off of the University of Valencia and to the Foundation for the Promotion of Health and Biomedical Research, where she investigated human gut microbiome alterations in chronic inflammatory diseases. She later worked as a postdoctoral researcher coordinating a clinical study on the effect of metabolic surgery on non-alcoholic steatohepatitis. Following the outbreak of the COVID-19 pandemic, she also worked on a project investigating the possible bi-directional relationship between COVID-19 and diabetes.

Her main research interests include the bioavailability and health effects of phytochemicals and natural dietary supplements; the effects of phytochemicals and dietary elements on chronic diseases involving oxidative stress, inflammation and gut microbiota alterations; gut health; and obesity management. She has 20 publications and an h-index of 10.

Charalampia Amerikanou

Dr Charalampia Amerikanou is a biologist with expertise in molecular nutrition and has many years of experience in investigating the effects of natural products on inflammation and oxidative stress. She has 37 publications in high-impact and indexed scientific journals and an h-index of 11. Her research interests include chronic inflammation, obesity and associated comorbidities, molecular mechanisms of inflammation and oxidative stress and microRNA profiling in response to dietary interventions. She graduated from the Biology Department of the Kapodistrian University of Athens, and she received her Master's degree in "Applications of Biology in Medicine". She obtained her PhD degree from the Department of Nutrition and Dietetics, Harokopio University of Athens, on a thesis entitled "Nutritional interventions in inflammatory diseases: investigating the molecular pathways". She has participated in several scientific projects related to inflammatory and oxidative stress processes as a Research Associate. Also, she was the project manager of a Marie Sklodowska-Curie Actions (MSCA) Research and Innovation Staff Exchange (RISE) program, entitled "Mastiha treatment for obese with NAFLD diagnosis". She has thorough knowledge of molecular biology and immunological techniques and applications, such as ELISA, PBMC, nucleic acid extraction and quantitation, PCR and others.

Marisa Colone

Since 1999, Marisa Colone has been working as a researcher at the Italian National Institute of Health. She is the author of about 60 publications in PubMed-indexed international journals. She is involved in studies of the relationship between the multidrug resistance and invasion potential of tumor cells. She has taken part in investigations via molecular biology and morphological, ultrastructural and immunocytochemical methods on the mechanisms of action at the subcellular level of several antitumoral and antimicrobic agents and on the mechanisms underlying the multidrug resistance phenomenon and on the chemosensitizing effects of natural compounds of different origin on cultured drug-resistant tumor cells and isolated microorganisms. She has studied new therapeutic strategies to treat male infertility, as well as nanomedicine, drug delivery and electron microscopy analysis.

In 2006, she received an award from AICC-CELLTOX to participate in the 3rd International Joint Meeting of AICC-CELLTOX at the 46th ETCS International Meeting (Verona, Italy). In 2007, she received the award for young researchers from the Italian Association of Cell Cultures (AICC), and in 2009, the award for best poster presented to the MC 2009 (Microscopy Conference 2009) in Graz (Austria). In 2009, she received a grant for participation in the Fifth National School Girse of EPR spectroscopy in Florence, (Italy). In 2011, she received an award for the best poster presented to the MC Microscopy Conference 2011 in Urbino, (Italy). Lastly, in 2011, she received an award for biological researchers from the Italian Society for Microscopical Sciences (SISM).

 *life*

Editorial

Plant-Derived Natural Products and Their Biomedical Properties: Recent Advances and Future Directions

Charalampia Amerikanou [1] and Efstathia Papada [2,*]

1 Department of Nutrition and Dietetics, School of Health Science and Education, Harokopio University, 17671 Athens, Greece; camer@hua.gr
2 Division of Medicine, University College London, London WC1E 6JF, UK
* Correspondence: e.papada@ucl.ac.uk

Citation: Amerikanou, C.; Papada, E. Plant-Derived Natural Products and Their Biomedical Properties: Recent Advances and Future Directions. *Life* **2023**, *13*, 2105. https://doi.org/10.3390/life13102105

Received: 16 October 2023
Accepted: 20 October 2023
Published: 23 October 2023

Nature has always been a source of inspiration and innovation to humanity. Throughout history, plants have provided us with not only sustenance but also a vast array of medicines, materials, and resources. In recent decades, there has been intense research interest in harnessing the biomedical potential of plant-derived natural products. To this end, the evolution of several methods for the extraction, isolation, and chemical characterization of plant-derived natural products has substantially contributed to the identification of their potential therapeutic properties. In this Special Issue, we explore recent advances in plant-derived natural products and their biomedical properties via intriguing in vitro and in vivo research articles and literature reviews.

Recently, Lantzouraki et al. showed that the extracts of two *Artemisia* species, *Artemisia arborescens* and *Artemisia inculta Delile*, from the Greek island of Crete exhibit an interesting profile of secondary metabolites [1]. More specifically, a high phenolic and terpenoid content and safe metal levels were reported in the aqueous–glycerolic extract of these two species. Furthermore, this extract offered a higher antioxidant potential compared to decoctions and methanolic extracts, suggesting its potential use not only in the food industry but also for dermocosmetic and pharmacological applications, given that glycerol increases the transdermal delivery of active substances [1]. On a similar note, the chemical compositions of different extracts of another Mediterranean plant have been evaluated. Ouahabi et al. determined the fatty acid content in the n-hexane extract and the phenolic content in the methanolic extract of the leaves of Moroccan *Pistacia lentiscus* (mastic tree), with the main component being linoleic acid in the first case and catechin in the second [2]. The extracts also exhibited significant antioxidant, antimicrobial, and antifungal activity, with in silico analyses revealing some interesting interactions between the identified components of the extracts and specific enzymes. An example interaction is between catechin and enoyl-acyl carrier protein reductase (FabI), a native *E. coli* enzyme, with potential inhibitory activity, supporting the potential use of mastic tree and its extracts in drug development [2]. *Euonymus laxiflorus* Champ. (ELC), a Vietnamese medicinal plant already known for its antioxidant and antidiabetic activity, has been investigated for its anti-acetylcholinesterase effect, an important mechanism for the management of Alzheimer's disease [3]. Interestingly, the ELC trunk bark extract featured a high phenolic and flavonoid content, and its in vitro anti-acetylcholinesterase activity was comparable to that of berberine chloride, a known acetylcholinesterase inhibitor. The compounds with the highest concentration included chlorogenic acid, epigallocatechin gallate, epicatechin, apigetrin, and quercetin, while docking-based simulations revealed their potential drug properties due to their significant binding energy to acetylcholinesterase [3].

Several animal studies have also emerged regarding the exploitation of the biomedical properties of plant-derived natural products. Recently, a rat model of diet-induced obesity was used to explore the effect of alcoholic and aqueous extracts of *Jatropha tanjorensis* (JT) and *Fraxinus micrantha* (FM) leaves on parameters related to metabolic syndrome [4]. A reduction in weight gain, food intake, serum glucose, and lipid profile was observed in rats treated with the extracts compared to the control high-fat-fed rats. Interestingly, the levels of antioxidant enzymes increased significantly after treatment with JT extract and showed similar levels to those achieved after treatment with orlistat, a common medicine used in obesity management [4]. Another study on rats investigated the gastroprotective effects of *Eurycoma longifolia* Jack (ELJ), a popular traditional herbal medicine of Southeast Asia [5]. The oral administration of a high dose of ELJ (1200 mg/kg) in carrageenan-induced rat paw edema showed a similar anti-antinociceptive and anti-inflammatory activity to aspirin (300 mg/kg) and reduced the rectal temperature of yeast-infected rats, suggesting central antipyretic effects. Finally, in the same animal model, treated with acidified ethanol for the development of gastric ulcers, different ELJ extract doses prevented gastric ulcer formation, demonstrating its gastroprotective properties [5].

Ocimum sanctum (Tulsi) is a well-known medicinal herb, commonly used in the Ayurvedic system for anxiety, cough, diarrhea, fever, vomiting, and other ailments [6]. Yadav et al. explored whether its extract administration to mice with a photothrombotic-ischemic-stroke-like injury could show an effect on their brain and the lipidomic profile of the brain and plasma. Untargeted lipidomic profiling with the Q-Exactive Mass Spectrometer revealed 77 upregulated and 33 downregulated lipid species in the brains of untreated lesioned mice compared to those treated with Tulsi. The most interesting finding was the increased presence of lysophosphatidylcholine (LPC) (16:1) in the brain of Tulsi-treated mice. This lysophospholipid is a major component of membranes, which implicates it in immune regulation after cerebral ischemia in rats, and it has also been found in Tulsi extract, suggesting a neuroprotective effect [6]. Another study evaluated the anticarcinogenic effects of hesperetin (HES), a naturally occurring flavone in *Citrus aurantium* L. (Rutaceae) fruit peel, alone and in combination with capecitabine (CAP), a chemotherapeutic drug, on 1,2 dimethylhydrazine (DMH)-induced colon carcinogenesis in Wistar rats [7]. The results showed that treatment with HES and/or CAP prevented histological cancerous changes in combination with a decrease in colon-Ki67 expression and serum carcinoembryonic antigen. Additionally, treatments with HES and/or CAP induced a significant reduction in serum lipid peroxides, and an increase in serum reduced glutathione, as well as the enhancement of colon-tissue superoxide dismutase, glutathione reductase, and glutathione-S-transferase activities. Interestingly, an increase in the mRNA expression of the anti-inflammatory IL-4, as well as the proapoptotic protein p53, in the colon tissues of the DMH-administered rats treated with HES and/or CAP was reported [7].

The therapeutic effects of aqueous extract (BCAE) and ethanolic extract (BCEE) obtained from the aerial parts of *Brocchia cinerea* (Delile) were evaluated by Agour and colleagues [8]. Firstly, the total polyphenol content (TPC), total flavonoid content (TF), and condensed tannin content (CT) were determined, and the results showed that BCEE had the highest content of polyphenols, flavonoids, and tannins. In vitro, DPPH, FRAP, and TAC were used to evaluate antioxidant efficacy, and BCEE demonstrated a strong antiradical activity against DPPH and a medium iron-reducing potential, while BCAE inhibited the growth of the antibiotic-resistant bacterium *P. aeruginosa*. The analgesic power was evaluated in vivo using the abdominal contortion model in mice, and the anti-inflammatory activity was assessed via the carrageenan-induced edema model in rats, while wound healing was evaluated in an experimental second-degree-burn model. BCAE exhibited significant pharmacological effects and analgesic efficacy and also contributed to the re-epithelialization of wounds [8]. Salamatullah investigated the chemical composition, as well as the antioxidant, anti-inflammatory, and analgesic properties, of a polyphenol-rich fraction from *Withania adpressa* Coss. ex Batt [9]. High-performance liquid chromatography (HPLC) was applied and revealed bioactive phenols including epicatechin, apigenin, luteolin, quercetin, caffeic acid, p-coumaric acid, and rosmarinic acid. This fraction showed anti-free-radical potency and a good total antioxidant capacity. In a rat model, the polyphenol-rich fraction strongly alleviated the inflammatory effect of carrageenan injected into the plantar fascia of rats and showed a good analgesic effect after heat stimulation.

Apart from the new research in vitro, and in vivo in animal models, several researchers around the world have attempted, through literature reviews, to gather and critically evaluate the current evidence on the biomedical properties of various plant-derived natural compounds. A recent review evaluated plagiochilins, a series of seco-aromadendrane-type sesquiterpenes isolated from leafy liverworts of the genus *Plagiochila* [10]. Currently, a total of 24 plagiochilins and many derivatives have been isolated and characterized, but there are limited studies on their pharmacological properties. Interestingly, plagiochilins A and C have demonstrated notable antiproliferative activity against cultured cancer cells by inhibiting the termination phase of cytokinesis. However, these compounds should be further evaluated for their antiproliferative potential [10]. Another recent review compiled the evidence regarding the biomedical properties of mogrol, a triterpene found in the fruits of the traditional Chinese medicinal plant *Siraitia grosvenorii*, and also highlighted the current research gaps and pointed out directions for future work [11]. Research has shown that mogrol could act as a therapeutic candidate with multiple pharmacological properties, including, but not limited to, neuroprotective, anticancer, anti-inflammatory, and antidiabetic effects. Although the molecular mechanisms behind these activities might involve several important targets, including AMPK, TNF-α, and NF-κB, there is still a gap in the literature concerning the exact mechanisms and targets [11].

Juglans regia Linn., a tree accounting for around 88% of the total walnut production globally, has been evaluated for the phytochemical content of its various components, including the bark, leaves, and fruit. Bhat and colleagues reviewed the scientific literature on the antimicrobial, antioxidant, antifungal, and anticancer properties of various compounds isolated from different solvents and different parts of *J. regia* and suggested that synthetic analogues and various extracts should be further assessed in a concentration-dependent manner to increase our understanding of these promising properties [12]. Last but not least, Davila and Papada critically evaluate the relevant evidence around the use of plant-derived natural products in the management of inflammatory bowel disease, with a specific focus on the clinical evidence so far for curcumin, Mastiha, *Boswellia serrata*, and *Artemisia absinthium*. As the results from human trials with relatively small sample sizes are very limited, the authors highlight the need to prioritize clinical trials focusing on phytochemical bioavailability, optimal doses, and safety data, which are essential for the inclusion of plant-derived natural compounds in the management pathways of chronic inflammatory conditions [13].

In conclusion, the research findings in this Special Issue, summarized above, shed further light on our knowledge and understanding of the effects of various plant-derived natural compounds on health and disease. In order for these compounds to be considered for use in established therapeutic regimes for chronic and acute pathological conditions, it is crucial to gain further insights into their exact mechanisms of action, interactions with drugs and nutrients, dosages, and safety considerations.

Author Contributions: Conceptualization, E.P.; Writing—original draft preparation, C.A. and E.P.; Writing—review and editing, C.A. and E.P. All authors have read and agreed to the published version of the manuscript.

Conflicts of Interest: The authors declare no conflict of interest.

References

1. Lantzouraki, D.Z.; Amerikanou, C.; Karavoltsos, S.; Kafourou, V.; Sakellari, A.; Tagkouli, D.; Zoumpoulakis, P.; Makris, D.P.; Kalogeropoulos, N.; Kaliora, A.C. *Artemisia arborescens* and *Artemisia inculta* from Crete; Secondary Metabolites, Trace Metals and In Vitro Antioxidant Activities. *Life* **2023**, *13*, 1416. [CrossRef]
2. Ouahabi, S.; Loukili, E.H.; Elbouzidi, A.; Taibi, M.; Bouslamti, M.; Nafidi, H.-A.; Salamatullah, A.M.; Saidi, N.; Bellaouchi, R.; Addi, M.; et al. Pharmacological Properties of Chemically Characterized Extracts from Mastic Tree: In Vitro and In Silico Assays. *Life* **2023**, *13*, 1393. [CrossRef] [PubMed]

3. Nguyen, V.B.; Wang, S.-L.; Phan, T.Q.; Doan, M.D.; Phan, T.K.P.; Phan, T.K.T.; Pham, T.H.T.; Nguyen, A.D. Novel Anti-Acetylcholinesterase Effect of *Euonymus laxiflorus* Champ. Extracts via Experimental and In Silico Studies. *Life* **2023**, *13*, 1281. [CrossRef] [PubMed]

4. Srivastava, S.; Virmani, T.; Haque, M.R.; Alhalmi, A.; Al Kamaly, O.; Alshawwa, S.Z.; Nasr, F.A. Extraction, HPTLC Analysis and Antiobesity Activity of *Jatropha tanjorensis* and *Fraxinus micrantha* on High-Fat Diet Model in Rats. *Life* **2023**, *13*, 1248. [CrossRef] [PubMed]

5. Subhawa, S.; Arpornchayanon, W.; Jaijoy, K.; Chansakaow, S.; Soonthornchareonnon, N.; Sireeratawong, S. Anti-Inflammatory, Antinociceptive, Antipyretic, and Gastroprotective Effects of *Eurycoma longifolia* Jack Ethanolic Extract. *Life* **2023**, *13*, 1465. [CrossRef]

6. Yadav, I.; Sharma, N.; Velayudhan, R.; Fatima, Z.; Maras, J.S. *Ocimum sanctum* Alters the Lipid Landscape of the Brain Cortex and Plasma to Ameliorate the Effect of Photothrombotic Stroke in a Mouse Model. *Life* **2023**, *13*, 1877. [CrossRef] [PubMed]

7. Hassan, A.K.; El-Kalaawy, A.M.; Abd El-Twab, S.M.; Alblihed, M.A.; Ahmed, O.M. Hesperetin and Capecitabine Abate 1,2 Dimethylhydrazine-Induced Colon Carcinogenesis in Wistar Rats via Suppressing Oxidative Stress and Enhancing Antioxidant, Anti-Inflammatory and Apoptotic Actions. *Life* **2023**, *13*, 984. [CrossRef] [PubMed]

8. Agour, A.; Mssillou, I.; El Barnossi, A.; Chebaibi, M.; Bari, A.; Abudawood, M.; Al-Sheikh, Y.A.; Bourhia, M.; Giesy, J.P.; Aboul-Soud, M.A.M.; et al. Extracts of *Brocchia cinerea* (Delile) Vis Exhibit In Vivo Wound Healing, Anti-Inflammatory and Analgesic Activities, and Other In Vitro Therapeutic Effects. *Life* **2023**, *13*, 776. [CrossRef] [PubMed]

9. Salamatullah, A.M. Antioxidant, Anti-Inflammatory, and Analgesic Properties of Chemically Characterized Polyphenol-Rich Extract from *Withania adpressa* Coss. ex Batt. *Life* **2022**, *13*, 109. [CrossRef] [PubMed]

10. Bailly, C. Discovery and Anticancer Activity of the Plagiochilins from the Liverwort Genus *Plagiochila*. *Life* **2023**, *13*, 758. [CrossRef] [PubMed]

11. Jaiswal, V.; Lee, H.-J. Pharmacological Activities of Mogrol: Potential Phytochemical against Different Diseases. *Life* **2023**, *13*, 555. [CrossRef] [PubMed]

12. Bhat, A.A.; Shakeel, A.; Rafiq, S.; Farooq, I.; Malik, A.Q.; Alghuthami, M.E.; Alharthi, S.; Qanash, H.; Alharthy, S.A. *Juglans regia* Linn.: A Natural Repository of Vital Phytochemical and Pharmacological Compounds. *Life* **2023**, *13*, 380. [CrossRef] [PubMed]

13. Davila, M.M.; Papada, E. The Role of Plant-Derived Natural Products in the Management of Inflammatory Bowel Disease—What Is the Clinical Evidence So Far? *Life* **2023**, *13*, 1703. [CrossRef] [PubMed]

Article

Artemisia arborescens and *Artemisia inculta* from Crete; Secondary Metabolites, Trace Metals and In Vitro Antioxidant Activities

Dimitra Z. Lantzouraki [1,†], Charalampia Amerikanou [2,†], Sotirios Karavoltsos [3], Vasiliki Kafourou [2], Aikaterini Sakellari [3], Dimitra Tagkouli [2], Panagiotis Zoumpoulakis [4], Dimitris P. Makris [5], Nick Kalogeropoulos [2] and Andriana C. Kaliora [2,*]

[1]   Institute of Chemical Biology, National Hellenic Research Foundation, 48 Vas. Constantinou Ave., 11635 Athens, Greece; dlantzouraki@gmail.com
[2]   Department of Nutrition and Dietetics, School of Health Science and Education, Harokopio University, 70 El. Venizelou Ave., 17676 Athens, Greece; amerikanou@windowslive.com (C.A.)
[3]   Laboratory of Environmental Chemistry, Department of Chemistry, National and Kapodistrian University of Athens, 15784 Athens, Greece; skarav@chem.uoa.gr (S.K.)
[4]   Department of Food Science and Technology, University of West Attica, Ag. Spyridonos, 12243 Egaleo, Greece
[5]   Department of Food Science & Nutrition, School of Agricultural Sciences, University of Thessaly, N. Temponera Street, 43100 Karditsa, Greece
*   Correspondence: akaliora@hua.gr or andrianakaliora@gmail.com; Tel.: +30-2109549226
†   These authors contributed equally to this work.

Citation: Lantzouraki, D.Z.; Amerikanou, C.; Karavoltsos, S.; Kafourou, V.; Sakellari, A.; Tagkouli, D.; Zoumpoulakis, P.; Makris, D.P.; Kalogeropoulos, N.; Kaliora, A.C. *Artemisia arborescens* and *Artemisia inculta* from Crete; Secondary Metabolites, Trace Metals and In Vitro Antioxidant Activities. *Life* **2023**, *13*, 1416. https://doi.org/10.3390/life13061416

Academic Editor: Stefania Lamponi

Received: 17 May 2023
Revised: 8 June 2023
Accepted: 16 June 2023
Published: 19 June 2023

Abstract: Background: Currently, the use of medicinal plants has increased. *Artemisia* species have been used in several applications, including medicinal use and uses in cosmetics, foods and beverages. *Artemisia arborescens* L. and *Artemisia inculta* are part of the Mediterranean diet in the form of aqueous infusions. Herein, we aimed to compare the secondary metabolites of the decoctions and two different extracts (methanolic and aqueous-glycerolic) of these two species, as well as their antioxidant capacity and trace metal levels. Methods: Total phenolic, total flavonoid, total terpenes, total hydroxycinnamate, total flavonol, total anthocyanin contents and antioxidant/antiradical activity were determined, and GC/MS analysis was applied to identify and quantify phenolics and terpenoids. Trace metals were quantified with ICP-MS. Results: Aqueous-glycerolic extracts demonstrated higher levels of total secondary metabolites, greater antioxidant potential and higher terpenoid levels than decoctions and methanolic extracts. Subsequently, the aqueous-glycerolic extract of a particularly high phenolic content was further analyzed applying targeted LC-MS/MS as the most appropriate analytic tool for the determination of the phenolic profile. Overall, twenty-two metabolites were identified. The potential contribution of infusions consumption to metal intake was additionally evaluated, and did not exceed the recommended daily intake. Conclusions: Our results support the use of these two species in several food, cosmetic or pharmaceutical applications.

Keywords: *Artemisia arborescens*; *Artemisia inculta*; phenolic compounds; terpenes; trace metals; antioxidant capacity

1. Introduction

To meet their treatment needs, nowadays most people rely on traditional herbs, and 60% of medicines in pharmacies are derived from medicinal plants [1]. The use of medicinal plants and their associated formulations is becoming more common throughout the world due to the fact that they are available, safe, effective and the subject of valuable traditional knowledge that can be used to prevent and treat a variety of diseases [2–4]. On the other hand, several high-priced medications that are routinely used have unpredictable and serious side effects. Thus, as a consequence of the increasing demand for new therapeutic strategies worldwide, it is crucial to investigate botanical plants in terms of their potential for safe treatments methods. For example, it has been established that bioactive phytochemical compounds detected in several botanic species could be used to prevent and treat diseases linked with oxidative stress, such as diabetes, cardiovascular diseases, different forms of cancer, rheumatoid arthritis or Alzheimer's disease [5].

Artemisia is a plant genus of the Asteraceae family with hundreds of species, mainly found in the drier climates of the Northern Hemisphere, with several culinary, beverage, aromatic and industrial uses [6]. For instance, on the island of Crete, Greece, *Artemisia arborescens* L. or *Arboreus absinth* and *Artemisia inculta Delile* constitute part of the Mediterranean diet in the form of aqueous infusions. *Artemisia* species are of great importance in traditional medicine, mentioned even in ancient sources for the treatment of fever, dysentery and hemorrhoids, as an antispasmodic or for calming of children [7]. In recent years, *Artemisia* species have attracted considerable research interest because of their chemical composition and biological activities [8]. The 2015 Nobel Prize in Physiology or Medicine was awarded to Professor Youyou Tu for her key contribution to the discovery of artemisinin, a new class of antimalarial drugs that have saved millions of lives and represents one of the significant contributions of China to global health. Different classes of secondary metabolites have been detected among the 260 *Artemisia* species, including lignans, sesquiterpenoids, flavonoids, coumarins, glycosides, caffeoylquinic acids, sterols and polyacetylenes [9,10]. Additionally, different species of *Artemisia* exhibit neuroprotective, antidepressant, cytotoxic, digestive and antimicrobial activities [11–14] as well as nephroprotective [15] or hepatoprotective [16] properties. Added to the above, the use of *Artemisia* species in cosmetic products has increased significantly, mostly due to their antibacterial or antioxidant properties [11].

The aforementioned significant health benefits of *Artemisia* species promote an increase in the consumption, and several other uses, of this plant. However, particular attention should be paid to the metal content of the plant material used, since certain metals' gradual accumulation in vital organs, combined with their incomplete excretion from the human organism, poses a serious health risk [17,18]. Among the metals most frequently examined in the literature, also studied here were Co, Fe, Mn and Zn, representing essential nutrients; Cr, Cu and Ni, which are essential, albeit exerting toxicity only at elevated concentrations; and Cd and Pb, being exclusively toxic with no beneficial properties even at low levels.

The research of *Artemisia* species from the island of Crete, Greece, the southernmost point of Europe, is limited. To the best of our knowledge, this is the first study that evaluates the main phytochemical compounds of *A. arborescens* and *A. inculta* from Crete following different extraction methodologies, as well as their antioxidant activities and trace metals. Additionally, another novelty lies in the proposed aqueous-glycerolic method as it yields products with high secondary metabolite contents, antioxidant capacity and acceptable levels of trace elements in terms of toxicity.

2. Materials and Methods

2.1. Chemicals, Standards and Solvents

Ferric chloride hexahydrate (FeCl$_3$ 6H$_2$O) of analytical grade was supplied from Acros Organics (Morris Plains, NJ, USA), and aluminum chloride (AlCl$_3$) from Fisher Scientific (Princeton, NJ, USA). Gallic acid, Folin–Ciocalteu's phenol reagent, rutin (quercetin 3-O-rutinoside), ascorbic acid, 2,2'-diphenyl-1-picrylhydrazyl (DPPH•) free radical, p-(dimethylamino)-cinnamaldehyde (DMAC), (+)-catechin, Trolox 6-hydroxy-2,5,7,8-tetramethylchroman-2-carboxylic acid and 2,4,6-tripyridyl-s-triazine (TPTZ) were purchased from Sigma Chemical Co. (St. Louis, MO, USA). Standard phenolic compounds—namely, 3,4,5-trihydroxybenzoic acid, trans-4-hydroxycinnamic acid, 3,4-dihydroxycinnamic acid, vanillin and quercetin—were purchased from Alfa Aesar (Karlsruhe, Germany), while (±)-naringenin and (±)-catechin were obtained from Sigma-Aldrich (Steinheim, Germany). Cinnamic acid, 4-hydroxybenzoic acid, nitric acid 65% supra pure and hydrogen peroxide 30% supra pure were purchased from Merck KGaA (Darmstadt, Germany), 2,6-di-tert-butyl-4-methylphenol was purchased from Acros Organics (Geel, Belgium) and

2-(4-hydroxyphenyl) ethanol was purchased from Fluka Analytical (Merck KGaA, Darmstadt, Germany). All solvents used were of GC, HPLC or MS grade and were purchased from Sigma Aldrich Co. (Gillingham, UK), Fisher Chemical (Loughborough, UK) and Merck KGaA (Darmstadt, Germany). Glycerol purchased from Oleon Corporate M&S (Ertvelde, Belgium) and used for extractions was 99 % pure. Formic acid of MS grade was purchased from LGC Standards (Wesel, Germany).

2.2. Sampling and Preparation

The plant material used in this study consisted of the aerial parts of *A. inculta* Delile and *A. arborescens*, which were provided by the Mediterranean Plant Conservation Unit, The Mediterranean Agronomic Institute of Chania (M.A.I.Ch., Chania, Crete, Greece), where voucher specimens were deposited (*A. inculta*: 9493 MAIC; *A. arborescens*: 9504 MAIC). The aerial parts, composed of foliage and stems, of *A. arborescens* and *A. inculta* were collected from the Almyrida area (Apokoronas region of Chania regional unit, Crete, Greece) and the island of Gavdos (regional unit of Chania, Crete, Greece), respectively. In particular, the different samples from the experts of M.A.I.Ch. were collected from fully grown shrubs during in October 2014 and October 2015 (mean values of temperature and daily rainfall in Almyrida area were 19.9 °C, 4.23 mm and 20.8 °C, 3.30 mm in October 2014 and October 2015, respectively; the corresponding data for Gavdos island were 16.0 °C, 2.56 mm in October 2014 and 16.3 °C, 3.11 mm in October 2015).

Stems of plant samples were discarded, while flowering tops and leaves were carefully washed in cold distilled water, drained and left to dry at room temperature in a dry dark chamber for 7 days. Dried samples were grounded to a fine powder in a mechanical grinder and stored in the dark at 4 °C until their further use within 4 months post-collection.

2.3. Preparation of Samples

Each sample of *A. arborescens* and *A. inculta* obtained from the two samplings was homogenized separately. Methodologies followed for the preparation of herbal decoctions or methanolic and aqueous-glycerolic extracts were identical for the two species, while each procedure was carried out in triplicate. All *Artemisia* samples prepared as indicated below were stored at −80 °C in darkness until analysis.

2.3.1. Decoctions

Decoctions were prepared by adding 3 g of dried herb to 200 mL [1:67 (*w/v*) material to solvent ratio] of bottled natural mineral water in a glass (Pyrex) boiling pot. The mixture was placed on a preheated heating plate, left at boiling temperature for 3 min, then at room temperature for 2 min, and finally filtered by a Buchner funnel under vacuum. An appropriate amount of water was added to maintain the final volume of 200 mL. Decoctions were freeze-dried for 120 h in a Cryodos freeze dryer (Telstar Industrial, Barcelona, Spain) and the dry residue was weighed. Furthermore, the residue of total salts per volume of the mineral water used was determined after freeze-drying for the correction of extractable yield values. Freeze-dried decoctions were stored at −80 °C in darkness and appropriately diluted prior to analysis (3 g dry residue of decoction per 200 mL distilled water).

2.3.2. Herbal Extracts

For the methanolic extracts, a classical extraction procedure was performed with a 1:100 (*w/v*) material to solvent ratio as follows. Approximately 0.5 g of dried herb was macerated in 50 mL of methanol, and the mixture was left in darkness under constant stirring at room temperature for 24 h. The crude extracts were then centrifuged at 3600 rpm for 10 min, and the supernatants were collected and evaporated to dryness using a rotary evaporator at 40 °C. Each dry residue was separately redissolved in 10 mL of methanol and the solvent was evaporated to dryness once more using a centrifugal concentrator (Speed Vac, Labconco Corporation, Kansas City, MO, USA) at 40 °C. Finally, the dried extracts were redissolved in 2 mL of methanol using an ultrasonic bath. The concentrated samples were preserved in the dark at $-80\ ^{\circ}$C and diluted to a final concentration of 1 g of extract dry residue per 100 mL of methanol for further analysis.

The glycerol–water extracts were prepared as described by Shehata et al. [19] with minor modifications. One (1) gram of dried plant sample was mixed with 125 mL of glycerol–water 9:1 (*w/v*) mixture, and the extraction took place on a magnetic stirrer hot plate at 80 ($\pm$1) $^{\circ}$C, under continuous stirring for 160 min. The extracts were then cooled in a water bath at room temperature, centrifuged at 3500 rpm for 30 min, and the supernatants were collected for further analysis.

2.4. Determination of Total Phenolic Content

Total phenolic content (TPC) of each sample preparation was determined by applying a micro method of Folin–Ciocalteu's colorimetric assay, based on the procedure described by Karakashov et al. [20]. Briefly, in a 1.5 mL tube, 20 µL of sample, standard solution or blank was added to 780 µL of distilled water and 50 µL of Folin–Ciocalteu reagent, mixed thoroughly and then allowed to stand for 1 min. Subsequently, 150 µL of saturated (20% *w/v*) aqueous sodium carbonate solution was added, and the mixture was vortexed and allowed to stand at room temperature in darkness for 60 min. The samples were transferred to a 96-well plate, and the absorbance was measured at 750 nm using an ELISA microplate reader (Power Wave XS2, Microplate Spectrophotometer, BioTekInstruments, Winooski, VT, USA). The TPC was expressed as mg of caffeic acid equivalents (CAE) per gram of dried *Artemisia* sp. using a standard curve within a range of 40–1000 mg·L^{-1} caffeic acid in assay solution (y = 0.0007x $-$ 0.0129, R^2 = 0.998).

2.5. Determination of Total Flavonoid Content (TFC)

For estimating the total flavonoid content (TFC) of decoctions or extract samples of the two *Artemisia* species, a previously published protocol was applied with some modifications [21]. In detail, an aliquot of 25 µL of sample was mixed with 30 µL of sodium nitrite (NaNO$_2$) aqueous solution 5 % (*w/v*), and the derived solution was incubated in 300 µL of ethanol–water 1:1 (*v/v*) for 5 min at room temperature. Afterwards, 150 µL of aluminum chloride hexahydrate solution (AlCl$_3$·6H$_2$O) 2% (*w/v*) in water was added and allowed to stand at room temperature for 5 min. After the addition of 200 µL of sodium hydroxide (NaOH) 1 M aqueous solution, the mixture was adjusted to a final volume of 1 mL with ethanol–water 1:1 (*v/v*). The absorbance was measured at 510 nm using a 96-well plate and an ELISA microplate reader, while the total flavonoid concentration was expressed as mg catechin equivalents (CE) per gram of dried *Artemisia* species. The range of the concentrations for catechin was 20–1000 mg·L^{-1} in assay solution (y = 0.0003x + 0.0048, R^2 = 0.998).

2.6. Determination of Phenolic Classes

The methodology employed by Galanakis et al. [22] was performed to determine different phenolic classes in the extracts and decoctions of *Artemisia* sp., namely, hydroxycinnamates, flavonols and anthocyanins. In short, 1 mL of each sample and 1 mL of aqueous ethanol (95% *v/v*) containing 0.1% (*v/v*) hydrochloric acid were mixed to a final volume of 10 mL with 2% (*v/v*) hydrochloric acid. The absorbance of the mixture was measured at 320, 360 and 520 nm to determine total hydroxycinnamate content (THCC) as mg of caffeic acid equivalents (CAE) per gram of dried *Artemisia* sp., total flavonol content (TFnolC) as mg of quercetin equivalents (QE) per gram of dried plant and total anthocyanin content (TAC) as µg of cyanidin equivalents (CNE) per gram of dried plant, respectively. Concentration ranges and equations of the corresponding standard curves of the abovementioned determinations were as follows: caffeic acid, 5–20 mg·L^{-1} of assay solution, y = 0.0686x $-$ 0.0100 (R^2 = 0.998); quercetin, 5–20 mg·L^{-1} of assay solution, y = 0.0444x $-$ 0.0290 (R^2 = 0.999); cyanidin chloride, 40–300 µg·L^{-1} of assay solution, y = 0.0009x $-$ 0.0153 (R^2 = 0.990).

2.7. Determination of Total Terpenes

A colorimetric assay method based on Fan and He [23] was used to estimate the content of total terpenic compounds (TTC). For each sample preparation, 200 μL were evaporated to dryness in a boiling water bath. The dry residue was re-diluted with 0.3 mL (5% w/v) vanillin in glacial acetic acid and 1 mL of perchloric acid solution. The mixture was heated for 45 min at 60 °C and then cooled in an ice-water bath to ambient temperature. The absorbance of assay solutions was measured at 548 nm following the addition of 5 mL glacial acetic acid. Ursolic acid was used as the standard compound within a range of 3–30 mg·L^{-1} of the assay solution (y = 0.0298x − 0.0664, R^2 = 0.988). The TTC of extracts and decoctions was expressed as mg of ursolic acid equivalents (UAE) per gram of dried *Artemisia* plant.

2.8. Assessment of Antioxidant Activity

The antioxidant activity was assessed by measuring the radical-scavenging activity and reducing antioxidant potential of *Artemisia* decoctions and extracts.

The antiradical power of tested *Artemisia* preparations was assessed as described in a previous study [24]. The DPPH assay provides an evaluation of the samples' potency to scavenge the 2,2′-diphenyl-1-picrylhydrazyl free radical, which was depicted as the concentration of Trolox equivalents (TE) per gram of dry herb, using a standard curve ranging from 0.050 to 1.2 mM of Trolox (y = 0.31504x + 0.00161, R^2 = 0.993). The absorbance was recorded at 515 nm twice, i.e., at 5 and 30 min, where the absorbance was stabilized at a minimum value.

The antioxidant power of *Artemisia* decoctions and extracts was evaluated based on the reduction of iron from ferric to ferrous form when being complexed with 2,4,6-tris(2-pyridyl)-s-triazine (TPTZ). Ferric Reducing/Antioxidant Power (FRAP) assay was carried out according to a previously published work [25]. For the construction of the standard curve (y = 0.0.83601x + 0.04754, R^2 = 0.999), 10–500 μM of standard solutions of L-ascorbic acid were prepared. The absorbance for samples, blanks and standards was measured until stabilization to a peak value at 620 nm, and the results were expressed as mg of L-ascorbic acid equivalents (AAE) per gram of dry herb.

In addition, inhibition of copper-induced lipid oxidation in total serum solubilized in phosphate buffer saline (PBS), using lag time as a criterion for antioxidative potency, was evaluated as a more biologically relevant assay to assess the antioxidant activity of the *Artemisia* samples. Venous blood was collected under sterile conditions from healthy humans, and serum was obtained after centrifugation at 3000 rpm at 4 °C for 10 min directly after collection. The study of the kinetics of copper-induced oxidation in 12-fold diluted serum was performed by monitoring the absorbance of lipid oxidation products at 245 nm using an ELISA reader (PowerWaveXS2, Microplate Spectrophotometer, BioTek, Winooski, VT, USA). At time point 0, CuSO$_4$ was added in the serum (20 μL) to a final concentration of 10^{-5} M in PBS. Copper-induced oxidation of lipids in serum leads to the formation of conjugated dienic hydroperoxides that absorb at 245 nm. The kinetics of oxidation was analyzed in terms of the lag time prior to oxidation and was expressed in seconds.

2.9. GC/MS Analysis of Phenolic Compounds and Terpenoids

Gas chromatography/mass spectrometry (GC/MS) analysis of phenolic and terpenic compounds was performed. An Agilent (Wallborn, Germany) HP series GC 6890N coupled with a HP 5973 MS detector (EI, 70 eV), split–splitless injector and an HP 7683 autosampler were used for the determination of phenolic and terpenic compounds of *Artemisia* decoctions and methanolic or hydroglycerolic extracts. An aliquot (1 μL) of the silylated samples was injected into the gas chromatograph at a split ratio of 1:20. Separations were achieved on a HP-5 MS capillary column (30 m × 0.25 mm × 250 μm), employing high purity helium at 0.6 mL/min as the carrier gas. The injector and transfer line temperatures were kept at 250 and 300 °C, respectively, and the oven temperature was kept initially at 70 °C for 5 min, then raised to 70–130 °C at 15 °C /min, then 130–160 °C at 4 °C/min, kept at 160 °C for 15 min and finally raised to 160–300 °C at 10 °C/min and kept at 300 °C for 15 min.

A selective ion monitoring (SIM) GC/MS method was applied for the detection of 17 phenolic compounds, 1 stilbene, 4 terpenic compounds and the internal standard based on the ±0.05 RT presence of target and qualifier ions of commercial standard compounds at the predetermined ratios. The target and qualifier ions used for the identification of compounds under concern are presented in Table S1. Identification of chromatographic peaks was made by comparing the retention times and ratios of two or three fragment ions of each phenolic or terpenic compound with those of commercial reference standards [26]. Quantification was carried out by constructing reference curves for each compound, based on a series of 9 standard mixtures of the phenolic and terpenic compounds containing the same quantity of internal standard as that of samples. Serving as the internal standard was 3-(4-hydroxyphenyl)-1-propanol.

2.10. Targeted LC-MS/MS Profiling of Phenolic Compounds in Glycerolic Extracts

Liquid chromatography–mass spectrometry (LC-MS) was employed for further investigation of phenolic compounds in the aqueous-glycerolic extracts of *Artemisia* sp. samples, as previously described [24,27]. HPLC-PDA-ESI-MS/MS paired with an in-house multiple reaction monitoring (MRM) spectral library was employed. Phenolic compound separation was carried out using a Thermo Scientific Surveyor Plus HPLC-PDA-ESI-MS/MS system (San José, CA, USA). The platform comprises a Thermo Scientific Surveyor HPLC Pump Plus, a Thermo Scientific Surveyor Autosampler Plus Lite and an LCQ FLEET mass spectrometer equipped with an Electrospray Ionization (ESI) Probe and an Ion Trap analyzer. Data were processed using the Xcalibur software program (version 2.1).

Prior to LC-MS analysis, the extracts were diluted as 1:2 (v/v) with a mixture of MeOH-H$_2$O 7:3 (v/v). The chromatographic separation of phenolics was carried out using a Finnigan Surveyor system and a Hypersil Gold Column (3 mm, 2.1 × 100 mm, Thermo, Palo Alto, Santa Clara, CA, USA) protected with a security guard cartridge (Hypersil Gold, 3 mm, 10 × 2.1 mm i.d.). The gradient mobile phase consisted of solvent A [water—0.5% (v/v) formic acid] and solvent B (acetonitrile). The flow rate was 0.3 mL·min^{-1} and the injection volume was 5.0 µL. The gradient elution program was initially 5% B, linear 5–9% B at 4 min, linear 9–15% B at 8 min, linear 15–18% B at 11 min, held constant for 1 min, linear 18–50% B at 15 min, held constant for 2 min, purging with 100% B during 6 min and re-equilibration of the column for 10 min.

Mass spectrometric analysis of sample solutions was operated in negative electrospray ionization (ESI) mode, and different collision energies were applied for tandem MS analysis. Mass spectrometer parameters for negative ion mode were as follows: source voltage, 4.0 kV; capillary voltage, −18 V; capillary temperature, 300 °C; sheath gas flow, 50 (arbitrary units); sweep gas flow, 20 (arbitrary units); full max ion time, 300 ms; and full micro scans, 3. MRM experiments were performed by specifying the deprotonated parent ion of each targeted compound for MS/MS fragmentation and the fragment ions were recorded. For the identification of each compound, the parent ion from the negative ionization mode as well as the characteristic fragments deriving from the fragmentation were used. Data-dependent scans for MS/MS analyses were carried out with the following conditions: collision energies, 15, 25, 30, 35 (arbitrary units); width, 1.00; repeat count, 2; repeat duration, 0.5 min; exclusion size list, 25; exclusion duration, 1.00 min; exclusion mass width, 3.00; and scanned mass range (m/z), 100–1600. The acquired MS/MS data were then compared with the in-house spectral libraries for the identification of the secondary metabolites. The identification was scored based on the similarity of fragmentation patterns between the acquired and the library spectra. The criteria we selected included MS2/MS3 fragment peak intensity ratios and isotope peak intensity ratios, among others. This acquisition scheme allowed the identification and characterization of not only the major (poly)phenolic compounds present in the studied extracts but also several low-level molecules [28–30].

2.11. Trace Metals Determination

All materials contacting the samples were soaked in dilute HNO_3 (Merck, Darmstadt, Germany) and rinsed thoroughly with ultrapure water of 18.2 MΩ cm (Millipore, Bedford, MA, USA). Class A volumetric glassware was used for preparing all solutions required. For trace metals determination, samples of both plants and prepared decoctions were wet-digested by adding HNO_3 65% supra pure (Merck) and H_2O_2 30% supra pure (Merck). Digestion was performed by a microwave digestion system (Anton Paar Multiwave GO Plus, Graz, Austria) and digested samples were subsequently diluted to a final volume of 25 mL [31]. The measurement of trace metals was carried out by Inductively Coupled Plasma Mass Spectrometry (ICP-MS), employing a Thermo Scientific ICAP Qc (Waltham, MA, USA) instrument, in a single collision cell mode, with kinetic energy determination (KED) using pure He. Matrix-induced signal suppressions and instrumental drift were corrected by internal standardization (45 Sc, 103 Rh).

Calculation of limits of detection (LODs) was performed by multiplying the standard deviation of seven replicate samples prepared at an approximately low concentration by 3.14 [32]. Calculated LODs in $\mu g\ g^{-1}$ referring to dry weight were equal to 0.003 for Cd, 0.004 for Co and Ni, 0.02 for Cr, Cu and Mn, 0.05 for Fe, 0.008 for Pb and 0.04 for Zn. For quality assurance purposes, a procedural blank was included in samples' analyses, in which no analytes were detected. For the verification of the accuracy and precision of the method, the certified reference material $ERM^{®}$-CD281 (rye grass) was analyzed, with calculated metal recoveries ranging between 90 and 110%.

Metal extractability from the herb towards the aqueous infusion was calculated from their corresponding metal contents, while also considering the solid residue per cup of infusion and the amount of 3 g of the plant used for infusion preparation:

$$\% \text{ Extraction Efficiency (EE)} = 100 \times (\text{Metal content infusion} \times \text{Solid residue}$$
$$\text{per cup})/(\text{Metal content plant tissue} \times 3)$$

3. Results and Discussion

3.1. Secondary Metabolites

Table 1 represents the analytical data for total phenolic, total flavonoid, total hydroxycinnamate, total flavonol, total anthocyanin and total terpenic contents of the decoctions and methanolic and aqueous-glycerolic extracts of *A. arborescens* and *A. inculta*. It is well-known that the three fundamental classes of bioactive compounds of *Artemisia* are flavonoids, phenolic acids and terpenes [33]. In total, in our samples, aqueous-glycerolic extracts of both species had higher levels of the above secondary metabolites compared to decoctions and methanolic extracts. The content of total (poly)phenolic compounds of the aqueous-glycerolic extracts of *A. arborescens* and *A. inculta* has been investigated before, being dependent on the concentration of glycerol and the liquid-to-solid ratio [19]. Different methanol, ethanol and acetonitrile extracts of A. absinthium contained TPC ranging from 659 to 1033 mg gallic acid equivalents/100 g dm (dry matter), and TFC ranging from 259 to 392 mg catechin equivalents/100 g dm [34]. Singh et al. [35] reported that TPC and TFC were higher in ethanolic extracts of *A.* absinthium compared to aqueous and chloroform extracts, suggesting the organic solvent (ethanol) is ideal to extract bioactive phenolic compounds. In our study, the aqueous-glycerol extract exhibited the greater potential to possess more polyphenols and terpenes.

In the study by Bourgou et al. [36], ethyl acetate fractions of *A. herba-alba* showed higher quantity of TPC (87.5 mg gallic acid equivalents /100 g dm) and TFC (96.5 mg QF/g dm) compared to the water fraction (TPC = 40 mg gallic acid equivalents/100 g dw, TFC = 60.6 mg QE/g dm).

Table 1. Secondary metabolites detected in *A. arborescens* and *A. inculta* extracts.

	A. arborescens			*A. inculta*		
	Decoction	Methanolic	Aqueous-Glycerolic	Decoction	Methanolic	Aqueous-Glycerolic
TPC (mg CAE·g^{-1} dm)	8.29 ± 0.27 [d]	7.3 ± 1.1 [d]	32.82 ± 0.50 [b]	9.74 ± 0.15 [c]	10.80 ± 0.90 [c]	36.0 ± 2.5 [a]
TFC (mg CE·g^{-1} dm)	6.05 ± 0.17 [c]	4.48 ± 0.37 [d]	19.57 ± 0.76 [a]	6.04 ± 0.72 [c]	7.0 ± 1.3 [c]	16.77 ± 0.36 [b]
THCC (mg CAE·g^{-1} dm)	0.0985 ± 0.0035 [e]	0.095 ± 0.017 [d,e]	0.310 ± 0.010 [b]	0.1117 ± 0.0012 [d]	0.1257 ± 0.0050 [c]	0.338 ± 0.013 [a]
TFnolC (mg QE·g^{-1} dm)	0.0859 ± 0.0025 [e]	0.100 ± 0.016 [d,e]	0.293 ± 0.015 [b]	0.0988 ± 0.0015 [d]	0.1383 ± 0.0048 [c]	0.370 ± 0.013 [a]
TAC (μg CNE·g^{-1} dm)	0.651 ± 0.032 [d]	1.424 ± 0.028 [b]	2.680 ± 0.082 [a]	0.462 ± 0.011 [e]	0.764 ± 0.058 [c]	2.61 ± 0.14 [a]
TTC (mg UAE·g^{-1} dm)	0.229 ± 0.029 [d]	1.859 ± 0.160 [a]	0.374 ± 0.021 [c]	0.215 ± 0.021 [d]	1.857 ± 0.346 [a]	0.438 ± 0.024 [b]

Total phenolic content was expressed as caffeic acid equivalents (CAE), total flavonoid content as catechin equivalents (CE), total hydroxycinnamate content as caffeic acid equivalents (CAE), total flavonol content as quercetin equivalents (QE), total anthocyanin content as cyanidin equivalents (CNE), and total terpenic content as ursolic acid equivalents (UAE) on a dry material basis for the decoctions, methanolic (MeOH), and glycerolic extracts of *A. arborescens* and *A. inculta*. Values are presented as mean (±standard deviation) (n = 3). TPC: total phenolic content, TFC: total flavonoid content, THCC: total hydroxycinnamate content, TFnolC: total flavonol content, TAC: total anthocyanin content, TTC: total terpenic content, dm: dry matter. [a–e] Means per row denoted by a common superscript letter are not significantly different according to Tukey's test at 5% level of significance.

3.2. Antioxidant Properties

The aqueous-glycerol extract exhibits greater antioxidant potential compared to decoctions and methanolic extracts, as shown in Table 2. More specifically, scavenging/antiradical activity, as assessed by DPPH assay, antioxidant power as assessed by FRAP assay and inhibition of copper-induced lipid oxidation in total serum were higher in the aqueous-glycerol extract. Several studies have proven the antioxidant properties of leaf extracts and essential oil of *Artemisia* species and have linked these effects with their components [37–40].

Table 2. Antioxidant potential of *A. arborescens* and *A. inculta* extracts.

	A. arborescens			*A. inculta*		
	Decoction	Methanolic	Aqueous-Glycerolic	Decoction	Methanolic	Aqueous-Glycerolic
Antiradical activity (mg TE·g^{-1} dm)	5.18 ± 0.50 [e]	7.524 ± 0.039 [d]	30.9 ± 1.2 [a]	5.68 ± 0.21 [e]	8.23 ± 0.40 [c]	27.8 ± 1.1 [b]
FRAP (mg AAE·g^{-1} dm)	3.45 ± 0.14 [e]	5.36 ± 0.44 [c]	21.77 ± 0.70 [a]	2.92 ± 0.25 [e]	4.31 ± 0.47 [d]	18.19 ± 0.27 [b]
TSO (sec)	1107.7 ± 8.1 [f]	627.8 ± 366.3 [a]	2007.25 ± 171.25 [f]	2120 ± 493.8 [b]	1400.3 ± 377.2 [f]	3850.25 ± 320.25 [c]

[a–f] Means per row denoted by a common superscript letter are not significantly different according to Tukey's test at 5% level of significance. Values are presented as mean (±standard deviation) (n = 3). TE: Trolox equivalents, FRAP: Ferric Reducing/Antioxidant Power, TSO: total serum oxidizability.

3.3. GC-MS Profiling

GC-MS analysis provided the composition of the predominant phenolic and terpenoid compounds in the studied *Artemisia* preparations (Table 3). In total, 22 compounds were identified. Most phenolic compounds were detected at higher levels or appeared only in the methanolic extract, whereas terpenoids were detected at higher levels or presented only in the aqueous-glycerol extract. Comparing the two species, a great variation was observed, with some phenolic compounds being higher in the methanolic extract or the aqueous-glycerol extract of *A. arborescens* or *A. inculta*, and some others being higher in the decoctions of both species. Ursolic acid was higher in *A. inculta* and in aqueous-glycerol extracts compared to methanolic ones, and oleanolic acid was higher in aqueous-glycerol extracts compared to methanolic ones, but higher in *A. inculta* compared to *A. arborescens* only in the aqueous-glycerol extract. Ursolic and oleanolic acids have been isolated from *A. indica*, showing modulatory effects on γ-Aminobutyric acid (GABA-A) receptors, demonstrating anxiolytic activity in mouse models, with no signs of acute toxicity [41]. Additionally, ursolic acid isolated from the methanolic extracts of *A. capillaris* inhibited the growth of both susceptible and resistant strains of Mycobacterium tuberculosis, exhibiting promising results against tuberculosis [42]. It is noteworthy that erythrodiol and uvaol, well-known for their antioxidant and anti-inflammatory activities [43,44], were identified only in the aqueous-glycerol extracts. To the best of our knowledge, this is the first time these two terpenoids are identified in an *Artemisia* extract.

Table 3. Composition of *A. arborescens* and *A. inculta* extracts samples assessed by GC-MS (expressed as µg per g of dry material).

Phenolic Compounds	Molecular Formula	A. arborescens			A. inculta		
		Decoction	Methanolic	Aqueous-Glycerolic	Decoction	Methanolic	Aqueous-Glycerolic
Caffeic acid	$C_9H_8O_4$	94.6 ± 7.5 [d]	121.9 ± 9.0 [c]	39.1 ± 4.5 [e]	382 ± 12 [a]	289.6 ± 6.7 [b]	19.1 ± 3.0 [f]
Chlorogenic acid	$C_{16}H_{18}O_9$	537.1 ± 8.9 [e]	5754 ± 70 [a]	1669.2 ± 1.3 [d]	2332 ± 179 [c]	5052 ± 56 [b]	285 ± 19 [f]
Chrysin	$C_{15}H_{10}O_4$	nd	3.89 ± 0.05 [b]	nd	nd	6.51 ± 0.13 [a]	nd
p-Coumaric acid	$C_9H_8O_3$	1.52 ± 0.13 [d]	2.97 ± 0.32 [c]	nd	37.78 ± 0.26 [b]	43.2 ± 3.5 [a]	nd
Ferulic acid	$C_{10}H_{10}O_4$	16.26 ± 0.59 [b]	2.54 ± 0.16 [d]	1.64 ± 0.05 [e]	25.3 ± 1.4 [a]	16.8 ± 1.4 [b]	3.94 ± 0.06 [c]
Gallic acid	$C_7H_6O_5$	nd	0.94 ± 0.02 [a]	nd	nd	0.67 ± 0.06 [a]	nd
p-Hydroxybenzoic acid	$C_7H_6O_3$	7.89 ± 0.16 [c]	1.16 ± 0.10 [e]	3.82 ± 0.46 [d]	45.4 ± 2.6 [a]	nd	23.99 ± 0.42 [b]
p-Hydroxyphenylacetic acid	$C_8H_8O_3$	nd	0.39 ± 0.05 [b]	nd	nd	0.44 ± 0.03 [b]	5.17 ± 0.68 [a]
Kaempferol	$C_{15}H_{10}O_6$	nd	1.15 ± 0.11 [b]	nd	nd	3.19 ± 0.18 [a]	nd
Naringenin	$C_{15}H_{12}O_5$	nd	3.07 ± 0.28 [c]	nd	15.84 ± 0.25 [b]	40.1 ± 2.7 [a]	36.5 ± 2.4 [a]
Phloretic acid	$C_9H_{10}O_3$	nd	nd	1.70 ± 0.09 [a]	nd	0.49 ± 0.01 [b]	1.02 ± 0.02 [c]
Protocatechuic acid	$C_7H_6O_4$	8.73 ± 0.80 [b]	6.44 ± 0.33 [b,c]	nd	21.2 ± 1.0 [a]	5.51 ± 0.61 [c]	nd
Quercetin	$C_{15}H_{10}O_7$	nd	7.21 ± 0.31 [c]	nd	14.93 ± 0.89 [b]	23.04 ± 0.54 [a]	nd
Resveratrol	$C_{14}H_{12}O_3$	nd	0.36 ± 0.04 [b]	1.62 ± 0.09 [a]	nd	0.24 ± 0.03 [b]	nd
Sinapic acid	$C_{11}H_{12}O_5$	nd	n[d]	30.1 ± 3.0	nd	nd	nd
Syringic acid	$C_9H_{10}O_5$	6.47 ± 0.51 [b]	2.96 ± 0.29 [c]	6.34 ± 0.47 [b]	10.39 ± 0.26 [a]	5.84 ± 0.37 [b]	7.38 ± 0.13 [b]
Tyrosol	$C_8H_{10}O_2$	nd	0.05 ± 0.01 [b]	nd	nd	0.14 ± 0.01 [a]	nd
Vanillic acid	$C_8H_8O_4$	6.66 ± 0.70 [d]	2.20 ± 0.21 [f]	3.06 ± 0.17 [e]	21.1 ± 1.1 [a]	13.96 ± 0.44 [b]	9.18 ± 0.58 [c]
Total Phenolic Compounds		679 ± 19 [e]	5911 ± 81 [a]	1756 ± 10 [d]	2906 ± 199 [c]	5502 ± 72 [b]	392 ± 27 [f]
Terpenoids							
Erythrodiol	$C_{30}H_{50}O_2$	nd	nd	487.8 ± 14 [a]	nd	nd	420 ± 30 [a]
Oleanolic acid	$C_{30}H_{48}O_3$	nd	8.54 ± 0.86 [c]	242.6 ± 18 [b]	nd	7.14 ± 0.68 [c]	480 ± 45 [a]
Ursolic acid	$C_{30}H_{48}O_3$	nd	14.24 ± 0.93 [c]	35.2 ± 5.0 [b]	nd	15.58 ± 0.63 [c]	82.4 ± 7.9 [a]
Uvaol	$C_{30}H_{50}O_2$	nd	nd	712.9 ± 18 [a]	nd	nd	584 ± 42 [b]
Total Terpenoids		nd	22.8 ± 1.8 [b]	1478 ± 56 [a]	nd	22.7 ± 1.3 [b]	1568 ± 126 [a]

nd: Not detected. [a–f] Means per row denoted by a common superscript letter are not significantly different according to Tukey's test at 5% level of significance. Values are presented as mean (±standard deviation) ($n = 3$).

3.4. HPLC-MS Profiling in Hydro-Glycolic Extracts

Based on the results reported for the majority of spectrophotometric assays, the glycerolic extracts were further analyzed by applying LC-MS as it is more suitable for the determination of a wider range of polar and semi-polar compounds, which are often present in plant glycerolic extracts [45]. Under this perspective, we proceeded with a targeted LC-MS method to separate and detect individual (poly)phenolic compounds in the glycerolic extracts of *Artemisia* in order to further investigate the phytochemical profile of the studied glycerol extracts. Tandem mass spectrometry (MS/MS) and built-in MRM spectral libraries were employed to confirm the identity of the compounds.

Table 4 demonstrates the phenolic compounds identified in the aqueous-glycerolic extracts of *A. arborescens* and *A. inculta*. Chlorogenic acid, isorhamnetin, kaempferol-3-O-glucoside and kaempferol-3-O-rutinoside were common in the composition of extracts from both *Artemisia* species. Slimestad et al. [46] identified chlorogenic acid, the ester of caffeic and quinic acid, in both leaves and stalks of *A. annua*, while the antimicrobial potency of extracts from wormwood (*A. gmelinii*) against Gram-positive bacteria and Candida spp. was partially attributed to chlorogenic acid, which dominated the ethanolic preparation from the aerial parts of the plant [47]. High yields of chlorogenic acid from sweet wormwood (*A. annua*) and tarragon (*A. dracunculus*) were obtained in ethanolic fractions that elicited strong radical-scavenging activity [48]. Further, isorhamnetin, naturally contained in Hippophaerhamnoides and Ginkgo biloba [49], was also detected in *A. Annua* [50]. This O-methylated flavonol can protect against atherosclerosis [51], also displaying significant anti-tumor activity [52]. Several kaempferol glycosides were also reported in infusions from the aerial parts of *A. copa Phil.* [53] while kaempferol-3-O-glucoside (astragalin) and its aglycone flavonol, i.e., 3,4′,5,7-tetrahydroxyflavone, were predominant in *A. annua* [50]. Kaempferol is a common dietary flavonoid that exhibits antioxidant and anti-inflammatory effects [54].

Table 4. Phenolic compounds identified in *Artemisia* spp. aqueous-glycerolic extracts with HPLC-ESI(–)-MS/MS(MRM) analysis.

Phenolic Compound	Molecular Formula	[M–H]—(m/z) [1,2]	MS [2] Product Ions (m/z)	A. arborescens	A. inculta
Caffeic acid hexoside	$C_{15}H_{18}O_9$	341.11	179, 161, 135		+
Chlorogenic acid	$C_{16}H_{18}O_9$	353.15	217, 191	+	+
Dihydrokaempferol 3-O-glucoside	$C_{21}H_{22}O_{11}$	449.09	287	+	
Dihydrokaempferol-3-O-rhamnoside (Engeletin)	$C_{21}H_{22}O_{10}$	433.00	269, 179, 151		+
Procyanidin B2	$C_{30}H_{26}O_{12}$	577.24	425	+	
Ellagic acid	$C_{14}H_6O_8$	301.06	301, 257, 229, 185		+
Ellagic acid-O-hexoside	$C_{20}H_{16}O_{13}$	463.11	301, 300, 283, 257, 229		+
Gallic acid derivative	not defined	243.27	169, 225, 151, 139, 125	+	
Hexose ester of protocatechuic acid	$C_{13}H_{15}O_9$	314.77	153	+	
p-Hydroxybenzoic acid	$C_7H_6O_3$	137.06	93	+	
Isorhamnetin	$C_{16}H_{12}O_7$	315.20	300, 301	+	+
Kaempferol-3-O-glucoside (Astragalin)	$C_{21}H_{20}O_{11}$	447.24	285, 255, 327	+	+
Kaempferol-3-O-rutinoside (Nictoflorin)	$C_{27}H_{30}O_{15}$	593.26	285	+	+
Phlorizin	$C_{21}H_{24}O_{10}$	435.20	297, 273, 167	+	
Pyrogallol	$C_6H_6O_3$	125.06	106, 97, 81	+	
Quercetin-3-O-glucuronide (Miquelianin)	$C_{21}H_{18}O_{13}$	477.26	301	+	
Quercetin-3-O-glucoside	$C_{21}H_{20}O_{12}$	463.19	301		+
Quercetin-O-xyloside	$C_{20}H_{18}O_{11}$	433.19	301	+	
Syringaldehyde	$C_9H_{10}O_4$	181.12	166		+
Syringetin-3-O-glucoside	$C_{23}H_{24}O_{13}$	507.25	345		+
Syringetin-hexoside	$C_{23}H_{24}O_{13}$	507.25	345, 327, 315		+
Valoneic acid bilactone	$C_{21}H_{10}O_{13}$	469.04	425, 407	+	

[1] Ions with relative abundance greater than 10% are shown; [2] [M–H]⁻: parent ion derived from molecular mass under negative electrospray ionization conditions; a positive identification for a phenolic compound in the glycerolic extracts is marked with the plus sign symbol (+).

Notably, the phenolic profile, as determined by LC-MS, greatly differentiated between the two *Artemisia* species. A total of 14 phenolic targets were present in *A. arborescens* glycerol extract, while dihydrokaempferol 3-O-glucoside, procyanidin B2, hexose ester of protocatechuic acid, p-hydroxybenzoic acid and quercetin-O-xyloside were among others that were not detected in *A. inculta*. Other researchers have reported simple phenolic compounds and flavonoids found in *A. arborescens* solvent fractions, namely p-coumaric and caffeic acids, chrysosplenol-D, casticin, eupatin, cirsilineol, chrysosplenetin and artemetin [55].

According to our results reported in Table 4, *A. inculta* glycerolic fraction investigation hit 12 positive results corresponding to MRM mass spectra of (poly)phenolic compounds included in the in-house library, such as caffeic acid hexoside, dihydrokaempferol-3-O-rhamnoside, ellagic acid, quercetin-3-O-glucoside and syringetin-hexoside. There is rather limited published data so far on the (poly)phenolic profile of *A. inculta*. However, a study by Younsi et al. [56] indicated chlorogenic acid and 1,4 dicaffeoylquinic acid as the major phenolic constituents in a methanolic extract from *A. inculta* leaves and flowers, while apigenin-6-C-glycosyl flavonoids and caffeoylquinic acids were also present. A recent study revealed that caffeic acids and C-glycosyl flavonoids, such as myricetin, prevail in the (poly)phenolic composition of different extracts from the specific *Artemisia* species [57]. Furthermore, Mohammed et al. [58] reported on significant levels of hydroquinone and 4-hydroxybenzoic acid in *A. inculta* extracts demonstrating antibacterial activity. To the best of our knowledge, ellagic acid, a hydroxybenzoic acid dimer, has not been previously reported in *A. inculta*; however, it was a main phenol in *A. aucheri* [59], while an ellagic acid derivative was detected in *A. argentea L' Hér* alcoholic extract [40]. Ellagic acid is abundant in various fruits such as pomegranate, strawberry, raspberry and blackberry. It is also found in nuts such as walnuts, certain trees such as oak and birch and some medicinal plants and herbs, including Terminalia chebula and Eucalyptus globulus [8]. Ellagic acid is considered a prominent bioactive compound due to its potential health-promoting properties. It has been shown to possess several properties such as antioxidant [60], anti-inflammatory [61] and cardioprotective activities [62].

3.5. Trace Metals

Levels of detected trace metals in *A. arborescens* and *A. inculta* samples are presented in Table 5. The samples differed in terms of their Cd, Co, Cr, Mn, Ni and Pb contents, with *A. arborescens* demonstrating higher concentrations in both the herbal tissue and infusion samples. The concentrations of Cu, Fe and Zn were comparable between the two species. Among the trace metals examined, Fe, Mn and Zn were present at higher levels in both the herbal tissues and corresponding infusions of *A. arborescens* and *A. inculta* samples. Fe, which is an essential element, represents the principal component in several enzymes and proteins and plays a crucial role in the transportation of oxygen to the tissues of the human body through hemoglobin [63], varied between 44.8 and 228 μg g^{-1}. Mn, which is also classified among essential elements, participates in enzymes and contributes to oxidative stress response [64], bone formation and metabolism of amino acids, cholesterol and carbohydrates [65], varied between 27.8 and 101 μg g^{-1}. Zn, which enhances body immunity and protection against several diseases, maintaining a crucial role in many enzymes and participating in metabolic reactions [66], was measured from 33.8 to 56.1 μg g^{-1}.

Table 5. Levels of trace metals in dry herbal tissues and prepared infusions (μg g^{-1}) of *A. arborescens* and *A. inculta* and extraction efficiency (% EE) of metals from the herb to the infusion.

	Cd	Co	Cr	Cu	Fe	Mn	Ni	Pb	Zn
A. arborescens									
Herbal tissue	0.621 ± 0.056	0.295 ± 0.027	2.42 ± 0.03	9.38 ± 1.02	228 ± 25	87.9 ± 10.0	22.6 ± 1.9	0.676 ± 0.056	56.1 ± 6.0
Infusion	0.254 ± 0.021	0.554 ± 0.049	1.12 ± 0.14	16.4 ± 1.88	44.8 ± 5.2	101 ± 11	41.9 ± 5.1	0.298 ± 0.031	52.6 ± 4.8
% EE	13.0	59.7	14.8	55.6	6.2	36.6	29.1	14.0	29.8
A. inculta									
Herbal tissue	0.064 ± 0.007	0.117 ± 0.011	0.781 ± 0.063	10.1 ± 0.9	175 ± 18	28.6 ± 3.4	1.47 ± 0.12	0.174 ± 0.016	33.8 ± 2.9
Infusion	0.059 ± 0.006	0.156 ± 0.013	1.04 ± 0.12	21.1 ± 1.9	45.8 ± 5.7	27.8 ± 2.2	4.77 ± 0.51	0.303 ± 0.036	
% EE	24.6	35.5	35.5	41.3	7.0	26.0	36.5	46.6	42.4

Cd: cadmium, Co: cobalt, Cr: chromium, Cu: copper, Fe: iron, Mn: manganese, Ni: nickel, Pb: lead, Zn: zinc.

Comparatively lower concentrations were measured for Ni (1.47–41.9 μg g^{-1}) and Cu (9.38–21.1 μg g^{-1}), which, although essential, may exhibit a toxic impact when detected at elevated concentrations. Despite the relatively limited data available, a beneficial role of Ni in physiological processes of animal species has been demonstrated, together with potential carcinogenic effects accompanying exposure to nickel compounds [67]. Cr concentrations measured in *Artemisia* tissue and infusion samples analyzed herein were equal to 0.781–2.42 μg g^{-1}. Even lower levels were detected for Co (0.117–0.554 μg g^{-1}), which is closely associated with the physiological role of vitamin B12 in the production and maintenance of red blood cells.

A relatively low content of the toxic elements Cd (0.059–0.621 μg g^{-1}) and Pb (0.174–0.676 μg g^{-1}) was determined in *Artemisia* tissue and infusion samples. Classified by the International Agency for Research on Cancer (IARC) as "carcinogenic to humans" [68] and ranked by the EU in category 1 [69], Cd has been characterized as responsible for renal tubular dysfunction, bone fragility and reproductive disorders following prolonged oral exposure. Concerning Pb, its well-demonstrated toxicity threatens both young children, with the central nervous system representing the target organ, as well as adults with the manifestation of chronic kidney disease and cardiovascular dysfunctions. The maximum permissible levels in raw plant materials, set at 0.3 μg g^{-1} for Cd and 10 μg g^{-1} for Pb by the World Health Organization [70], were exceeded only in the case of Cd measured in the *A. arborescens* tissue sample (0.621 μg g^{-1}).

Values detected in *A. arborescens* are similar to those reported for various other *Artemisia* species (in μg g^{-1} per dry weight of herb) for Cd (0.05–0.75), Cu (5.9–16.9), Fe (79.1–209.3), Mn (47.7–75.2) and Zn (35.2–58.6), whereas they were lower for Pb (1.25–2.08) [71]. Begaa et al. [72] reported similar values for Co (0.27–0.30) and Cr (0.74–1.50) for A. campestris and *A. herba-alba*, while reporting lower values for Zn (13–18) and higher values for Fe (617–631). Values comparable to these of the present work were recently presented by Ait Bouzid et al. [73] for *A. herba-alba* samples as follows: Cd (0.02 $\pm$ 0.01), Cu (6.6 $\pm$ 0.5), Fe (499 $\pm$ 40), Mn (80.3 $\pm$ 6.5), Pb (1.50 $\pm$ 0.03) and Zn (22.5 $\pm$ 1.8). As regards other herbal species consumed in the form of infusions, comparable levels (in μg g^{-1} per dry weight of herb) of Cd (0.01–0.39), Cr (0.27–2.45) and Ni (2.70–13.41) as well as significantly higher levels of Cu (7.73–63.71) and Pb (0.48–10.57) were detected in a Chinese tea [74]. Similar results (in μg g^{-1}) for Cd (0.16–0.68), Cu (4.19–9.49), Fe (79.4–522) and Pb (0.73–1.51) were reported by Kalny et al. [75] for *Taraxacum officinale* (dandelion), *Betula* sp. (birch) and *Crataegus* sp. (hawthorn), commonly used for tea preparation.

Herbal infusions are taken orally and ingested in our digestive system. The element fraction actually retained in the human body following consumption is determined by the levels of elements extracted in the infusion. The extraction efficiency of trace elements is further dependent on the plant species, the organic matrix composition of the infusion prepared and element incorporation therein, in the form of either different covalent species or coordination complexes. In addition to the trace elements content of the initial herbal tissue, the corresponding infusion provides, hence, valuable information [76]. Although the species of *Artemisia* examined differed in the extractability order of trace elements transferred from the herbal tissue towards the infusion, Cu was significantly extracted in both cases (55.6 and 41.3% for *A. arborescens* and *A. inculta*, respectively) followed by Co (59.7 and 35.5%), Zn (29.8 and 42.4%), Cr (14.8 and 35.5%), Mn (36.6 and 26.0%), Ni (29.1 and 36.5%) and Pb (14.0 and 46.6%) which migrated moderately, while Cd (13.0 and 24.6%) and Fe (6.2 and 7.0%) were poorly extracted (Table 5). According to Matsuura et al. [77] differences characterizing the extraction efficiencies of transition metals are difficult to explain, being related to their ionic and covalent features.

To estimate the contribution of *Artemisia* infusion consumption to metal intake, a daily consumption of 2 cups (200 mL per cup) and a body weight of 65 kg were assumed. Metal concentrations expressed per cup for *A. arborescens* and *A. inculta* were, respectively, equal to 0.239 and 0.047 for Cd, 0.523 and 0.125 for Co, 1.06 and 0.833 for Cr, 15.5 and 12.6 for Cu, 42.3 and 36.7 for Fe, 95.6 and 22.3 for Mn, 19.5 and 1.60 for Ni, 0.281 and 0.242 for Pb and 49.6 and 42.9 for Zn. Potential intake of inadequate Fe levels might be responsible for a gradual reduction of Fe stores, further leading to Fe deficiency, a threat mainly to women. A Recommended Daily Intake (RDI) for Fe was set at 8–18 mg·day^{-1} [78]. Due to a lack of adequate data, no upper limit (UL) has been set for Mn so far, while its Adequate Intake (AI) was set at 5–5.5 mg·day^{-1} [79]. For Zn the RDI was set at 8–14 mg·day^{-1}; however, due to the negative impact an excessive Zn intake might provoke, its UL has been set at 5–40 mg·day^{-1} [79]. For Fe, Mn and Zn, a 2-cup daily consumption of *Artemisia* infusions contributed to less than 3% of the corresponding lower bounds. Regarding the potentially toxic metals examined, a Tolerable Daily Intake (TDI) for Ni was recently established equal to 13 $\mu g \cdot kg^{-1}$ bw·day^{-1} [80] and due to lack of adequate evidence, EFSA [78] adopted a TDI of 300 $\mu g\ kg^{-1}$ bw·day^{-1} for Cr. Based on the classification of Co(II) compounds as "possibly carcinogenic to humans" [81], a TDI equal to 1.6–8 $\mu g \cdot kg^{-1}$ bw·day^{-1} was set [82]. In all cases of Ni, Co and Cr, daily consumption contributed to an intake not exceeding 5%. A Provisional Tolerable Weekly Intake (PTWI) of 2.5 $\mu g \cdot kg^{-1}$ bw·week^{-1} has been set for Cd by EFSA [83], while, due to Pb toxicity, a PTWI has been set at 25 $\mu g \cdot kg^{-1}$ bw·week^{-1} in 1986 by JECFA. The latter is a health guidance value that, however, has been withdrawn and not replaced so far [84]. In both cases, *Artemisia* infusions contributed at a percentage not exceeding 2%.

4. Conclusions

The results of our study highlight the potential use of the investigated *Artemisia* species not only in the nutrition and food industry, but also in the development of dermo-cosmetic applications, as glycerol is well known for its ability to increase the transdermal delivery of active substances. The latter should be seen from the perspective of the increased consumer demand for plant-derived substances in cosmetology, as well as for more green and sustainable products in general.

Supplementary Materials: The following supporting information can be downloaded at: https://www.mdpi.com/article/10.3390/life13061416/s1, Table S1: Target and qualifier ions for the trimethylsilyl ethers (TMS) of simple phenols, stilbenes, terpenic compounds, and the internal standard (IS).

Author Contributions: D.Z.L.: data curation; investigation; writing—original draft. C.A.: investigation, writing—original draft. S.K.: investigation; data curation; writing—original draft. V.K.: investigation. A.S.: investigation, writing—original draft. D.T.: investigation. P.Z.: methodology; supervision. D.P.M.: conceptualization. N.K.: conceptualization; methodology; supervision. A.C.K.: investigation; data curation; supervision; writing—original draft. All authors have read and agreed to the published version of the manuscript.

Funding: This research did not receive any specific grant from funding agencies in the public, commercial or not-for-profit sectors.

Institutional Review Board Statement: Not applicable.

Informed Consent Statement: Not applicable.

Data Availability Statement: Data will be made available on request from the corresponding author.

Conflicts of Interest: The authors declare no conflict of interest.

References

1. Leonti, M.; Casu, L. Traditional medicines and globalization: Current and future perspectives in ethnopharmacology. *Front. Pharmacol.* **2013**, *4*, 92. [CrossRef] [PubMed]
2. Alami Merrouni, I.; Elachouri, M. Anticancer medicinal plants used by Moroccan people: Ethnobotanical, preclinical, phytochemical and clinical evidence. *J. Ethnopharmacol.* **2021**, *266*, 113435. [CrossRef] [PubMed]
3. Bussmann, R.W.; Glenn, A. Medicinal plants used in Northern Peru for reproductive problems and female health. *J. Ethnobiol. Ethnomed.* **2010**, *6*, 30. [CrossRef] [PubMed]
4. Ali-Shtayeh, M.S.; Jamous, R.M.; Al-Shafie, J.H.; Elgharabah, W.A.; Kherfan, F.A.; Qarariah, K.H.; Khdair, I.S.; Soos, I.M.; Musleh, A.A.; Isa, B.A.; et al. Traditional knowledge of wild edible plants used in Palestine (Northern West Bank): A comparative study. *J. Ethnobiol. Ethnomed.* **2008**, *4*, 13. [CrossRef]
5. Hassan, W.; Noreen, H.; Rehman, S.; Gul, S.; Amjad Kamal, M.; Paul Kamdem, J.; Zaman, B.; BT da Rocha, J. Oxidative Stress and Antioxidant Potential of One Hundred Medicinal Plants. *Curr. Top. Med. Chem.* **2017**, *17*, 1336–1370. [CrossRef]
6. Watson, B.; Kennel, E. *Artemisia* spp. Available online: https://www.herbsociety.org (accessed on 19 March 2022).
7. Feng, X.; Cao, S.; Qiu, F.; Zhang, B. Traditional application and modern pharmacological research of *Artemisia annua* L. *Pharmacol. Ther.* **2020**, *216*, 107650. [CrossRef]
8. Sharifi-Rad, J.; Herrera-Bravo, J.; Semwal, P.; Painuli, S.; Badoni, H.; Ezzat, S.M.; Farid, M.M.; Merghany, R.M.; Aborehab, N.M.; Salem, M.A.; et al. *Artemisia* spp.: An Update on Its Chemical Composition, Pharmacological and Toxicological Profiles. *Oxidative Med. Cell. Longev.* **2022**, *2022*, 5628601. [CrossRef]

9. Watson, L.E.; Bates, P.L.; Evans, T.M.; Unwin, M.M.; Estes, J.R. Molecular phylogeny of Subtribe Artemisiinae (Asteraceae), including *Artemisia* and its allied and segregate genera. *BMC Evol. Biol.* **2002**, *2*, 17. [CrossRef]

10. Algieri, F.; Rodriguez-Nogales, A.; Rodriguez-Cabezas, M.E.; Risco, S.; Ocete, M.A.; Galvez, J. Botanical Drugs as an Emerging Strategy in Inflammatory Bowel Disease: A Review. *Mediat. Inflamm.* **2015**, *2015*, 179616. [CrossRef]

11. Ekiert, H.; Klimek-Szczykutowicz, M.; Rzepiela, A.; Klin, P.; Szopa, A. *Artemisia* Species with High Biological Values as a Potential Source of Medicinal and Cosmetic Raw Materials. *Molecules* **2022**, *27*, 6427. [CrossRef]

12. Jaradat, N.; Qneibi, M.; Hawash, M.; Al-Maharik, N.; Qadi, M.; Abualhasan, M.N.; Ayesh, O.; Bsharat, J.; Khadir, M.; Morshed, R.; et al. Assessing *Artemisia arborescens* essential oil compositions, antimicrobial, cytotoxic, anti-inflammatory, and neuroprotective effects gathered from two geographic locations in Palestine. *Ind. Crop. Prod.* **2022**, *176*, 114360. [CrossRef]

13. Russo, A.; Bruno, M.; Avola, R.; Cardile, V.; Rigano, D. Chamazulene-Rich *Artemisia arborescens* Essential Oils Affect the Cell Growth of Human Melanoma Cells. *Plants* **2020**, *9*, 1000. [CrossRef] [PubMed]

14. Zeng, Z.-W.; Chen, D.; Chen, L.; He, B.; Li, Y. A comprehensive overview of Artemisinin and its derivatives as anticancer agents. *Eur. J. Med. Chem.* **2022**, *247*, 115000. [CrossRef]

15. Dhibi, S.; Bouzenna, H.; Samout, N.; Tlili, Z.; Elfeki, A.; Hfaiedh, N. Nephroprotective and antioxidant properties of *Artemisia arborescens* hydroalcoholic extract against oestroprogestative-induced kidney damages in rats. *Biomed. Pharmacother.* **2016**, *82*, 520–527. [CrossRef] [PubMed]

16. Dhibi, S.; Ettaya, A.; Elfeki, A.; Hfaiedh, N. Protective effects of *Artemisia arborescens* essential oil on oestroprogestative treatment induced hepatotoxicity. *Nutr. Res. Pract.* **2015**, *9*, 466. [CrossRef]

17. Ijomone, O.M.; Ifenatuoha, C.W.; Aluko, O.M.; Ijomone, O.K.; Aschner, M. The aging brain: Impact of heavy metal neurotoxicity. *Crit. Rev. Toxicol.* **2020**, *50*, 801–814. [CrossRef]

18. Fu, Z.; Xi, S. The effects of heavy metals on human metabolism. *Toxicol. Mech. Methods* **2020**, *30*, 167–176. [CrossRef]

19. Shehata, E.; Grigorakis, S.; Loupassaki, S.; Makris, D.P. Extraction optimisation using water/glycerol for the efficient recovery of polyphenolic antioxidants from two *Artemisia* species. *Sep. Purif. Technol.* **2015**, *149*, 462–469. [CrossRef]

20. Karakashov, B.; Grigorakis, S.; Loupassaki, S.; Makris, D.P. Optimisation of polyphenol extraction from Hypericum perforatum (St. John's Wort) using aqueous glycerol and response surface methodology. *J. Appl. Res. Med. Aromat. Plants* **2015**, *2*, 1–8. [CrossRef]

21. Mitić, S.S.; Paunović, D.Đ.; Pavlović, A.N.; Tošić, S.B.; Stojković, M.B.; Mitić, M.N. Phenolic profiles and total antioxidant capacity of marketed beers in Serbia. *Int. J. Food Prop.* **2013**, *17*, 908–922. [CrossRef]

22. Galanakis, C.M.; Tornberg, E.; Gekas, V. Recovery and preservation of phenols from olive waste in ethanolic extracts. *J. Chem. Technol. Biotechnol.* **2010**, *85*, 1148–1155. [CrossRef]

23. Fan, J.-P.; He, C.-H. Simultaneous quantification of three major bioactive triterpene acids in the leaves of diospyros kaki by high-performance liquid chromatography method. *J. Pharm. Biomed. Anal.* **2006**, *41*, 950–956. [CrossRef]

24. Lantzouraki, D.Z.; Sinanoglou, V.J.; Zoumpoulakis, P.G.; Glamočlija, J.; Ćirić, A.; Soković, M.; Heropoulos, G.; Proestos, C. Antiradical–antimicrobial activity and phenolic profile of pomegranate (*Punica granatum* L.) juices from different cultivars: A comparative study. *RSC Adv.* **2015**, *5*, 2602–2614. [CrossRef]

25. Lantzouraki, D.Z.; Tsiaka, T.; Soteriou, N.; Asimomiti, G.; Spanidi, E.; Natskoulis, P.; Gardikis, K.; Sinanoglou, V.J.; Zoumpoulakis, P. Antioxidant profiles of *Vitis vinifera* L. and *Salvia triloba* L. leaves using high-energy extraction methodologies. *J. AOAC Int.* **2020**, *103*, 413–421. [CrossRef]

26. Kaliora, A.C.; Batzaki, C.; Christea, M.G.; Kalogeropoulos, N. Nutritional evaluation and functional properties of traditional composite salad dishes. *LWT—Food Sci. Technol.* **2015**, *62*, 775–782. [CrossRef]

27. Lantzouraki, D.Z.; Sinanoglou, V.J.; Tsiaka, T.; Proestos, C.; Zoumpoulakis, P. Total phenolic content, antioxidant capacity and phytochemical profiling of grape and pomegranate wines. *RSC Adv.* **2015**, *5*, 101683–101692. [CrossRef]

28. van der Laan, T.; Boom, I.; Maliepaard, J.; Dubbelman, A.-C.; Harms, A.C.; Hankemeier, T. Data-independent acquisition for the quantification and identification of metabolites in plasma. *Metabolites* **2020**, *10*, 514. [CrossRef] [PubMed]

29. Guo, J.; Huan, T. Comparison of full-scan, data-dependent, and data-independent acquisition modes in liquid chromatography–mass spectrometry based untargeted metabolomics. *Anal. Chem.* **2020**, *92*, 8072–8080. [CrossRef]

30. Davies, V.; Wandy, J.; Weidt, S.; van der Hooft, J.J.; Miller, A.; Daly, R.; Rogers, S. Rapid development of improved data-dependent acquisition strategies. *Anal. Chem.* **2021**, *93*, 5676–5683. [CrossRef] [PubMed]

31. Grigoriou, C.; Costopoulou, D.; Vassiliadou, I.; Karavoltsos, S.; Sakellari, A.; Bakeas, E.; Leondiadis, L. Polycyclic aromatic hydrocarbons and trace elements dietary intake in inhabitants of Athens, Greece, based on a duplicate portion study. *Food Chem. Toxicol.* **2022**, *165*, 113087. [CrossRef] [PubMed]

32. U.S. Environmental Protection Agency (USEPA). Guidelines establishing test procedures for the analysis of pollutants (App. B, Part 136, definition and procedures for the determination of the method detection limit). In *U.S. Code of Federal Regulations*; U.S. Government Publishing Office: Washington, DC, USA, 1997; pp. 265–267.

33. Anibogwu, R.; Jesus, K.D.; Pradhan, S.; Pashikanti, S.; Mateen, S.; Sharma, K. Extraction, Isolation and Characterization of Bioactive Compounds from *Artemisia* and Their Biological Significance: A Review. *Molecules* **2021**, *26*, 6995. [CrossRef] [PubMed]

34. Ghafoori, H.; Sariri, R.; Naghavi, M.R. Study of effect of extraction conditions on the biochemical composition and antioxidant activity of *Artemisia* Absinthium by HPLC and TLC. *J. Liq. Chromatogr. Relat. Technol.* **2014**, *37*, 1558–1567. [CrossRef]
35. Singh, R.; Verma, P.; Singh, G. Total phenolic, flavonoids and tannin contents in different extracts of *Artemisia* absinthium. *J. Intercult. Ethnopharmacol.* **2012**, *1*, 101. [CrossRef]
36. Bourgou, S.; BettaiebRebey, I.; Mkadmini, K.; Isoda, H.; Ksouri, R.; Ksouri, W.M. LC-ESI-TOF-MS and GC-MS profiling of *Artemisia herba-alba* and evaluation of its bioactive properties. *Food Res. Int.* **2017**, *99*, 702–712. [CrossRef] [PubMed]
37. Pandey, A.K.; Singh, P. The Genus *Artemisia*: A 2012–2017 Literature Review on Chemical Composition, Antimicrobial, Insecticidal and Antioxidant Activities of Essential Oils. *Medicines* **2017**, *4*, 68. [CrossRef]
38. Carvalho, I.S.; Cavaco, T.; Brodelius, M. Phenolic composition and antioxidant capacity of six *Artemisia* species. *Ind. Crop. Prod.* **2011**, *33*, 382–388. [CrossRef]
39. Ornano, L.; Venditti, A.; Ballero, M.; Sanna, C.; Quassinti, L.; Bramucci, M.; Lupidi, G.; Papa, F.; Vittori, S.; Maggi, F.; et al. Chemopreventive and Antioxidant Activity of the Chamazulene-Rich Essential Oil Obtained from *Artemisia arborescens* L. Growing on the Isle of La Maddalena, Sardinia, Italy. *Chem. Biodivers.* **2013**, *10*, 1464–1474. [CrossRef]
40. Gouveia, S.; Castilho, P.C. Antioxidant potential of *Artemisia* argentea L'Hér alcoholic extract and its relation with the phenolic composition. *Food Res. Int.* **2011**, *44*, 1620–1631. [CrossRef]
41. Khan, I.; Karim, N.; Ahmad, W.; Abdelhalim, A.; Chebib, M. GABA-A Receptor Modulation and Anticonvulsant, Anxiolytic, and Antidepressant Activities of Constituents from *Artemisia indica* Linn. *Evid. Based Complement. Altern. Med.* **2016**, *2016*, 1215393. [CrossRef]
42. Jyoti, M.A.; Nam, K.-W.; Jang, W.S.; Kim, Y.-H.; Kim, S.-K.; Lee, B.-E.; Song, H.-Y. Antimycobacterial activity of methanolic plant extract of *Artemisia* capillaris containing ursolic acid and hydroquinone against Mycobacterium tuberculosis. *J. Infect. Chemother.* **2016**, *22*, 200–208. [CrossRef]
43. Peñas-Fuentes, J.L.; Siles, E.; Rufino-Palomares, E.E.; Pérez-Jiménez, A.; Reyes-Zurita, F.J.; Lupiáñez, J.A.; Fuentes-Almagro, C.; Peragón-Sánchez, J. Effects of Erythrodiol on the Antioxidant Response and Proteome of HepG2 Cells. *Antioxidants* **2021**, *11*, 73. [CrossRef] [PubMed]
44. Carmo, J.; Cavalcante-Araújo, P.; Silva, J.; Ferro, J.; Correia, A.C.; Lagente, V.; Barreto, E. Uvaol Improves the Functioning of Fibroblasts and Endothelial Cells and Accelerates the Healing of Cutaneous Wounds in Mice. *Molecules* **2020**, *25*, 4982. [CrossRef]
45. López-Ruiz, R.; Romero-González, R.; Garrido Frenich, A. Metabolomics approaches for the determination of multiple contaminants in food. *Curr. Opin. Food Sci.* **2019**, *28*, 49–57. [CrossRef]
46. Slimestad, R.; Johny, A.; Thomsen, M.G.; Karlsen, C.R.; Rosnes, J.T. Chemical profiling and biological activity of extracts from nine Norwegian medicinal and aromatic plants. *Molecules* **2022**, *27*, 7335. [CrossRef]
47. Mamatova, A.S.; Korona-Glowniak, I.; Skalicka-Woźniak, K.; Józefczyk, A.; Wojtanowski, K.K.; Baj, T.; Sakipova, Z.B.; Malm, A. Phytochemical composition of wormwood (*Artemisia gmelinii*) extracts in respect of their antimicrobial activity. *BMC Complement. Altern. Med.* **2019**, *19*, 288. [CrossRef]
48. Minda, D.; Ghiulai, R.; Banciu, C.D.; Pavel, I.Z.; Danciu, C.; Racoviceanu, R.; Soica, C.; Budu, O.D.; Muntean, D.; Diaconeasa, Z.; et al. Phytochemical profile, antioxidant and wound healing potential of three *Artemisia* species: In vitro and in OVO evaluation. *Appl. Sci.* **2022**, *12*, 1359. [CrossRef]
49. Gong, G.; Guan, Y.-Y.; Zhang, Z.-L.; Rahman, K.; Wang, S.-J.; Zhou, S.; Luan, X.; Zhang, H. Isorhamnetin: A review of Pharmacological Effects. *Biomed. Pharmacother.* **2020**, *128*, 110301. [CrossRef] [PubMed]
50. Ferreira, J.F.S.; Luthria, D.L.; Sasaki, T.; Heyerick, A. Flavonoids from *Artemisia annua* L. as antioxidants and their potential synergism with artemisinin against malaria and cancer. *Molecules* **2010**, *15*, 3135–3170. [CrossRef] [PubMed]
51. Luo, Y.; Sun, G.; Dong, X.; Wang, M.; Qin, M.; Yu, Y.; Sun, X. Isorhamnetin attenuates atherosclerosis by inhibiting macrophage apoptosis via PI3K/AKT activation and HO-1 induction. *PLoS ONE* **2015**, *10*, e0120259. [CrossRef]
52. Wei, J.; Su, H.; Bi, Y.; Li, J.; Feng, L.; Sheng, W. Anti-proliferative effect of isorhamnetin on Hela cells through inducing G2/M cell cycle arrest. *Exp. Ther. Med.* **2018**, *15*, 3917–3923. [CrossRef]
53. Larrazábal-Fuentes, M.J.; Fernández-Galleguillos, C.; Palma-Ramírez, J.; Romero-Parra, J.; Sepúlveda, K.; Galetovic, A.; González, J.; Paredes, A.; Bórquez, J.; Simirgiotis, M.J.; et al. Chemical profiling, antioxidant, anticholinesterase, and antiprotozoal potentials of *Artemisia* Copa phil. (Asteraceae). *Front. Pharmacol.* **2020**, *11*, 1911. [CrossRef]
54. Alam, W.; Khan, H.; Shah, M.A.; Cauli, O.; Saso, L. Kaempferol as a dietary anti-inflammatory agent: Current therapeutic standing. *Molecules* **2020**, *25*, 4073. [CrossRef]
55. Araniti, F.; Gullì, T.; Marrelli, M.; Statti, G.; Gelsomino, A.; Abenavoli, M.R. *Artemisia arborescens* L. Leaf Litter: Phytotoxic activity and phytochemical characterization. *Acta Physiol. Plant.* **2016**, *38*, 128. [CrossRef]
56. Younsi, F.; Trimech, R.; Boulila, A.; Ezzine, O.; Dhahri, S.; Boussaid, M.; Messaoud, C. Essential oil and phenolic compounds of *Artemisia herba-alba* (asso.): Composition, Antioxidant, antiacetylcholinesterase, and antibacterial activities. *Int. J. Food Prop.* **2015**, *19*, 1425–1438. [CrossRef]
57. Dhifallah, A.; Selmi, H.; Ouerghui, A.; Sammeri, H.; Aouini, D.; Rouissi, H. Comparative study of phenolic compounds and antiradical activities of four extracts of Tunisian *Artemisia herba Alba*. *Pharm. Chem. J.* **2022**, *56*, 226–232. [CrossRef]

58. Mohammed, M.J.; Anand, U.; Altemimi, A.B.; Tripathi, V.; Guo, Y.; Pratap-Singh, A. Phenolic composition, antioxidant capacity and antibacterial activity of white wormwood (*Artemisia herba-alba*). *Plants* **2021**, *10*, 164. [CrossRef]

59. Mehdizadeh, A.; Karimi, E.; Oskoueian, E. Nano-liposomal encapsulation of *Artemisia aucheri* phenolics as a potential phytobiotic against campylobacter jejuni infection in mice. *Food Sci. Nutr.* **2022**, *10*, 3314–3322. [CrossRef] [PubMed]

60. Alfei, S.; Marengo, B.; Zuccari, G. Oxidative stress, antioxidant capabilities, and bioavailability: Ellagic acid or urolithins? *Antioxidants* **2020**, *9*, 707. [CrossRef]

61. Gil, T.-Y.; Hong, C.-H.; An, H.-J. Anti-inflammatory effects of ellagic acid on keratinocytes via MAPK and stat pathways. *Int. J. Mol. Sci.* **2021**, *22*, 1277. [CrossRef] [PubMed]

62. Jordão, J.; Porto, H.; Lopes, F.; Batista, A.; Rocha, M. Protective effects of ellagic acid on cardiovascular injuries caused by hypertension in rats. *Planta Med.* **2017**, *83*, 830–836. [CrossRef] [PubMed]

63. Arzani, A.; Zeinali, H.; Razmjo, K. Iron and magnesium concentrations of mint accessions (*Mentha* spp.). *Plant Physiol. Biochem.* **2007**, *45*, 323–329. [CrossRef]

64. Michalke, B.; Fernsebner, K. New insights into manganese toxicity and speciation. *J. Trace Elem. Med. Biol.* **2014**, *28*, 106–116. [CrossRef]

65. Gimou, M.-M.; Pouillot, R.; Charrondiere, U.R.; Noël, L.; Guérin, T.; Leblanc, J.-C. Dietary exposure and health risk assessment for 14 toxic and essential trace elements in Yaoundé: The Cameroonian total diet study. *Food Addit. Contam.* **2014**, *31*, 1064–1080. [CrossRef]

66. Rose, M.; Baxter, M.; Brereton, N.; Baskaran, C. Dietary exposure to metals and other elements in the 2006 UK Total Diet Study and some trends over the last 30 years. *Food Addit. Contam.* **2010**, *27*, 1380–1404. [CrossRef]

67. International Agency for Research on Cancer. Chromium, nickel and welding. *IARC Monogr. Eval. Carcinog. Risks Hum.* **1990**, *49*, 647–648.

68. International Agency for Research on Cancer. Beryllium, cadmium, mercury, and exposures in the glass manufacturing industry. *IARC Monogr. Eval. Carcinog. Risks Hum.* **1993**, *19*, 360–363.

69. Pennington, J. Commission Directive 2004/ 73/EC, 29th Time Council Directive 67/548EEC. *OJEC* **2004**, *152*, 1–311. Available online: https://op.europa.eu/en/publication-detail/-/publication/7bf52e98-781e-4c0a-aec9-3795a32f5984 (accessed on 5 May 2023).

70. WHO. *Quality Control Methods for Medicinal Plant Materials*; World Health Organization: Geneva, Switzerland, 1998; Available online: https://www.who.int/docs/default-source/medicines/norms-and-standards/guidelines/quality-control/quality-control-methods-for-medicinal-plant-materials.pdf?sfvrsn=b451e7c6_0 (accessed on 15 March 2023).

71. Chizzola, R.; Michitsch, H.; Franz, C. Monitoring of metallic micronutrients and heavy metals in herbs, spices and medicinal plants from Austria. *Eur. Food Res. Technol.* **2003**, *216*, 407–411. [CrossRef]

72. Begaa, S.; Messaoudi, M.; Benarfa, A. Statistical Approach and Neutron Activation Analysis for Determining Essential and Toxic Elements in Two Kinds of Algerian *Artemisia* Plant. *Biol. Trace Elem. Res.* **2020**, *199*, 2399–2405. [CrossRef] [PubMed]

73. Ait Bouzid, H.; Oubannin, S.; Ibourki, M.; Bijla, L.; Hamdouch, A.; Sakar, E.H.; Harhar, H.; Majourhat, K.; Koubachi, J.; Gharby, S. Comparative evaluation of chemical composition, antioxidant capacity, and some contaminants in six Moroccan medicinal and Aromatic. *Biocatal. Agric. Biotechnol.* **2022**, *47*, 102569. [CrossRef]

74. Zhong, W.-S.; Ren, T.; Zhao, L.-J. Determination of Pb (Lead), Cd (Cadmium), Cr (Chromium), Cu (Copper), and Ni (Nickel) in Chinese tea with high-resolution continuum source graphite furnace atomic absorption spectrometry. *J. Food Drug Anal.* **2016**, *24*, 46–55. [CrossRef] [PubMed]

75. Kalny, P.; Fijałek, Z.; Daszczuk, A.; Ostapczuk, P. Determination of selected microelements in polish herbs and their infusions. *Sci. Total Environ.* **2007**, *381*, 99–104. [CrossRef] [PubMed]

76. Pohl, P.; Bielawska-Pohl, A.; Dzimitrowicz, A.; Greda, K.; Jamroz, P.; Lesniewicz, A.; Szymczycha-Madeja, A.; Welna, M. Understanding element composition of medicinal plants used in herbalism—A case study by analytical atomic spectrometry. *J. Pharm. Biomed. Anal.* **2018**, *159*, 262–271. [CrossRef] [PubMed]

77. Matsuura, H.; Hokura, A.; Katsuki, F.; Itoh, A.; Haraguchi, H. Multielement Determination and Speciation of Major-to-Trace Elements in Black Tea Leaves by ICP-AES and ICP-MS with the Aid of Size Exclusion Chromatography. *Anal. Sci.* **2001**, *17*, 391–398. [CrossRef] [PubMed]

78. NDA. Scientific Opinion on Dietary Reference Values for chromium. *EFSA J.* **2014**, *12*, 3845. [CrossRef]

79. National Health and Medical Research Council. *Nutrient Reference Values for Australia and New Zealand Including Recommended Dietary Intakes*; National Health and Medical Research Council: Canberra, Australia, 2006.

80. Schrenk, D.; Bignami, M.; Bodin, L.; Chipman, J.K.; del Mazo, J.; Grasl-Kraupp, B.; Hogstrand, C.; Hoogenboom, L.; Leblanc, J.; Nebbia, C.S.; et al. Update of the risk assessment of nickel in food and drinking water. *EFSA J.* **2020**, *18*, e06268. [CrossRef]

81. International Agency for Research on Cancer. *IARC Monographs on the Evaluation of Carcinogenic Risks to Humans*; Cobalt and Cobalt Compounds 52: Lyon, France, 1991.

82. AFSSA (Agence Française de Sécurité Sanitaire des Aliments/French Agency for Food Safety). Opinion of the French Food Safety Agency on a Request for Scientific and Technical Support Regarding the Migration of Cobalt from Porcelain Oven-Dishes Intended to Come in Contact with Food. Available online: https://www.anses.fr/Documents/MCDA2010sa0095EN.pdf (accessed on 11 May 2010).

83. EFSA. Opinion of the Scientific Panel on contaminants in the food chain [CONTAM] related to cadmium as undesirable substance in animal feed. *EFSA J.* **2004**, *2*, 72. [CrossRef]
84. Joint FAO/WHO Expert Committee on Food Additives; Meeting and World Health Organization. *Safety Evaluation of Certain Contaminants in Food*; Food and Agriculture Organization of the United Nations: Rome, Italy, 2011.

Article

Pharmacological Properties of Chemically Characterized Extracts from Mastic Tree: In Vitro and In Silico Assays

Safae Ouahabi [1,*], El Hassania Loukili [1], Amine Elbouzidi [2], Mohamed Taibi [2], Mohammed Bouslamti [3], Hiba-Allah Nafidi [4], Ahmad Mohammad Salamatullah [5], Nezha Saidi [1], Reda Bellaouchi [6], Mohamed Addi [2], Mohamed Ramdani [1], Mohammed Bourhia [7,*] and Belkheir Hammouti [1]

[1] Laboratory of Applied and Environmental Chemistry (LCAE), Faculty of Sciences, Mohammed First University, B.P. 717, Oujda 60000, Morocco

[2] Laboratoire d'Amélioration des Productions Agricoles, Biotechnologie et Environnement (LAPABE), Faculté des Sciences, Université Mohammed Premier, Oujda 60000, Morocco

[3] Laboratories of Natural Substances, Pharmacology, Environment, Modeling, Health and Quality of Life (SNAMOPEQ), Faculty of Sciences, Sidi Mohamed Ben Abdellah University, Fez 30000, Morocco

[4] Department of Food Science, Faculty of Agricultural and Food Sciences, Laval University, Quebec City, QC G1V 0A6, Canada

[5] Department of Food Science & Nutrition, College of Food and Agricultural Sciences, King Saud University, P.O. Box 2460, Riyadh 11451, Saudi Arabia

[6] Laboratory of Bioresources, Biotechnology, Ethnopharmacology and Health, Faculty of Sciences, Mohammed First University, Boulevard Mohamed VI, B.P. 717, Oujda 60000, Morocco

[7] Department of Chemistry and Biochemistry, Faculty of Medicine and Pharmacy, Ibn Zohr University, Laayoune 70000, Morocco

* Correspondence: ouahabi.safae@ump.ac.ma (S.O.); bourhia.mohammed@gmail.com (M.B.)

Citation: Ouahabi, S.; Loukili, E.H.; Elbouzidi, A.; Taibi, M.; Bouslamti, M.; Nafidi, H.-A.; Salamatullah, A.M.; Saidi, N.; Bellaouchi, R.; Addi, M.; et al. Pharmacological Properties of Chemically Characterized Extracts from Mastic Tree: In Vitro and In Silico Assays. *Life* **2023**, *13*, 1393. https://doi.org/10.3390/life13061393

Academic Editors: Efstathia Papada, Charalampia Amerikanou and Marisa Colone

Received: 16 May 2023
Revised: 5 June 2023
Accepted: 8 June 2023
Published: 14 June 2023

Abstract: The mastic tree, scientifically known as *Pistacia lentiscus*, which belongs to the Anacardiaceae family, was used in this study. The aim of this research was to analyze the chemical composition of this plant and assess its antioxidant and antibacterial properties using both laboratory experiments and computer simulations through molecular docking, a method that predicts the binding strength of a small molecule to a protein. The soxhlet method (SE) was employed to extract substances from the leaves of *P. lentiscus* found in the eastern region of Morocco. Hexane and methanol were the solvents used for the extraction process. The n-hexane extract was subjected to gas chromatography-mass spectrometry (GC/MS) to identify its fatty acid content. The methanolic extract underwent high-performance liquid chromatography with a diode-array detector (HPLC-DAD) to determine the presence of phenolic compounds. Antioxidant activity was assessed using the DPPH spectrophotometric test. The findings revealed that the main components in the n-hexane extract were linoleic acid (40.97 ± 0.33%), oleic acid (23.69 ± 0.12%), and palmitic acid (22.83 ± 0.10%). Catechin (37.05 ± 0.15%) was identified as the predominant compound in the methanolic extract through HPLC analysis. The methanolic extract exhibited significant DPPH radical scavenging, with an IC50 value of 0.26 ± 0.14 mg/mL. The antibacterial activity was tested against *Staphylococcus aureus*, *Listeria innocua*, and *Escherichia coli*, while the antifungal activity was evaluated against *Geotrichum candidum* and *Rhodotorula glutinis*. The *P. lentiscus* extract demonstrated notable antimicrobial effects. Additionally, apart from molecular docking, other important factors, such as drug similarity, drug metabolism and distribution within the body, potential adverse effects, and impact on bodily systems, were considered for the substances derived from *P. lentiscus*. Scientific algorithms, such as Prediction of Activity Spectra for Substances (PASS), Absorption, Distribution, Metabolism, Excretion (ADME), and Pro-Tox II, were utilized for this assessment. The results obtained from this research support the traditional medicinal usage of *P. lentiscus* and suggest its potential for drug development.

Keywords: *P. lentiscus*; GC/MS; HPLC-DAD; antioxidant activity; antibacterial activity; antifungal activity; extract; molecular docking

1. Introduction

The Anacardiaceae family encompasses the *Pistacia* genus, which comprises various plant species of notable significance in their food, medicinal, and ornamental properties [1]. The genus includes around 20 species, ranging from evergreen or deciduous trees, shrubs, and small trees, standing between 5 and 15 m in height [2,3]. *Pistacia* is widely distributed across regions such as Africa, Southern Europe, Asia, and North America. It mainly thrives in the Mediterranean region, where favorable humidity conditions exist for its growth [4]. *Pistacia* trees are dioecious, meaning they have male and female flowers growing on different trees [5]. Recently, the pharmaceutical industry has been interested in the ethnomedicinal and biological potentials of the *Pistacia* genus [1]. While the phytochemical composition of the genus has been widely studied, new research has been focusing on the therapeutic effects of its extracts for enhancing health [6–14].

The mastic tree, or *P. lentiscus*, is a shrub species belonging to the *Pistacia* genus. This evergreen bush has a characteristic scent and green leaves, growing in Mediterranean and Middle Eastern regions and reaching heights between 1 and 8 m [15]. Since ancient times, mastic tree extracts have been employed in folk medicine for their anti-inflammatory, antiseptic, and disease-treating properties, such as treating gastralgia and dyspepsia [16]. The aerial part of the mastic tree has been utilized as a stimulant and diuretic to treat hypertension [17]. Mastic tree products are widely used in the food industry due to their secondary metabolites, such as flavonoids, polyphenols, and phenolic acids. Recently, mastic tree extracts have been reported to exhibit antioxidant activity [17–20], and their use in cosmetic products has also been documented [21]. Additionally, this species has been shown to display antihepatotoxic, antibacterial, and antiproliferative properties in colon cancer cells [16,22,23].

This investigation aimed to analyze the chemical constituents of *P. lentiscus* leaves (Figure 1) from the eastern region of Morocco. Furthermore, the extracts' potential biological activities were assessed, explicitly focusing on the methanolic's antimicrobial (antibacterial and antifungal) properties and antioxidant potential as measured by the DPPH free radical scavenging assay method. In addition, computational methods (molecular docking) were employed to investigate the principal compounds and their interactions, seeking to elucidate the underlying mechanisms.

Figure 1. Leaves of *P. lentiscus*.

2. Materials and Methods

2.1. Chemicals and Reagents

N-hexane and methanol were purchased from Merck (Darmstadt, Germany). 1,1-Diphenyl-2-picrylhydrazyl (DPPH•), Phenolic standards: Catechin, 4-hydroxybenzoic acid, *p*-Coumaric acid, naringin, quercetin, p-Coumaric acid, and luteolin were purchased from Merck and Carl Roth GmbH (Karlsruhe, Germany). All other chemicals used were of analytical grade.

2.2. Collection of Plant Material

Leaves of *P. lentiscus* were collected in March 2019 from Ahfir, located in the eastern region of Morocco. Subsequently, the leaves were meticulously cleansed and rinsed multiple times with distilled water, followed by air-drying in a well-ventilated location shielded from light and direct sunlight for 48 h. After that, lyophilization was carried out, and the leaves were ground into powder before extraction.

2.3. Soxhlet Extraction

A Soxhlet extraction apparatus consisting of a condenser, a Soxhlet chamber, and an extraction flask was used. In the extraction flask, 32 g of powder from dried mastic leaves were placed into an extraction thimble with 300 mL of the selected solvent (hexane and methanol). The period for the Soxhlet extractions experiments was chosen to be the time needed for the solvent to become colorless. After evaporation under vacuum, the extracts obtained were named HEPL and MEPL, respectively.

2.4. Determination of Extraction Yield

The extracts were stored in a refrigerated environment for preservation purposes. The extraction yield, expressed as a percentage, was calculated by dividing the weight of the obtained extract (*M extract*) by the weight of the dried initial sample (*M dry matter*) used for the extraction process. This calculation was carried out following the formula specified in the corresponding equation:

$$Extraction\ yield\ (\%) = \left[\frac{M\ extract\ (g)}{M\ dry\ matter\ (g)}\right] \times 100$$

2.5. Fatty Acid GC-MS Analysis of P. lentiscus Extracts

A modified version of the Fatty Acid GC-MS Analysis of *P. lentiscus* Extracts to analyze the fatty acid content of the hexane extract of *P. lentiscus* protocol as described by Loukili et al. [24] was employed. The BPX25 capillary column coupled to a QP2010 MS from Kyoto, Japan, was utilized in identifying and separating the fatty acids. Pure helium gas was the carrier gas at a constant flow rate of 3 mL/min. The temperature of the injection, ion source, and interface was maintained at 250 °C, while the temperature of the column oven was gradually increased from 50 °C to 250 °C at a rate of 10 °C/min. The ionization of sample components was performed in the EI mode (70 eV) with a mass range scanned of 40–300 *m/z*. The extract, diluted in n-hexane, was injected in a split mode with a volume of 1 µL, and the sample was analyzed in triplicate. The compounds were identified by comparing their retention times with authentic standards and their mass spectral fragmentation patterns with those stored in databases or on the National Institute of Standards and Technology (NIST). LabSolutions software, version 2.5, was utilized for data collection and processing.

2.6. Identification of Phenolic Compounds by HPLC-DAD

The HPLC/DAD Waters Corporation USA was employed to analyze the methanol extract. The separation module of the liquid chromatography (Waters; e2695) was coupled with a diode array detector (Waters 2998; PDA), and Empower software was used for data processing. A mobile phase gradient mode was applied on the AC18 column (5 µm, 4.6 mm × 250 mm), and the resulting chromatogram was captured using wavelengths in the 254–300 nm range. The mobile phase was composed of solvent A (ultrapure water/acetic acid, 2% *v/v*) and solvent B (acetonitrile) with varying compositions over time: 0–5 min: 95% A and 5% B; 25–30 min: 65% A and 35% B; 35–40 min: 30% A and 70% B; 40–45 min: 95% A and 5% B. The sample (20 µL) was injected, and the flow rate was set at 0.9 mL/min. Standard polyphenolic compounds like catechin, p-Coumaric acid, 4-hydroxybenzoic acid, Coumaric acid, quercetin, luteolin, and naringenin were utilized to identify the peaks by matching their retention times and UV spectra.

2.7. Antioxidant Activity

2.7.1. Determination of Antioxidants by DPPH Radical Scavenging Activity

The objective of the present investigation was to evaluate the antioxidant potential of the methanolic extract derived from the foliage of *P. lentiscus*. The antioxidant activity was determined using the DPPH radical scavenging assay, wherein ascorbic acid was used as a reference compound. In brief, 0.8 mL of the sample or standard solution at different concentrations (0.5, 0.4, 0.3, and 0.1 mg/mL) was mixed with 2 mL of DPPH• solution (4 mg of DPPH• dissolved in 200 mL of ethanol) and then mixed manually. Following this, the samples were kept in the dark at room temperature for 30 min, and their absorbance was measured at λ of 517 nm with a reference blank. This procedure was performed in triplicate to ensure reproducibility, and the reduction in DPPH• absorbance was recorded. Subsequently, the percentage of DPPH• radical scavenging inhibition by the sample was calculated using the following equation:

$$Inhibition~(\%) = \left[\frac{(Ab - As)}{Ab} \right] \times 100$$

The determination of the extract concentration leading to 50% *inhibition* (IC50) involved the plotting of the inhibition percentage against the extract concentration on a graph. The absorbance of the blank (*Ab*) and that of the positive control or sample (*As*) were utilized in this process.

2.7.2. β-Carotene Bleaching Assay

The antioxidant activity of *P. lentiscus* extracts was evaluated using the β-Carotene Bleaching assay. The emulsion was prepared by dissolving 2 mg of β-carotene, 20 mg of linoleic acid, and 200 mg of Tween 80 in 10 mL of chloroform. The solution was evaporated at 40 °C under a vacuum, and 100 mL of distilled water was added while stirring vigorously. The emulsion was mixed with either the extract or a reference antioxidant (BHA) at a concentration of 1 mg/mL in separate test tubes to evaluate the antioxidant activity. The absorbance at 470 nm was measured using a 96-well microplate reader at two different times: immediately after adding the emulsion (t0) and after 2 h. The volume of the emulsion used was 0.2 mL.

2.8. Antimicrobial Activity

2.8.1. Bacterial Strains

In this investigation, three bacterial strains, namely *Escherichia coli* (ATCC 10536), *Staphylococcus aureus* (ATCC 6538), and *Listeria innocua* (ATCC 49.189) were utilized as model organisms to assess the effectiveness of *P. lentiscus* extract as a bacterial growth inhibitor. The bacterial strains were obtained from the Microbiology and Microbial Biotechnology Laboratory of the Faculty of Science in Oujda, Morocco. The Mueller Hinton agar medium was used to cultivate bacterial cultures, which were kept at 37 °C for 24 h. The concentration of bacterial cells was determined to be 106 cells/mL using a UV-visible spectrophotometer at a wavelength of 620 nm before assessing the inhibitory effects of the *P. lentiscus* extract.

2.8.2. Agar-Diffusion Method

To qualitatively evaluate the antimicrobial efficacy of the substance against microbial strains, the agar diffusion method, a widely accepted technique for assessing susceptibility to various microbial strains, was used. The National Clinical Laboratory Standards Committee's guidelines were adhered to while carrying out the method. This method efficiently evaluates a substance's capacity to inhibit bacterial strains' proliferation [25]. The process included the addition of prepared bacterial inoculum to Petri dishes containing agar growth medium (MHA), where wells were made using a Pasteur pipette filled with 50 μL of the extract being tested. These plates were then incubated at 37 °C for 48 h, and the size of the inhibitory zone was measured to determine antimicrobial activity. Each test was conducted in triplicate to ensure the results' accuracy.

2.8.3. Determination of MIC and MBC

In assessing the effectiveness of antimicrobial agents, the determination of minimum inhibitory concentration (MIC) is critical. The present study utilized the resazurin microtiter assay to evaluate the MIC of *P. lentiscus* extract [26]. In this assay, a colorimetric indicator called resazurin is employed. It is reduced by metabolically active cells, which results in a color shift from blue to pink. Each well of a 96-well microplate contained different concentrations of the antimicrobial agent, and a standardized inoculum of the test bacteria was added to each well during the assay. The microplates were then incubated at 37 °C for 24 h and added resazurin to each well. A further 4–6 h of incubation occurred until a color change was observed. The MIC was determined as the lowest concentration of the antimicrobial agent that resulted in no color change, indicating the absence of viable bacteria. Controls were included in each microplate to confirm the accuracy of the results. The MBC was determined by inoculating a volume of 3 μL was taken from the negative wells as a sample, which was plated onto Mueller Hinton Agar medium plates and incubated at 37 °C for 24 h. The extract's lowest concentration that did not result in bacterial growth determined the MBC. The experiment was repeated in triplicate to ensure reproducibility.

2.9. Antifungal Activity

2.9.1. Selection of Fungal Strains for Antifungal Activity Testing Using *P. Lentiscus* Extract

In this study, two distinct fungal strains, namely *Rhodotorula glutinis* (ON 209167) and *Geotrichum candidum*, were selected to evaluate the antifungal potential of *P. lentiscus* extract. The strains were obtained from the Microbiology and Biotechnology Laboratory of the Faculty of Sciences in Oujda, Morocco.

2.9.2. Agar Diffusion Method

Culture conditions for the two fungal strains, *G. candidum* and *R. glutinis*, were optimized before testing the antifungal activity of *P. lentiscus* extract. *G. candidum* was cultured on potato dextrose agar medium for seven days at 25 °C, and the resulting spore concentration was adjusted to 2×10^6 spores/mL using the Thoma cell hemocytometer. *R. glutinis*, on the other hand, was cultured on Yeast Extract Peptone Dextrose for 48 h at 25 °C, and the cell concentration was measured and adjusted to 10^6 cells/mL. To test the antifungal activity of the *P. lentiscus* extract, the agar diffusion method, as used for bacteria, was employed with a slight modification in the culture medium (Yeast Extract Glucose). This method is widely used for testing susceptibility to different fungal strains and was performed according to the guidelines established by the National Clinical Laboratory Standards Committee [25].

2.9.3. Determination of MIC and MBC

The determination of MIC is crucial in evaluating the efficacy of antifungal agents. In this investigation, the resazurin microtiter test was utilized to determine the MIC of *P. lentiscus* extract [27]. The test was carried out in 96-well microplates, each containing a concentration range of 16% to 0.25%. A standardized inoculum of the fungal strain was added to each well, followed by incubation at 25 °C for 48 h. After incubation, resazurin was added, and the microplates were further incubated for 2 h until a color change from blue to pink was observed. The MIC was defined as the lowest concentration of the antifungal agent that did not result in a color change, indicating the absence of viable fungi. Positive and negative controls were included in each microplate to confirm the accuracy of the results. The MBC was determined by inoculating a 3 μL sample from the negative wells onto YEG and PDA medium plates, which were then incubated at 25 °C for 48 to 72 h. The MFC corresponded to the extract's lowest concentration, which did not result in any observable growth.

2.10. ADME and Toxicity Prediction

To determine the pharmacokinetic characteristics of the compounds being studied, this research assessed their profile in terms of Absorption, distribution, metabolism, and excretion (ADME). Computational tools, like SwissADME and pkCSM web servers, were utilized to forecast these characteristics. The examination assessed various physical and chemical traits of the compounds, their similarity to drugs, and their pharmacokinetic properties. These properties included their capacity to penetrate cell membranes, interact with transporters and enzymes responsible for drug absorption and elimination, and their metabolic stability [28–30]. Moreover, the toxicity levels of the molecules were estimated using the Protox II online tool [30]. This tool employs a statistical algorithm that compares the chemical structure of a substance to a vast database of toxic compounds to predict the probability of the substance causing harmful effects or toxicity to both humans and other living organisms. By utilizing the Protox II tool, significant details concerning the toxicity class, LD50 values, and toxicological endpoints like immunotoxicity, mutagenicity, cytotoxicity, hepatotoxicity, and carcinogenicity were obtained. As a result, these advanced computational techniques and instruments provided valuable comprehension of the potential therapeutic uses and risks of toxicity associated with the identified compounds.

2.11. PASS Prediction

To evaluate the pharmacological activity and potential toxicity risks of the primary chemical constituents present in Artemisia species' essential oils, we employed a set of sophisticated computational methods and tools [31]. Our methodology comprised the utilization of the PASS (Prediction of Activity Spectra for Substances) technique, which involves a statistical calculation that compares the chemical composition of a particular compound with an extensive database of bioactive compounds. This process allows for the forecast of the biological outcomes of these compounds, including enzyme inhibition, receptor binding, and alteration of metabolic pathways [32,33]. With the use of this approach, it becomes possible to predict the probability of a substance exhibiting specific activities. The first step involved converting the substances to SMILES format through ChemDraw. The converted compounds were then analyzed using the PASS web application, which provides insight into the possible activity (Pa) and inactivity (Pi) of drug-like compounds before the evaluation.

2.12. Molecular Docking Analysis

To explore the possible therapeutic characteristics of *P. lentiscus* extract, we employed molecular docking techniques to anticipate the antioxidant, antibacterial, and antifungal features of the seven phytocompounds present in the extract (Figure 2, Table 1). Our research approach was based on established methods previously described in the literature [34–39]. We acquired the three-dimensional (3D) configurations of the molecules from PubChem in March 2023 and converted them into a "pdb" file via the PyMol program. To examine how these plant compounds interact with specific proteins, we acquired the protein structures from the Protein Data Bank website using their distinctive PDB IDs. The protein structures were subjected to standard procedures, including the removal of inhibitors, water molecules, and ions and the addition of polar H-bonds and Kollmann charges to improve the accuracy of the protein structures. For automated docking experiments, we used AutoDock Vina v1.5.6 software and AutoGrid to generate grid maps that displayed the interaction energy between the ligands and target proteins during the docking process [40]. We expanded the search space grid box to enhance the accuracy of the docking procedure (Table 2). The ligand complex binding energies (ΔG) were denoted in Kcal/mol, and we created two-dimensional (2D) diagrams with the Discovery Studio 4.1 program. The interactions were then subject to further analysis.

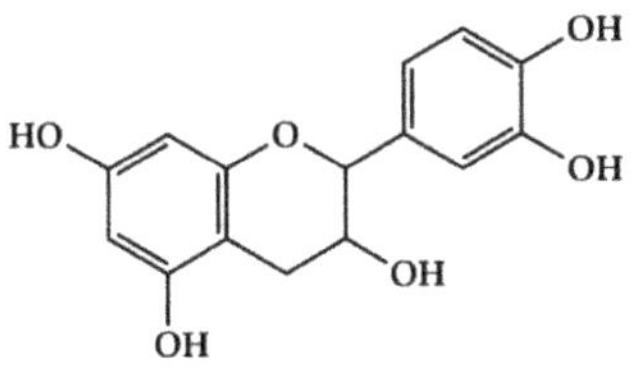

Figure 2. The molecular structure of the components found in *P. lentiscus* samples.

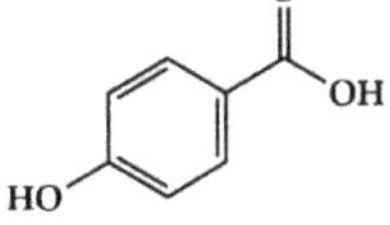

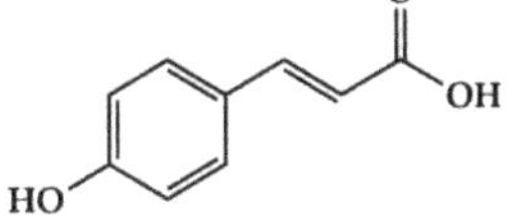

Table 1. Characteristics of the chosen small molecules for the docking study and their drug-likeness properties.

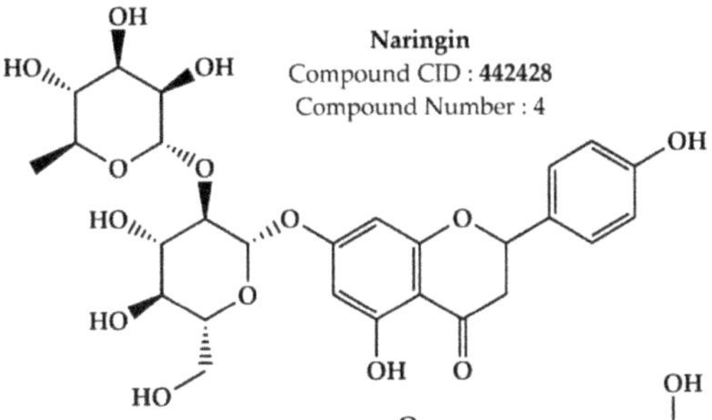

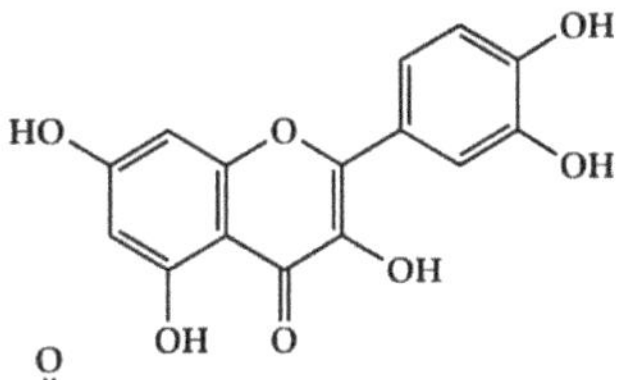

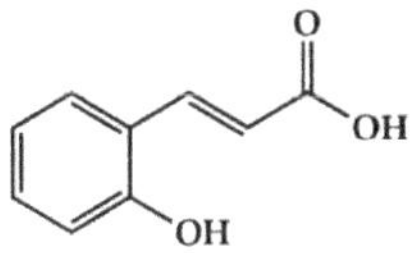

Molecules	MW * (g/mol)	TPSA (Å^2)	H-Bonds		Rotatable Bonds	Lipinski's Rule of Five (Violations)	Ghose Filter (Violations)	Veber Filter (Violations)
			Acceptors	Donors				
Catechin	290.27	110.38	6	5	1	Yes	Yes	Yes
4-Hydroxybenzoic acid	138.12	57.53	3	1	1	Yes	No (3 violations: MR < 40, MW < 160, number of atoms < 20)	Yes
p-Coumaric acid	164.16	57.53	3	1	2	Yes	Yes	Yes
Naringin	580.53	225.06	14	8	6	No (3 violations: MW > 500, N or O > 10, NH or OH > 5.)	No (4 violations: MW > 480, WLOGP < −0.4, MR > 130, number of atoms > 70)	No (1 violation: TPSA > 140)
Quercetin	302.24	131.36	7	5	1	Yes	Yes	Yes
o-coumaric acid	164.16	57.53	3	1	2	Yes	Yes	Yes
Luteolin	286.24	111.13	6	3	1	No (3 violations: MW > 500, N or O > 10, NH or OH > 5.)	No (4 violations: MW > 480, WLOGP < −0.4, MR > 130, number of atoms > 70)	No (1 violation: TPSA > 140)

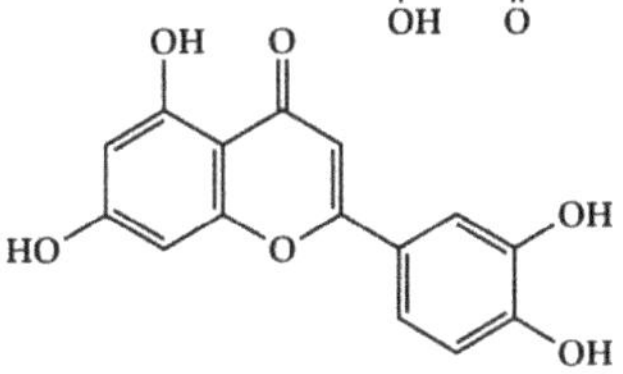

* MW: molecular weight; MR: molar refractivity.

Table 2. Molecular modeling targets and grid box characteristics.

Proteins/PDB IDs	Native Ligand	Grid Box Size (x, y, z)/Center (x, y, z)	Reference
DNA Gyrase Topoisomerase II (*E. coli*)/1KZN	Clorobiocin	(40, 40, 40)/(19.528, 19.500, 43.031)	[38]
Enoyl-Acyl Carrier Reductase Protein/3GNS	Triclosan	(40, 40, 40)/(−14.280, 0.562, −21.462)	[35,36]
Cytochrome P450 14 Alpha-Sterol Demethylase/1EA1	Fluconazole	(40, 40, 40)/(17.702, −3.978, 67.221)	[39,41]
N-Myristoyl Transferase/1IYL	Fluconazole	(40, 40, 40)/(−11.256, 49.991, 1.040)	[39,41]
Lipoxygenase/1N8Q	Protocatechuic Acid	(40, 40, 40)/(22.455, 1.293, 20.362)	[42]
CYP2C9/1OG5	Warfarin	(12.387, 11.653, 11.654)/(−19.823, 86.686, 38.275)	[42]

3. Results

3.1. Extraction Yield

The extraction yields of *P. lentiscus* extracts are presented in Table 3. To extract dried leaves of *P. lentiscus*, the efficiency of various solvents was examined, and the findings demonstrate that methanol (17.5 ± 0.05%) had the highest yield among the solvents tested. In contrast, hexane exhibited a low extraction yield (6.37 ± 0.13%).

Table 3. Extraction yield.

Extracts	Yield (%)	
	This Work	**Literature**
HEPL [1]	6.37 ± 0.13	2.00 ± 0.10
MEPL [2]	17.5 ± 0.05	13.10 ± 0.91

[1]: Hexanic extract of *P. lentiscus* leaves (HEPL), [2]: Methanolic extract of *P. lentiscus* leaves (MEPL).

3.2. Fatty Acid Analysis

The hexanoic extract from dried mastic leaves of *P. lentiscus* was analyzed by Gas chromatography coupled with mass spectrometry (GC–MS) to identify the profile of volatile compounds (Figure 3). The results are given in Table 4. The research findings showed that *P. lentiscus* contains a variety of fatty acids, with a notably high concentration of C18 USFA and C16 SFA (Figure 4), such as C18:2, C18:1, and C16:0. Specifically, the analysis identified six compounds, including C18:2 linoleic acid (making up 40.9% of the fatty acids), C18:1 oleic acid (23.6%), C16:0 palmitic acid (22.8%), C14:0 myristic acid (6.3%), D-Limonene (3.56%), and 10-methyl-Heptadecanoic acid (1.29%) like are shown in Table 4.

3.3. HPLC-DAD Analysis

The phytochemical study of the methanolic extract obtained from *P. lentiscus* was assessed using HPLC/DAD (high-performance liquid chromatography/diode-array detector). To determine the components within the extract, their retention times and UV spectra were compared to those of the corresponding standards. Figure 5 presents the HPLC chromatogram, which shows the identified polyphenolics, while Figure 6 exhibits the chemical structure of the main components found in the extract.

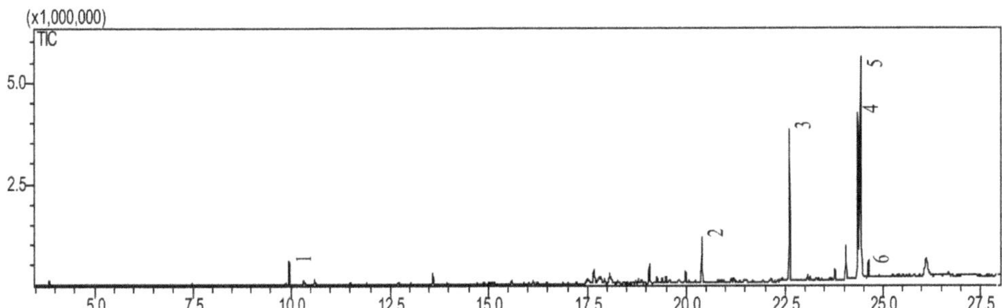

Figure 3. Chromatogram GC/MS of hexanic extract from *P. lentiscus.* (1) D-Limonene, (2) Myristic Acid (C14:0), (3) Palmitic Acid (C16:0), (4) Oleic Acid (C18:1), (5) Linoleic Acid (C18:2), and (6) 10-methyl-Heptadecanoic acid.

Table 4. Chemical composition of hexane extract from *P. lentiscus.* [a]: Unsaturated Fatty Acids (UFA); [b]: Saturated Fatty Acids (SFA); [c]: ratio UFA/SFA.

Compounds	TR (min)	HEPL (%)
D-Limonene	9.870	3.57 ± 0.12
Myristic Acid (C14:0)	20.392	6.39 ± 0.15
Palmitic Acid (C16:0)	22.608	22.84 ± 0.20
Oleic Acid (C18:1)	24.433	23.70 ± 0.13
Linoleic Acid (C18:2)	24.600	40.97 ± 0.11
10-methyl-Heptadecanoic acid	24.625	1.29 ± 0.01
UFA [a]	64.670	
SFA [b]	30.516	
UFA/SFA [c]	2.120	

Figure 4. Main compounds of hexanic extract: (**a**) linoleic acid (C18:2), (**b**) palmitic acid (C16:0).

The methanolic extract was found to contain several phenolic compounds, detectable at 254 nm as shown in Table 5. These compounds' highest concentration was catechin, accounting for approximately 37% of the total compounds detected. Other notable compounds present were 4-hydroxybenzoic acid and Coumaric acid, which accounted for around 17% of the total, followed by naringin at 8.9%, p-Coumaric acid at 7.1%, quercetin at 6.7%, and luteolin at 4.8%.

3.4. Antioxidant Activity

To assess the antioxidant potential of both the methanolic extract of *P. lentiscus* and the standard (ascorbic acid), the DPPH radical scavenging assay and β-carotene bleaching assay were conducted. The first test measures the reduction of the DPPH radical, which is reflected in a change in color from purple (DPPH•) to yellow (DPPH-H) upon adding an antioxidant. The test used a spectrophotometer to measure the absorbance at 515 nm, while the second test, the β-carotene bleaching assay, is used to evaluate a substance's antioxidant capacity. In this assay, β-carotene is mixed with linoleic acid and exposed to oxidative conditions. Antioxidants within a test substance can prevent or slow the degradation of β-carotene and linoleic acid, which results in a decreased rate of color fading, indicating a higher antioxidant capacity. The outcomes of the assay are presented in Table 6.

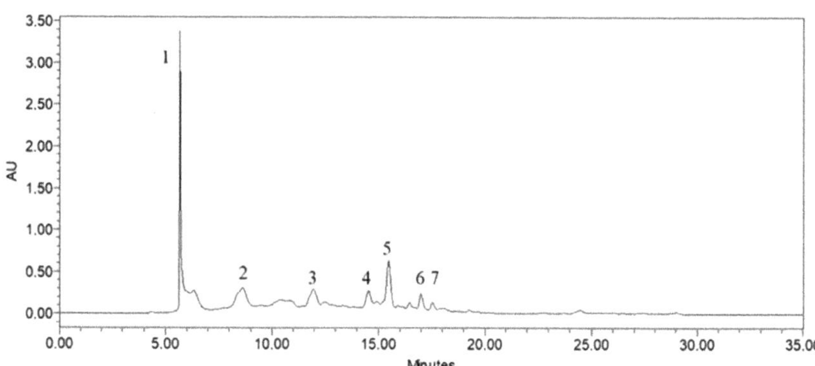

Figure 5. HPLC profile of the methanolic extract of *P. lentiscus*.

Catechin   **4-hydroxybenzoic acid**   **Coumaric Acid**

Figure 6. Most abundant compounds detected of *P. lentiscus* leaves methanolic extract.

Table 5. Phenolic compound of methanolic extract of *P. lentiscus* leaves.

Compounds	T_R (min)	MEPL (%)
Catechin	6.077	37.045 ± 0.15
4-hydroxybenzoic acid	8.527	17.763 ± 0.21
p-Coumaric acid	10.418	7.107 ± 0.10
Naringin	11.965	8.914 ± 0.09
Quercetin	14.539	6.739 ± 0.06
Coumaric Acid	15.473	17.589 ± 0.11
Luteolin	16.978	4.839 ± 0.02

Based on current research, the methanolic extract derived from *P. lentiscus* leaves exhibited potent free radical scavenging ability, as indicated by its IC_{50} value of 0.26 ± 0.13 mg/mL. This value is comparable to that of the reference standard, ascorbic acid, which had an IC_{50} value of 0.15 ± 0.11 mg/mL. These findings align with those reported by Zitouni et al. [43], who reported an IC_{50} value of 0.166 mg/mL for *P. lentiscus* from Algeria, and by Hemma et al. [44], who reported an IC_{50} value of 0.121 ± 0.001 mg/mL. Additionally, the extract demonstrated high β-carotene bleaching activity with an IC_{50} value of 0.19 ± 0.26 mg/mL, similar to the reference antioxidant BHA. In contrast, the hexanic extract exhibited IC_{50} values higher than the reference antioxidants, ascorbic acid, and BHA.

Table 6. Antioxidant activity of methanolic extract of *P. lentiscus*.

Antioxidant Activity	DPPH	β-Carotene
	Inhibitory Concentration 50 (μg/mL)	
HEPL	0.58 ± 0.73	0.64 ± 0.5
MEPL	0.26 ± 0.13	0.19 ± 0.26
Ascorbic Acid	0.15 ± 0.11	
BHA	0.09 ± 0.15	

3.5. Antibacterial Activity

The objective of this study was to investigate the antibacterial effects of *P. lentiscus*, a medicinal and aromatic plant, on Gram-positive bacteria (*Staphylococcus aureus* and *Listeria innocua*) and a Gram-negative bacterium (*Escherichia coli*). The diameters of the inhibition halos surrounding the discs were measured using a graduated ruler (Table 7) to determine the aortograms, and the minimum inhibitory concentration (MIC) and minimum bactericidal concentration (MBC) of the extract were assessed using the microdilution technique. The study revealed that *P. lentiscus* extracts had different antibacterial properties against the bacterial strains tested, with growth inhibition zone diameters ranging from 7 to 20 mm. *Staphylococcus aureus* exhibited the largest inhibition zone diameter (IZ) of 20 mm, while *E. coli* exhibited the smallest IZ of 7 mm. Specifically, a concentration of 2% was shown to be effective against *Staphylococcus aureus*. In addition, the extract inhibited the growth of *Listeria innocua* at concentrations of 4% and 8%.

Table 7. Exploring the MIC and the MBC concentrations of *P. lentiscus* extract.

Bacterial Strains	Inhibition Zone (mm)		*P. lentiscus* Extract		
	P. lentiscus Extract	Gentamicin (1 mg/mL)	MIC (%)	MBC (%)	MBC/MIC
S. aureus	20.00	19.50	2	4	2
L. innocua	15.00	21.50	4	8	2
E. coli	07.00	20.50	>16	>16	-

3.6. Antifungal Activity

The extract exhibited moderate antifungal activity against the tested fungal species, with growth inhibition zones of 22 mm and 12 mm observed for *R. glutinis* and *G. candidum*, respectively (Table 8). The extract demonstrated a minimum inhibitory concentration value of 8% and a minimum fungicidal concentration of approximately 16% against *G. candidum*.

Table 8. Exploring the MIC and the MFC concentrations of *P. lentiscus* extract against fungal strains.

Fungi Strains	Inhibition Zone (mm)		*P. lentiscus* Extract		
	P. lentiscus Extract	Cycloheximide (1 mg/mL)	MIC (%)	MFC (%)	MFC/MIC
G. candidum	12.00	23.00	8	16	2
R. glutinis	22.00	21.00	>16	>16	-

3.7. Physiochemical and ADME Prediction Analysis

The results of the molecular analysis and the drug-likeness analysis of the chosen molecules are summarized in Table 9. The properties listed include molecular weight

(MW), topological polar surface area (TPSA), hydrogen bond donors (H-Bonds), hydrogen bond acceptors, and the number of rotatable bonds. Additionally, the table shows whether the molecules violated Lipinski's rule of five [45], Ghose filter [46], and Veber filter [47]. Overall, the chosen small molecules have a range of properties, and some violate certain drug-likeness filters. Catechin, quercetin, and *o*-Coumaric acid have good drug-likeness properties as they do not violate any of the filters.

Table 9. Pharmacokinetic properties of the identified compounds in *P. lentiscus* extract.

Prediction	1	2	3	4	5	6	7
ADME Prediction **Absorption Parameters**							
Bioavailability score	0.55	0.85	0.55	0.17	0.55	0.85	0.55
Water Solubility (log mol/L)	−3.117	−1.877	−2.378	−2.919	−3.221	−1.56	−3.294
Caco-2 Permeability	−0.283	1.151	1.21	−0.658	−0.057	1.158	0.286
Intestinal Absorption (%)	68.82	83.96	93.49	25.79	75.34	91.11	82.17
Distribution							
Class of solubility	Soluble	Soluble	Soluble	Soluble	Soluble	Soluble	Soluble
Log K_p (cm/s)	−7.82	−6.02	−6.26	−10.15	−7.05	−5.86	−6.25
VDss (log L/kg)	1.027	−1.557	−1.151	0.619	−0.03	−0.406	−0.173
BBB Permeability	No	Yes	Yes	No	No	Yes	No
Metabolism							
CYP2D6, and CYP3A4 Substrate	No	No	No	No	No	No	No
CYP2D6, and CYP3A4 Inhibitors	No	No	No	No	No	No	Yes
Excretion							
Total Clearance log (mL/min/kg)	0.183	0.593	0.662	0.318	0.484	0.746	0.568
Renal OCT2 Substrate	No	No	No	No	No	No	No

In contrast, naringin and luteolin violate multiple filters and are less likely to be considered potential drugs. 4-Hydroxybenzoic acid violates the Ghose filter but is accepted by the Lipinski and Veber filters. The molecular weight of the molecules ranges from 138.12 g/mol for 4-hydroxybenzoic acid to 580.53 g/mol for naringin, with the majority of molecules having a molecular weight of less than 350 g/mol. The TPSA values range from 57.53 Å^2 for 4-hydroxybenzoic acid and p-Coumaric acid to 225.06 Å^2 for naringin, with the majority of molecules having TPSA values less than 140 Å^2.

Table 9 presents the pharmacokinetic properties of identified compounds in *P. lentiscus* extract, which could provide insights into their potential use as therapeutic agents. The table includes absorption parameters, such as bioavailability score, water solubility, Caco-2 permeability, and intestinal absorption; distribution parameters, such as solubility class, skin permeation (log Kp), the Volume of Distribution at steady-state (VDss), and Blood–Brain Barrier (BBB) permeability, metabolism parameters such as CYP2D6 and CYP3A4 substrate and inhibitor, and excretion parameters such as total clearance and renal Organic Cation Transporter 2 (OCT2) substrate. In terms of absorption parameters, the bioavailability score ranges from 0.17 to 0.85, indicating that some compounds may be well absorbed while others may have lower bioavailability (Figure 7). The Caco-2 permeability values suggest that most of the compounds have moderate to high permeability, with one compound having low permeability. The intestinal absorption percentages range from 25.79% to 93.49%, with most compounds having high absorption rates. Regarding distribution parameters, all compounds are soluble, and the log Kp values suggest that they are likely to distribute into tissues. The VDss values suggest that some compounds may have limited tissue distribution while others may distribute widely. Additionally, BBB permeability indicates that most compounds are likely to cross the BBB. The metabolism parameters suggest that none of the compounds are substrates or inhibitors of CYP2D6 or CYP3A4, which are important enzymes involved in drug metabolism. Finally, the excretion parameters suggest that most compounds are eliminated primarily via hepatic clearance, with total clearance values varying from 0.183 to 0.746 mL/min/kg. None of the compounds are substrates of renal OCT2. Overall, these results provide insight into the pharmacokinetic properties of the identified compounds in *P. lentiscus* extract, which could be useful for predicting their potential efficacy and safety in various applications.

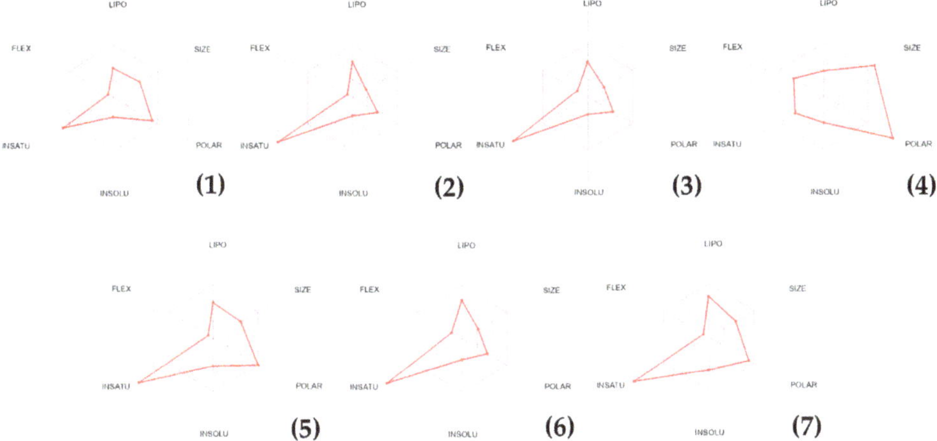

Figure 7. Bioavailability radars of *P. lentiscus* extract's molecules. (**1**) Catechin, (**2**) 4-Hydroxybenzoic acid, (**3**) p-Coumaric acid, (**4**) naringin, (**5**) quercetin, (**6**) o-Coumaric acid, (**7**) luteolin.

The BOILED-Egg model is based on the molecule's lipophilicity (WLOGP) and polarity (TPSA), which are two essential characteristics that play a critical role in determining a molecule's behavior in the body [48]. The BOILED-Egg model visually represents these characteristics, with the white area indicating molecules likely to be absorbed by the intestines and the yellow area indicating high potential for the molecule to penetrate the blood–brain barrier. In a recent study on *P. lentiscus* phytocompounds, the BOILED-Egg model was used to evaluate their potential to be absorbed by the intestines and penetrate the blood-brain barrier (Figure 8). Three compounds, 4-Hydroxybenzoic acid, p-Coumaric acid, and O-Coumaric acid, have a high ability to be absorbed by the intestines and penetrate the BBB while also being non-substrates of the P-glycoprotein (P-GP), which is a protein that can prevent certain molecules from entering the brain. On the other hand, compounds 1, 5, and 7 (catechin, quercetin, and luteolin, respectively) were found to have a poor ability to penetrate the brain endothelial cells. P-GP was found to be able to transport catechin as a substrate, meaning it may have difficulty entering the brain due to P-GP's actions. However, the other two compounds were identified as non-substrates of P-GP, indicating they may still have some potential to penetrate the blood-brain barrier. One compound, naringin (4), was found to be out of range due to its high TPSA value. This indicates that naringin may not be an effective candidate for penetrating the blood–brain barrier due to its high polarity.

3.8. Toxicity Prediction Using Pro-Tox II Webserver

The Pro Tox-II server was used to examine the toxicity profile of chemicals derived from *P. lentiscus*, and Table 10 provides a summary of the results. Based on the findings, none of the chemicals derived from *P. lentiscus* were found to have the potential to induce hepatotoxicity or cytotoxicity., indicating that they were all rather safe to use. Four of the studied substances, notably *p*-Coumaric acid, quercetin, *o*-Coumaric acid, and luteolin, were shown to be possibly carcinogenic in terms of their ability to induce cancer. Nonetheless, it was found that the likelihood of their occurrence was less than 0.68, indicating a negligibly low propensity to cause cancer. Contrarily, the immunotoxicity of naringin (4) was determined with a prediction probability of 0.99, demonstrating the substance's potential to cause immunotoxicity. Regarding the mutagenicity of the tested compounds, only two compounds, quercetin (5) and luteolin (7), were identified as being potentially mutagenic with a probability of 0.51. These findings highlight the importance of assessing natural compounds' safety and toxicity profiles before utilizing them for various applications, particularly in medicine and the food industry, to ensure their safe and effective use.

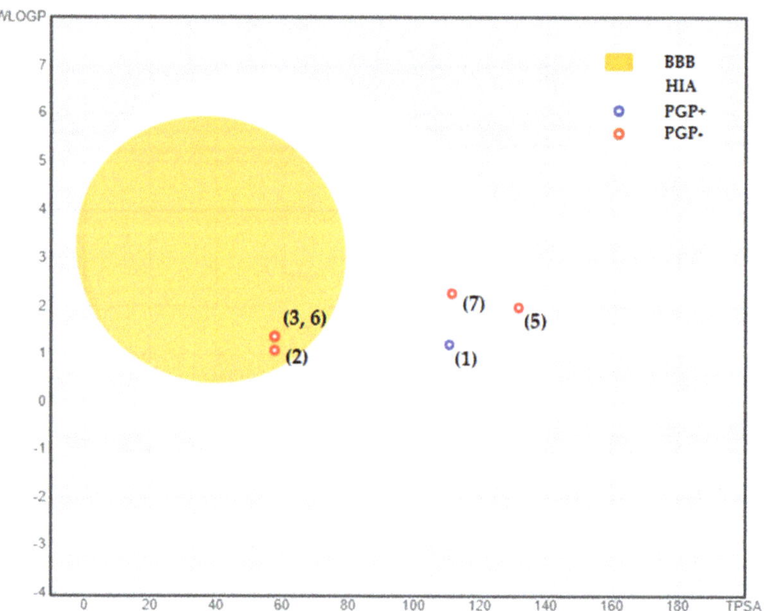

Figure 8. Boiled-egg model of the identified molecules in *P. lentiscus* extract. (1) Catechin, (2) 4-Hydroxybenzoic acid, (3) p-Coumaric acid, (5) quercetin, (6) o-Coumaric acid, (7) luteolin. Remark: naringin (4) was found to be out of range.

Table 10. The toxicological characteristics of compounds derived from the methanolic extract of *P. lentiscus* leaves were evaluated using Pro-Tox II. (1) Catechin, (2) 4-Hydroxybenzoic acid, (3) p-Coumaric acid, (4) naringin, (5) quercetin, (6) o-Coumaric acid, (7) luteolin.

	Predicted LD$_{50}$ (mg/kg)	Class	Hepatotoxicity		Carcinogenicity		Immunotoxicity		Mutagenicity		Cytotoxicity	
			Predi. *	Prob.	Predi.	Prob.	Predi.	Prob.	Predi.	Prob.	Predi.	Prob.
1	10,000	VI	In.	0.72	In.	0.51	In.	0.96	In.	0.55	In.	0.84
2	2200	V	In.	0.52	In.	0.51	In.	0.99	In.	0.99	In.	0.86
3	2850	V	In.	0.52	Ac.	0.50	In.	0.91	In.	0.93	In.	0.81
4	2300	V	In.	0.81	In.	0.80	Ac.	0.99	In.	0.73	In.	0.66
5	159	III	In.	0.69	Ac.	0.68	In.	0.87	Ac.	0.51	In.	0.99
6	2850	V	In.	0.52	Ac.	0.50	In.	0.91	In.	0.93	In.	0.81
7	3919	V	In.	0.69	Ac.	0.68	In.	0.97	Ac.	0.51	In.	0.99

* Predi.: Prediction, Prob: Probability, In.: Inactive, Ac.: Active.

3.9. PASS Prediction and Molecular Docking Analysis

3.9.1. PASS Prediction

Prediction of Activity Spectra for Substances (PASS) involves a series of algorithms that utilize structural data about the molecule to predict its activity against a wide range of biological targets, including enzymes, receptors, and ion channels [49,50]. By utilizing a vast database of known active and inactive chemicals, statistical models are developed that can accurately predict the activity of newly developed compounds [49]. In the early stages of drug discovery, PASS analysis is a widely used tool to prioritize compounds for further testing and to guide the design of new compounds with optimal activity profiles. As depicted in Table 11, PASS predicted activity (P act) and predicted inactivity (P ina) values for seven different compounds, namely catechin, 4-Hydroxybenzoic acid, p-coumaric acid, naringin, quercetin, o-Coumaric acid, and luteolin, were assessed across three categories of biological activity, namely antioxidant, antibacterial, and antifungal. For the antioxidant activity, it appears that compounds 1, 4, 5, and 7 (catechin, naringin, quercetin, and luteolin) have relatively high P act values (above 0.75), suggesting they may have antioxidant activity. Compounds 2, 3, and 6 (4-Hydroxybenzoic acid, *p*-Coumaric acid, and *o*-Coumaric acid) have lower P act values, indicating a lower likelihood of antioxidant activity. For the antibacterial and antifungal activities, the predictions revealed a potent antifungal activity for compound (4) Naringin with a P act = 0.816.

Table 11. PASS prediction of the identified phytocompounds. (1) Catechin, (2) 4-Hydroxybenzoic acid, (3) *p*-Coumaric acid, (4) naringin, (5) quercetin, (6) *o*-Coumaric acid, (7) luteolin. P act: probable activity; P ina: probable inactivity.

Prediction	1	2	3	4	5	6	7
	PASS Prediction (P act/P ina)						
Antioxidant	0.810/0.003	0.320/0.020	0.553/0.005	0.851/0.003	0.872/0.003	0.553/0.005	0.775/0.004
Antibacterial	0.320/0.053	0.384/0.034	0.343/0.045	0.669/0.005	0.387/0.033	0.343/0.045	0.388/0.033
Antifungal	0.552/0.023	0.384/0.053	0.451/0.039	0.816/0.004	0.490/0.032	0.451/0.039	0.520/0.027

3.9.2. Molecular Docking Results

Molecular docking is a commonly used methodology employed mainly in drug design that is based on molecular structure. Its development in the 1980s has been instrumental in drug discovery, as it enables the prediction of small-molecule ligand placement in target binding sites and can help stabilize ligand–receptor complexes through the investigation of molecular events, such as ligand-protein binding [51,52].

The docking scores, represented as binding energies (in kcal/mol), of the seven identified compounds from *P. lentiscus* against six proteins with antibacterial, antifungal, and antioxidant properties are shown in Table 12. The docking scores of the native ligands are also provided as a reference point. A heat map is used to visualize the data, with red indicating a stronger binding affinity than the native ligand, yellow indicating a similar affinity, and green indicating a weaker affinity.

Molecular docking can help stabilize ligand–receptor complexes. In this study, we have used molecular docking to investigate how *P. lentiscus* extract components work. The resulting binding affinity values indicate whether the molecule has a higher or lower affinity for the target compared to a known inhibitor. The interactions of these components have been investigated with specific enzymatic proteins, including lipoxygenase-3 (PDB ID: 1N8Q) [53,54] and cytochrome P450 (PDB ID: 1OG5) [53,55], which are known target receptors for antioxidant chemicals. For proteins known to have bactericidal/bacteriostatic activity, DNA Gyrase Topoisomerase II and Enoyl-Acyl Carrier Protein Reductase were selected as therapeutic targets (PDB IDs: 1KZN and 3GNS, respectively) [36,38,53]. Furthermore, the investigation focused on two proteins associated with antifungal activity, namely Cytochrome P450 14 alpha-sterol Deme-thylase (PDB ID: 1EA1) and N-Myristoyl transferase (PDB ID: 1IYL) [53].

Table 12. Heatmap displaying the docking scores (affinity values in kcal/mol) of components from *P. lentiscus*.

N°	Compounds	Antibacterial Proteins		Antifungal Proteins		Antioxidant Proteins	
		1KZN	3GNS	1EA1	1IYL	1N8Q	1OG5
		Docking Scores (Kcal/mol) *					
-	Native Ligand	−9.6	−6.2	−5.8	−5.8	−6	−6.6
1	Catechin	−8.3	−7.1	−6.6	−7	−6.3	−8.3
2	4-Hydroxybenzoic acid	−5.6	−5.5	−4.5	−4.7	−5.4	−5.3
3	*p*-coumaric acid	−5.8	−5.3	−4.6	−5.4	−5.7	−6.2
4	Naringin	−9.3	−8.1	−9.1	−7.7	−5.6	−7.8
5	Quercetin	−8.3	−7	−7.1	−7.1	−8.3	−8.8
6	*o*-coumaric acid	−6.1	−5.7	−4.7	−5.6	−6.3	−5.9
7	Luteolin	−8.9	−7.2	−7.3	−7.2	−6.3	−8.8

* Each column is assigned a color scale ranging from red, which represents the docking score of the native ligand (ΔG), to green, which represents a higher binding affinity than the native ligand (ΔG + 4 kcal/mol). The midpoint is represented by the color yellow.

The bacterial enzyme DNA gyrase topoisomerase II (PDB: 1KZN) is ubiquitously present in all bacterial species and regulates the topological state of bacterial DNA. In this study, we identified the enzyme DNA gyrase topoisomerase II, which consists of two subunits, GyrA and GyrB, as the target protein for binding. The compounds extracted from *P. lentiscus* exhibited weak binding affinity with this protein, except for naringin, which demonstrated a binding score ranging from −5.6 to −9.3 kcal/mol, with a score of −9.3 kcal/mol. When protein 1KZN was docked with its natural ligand clorobiocin, the outcome demonstrated strong inhibitory potential, as evidenced by the docking score of −9.6 kcal/mol. The ligand established two conventional H-bonds with the amino acid residues of the active site, THR165, and ASP73, which was mentioned in the reference [53]. Our docking analysis revealed the presence of four potent inhibitors of FabI activity (PDB ID: 3GNS), namely quercetin, catechin, luteolin, and naringin, with binding scores of −7, −7.1, −7.2, and −8.1 kcal/mol, respectively, compared to triclosan, a well-known FabI inhibitor, with a score of −6.2 kcal/mol.

The formation and upkeep of bacterial cell membranes require fatty acid biosynthesis, which involves a sequence of enzyme-catalyzed reactions that transform fundamental components into long, unsaturated fatty acids, constituting a significant part of these structures. The final stage of this process is the reduction of the double bond in an intermediate molecule referred to as an enoyl-acyl carrier protein (ACP) derivative [56]. The enzyme FabI, also known as enoyl-ACP reductase, is responsible for facilitating the reduction process in the final stage of fatty acid biosynthesis. It plays an essential role in the elongation phase of this process, which is necessary for the cycle's completion [57]. FabI has been identified in crystal structures of various bacteria, including *Escherichia coli* and *Staphylococcus aureus* [58].

The N-terminal glycine of various proteins in eukaryotic organisms is bound by N-Myristoyl Transferase (NMT), which utilizes myristoyl-CoA as a substrate and attaches myristate fatty acid [59]. NMT is essential for various crucial biological processes, including cell death, signal transduction, and the proliferation of fungal pathogens [60]. The study revealed that catechin, naringin, quercetin, and luteolin are potent inhibitors of the second lipoxygenase member, CYP2C9, with binding affinity values lower than −6.6 kcal/mol, the value of the ligand warfarin.

The biosynthesis of sterols in fungi depends heavily on the enzyme CYP51s, also referred to as Cytochrome P450 14α-Sterol Demethylase. This enzyme is responsible for producing intermediary compounds required for the formation of ergosterol, making it a critical component of the process [61]. Due to its indispensable role in the formation of sterols, CYP51s have become a prime aim for antifungal drugs. In this study, molecular docking analysis was carried out using the crystal structure of CYP51s (with a PDB ID of 1EA1) as a basis. The results of the analysis indicate that the molecules quercetin, catechin, luteolin, and naringin possess strong inhibitory potential with docking scores of −7.1, −6.6, −7.3, and −9.1 kcal/mol, respectively.

The current study also shows that the same four molecules, quercetin, catechin, luteolin, and naringin, exhibit potent inhibitory potential against NMT, with binding affinities higher than that of Fluconazole, a known antifungal drug with a docking score of −5.8 kcal/mol.

Lipoxygenases are a group of enzymes that use a redox mechanism to catalyze the oxidation of polyunsaturated fatty acids, producing hydroperoxide, which is an oxygen-centered radical that can be involved in the development of severe diseases.

In this study, two proteins were selected, namely lipoxygenase (1N8Q) and cytochrome P450 (1OG5). The study identified catechin, O-Coumaric acid, luteolin, and quercetin as potent inhibitors of lipoxygenase (PDB ID: 1N8Q), with binding affinity scores of −6.3, −6.3, −6.3, and −8.3 kcal/mol, respectively, compared to the inhibitor Protocatechuic Acid, which had a docking score of −6.0 kcal/mol.

4. Discussion

In this research, extracts were prepared from the leaves of *P. lentiscus*. Several studies [62,63] have confirmed that the yield depends on the polarity of the solvent used, which is consistent with our study. In line with Ariana Bampouli et al.'s study [64], our results demonstrate that hexane is not an appropriate solvent for extracting metabolites from mastic tree leaves due to its low extraction efficiency. On the other hand, ethanol or methanol as solvents led to significantly higher yields. Generally, non-polar solvents like hexane extract non-polar compounds like carotenes, terpenes, and lipids, while polar compounds like phenolic compounds are typically extracted using water, ethanol, or methanol [65].

The chemical composition of H.E was determined by gas chromatography-mass spectrometry (GC-MS). The primary compound detected in the *P. lentiscus* extract was linoleic acid, which is a precursor for arachidonic acid biosynthesis and serves as a substrate for synthesizing eicosanoids. According to previous research, linoleic acid has potential cholesterol-lowering effects and can be beneficial for the skin, making it a widely used ingredient in the cosmetic industry [66,67]. The *P. lentiscus* extract is rich in unsaturated fatty acids (UFA), specifically C18:1 (MUFA) and C18:2 (PUFA), accounting for around 64.67% of its content. These levels suggest that the extract is resistant to oxidation and could serve as a valuable source of MUFA and PUFA in the diet. MUFA is essential for its nutritional implications and impact on the extract's oxidative stability. Previous studies have shown that a diet high in MUFA may be a healthier alternative to a low-fat diet, as it can lower blood cholesterol levels, modulate immune function, decrease the susceptibility of LDL to oxidation, and enhance the fluidity of HDL [68]. Furthermore, including PUFA in the diet is crucial for the structure and function of membrane proteins, such as receptors, enzymes, and active transport molecules [69].

The major phenolic compounds of the M.E were determined through HPLC analysis. It revolves around the presence of catechins as major compounds. Catechins are flavan-3-ols, a natural polyphenolic compound belonging to the flavonoid family, found in numerous plants. These compounds possess various biochemical properties and exhibit antioxidant activity, which can prevent diseases such as cancer, cardiovascular, and neurodegenerative diseases by reducing oxidative stress [70].

The antioxidant activity was analyzed using two methods (the DPPH radical scavenging method and the β-carotene bleaching assay). The potent scavenging activity of the methanolic extract from *P. lentiscus* leaves is believed to be attributed to its high concentration of phenolic compounds, particularly catechin. The present discovery corresponds with the investigation conducted by Luo and colleagues [71], where they proposed that dissimilarities in the number of phenolic compounds among diverse plant species can elucidate the discrepancies in their antioxidant activity.

The antibacterial and antifungal activities of *P. lentiscus* methanolic extract were analyzed. The antibacterial potency of extracts from medicinal and aromatic plants depends on the diversity of the molecules present [72]. These results suggest that moderate concentrations of the extract may efficiently inhibit the growth and spread of the tested bacterial strains. The antibacterial activity of *P. lentiscus* is attributed to its chemical composition, particularly the presence of phenolic compounds [73]. Catechin, the primary compound found in this plant, is thought to be responsible for the antibacterial effects [74], while naringin and other compounds can also contribute to the activity against Gram-positive bacteria [75]. The activity of the extract against Gram-negative bacteria is comparatively lower than that against Gram-positive bacteria due to the differences in the composition of their cell walls. In Gram-positive microorganisms, the resistance is attributed to the presence of a hydrophilic barrier that hinders diffusion through the outer cell membrane. However, in Gram-negative bacteria, the inhibition of activity is primarily based on the direct interaction between the extract's hydrophobic components and the phospholipids present in the cell membrane. This interaction can lead to structural damage and complete rupture of the cell membranes [50].

The growth-inhibitory properties of *P. lentiscus* extract against these fungi are attributed to its chemical composition. The antifungal potential of plant-derived substances has been widely investigated through in vitro studies [76]. This moderate activity is a result of several factors acting together to achieve the antifungal effect. The extract's chemical composition and its richness in bioactive constituents are considered to be important determinants of antifungal activity. Notably, the presence of flavonoids is responsible for the extract's antifungal properties by disrupting the fungal cell membrane, degrading the mitochondrial function, inhibiting electron transport, and altering mitochondrial ATPase activity [77]. On the other hand, the extract was found to be inactive against molds, such as *R. glutinis*, which could be attributed to the inherent resistance of molds to the components present in the sample.

A drug needs to exhibit high selectivity and minimal side effects [78]. However, despite being effective, many potential treatments do not make it through the later stages of testing due to their high rate of attrition caused by unavoidable side effects [78]. These side effects can arise due to the complexity of the biological system and can be difficult to predict. To reduce the risk of drug candidate failure, in silico studies can be conducted to assess drug-likeness and ADME properties of potential molecules [45,46]. After the initial screening process, the identified molecules undergo a thorough analysis to determine their drug-like properties and ADME profiles. This comprehensive analysis can help to identify potential issues and minimize the risk of a drug candidate failing in the later stages of testing.

Lipinski's rule of five is a widely used rule to predict drug-likeness, and violations of this rule suggest that the molecule may have poor Absorption or permeation in the body. The Ghose and Veber filters are other commonly used filters to predict drug-likeness. Among the molecules listed, catechin, 4-Hydroxybenzoic acid, *p*-Coumaric acid, quercetin, and o-Coumaric acid have good drug-like properties, with no violations of any filters. These molecules may have the potential for drug development. Naringin and luteolin have multiple violations of drug-likeness filters, indicating that they may have poor pharmacokinetic properties and are less likely to be developed as drugs. However, it is important to note that the results of these filters are not absolute and do not entirely rule out the possibility of these molecules being developed as drugs.

The BOILED-Egg model is an effective and convenient tool to evaluate the potential for a substance to be absorbed by the intestines and to penetrate the blood–brain barrier [48]. It was able to provide valuable insights into the potential of *P. lentiscus* phytocompounds to be absorbed by the intestines and penetrate the BBB, which can be useful for further drug development and therapeutic applications.

Prediction of Activity Spectra for Substances (PASS) is a computational approach extensively utilized in drug research to predict the biological activity of chemical compounds [49]. The results of this analysis indicate that some compounds have relatively modest to high P act values for one category but not for the other, which may suggest that any noticed biological activity may be attributed to a potential synergy.

Molecular docking, a widely used methodology in drug design, was employed in this study to investigate the interactions between the components of *P. lentiscus* extract and specific enzymatic proteins. The binding affinity values obtained from the docking analysis provided insights into the potential inhibitory activity of the compounds. For DNA Gyrase Topoisomerase II, most compounds from *P. lentiscus* showed weak binding affinity, except for naringin, which demonstrated strong binding affinity. The natural ligand clorobiocin exhibited potent inhibitory potential, establishing two conventional H-bonds with the amino acid residues of the active site. In the case of Enoyl-Acyl Carrier Protein Reductase (FabI), four compounds from *P. lentiscus*, namely quercetin, catechin, luteolin, and naringin, showed potent inhibitory activity, suggesting their potential as FabI inhibitors. These compounds displayed higher binding scores compared to the well-known FabI inhibitor Triclosan. N-Myristoyl Transferase (NMT) inhibition was observed with catechin, naringin, quercetin, and luteolin, indicating their potential as inhibitors for this enzyme. These compounds exhibited higher binding affinities than the antifungal drug Fluconazole.

The results also revealed the strong inhibitory potential of quercetin, catechin, luteolin, and naringin against Cytochrome P450 14α-Sterol Demethylase (CYP51s), an enzyme crucial for sterol biosynthesis in fungi. These compounds displayed significant binding scores, indicating their potential as antifungal agents targeting CYP51s. Lastly, lipoxygenase-3 (1N8Q) inhibition was observed with catechin, O-coumaric acid, luteolin, and quercetin, suggesting their potential as lipoxygenase inhibitors. These compounds showed higher binding affinities compared to the inhibitor Protocatechuic Acid. Overall, the results of this study highlight the potential of the identified compounds from *P. lentiscus* extract as inhibitors for enzymes involved in antibacterial, antifungal, and antioxidant activities. Further investigations and experimental validations are warranted to confirm their efficacy and suitability as therapeutic agents.

5. Conclusions

Our research aims to improve the extracts derived from *P. lentiscus* in the Ahfir region of eastern Morocco. Through the analysis of the hexanic extract using GC/MS, we have identified linoleic acid as the major fatty acid compound, comprising approximately 40% of the extract. The unsaturated fatty acid to saturated fatty acid ratio is 2.12%, indicating a high content of unsaturated fatty acids, which are valuable in the food industry. The chemical composition of the methanolic extract was determined using HPLC-DAD, revealing that catechin is the predominant compound, accounting for around 37% of the extract. The methanolic extract demonstrated antioxidant activity close to that of ascorbic acid when assessed using the DPPH radical scavenging method. However, as these extracts are crude and contain various compounds, purified compounds may exhibit comparable activity to ascorbic acid. Our findings indicate that *P. lentiscus* extracts possess significant antimicrobial properties, as demonstrated by both in vitro and in silico assays, suggesting their potential as a phytomedicine. The in silico assays supported the experimental data and provided insights into the molecular interactions between the tested drugs and specific enzymatic proteins, indicating substantial binding affinities. In conclusion, our research has shown that the studied products exhibit excellent antioxidative and antibacterial properties, positioning them as potential candidates for various biotechnological applications.

Author Contributions: Conceptualization, writing—original draft preparation: S.O. and E.H.L.; methodology, validation: A.E.; software, investigation: M.T. and M.B. (Mohammed Bouslamti); visualization: H.-A.N.; data curation, E.H.L., A.M.S. and R.B.; writing—review and editing, M.B. (Mohammed Bourhia) and N.S.; validation, writing—review and editing, M.A., M.R. and B.H. supervision, project administration. All authors have read and agreed to the published version of the manuscript.

Funding: This work was funded by the Deputyship for Research & innovation, "Ministry of Education" in Saudi Arabia for funding this research (IFKSUOR3-410-1).

Institutional Review Board Statement: Not applicable.

Informed Consent Statement: Not applicable.

Data Availability Statement: Not applicable.

Acknowledgments: The authors extend their appreciation to the Deputyship for Research & innovation, "Ministry of Education" in Saudi Arabia for funding this research (IFKSUOR3-410-1).

Conflicts of Interest: The authors declare no conflict of interest.

References

1. Rauf, A.; Patel, S.; Uddin, G.; Siddiqui, B.; Bashir, A.; Muhammad, N.; Mabkhot, Y.; Ben Hadda, T. Phytochemical, Ethnomedicinal Uses and Pharmacological Profile of Genus Pistacia. *Biomed. Pharmacother.* **2017**, *86*, 393–404. [CrossRef] [PubMed]
2. Bozorgi, M.; Memariani, Z.; Mobli, M.; Surmaghi, M.; Ardekani, M.; Rahimi, R. Five Pistacia Species (*P. vera, P. atlantica, P. terebinthus, P. khinjuk,* and *P. lentiscus*): A Review of Their Traditional Uses, Phytochemistry, and Pharmacology. *Sci. World J.* **2013**, *2013*, 219815. [CrossRef]
3. Kafkas, S.; Khodaeiaminjan, M.; Guney, M.; Kafkas, E. Identification of Sex-Linked SNP Markers Using RAD Sequencing Suggests ZW/ZZ Sex Determination in *Pistacia vera* L. *BMC Genom.* **2015**, *16*, 98. [CrossRef]
4. Munné-Bosch, S.; Penuelas, J. Photo- and Antioxidative Protection During Summer Leaf Senescence in *Pistacia lentiscus* L. Grown under Mediterranean Field Conditions. *Ann. Bot.* **2003**, *92*, 385–391. [CrossRef]
5. Rodríguez González, Á.; Da Silva, F.; Gutiérrez, S.; Casquero, P. *Lippia* spp. Essential Oil as a Control Agent against *Acanthoscelides obtectus*, an Insect Pest in *Phaseolus vulgaris* Beans. In Proceedings of the 1st International Electronic Conference on Plant Science, Sciforum.net, 1–15 December 2020; MDPI: Basel, Switzerland, 2020; p. 8767.
6. Abidi, A.; Aissani, N.; Sebai, H.; Serairi, R.; Kourda, N.; Khamsa, S. Protective Effect of *Pistacia lentiscus* Oil Against Bleomycin-Induced Lung Fibrosis and Oxidative Stress in Rat. *Nutr. Cancer* **2017**, *69*, 490–497. [CrossRef] [PubMed]
7. Gündoğdu, M.; Akdeniz, F.; Öztürkkan, F.; Demirci, S.; Adiguzel, V. A Promising Method for Recovery of Oil and Potent Antioxidant Extracts from *Pistacia khinjuk* Stocks Seeds. *Ind. Crop. Prod.* **2016**, *83*, 515–521. [CrossRef]
8. Naouar, M.; Zouiten-Mekki, L.; Charfi, L.; Boubaker, J.; Filali, A. Preventive and Curative Effect of *Pistacia lentiscus* Oil in Experimental Colitis. *Biomed. Pharmacother.* **2016**, *83*, 577–583. [CrossRef]
9. Andreadou, I.; Paraschos, S.; Efentakis, P.; Magiatis, P.; Kaklamanis, L.; Halabalaki, M.; Skaltsounis, L.; Iliodromitis, E. "*Pistacia lentiscus* L." Reduces the Infarct Size in Normal Fed Anesthetized Rabbits and Possess Antiatheromatic and Hypolipidemic Activity in Cholesterol Fed Rabbits. *Phytomedicine* **2016**, *23*, 1220–1226. [CrossRef]
10. Beghlal, D.; El Bairi, K.; Marmouzi, I.; Haddar, L.; Mohammed, B. Phytochemical, Organoleptic and Ferric Reducing Properties of Essential Oil and Ethanolic Extract from *Pistacia lentiscus* (L.). *Asian Pac. J. Trop. Dis.* **2016**, *6*, 305–310. [CrossRef]
11. Kalogeropoulos, N.; Chiou, A.; Ioannou, M.; Karathanos, V. Nutritional Evaluation and Health Promoting Ac-tivities of Nuts and Seeds Cultivated in Greece. *Int. J. Food Sci. Nutr.* **2013**, *64*, 757–767. [CrossRef]
12. Gazioğlu, I.; Koseoglu Yilmaz, P.; Haşimi, N.; Kilinc, E.; Tolan, V.; Kolak, U. In vitro Biological Activities and Fatty Acid Profiles of *Pistacia terebinthus* Fruits and *Pistacia khinjuk* Seeds. *Nat. Prod. Res.* **2014**, *29*, 444–446. [CrossRef]
13. Hadjimbei, E.; Botsaris, G.; Goulas, V.; Gekas, V. Health-Promoting Effects of *Pistacia* Resins: Recent Advances, Challenges, and Potential Applications in the Food Industry. *Food Rev. Int.* **2014**, *31*, 1–12. [CrossRef]
14. Yemmen, M.; Landoulsi, A.; Hamida, J.; Mégraud, F.; Trabelsi Ayadi, M. Antioxidant Activities, Anticancer Activity and Polyphenolics Profile, of Leaf, Fruit and Stem Extracts of *Pistacia lentiscus* from Tunisia. *Cell. Mol. Biol.* **2017**, *63*, 87–95. [CrossRef] [PubMed]
15. Iauk, L.; Ragusa, S.; Rapisarda, A.; Franco, S.; Nicolosi, V.M. In vitro Antimicrobial Activity of *Pistacia lentiscus* L. Extracts: Preliminary Report. *J. Chemother.* **1996**, *8*, 207–209. [CrossRef] [PubMed]
16. Ljubuncic, P.; Song, H.; Cogan, U.; Azaizeh, H.; Bomzon, A. The Effects of Aqueous Extracts Prepared from the Leaves of *Pistacia lentiscus* in Experimental Liver Disease. *J. Ethnopharmacol.* **2005**, *100*, 198–204. [CrossRef]
17. Gardeli, A.; Papageorgiou, V.; Mallouchos, A.; Kibouris, T.; Komaitis, M. Essential Oil Composition of *Pistacia lentiscus* L. and *Myrtus communis* L.: Evaluation of Antioxidant Capacity of Methanolic Extracts. *Food Chem.* **2008**, *107*, 1120–1130. [CrossRef]
18. Andrikopoulos, N.; Kaliora, A.; Assimopoulou, A.; Papapeorgiou, V. Biological Activity of Some Naturally Occurring Resins, Gums, and Pigments Against In Vitro LDL Oxidation. *Phytother. Res. PTR* **2003**, *17*, 501–507. [CrossRef]

19. Barra, A.; Coroneo, V.; Dessi, S.; Cabras, P.; Angioni, A. Characterization of the Volatile Constituents in the Essential Oil of *Pistacia lentiscus* L. from Different Origins and Its Antifungal and Antioxidant Activity. *J. Agric. Food Chem.* **2007**, *55*, 7093–7098. [CrossRef]

20. Belyagoubi-Benhammou, N.; Bekkara, F.; Kadifkova Panovska, T. Antioxidant and Antimicrobial Activities of the *Pistacia lentiscus* and *Pistacia atlantica* Extracts. *Afr. J. Pharm. Pharmacol.* **2008**, *2*, 022–028.

21. D'Auria, M.; Racioppi, R. Characterization of the Volatile Fraction of Mastic Oil and Mastic Gum. *Nat. Prod. Res.* **2020**, *36*, 3460–3463. [CrossRef]

22. Rauha, J.-P.; Remes, S.; Heinonen, M.; Hopia, A.; Kähkönen, M.; Kujala, T.; Pihlaja, K.; Vuorela, H.; Vuorela, P. Antimicrobial Effects of Finnish Plant Extracts Containing Flavonoids and Other Phenolic Compounds. *Int. J. Food Microbiol.* **2000**, *56*, 3–12. [CrossRef] [PubMed]

23. Balan, K.; Prince, J.; Han, Z.; Dimas, K.; Hadzopoulou-Cladaras, M.; Wyche, J.H.; Sitaras, N.M.; Pantazis, P. Antiproliferative Activity and Induction of Apoptosis in Human Colon Cancer Cells Treated in vitro with Constituents of a Product Derived from *Pistacia lentiscus* L. Var. Chia. *Phytomedicine* **2007**, *14*, 263–272. [CrossRef] [PubMed]

24. Loukili, E.H.; Bouchal, B.; Bouhrim, M.; Abrigach, F.; Genva, M.; Kahina, Z.; Bnouham, M.; Bellaoui, M.; Hammouti, B.; Addi, M.; et al. Chemical Composition, Antibacterial, Antifungal and Antidiabetic Activities of Ethanolic Extracts of *Opuntia dillenii* Fruits Collected from Morocco. *J. Food Qual.* **2022**, *2022*, 9471239. [CrossRef]

25. CLSI. *M27-A3: Reference Method for Broth Dilution Antifungal Susceptibility Testing of Yeasts*; Approved Standard—Third Edition; Clinical and Laboratory Standards Institute: Wayne, PA, USA, 2008.

26. Agour, A.; Mssillou, I.; Mechchate, H.; Es-safi, I.; Allali, A.; El Barnossi, A.; Kamaly, O.; Alshawwa, S.; Moussaoui, A.; Bari, A.; et al. *Brocchia cinerea* (Delile) Vis. Essential Oil Antimicrobial Activity and Crop Protection against Cowpea Weevil *Callosobruchus maculatus* (Fab.). *Plants* **2022**, *11*, 583. [CrossRef]

27. Mssillou, I.; Agour, A.; Hamamouch, N.; Badiaa, L.; Derwich, E. Chemical Composition and In vitro Antioxidant and Antimicrobial Activities of *Marrubium vulgare* L. *Sci. World J.* **2021**, *2021*, 7011493. [CrossRef]

28. Daina, A.; Michielin, O.; Zoete, V. SwissADME: A Free Web Tool to Evaluate Pharmacokinetics, Drug-Likeness and Medicinal Chemistry Friendliness of Small Molecules. *Sci. Rep.* **2017**, *7*, 42717. [CrossRef]

29. Pires, D.E.V.; Blundell, T.L.; Ascher, D.B. PkCSM: Predicting Small-Molecule Pharmacokinetic and Toxicity Properties Using Graph-Based Signatures. *J. Med. Chem.* **2015**, *58*, 4066–4072. [CrossRef]

30. Kandsi, F.; Lafdil, F.Z.; Elbouzidi, A.; Bouknana, S.; Miry, A.; Addi, M.; Conte, R.; Hano, C.; Gseyra, N. Evaluation of Acute and Subacute Toxicity and LC-MS/MS Compositional Alkaloid Determination of the Hydroethanolic Extract of *Dysphania ambrosioides* (L.) Mosyakin and Clemants Flowers. *Toxins* **2022**, *14*, 475. [CrossRef]

31. Alam, A.; Jawaid, T.; Alam, P. In vitro Antioxidant and Anti-Inflammatory Activities of Green Cardamom Essential Oil and Silico Molecular Docking of Its Major Bioactives. *J. Taibah Univ. Sci.* **2021**, *15*, 757–768. [CrossRef]

32. Linde, G.A.; Gazim, Z.C.; Cardoso, B.K.; Jorge, L.F.; Tešević, V.; Glamočlija, J.; Soković, M.; Colauto, N.B. Antifungal and Antibacterial Activities of Petroselinum Crispum Essential Oil. *Genet. Mol. Res.* **2016**, *15*, 3. [CrossRef]

33. Filimonov, D.A.; Lagunin, A.A.; Gloriozova, T.A.; Rudik, A.V.; Druzhilovskii, D.S.; Pogodin, P.V.; Poroikov, V.V. Prediction of the Biological Activity Spectra of Organic Compounds Using the PASS Online Web Resource. *Chem. Heterocycl. Compd.* **2014**, *50*, 444–457. [CrossRef]

34. Cosconati, S.; Forli, S.; Perryman, A.L.; Harris, R.; Goodsell, D.S.; Olson, A.J. Virtual Screening with AutoDock: Theory and Practice. *Expert Opin. Drug Discov.* **2010**, *5*, 597–607. [CrossRef] [PubMed]

35. Priyadarshi, A.; Kim, E.E.; Hwang, K.Y. Structural Insights into Staphylococcus Aureus Enoyl-ACP Reductase (FabI), in Complex with NADP and Triclosan. *Proteins Struct. Funct. Bioinform.* **2010**, *78*, 480–486. [CrossRef]

36. Trevisan, D.A.C.; da Silva, P.V.; Farias, A.B.P.; Campanerut-Sá, P.A.Z.; Ribeiro, T.; Faria, D.R.; de Mendonça, P.S.B.; de Mello, J.C.P.; Seixas, F.A.V.; Mikcha, J.M.G. Antibacterial Activity of Barbatimão (Stryphnodendron adstringens) against Staphylococcus Aureus: In vitro and Silico Studies. *Lett. Appl. Microbiol.* **2020**, *71*, 259–271. [CrossRef] [PubMed]

37. Ebrahimipour, S.Y.; Sheikhshoaie, I.; Simpson, J.; Ebrahimnejad, H.; Dusek, M.; Kharazmi, N.; Eigner, V. Antimicrobial Activity of Aroylhydrazone-Based Oxido Vanadium (v) Complexes: In vitro and in silico Studies. *New J. Chem.* **2016**, *40*, 2401–2412. [CrossRef]

38. Tittal, R.K.; Yadav, P.; Lal, K.; Kumar, A. Synthesis, Molecular Docking and DFT Studies on Biologically Active 1, 4-Disubstituted-1, 2, 3-Triazole-Semicarbazone Hybrid Molecules. *New J. Chem.* **2019**, *43*, 8052–8058.

39. Salaria, D.; Rolta, R.; Patel, C.N.; Dev, K.; Sourirajan, A.; Kumar, V. In vitro and Silico Analysis of Thymus Serpyllum Essential Oil as Bioactivity Enhancer of Antibacterial and Antifungal Agents. *J. Biomol. Struct. Dyn.* **2021**, *40*, 10383–10402. [CrossRef]

40. Trott, O.; Olson, A.J. AutoDock Vina: Improving the Speed and Accuracy of Docking with a New Scoring Function, Efficient Optimization, and Multithreading. *J. Comput. Chem.* **2010**, *31*, 455–461. [CrossRef]

41. Rolta, R.; Salaria, D.; Kumar, V.; Patel, C.N.; Sourirajan, A.; Baumler, D.J.; Dev, K. Molecular Docking Studies of Phytocompounds of Rheum Emodi Wall with Proteins Responsible for Antibiotic Resistance in Bacterial and Fungal Pathogens: In silico Approach to Enhance the Bio-Availability of Antibiotics. *J. Biomol. Struct. Dyn.* **2022**, *40*, 3789–3803. [CrossRef]

42. Rădulescu, M.; Jianu, C.; Lukinich-Gruia, A.T.; Mioc, M.; Mioc, A.; Șoica, C.; Stana, L.G. Chemical Composition, in vitro and in silico Antioxidant Potential of *Melissa officinalis* subsp. *officinalis* Essential Oil. *Antioxidants* **2021**, *10*, 1081. [CrossRef]

43. Asma, T.; Bariza, Z.; Noui, Y. Algerian Prickly Pear (*Opuntia ficusindica* L.) Physicochemical Characteristics. *Int. J. Sci. Res.* **2017**, *5*, 14–17.
44. Rym, H.; Belhadj, S.; Celia, O.; Fairouz, S. Antioxidant activity of *Pistacia lentiscus* methanolic extracts. *Rev. Agrobiol.* **2018**, *8*, 845–852.
45. Lipinski, C.A. Lead-and Drug-like Compounds: The Rule-of-Five Revolution. *Drug Discov. Today Technol.* **2004**, *1*, 337–341. [CrossRef]
46. Ghose, A.K.; Viswanadhan, V.N.; Wendoloski, J.J. A Knowledge-Based Approach in Designing Combinatorial or Medicinal Chemistry Libraries for Drug Discovery. 1. A Qualitative and Quantitative Characterization of Known Drug Databases. *J. Comb. Chem.* **1999**, *1*, 55–68. [CrossRef] [PubMed]
47. Veber, D.F.; Johnson, S.R.; Cheng, H.-Y.; Smith, B.R.; Ward, K.W.; Kopple, K.D. Molecular Properties That Influence the Oral Bioavailability of Drug Candidates. *J. Med. Chem.* **2002**, *45*, 2615–2623. [CrossRef] [PubMed]
48. Daina, A.; Zoete, V. A BOILED-Egg To Predict Gastrointestinal Absorption and Brain Penetration of Small Molecules. *ChemMedChem* **2016**, *11*, 1117–1121. [CrossRef] [PubMed]
49. Lagunin, A.; Stepanchikova, A.; Filimonov, D.; Poroikov, V. PASS: Prediction of Activity Spectra for Biologically Active Substances. *Bioinformatics* **2000**, *16*, 747–748. [CrossRef]
50. Baskin, I.I. Machine Learning Methods in Computational Toxicology. *Comput. Toxicol. Methods Protoc.* **2018**, *1800*, 119–139.
51. Ferreira, L.G.; Dos Santos, R.N.; Oliva, G.; Andricopulo, A.D. Molecular Docking, and Structure-Based Drug Design Strategies. *Molecules* **2015**, *20*, 13384–13421. [CrossRef]
52. López-Vallejo, F.; Caulfield, T.; Martínez-Mayorga, K.; A Giulianotti, M.; Nefzi, A.; A Houghten, R.; L Medina-Franco, J. Integrating Virtual Screening and Combinatorial Chemistry for Accelerated Drug Discovery. *Comb. Chem. High Throughput Screen.* **2011**, *14*, 475–487. [CrossRef]
53. Kandsi, F.; Elbouzidi, A.; Lafdil, F.Z.; Meskali, N.; Azghar, A.; Addi, M.; Hano, C.; Maleb, A.; Gseyra, N. Antibacterial and Antioxidant Activity of *Dysphania ambrosioides* (L.) Mosyakin and Clemants Essential Oils: Experimental and Computational Approaches. *Antibiotics* **2022**, *11*, 482. [CrossRef]
54. Semidalas, C.; Semidalas, E.; Matsoukas, M.T.; Nixarlidis, C.; Zoumpoulakis, P. In silico Studies Reveal the Mechanisms behind the Antioxidant and Anti-Inflammatory Activities of Hydroxytyrosol. *Med. Chem. Res.* **2016**, *25*, 2498–2511. [CrossRef]
55. Elbouzidi, A.; Ouassou, H.; Aherkou, M.; Kharchoufa, L.; Meskali, N.; Baraich, A.; Mechchate, H.; Bouhrim, M.; Idir, A.; Hano, C.; et al. LC–MS/MS Phytochemical Profiling, Antioxidant Activity, and Cytotoxicity of the Ethanolic Extract of *Atriplex halimus* L. against Breast Cancer Cell Lines: Computational Studies and Experimental Validation. *Pharmaceuticals* **2022**, *15*, 1156. [CrossRef] [PubMed]
56. Payne, D.J.; Miller, W.H.; Berry, V.; Brosky, J.; Burgess, W.J.; Chen, E.; DeWolf, W.E.; Fosberry, A.P.; Greenwood, R.; Head, M.S.; et al. Discovery of a Novel and Potent Class of FabI-Directed Antibacterial Agents. *Antimicrob. Agents Chemother.* **2002**, *46*, 3118–3124. [CrossRef] [PubMed]
57. Rafi, S.; Novichenok, P.; Kolappan, S.; Zhang, X.; Stratton, C.F.; Rawat, R.; Kisker, C.; Simmerling, C.; Tonge, P.J. Structure of Acyl Carrier Protein Bound to FabI, the FASII Enoyl Reductase from Escherichia Coli. *J. Biol. Chem.* **2006**, *281*, 39285–39293. [CrossRef] [PubMed]
58. Heath, R.J.; Rock, C.O. Enoyl-Acyl Carrier Protein Reductase (FabI) Plays a Determinant Role in Completing Cycles of Fatty Acid Elongation in *Escherichia Coli* (∗). *J. Biol. Chem.* **1995**, *270*, 26538–26542. [CrossRef]
59. Trzaskos, J.M.; Fischer, R.T.; Favata, M.F. Mechanistic Studies of Lanosterol C-32 Demethylation. Conditions Which Promote Oxysterol Intermediate Accumulation during the Demethylation Process. *J. Biol. Chem.* **1986**, *261*, 16937–16942. [CrossRef] [PubMed]
60. Wright, M.H.; Heal, W.P.; Mann, D.J.; Tate, E.W. Protein Myristoylation in Health and Disease. *J. Chem. Biol.* **2010**, *3*, 19–35. [CrossRef]
61. Raju, R.V.S.; Datla, R.S.S.; Moyana, T.N.; Kakkar, R.; Carlsen, S.A.; Sharma, R.K. N-Myristoyltransferase. *Mol. Cell. Biochem.* **2000**, *204*, 135–155. [CrossRef]
62. Kaneria, M.; Chanda, S. Evaluation of Antioxidant and Antimicrobial Properties of *Manilkara zapota* L. (Chiku) Leaves by Sequential Soxhlet Extraction Method. *Asian Pac. J. Trop. Biomed.* **2012**, *2*, S1526–S1533. [CrossRef]
63. Yang, D.; Wang, Q.; Ke, L.; Jiang, J.; Ying, T. Antioxidant Activities of Various Extracts of Lotus (Nelumbo Nuficera Gaertn) Rhizome. *Asia Pac. J. Clin. Nutr.* **2007**, *16* (Suppl. S1), 158–163. [PubMed]
64. Bampouli, A.; Kyriakopoulou, K.; Papaefstathiou, G.; Louli, V.; Krokida, M.; Magoulas, K. Comparison of Different Extraction Methods of *Pistacia lentiscus* Var. Chia Leaves: Yield, Antioxidant Activity and Essential Oil Chemical Composition. *J. Appl. Res. Med. Aromat. Plants* **2014**, *1*, 81–91. [CrossRef]
65. Vázquez, E.; Garcia-Risco, M.; Jaime, L.; Reglero, G.; Fornari, T. Simultaneous Extraction of Rosemary and Spinach Leaves and Its Effect on the Antioxidant Activity of Products. *J. Supercrit. Fluids* **2013**, *82*, 138–145. [CrossRef]
66. Manoko, M.L.K.; van Weerden, G.M.; van Berg, R.G.; Mariani, C. A New Tetraploid Species of *Solanum* L. Sect. *Solanum* (Solanaceae) from Tanzania. *PhytoKeys* **2012**, *16*, 65–74. [CrossRef] [PubMed]
67. Letawe, C.; Boone, M.; Pierard, G. Digital Image Analysis of the Effect of Topically Applied Linoleic Acid on Acne microcomedones. *Clin. Exp. Dermatol.* **1998**, *23*, 56–58. [CrossRef]

68. Villa, B.; Calabresi, L.; Chiesa, G.; Risè, P.; Galli, C.; Sirtori, C. Omega-3 Fatty Acid Ethyl Esters Increase Heart Rate Variability in Patients with Coronary Disease. *Pharmacol. Res. Off. J. Ital. Pharmacol. Soc.* **2002**, *45*, 475. [CrossRef]

69. Vafeiadou, K.; Weech, M.; Altowaijri, H.; Mihaylova, R.; Todd, S.; Yaqoob, P.; Jackson, K.; Lovegrove, J. The Effects of Substitution of Dietary Saturated Fatty Acids with Either Monounsaturated Fatty Acids or N-6 Polyun-saturated Fatty Acids on Measures of Endothelial Function, Arterial Stiffness, and Blood Pressure: Results from the DIVAS Study. *Proc. Nutr. Soc.* **2014**, *73*, E12. [CrossRef]

70. Bernatoniene, J.; Kopustinskiene, D. The Role of Catechins in Cellular Responses to Oxidative Stress. *Molecules* **2018**, *23*, 965. [CrossRef]

71. Cai, Y.; Luo, Q.; Sun, M.; Corke, H. Antioxidant Activity and Phenolic Compounds of 112 Traditional Chinese Medicinal Plants Associated with Anticancer. *Life Sci.* **2004**, *74*, 2157–2184. [CrossRef]

72. Taibi, M.; Elbouzidi, A.; OU-yahia, D.; Dalli, M.; Bellaouchi, R.; Tikent, A.; Roubi, M.; Gseyra, N.; Asehraou, A.; Hano, C.; et al. Assessment of the Antioxidant and Antimicrobial Potential of Ptychotis Verticillata Duby Essential Oil from Eastern Morocco: An In vitro and Silico Analysis. *Antibiotics* **2023**, *12*, 655. [CrossRef]

73. Milia, E.; Bullitt, S.; Mastandrea, G.; Szotakova, B.; Schoubben, A.; Langhansova, L.; Quartu, M.; Bortone, A.; Eick, S. *Pistacia lentiscus*: From Phytopharmacology to Scientific Explanations on Its Anti-Inflammatory and An-Timicrobial Capacity. *Antibiotics* **2021**, *10*, 425. [CrossRef] [PubMed]

74. Céliz, G.; Daz, M.; Audisio, M.C. Antibacterial Activity of Naringin Derivatives against Pathogenic Strains. *J. Appl. Microbiol.* **2011**, *111*, 731–738. [CrossRef] [PubMed]

75. Gachkar, L.; Yadegari, D.; Rezaei, M.B.; Taghizadeh, M.; Astaneh, S.A.; Rasooli, I. Chemical and Biological Characteristics of *Cuminum cyminum* and *Rosmarinus officinalis* Essential Oils. *Food Chem.* **2007**, *102*, 898–904. [CrossRef]

76. Giordani, R.; Kaloustian, J. Action Anticandidosique Des Huiles Essentielles: Leur Utilisation Concomitante Avec Des Médicaments Antifongiques. *Phytothérapie* **2006**, *4*, 121–124. [CrossRef]

77. Slimani, I.; Nassiri, L.; Boukil, A.; Bouiamrine, E.H.; Bachiri, L.; Bammou, M.; Ibijbijen, J. Inventaire Des Plantes Aromatiques et Médicinales Du Site d'intérêt Biologique et Écologique de Jbel Zerhoun, Région Meknès Tafilalet. *Afr. Sci.* **2016**, *12*, 393–409.

78. Selick, H.E.; Beresford, A.P.; Tarbit, M.H. The Emerging Importance of Predictive ADME Simulation in Drug Discovery. *Drug Discov. Today* **2002**, *7*, 109–116. [CrossRef]

Article

Novel Anti-Acetylcholinesterase Effect of *Euonymus laxiflorus* Champ. Extracts via Experimental and In Silico Studies

Van Bon Nguyen [1,*], San-Lang Wang [2,3,*], Tu Quy Phan [4], Manh Dung Doan [1], Thi Kim Phung Phan [4], Thi Kim Thu Phan [5], Thi Huyen Thoa Pham [5] and Anh Dzung Nguyen [1]

[1] Institute of Biotechnology and Environment, Tay Nguyen University, Buon Ma Thuot 630000, Vietnam; dmdung@ttn.edu.vn (M.D.D.); nadzung@ttn.edu.vn (A.D.N.)
[2] Department of Chemistry, Tamkang University, New Taipei City 25137, Taiwan
[3] Life Science Development Center, Tamkang University, New Taipei City 25137, Taiwan
[4] Faculty of Medicine and Pharmacy, Tay Nguyen University, Buon Ma Thuot 630000, Vietnam; phantuquy@ttn.edu.vn (T.Q.P.); ptkphung@ttn.edu.vn (T.K.P.P.)
[5] Department of Science and Technology, Tay Nguyen University, Buon Ma Thuot 630000, Vietnam; ptkthu@ttn.edu.vn (T.K.T.P.); pththoa@ttn.edu.vn (T.H.T.P.)
* Correspondence: nvbon@ttn.edu.vn (V.B.N.); sabulo@mail.tku.edu.tw (S.-L.W.)

Citation: Nguyen, V.B.; Wang, S.-L.; Phan, T.Q.; Doan, M.D.; Phan, T.K.P.; Phan, T.K.T.; Pham, T.H.T.; Nguyen, A.D. Novel Anti-Acetylcholinesterase Effect of *Euonymus laxiflorus* Champ. Extracts via Experimental and In Silico Studies. *Life* **2023**, *13*, 1281. https://doi.org/10.3390/life13061281

Academic Editors: Marisa Colone, Charalampia Amerikanoua and Efstathia Papada

Received: 26 April 2023
Revised: 22 May 2023
Accepted: 28 May 2023
Published: 30 May 2023

Abstract: Alzheimer's disease (AD) is the most common form of dementia, which is recorded as a global health issue. Natural acetylcholinesterase inhibitors (AChEIs) are considered a helpful therapy for the management of symptoms of patients with mild-to-moderate AD. This work aimed to investigate and characterize *Euonymus laxiflorus* Champ. (ELC) as a natural source of AChEIs compounds via in vitro and virtual studies. The screening parts used, including the leaves, heartwood, and trunk bark of ELC, revealed that the trunk bark extract possessed the highest activity, phenolics and flavonoid content. The in vitro anti-Alzheimer activity of ELC trunk bark was notably reclaimed for the first time with comparable effect (IC_{50} = 0.332 mg/mL) as that of a commercial AChEI, berberine chloride (IC_{50} = 0.314 mg/mL). Among various solvents, methanol was the most suitable to extract ELC trunk bark with the highest activity. Twenty-one secondary metabolites (**1–21**) were identified from ELC trunk bark extract, based on GCMS and UHPLC analyses. Of these, 10 volatile compounds were identified from this herbal extract for the first time. One phenolic (**11**) and seven flavonoid compounds (**15–21**) were also newly found in this herbal extract. Of the identified compounds, chlorogenic acid (**11**), epigallocatechin gallate (**12**), epicatechin (**13**), apigetrin (**18**), and quercetin (**20**) were major compounds with a significant content of 395.8–2481.5 μg/g of dried extract. According to docking-based simulation, compounds (**11–19**, and **21**) demonstrated more effective inhibitory activity than berberine chloride, with good binding energy (DS values: −12.3 to −14.4 kcal/mol) and acceptable RMSD values (0.77–1.75 Å). In general, these identified compounds processed drug properties and were non-toxic for human use, based on Lipinski's rule of five and ADMET analyses.

Keywords: Alzheimer; acetylcholinesterase inhibitors; *Euonymus laxiflorus* Champ.; bioactive compounds; medicinal plants

1. Introduction

Alzheimer's disease (AD) is the most common form of dementia. This disease has been recorded as a global health issue, with the number of patients worldwide reaching approximately 50 million; the number of cases of AD may double every 5 years, and it is estimated to reach 152 million by 2050 [1,2]. AD significantly affects individuals, the families of the patients, and the global economy, with the annual incurred costs estimated to reach about USD 1 trillion worldwide [1]. Up to date, no therapies may completely cure AD. However, some current therapies may significantly help to manage the symptoms, provide temporary relief, and increase the quality of life of patients with AD [3]. Among the therapies, the utilization of herbs for the management of AD has been considered since they are natural sources of bioactive compounds and are less toxic [4–10].

Vietnam, a tropical country, has been ranked as the sixteenth most biodiverse worldwide, with more than 10,000 herbal species recorded and 4000 species in use as traditional medicine [11]. Thus, studies based on the investigation of the biological effects and chemical profiles of medicinal plants from Vietnam have received increasing interest in recent years [12–17]. However, only a few studies and new records of herbs with potential anti-Alzheimer effects of this biodiverse area have been reported [15–17]; as such, the discovery of medicinal plants with biological activities related to Alzheimer's drugs from Vietnam has been received with great interest.

Euonymus laxiflorus Champ. (ELC), a medicinal plant, is wildly grown in Vietnam, India, Cambodia, Myanmar, and China [12,14,18]. ELC has been found to be a rich source of bioactivities, including antioxidant and anti-nitric oxide (anti-NO) properties [19], a potent anti-enzyme targeting anti-diabetes, such as anti-α-glucosidase and anti-α-amylase [12,14], and a significant reduction in the blood glucose of normal and diabetic rats with fewer or no side effects [20,21]. Some target compounds of α-glucosidase, α-amylase, antioxidant, and the anti-NO effect were purified and identified in our earlier works [13,19,22]. However, no data are available on anti-acetylcholinesterase (AChE) targeting anti-Alzheimer, as well as the concerning constituents of this herb.

As a part of our ongoing objective to investigate novel bioactivity and the active constituents of the herbal ELC for the development of this herb as a natural and functional food and/or as drugs, in this investigation, we collected this herb from Dak Lak province of Vietnam and determined the bioactivities. This study is the first to report the anti-AChE activity of ELC. Further screening of the most active part used, including the heartwood, trunk bark, leaves of ELC, and the chemical profiles, the interaction of active molecules toward the targeting enzyme acetylcholinesterase, as well as Lipinski's rule of five and ADMET-based pharmacokinetics and pharmacology analyses via computational study are also presented in this study. All steps of our work are summarized in Scheme 1.

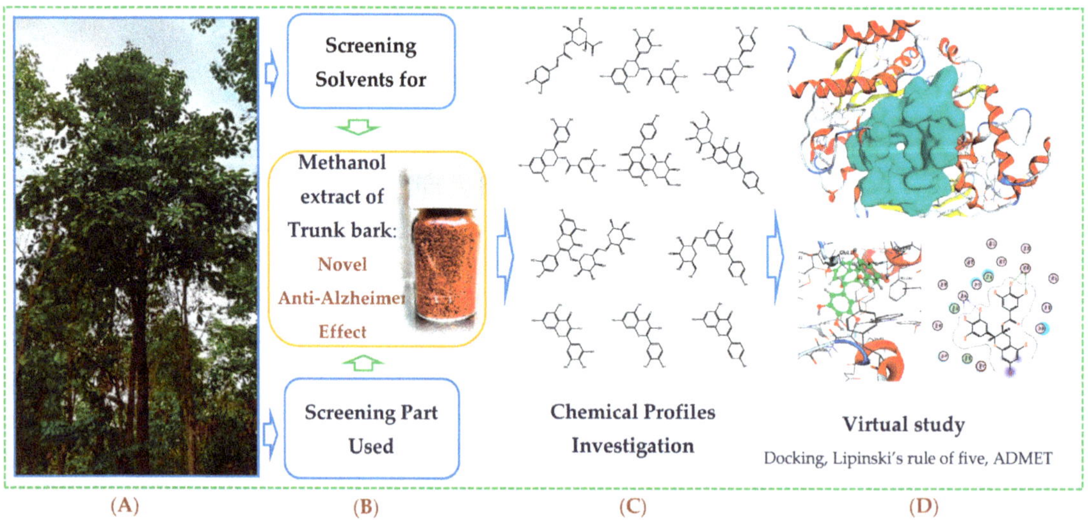

Scheme 1. A schematic of this study. (**A**) The samples of *Euonymus laxiflorus* Champ. (**B**) were collected for screening extraction solvents and functional part used. (**C**) The most bioactive extract was investigated for its chemical profiles, (**D**) then the virtual study was conducted for the prediction of active compounds and drug discovery.

2. Materials and Methods

2.1. Materials

The samples of some parts used, such as leaves, heartwood, and trunk bark of ELC, were collected in Yok Don National Park, Dak Lak Province, Vietnam, in 2022. The dried herbal parts were packed in PE bags and stored at -30 °C before extraction. Acetylcholinesterase and berberine chloride were acquired from Sigma Aldrich (St. Louis, MO, USA). The highest-grade solvents and common chemical agents available were used in this study.

2.2. Chemical Methods

2.2.1. Preparation of Herbal Extracts

Herbal extracts were prepared using the previously reported process [12,13]. The herbal samples were extracted using different solvents, including methanol, butanol, ethyl acetate, *n*-hexane, and water. The solvent was mixed with the plant sample for 24 h in a ratio of 100 mL/10 g in a conical glass flask using a shaker; then, the solution was filtered via a filter paper (no. 1, Whatman International Ltd., Maidstone, UK) at room temperature. The sample residues were further extracted twice with 200 mL using the same solvent with the same conditions described above. These three solutions were combined and concentrated at 60 °C under a vacuum using a rotary evaporator (IKA, Staufen, Germany). The extracts were stored at -30 °C before further use [12,13].

2.2.2. Gas Chromatography-Mass Spectrometry (GCMS) Analysis

The herbal extract was dissolved in methanol (MeOH), then purified by solid-phase extraction using the QuEChERS method. For further analysis, GC (Thermo Trace GC Ultra, USA) and ITQ900 (Thermo, Waltham, MA, USA) were conducted. A TG-SQC capillary column (30 m $\times$ 0.25 mm $\times$ 0.25 µm) was utilized for the GCMS analysis. Helium (99.999%), a carrier gas, was set at a constant flow rate of 1 mL/min. The sample solution (1 µL) was injected in a split ratio of 10:1. The temperature of the injector and the ion source were set at 250 and 230 °C, respectively. The temperature program of the oven was set at 70 °C (isothermal for 2 min) and increased up to 280 °C with an increasing speed of 15 °C/min, ending with a 10 min isothermal at 280 °C. MS data were at 70 eV, a scanning interval time of 0.5 s, and for fragments from 50 to 650 Da. The compounds were identified via comparison with reported compounds using compounds data of the Mass Spectra Library (NIST 17.L and Wiley).

2.2.3. UHPLC Analysis

The herbal extracts were dissolved in MeOH at 10 mg/mL and then filtered using a 0.45 µm polyvinylidene fluoride membrane filter (Millipore Sigma, Billerica, MA, USA). An amount of two microliters of the extract solution was injected into the UHPLC system (Thermo Ultimate 3000). The constituents in the sample were separated via a column (Hypersil GOLD aQ, 3 µm, 150 $\times$ 2.1 mm), which was maintained at a temperature of 30 °C. A mobile phase consisting of MeOH and 0.1% phosphoric acid in water was used, and the mobile phase program was set at 5% MeOH (0.0–0.5 min), 5–30% MeOH (0.5–8.0 min), 30–45% MeOH (8.0–13 min), 45–65% MeOH (13.0–18.0 min), 65–95% MeOH (18.0–22.0 min), and 95–5% MeOH (22.0–23.0 min). The flow was set at 0.2 mL/min, and the constituents were detected at 265 nm.

2.3. Enzyme Inhibition Assay

AChE was used to carry out the inhibitory assay using Ellman's method [23] with modifications. A 0.05 M phosphate buffer, pH 8.0, was used in this protocol. Additionally, 60 µL of the herbal extract was mixed with the same volume of 0.5 mM enzyme solution and 120 µL of phosphate buffer, and kept at 25 °C for 15 min in a flat-bottom 96-well plate. Agent 5, 5′-dithiobis 2 nitrobenzoic acid (DTNB, 30 µL of 0.003 M) and 40 µL of 0.002 M acetylthiocholine iodide (ATCI) were added into the mixture to start the reaction. The reaction was maintained at 25 °C for 10 min before the absorbance was measured at the wavelength of 415 nm. For the control group, 60 μL of phosphate buffer was used instead of 60 μL of the herbal extract solution and the same condition was used for the calculation of enzyme inhibition activity using the following equation:

$$AChE\ inhibition\ (\%) = (A_C - A_E)/A_C \times 100,$$

where A_E is the absorbance value measured at 415 nm of the reaction containing herbal extract and enzyme, while A_C is the absorbance value at 415 nm of the reaction containing enzyme and phosphate buffer instead of the herbal extract. The samples and berberine chloride (commercial AChE inhibitor) were dissolved in dimethylsulfoxide and eluted at various concentrations using a 0.05 M phosphate buffer and then used for the tests.

2.4. Virtual Study Methods

2.4.1. Docking Simulation

Virtual analysis was conducted using the MOE-2015.10 software to predict the active metabolite-concerning inhibition against the targeting enzyme. This virtual protocol was performed following the three typical steps mentioned in the previous reports [24,25].

- ✓ Preparation of protein structure: The structure data of enzyme AChE were obtained from the Research Collaboratory for Structural Bioinformatics Protein Data Bank for the preparation of their three-dimensional structures through the use of MOE-2015.10. The most active sites on the enzyme were found using the site finder function of the MOE software. A virtual pH 7 was used for the preparation of enzyme structure.
- ✓ Preparation of ligand structures: The structures of identified compounds (ligands) from the herbal extracts and commercial inhibitors were prepared using the Chem-BioOffice 2018 software and further optimized using the MOE software. The following parameters were required to prepare the structure of ligands: virtual pH 7, force field MMFF94x; R-Field 1:80; cutoff, rigid water molecules, space group p1, cell size 10, 10, 10; cell shape 90, 90, 90; gradient 0.01 RMS kcal·mol^{-1}A^{-2}.
- ✓ Docking ligands into enzymes and the obtained out-put data: The prepared ligands were docked into the active site of AChE using the MOE software. As major out-put for analysis, DS value, RMSD value, linkage types, compositions of amino acids, and the linkages' distances were harvested.

2.4.2. The Lipinski Rule of Five and ADMET Analyses

The online software accessed at (http://www.scfbioiitd.res.in/software/drugdesign/lipinski.jsp (accessed on 15 May 2023)) was utilized for the Lipinski Rule of Five performance. A web tool called SwissADME (http://www.swissadme.ch/ (accessed on 15 May 2023)) was used for the prediction of some pharmacokinetic parameters of ADMET. The out-put data of theoretical interpretations of pharmacokinetic parameters have been previously described [26] and used as a public reference online, accessed at (http://biosig.unimelb.edu.au/pkcsm/theory (accessed on 15 May 2023)).

3. Results and Discussion

3.1. New Records of the Potential Anti-Acetylcholinesterase Effect of Euonymus laxiflorus Champ

To evaluate ELC as a source of anti-Alzheimer drugs, the ELC trunk bark (ELCTB) was extracted with methanol, butanol, ethyl acetate, *n*-hexane, and water, and then tested for their inhibition against AChE, a key enzyme in the discovery of anti-Alzheimer drugs [27]. The biological effect was recorded and shown under the IC_{50} value, which is the sample concentration that reduces 50% of enzymatic effect; the lower the value a sample yields, the greater the effect it displays. The results are shown in Table 1, and Figure A1 shows that the MeOH extract of ELCTB could most efficiently inhibit AChE with a low IC_{50} value (0.323 mg/mL), comparable to that of a commercial inhibitor compound, berberine chloride

(IC_{50} = 0.314 mg/mL). This meOH extract was also found to inhibit AChE with a high max inhibition value of 97.0%, while other extracts showed a max effect of less than 80.1%.

Table 1. Acetylcholinesterase inhibitory effect of ELC trunk bark extracted by various solvents.

No.	ELCTB Extracted by Various Solvents	IC_{50} (mg/mL)	Max Inhibition (%)
1	Methanol extract	0.323 ± 0.021	97.0 ± 2.3%
2	Butanol extract	0.781 ± 0.042	80.1 ± 2.4%
3	Ethyl acetate extract	1.232 ± 0.098	65.2 ± 3.1%
4	*n*-Hexane extract	1.530 ± 0.101	57.3 ± 1.5%
5	Water extract	0.662 ± 0.032	79.1 ± 2.6%
6	Berberine chloride	0.314 ± 0.032	98.2 ± 1.8%

To screen for the most functional part to be used, the heartwood, trunk bark, and leaves of ELC were also extracted by meOH and used for the activity tests. The trunk bark MeOH extract showed the most AChE inhibitory activity with the smallest IC_{50} values of 0.336 mg/mL (Table 2 and Figure A2). This part used was also rich in total content of polyphenol (567.3 GAE/g dry extracts) and flavonoid (335.2 QE/g dry extracts). Thus, the MeOH extract of the trunk bark was further examined.

Table 2. Acetylcholinesterase-inhibitory activity, the total content of polyphenol, and flavonoid of different parts of *Euonymus laxiflorus* Champ.

Parts Used of *Euonymus laxiflorus* Champ. Extracted by MeOH	IC_{50} (mg/mL)	Total Content of Polyphenol (mg GAE/g Dry Extracts)	Total Content of Flavonoid (mg QE/g Dry Extracts)
Heartwood MeOH extract	1.326 ± 0.103	107.2 ± 22.1	98.0 ± 7.8
Trunk bark MeOH extract	0.336 ± 0.029	567.3 ± 27.2	335.2 ± 10
Leaves MeOH extract	0.981 ± 0.045	227.2 ± 15.1	101.3 ± 8.9
Berberine chloride	0.327 ± 0.031	-	-

Recently, ELCTB has been found showing several medical effects, including antioxidant, anti-NO, anti-α-glucosidase, anti-α-amylase, and antidiabetic effects, and ELCTB has also been used by folks for the treatment of several diseases [12,14,19–21]. However, the potential AChE inhibitory effect of ELC was newly recorded in this study; as such, the findings herein contributed to enriching the catalog of medical effects of this plant.

Clinical trials using herb extracts in AD treatment were conducted in some works. Some herbal extracts in China achieved good effects in clinical trials for treating dementia [28]. *Ginkgo biloba* extracts were trialed widely in AD treatment and trials were conducted for almost 3–6 months [29]. In the Ginkgo One Tablet A Day (GOTADAY) trial, the dose of the one-daily drug containing 240 mg of *G. biloba* extract showed a significant improvement in neuropsychiatric symptoms and cognition for dementia patients [30]. Overall, other reports of this herb showed non-potential or unreliable effects in trials [29]. Furthermore, a patient group using a high dose of *Panax ginseng* recorded significant results based on clinical dementia rating and Alzheimer disease assessment scale when compared with the control [31]. In Iran, *Salvia officinalis* extract was tested with a fixed dose of 60 drops/day for mild-to-moderate Alzheimer's patients in a 4-month period. It supported improved cognition after 16 weeks of treatment; however, some of its side effects need further confirmation [32]. Based on some of the above literature, using herb extracts as AD treatment agents is also considered a promising orientation. ELC extracts can also be suggested as a potential source for in-depth trials. This herb was demonstrated non-toxic for normal cells and mouse models and it has also been used as a traditional medicine by ethnic minorities in Vietnam for a long time [22].

3.2. The Chemical Profile of the MeOH Extract of Euonymus laxiflorus Champ. Trunk Bark

GCMS and UHPLC analyses were used for the identification of the chemical profile of ELCTB MeOH extract. Based on GCMS analysis using compounds data from the MS Library (NIST 17.L and Wiley), 10 volatile compounds (symbolized as **1–10**) in ELCTB were detected and identified (Figure 1, Table 3), including 3-hydroxydecanoic acid (**1**) [33], propane, 1,1-dipropoxy- (CAS) (**2**), (2-(2-butoxyisopropoxy)-2-isopropanol (**3**), p-xylene (**4**) [34], styrene (**5**) [35], oxalic acid, heptyl propyl ester (**6**) [36], sulfurous acid, isobutyl pentyl ester (**7**) [37], 2-phenylethyl allyl ether (**8**), 12-hydroxyalliacolide (**9**) [38], and 7-ethyl-quinoline (**10**). Of these volatile compounds, 3-hydroxydecanoic acid (**1**) was found as a major volatile compound in ELCTB extract in a significantly high amount (48.63% area), followed by p-xylene (**4**), styrene (**5**), oxalic acid, heptyl propyl ester (**6**), 12-hydroxyalliacolide (**9**), and 7-ethyl-quinoline (**10**) with recoded area in the range of 4.14–11.63%; other volatiles were present in a minor amount (0.38–2.91% area). The GC profiles of these volatile compounds are presented in Figure A3.

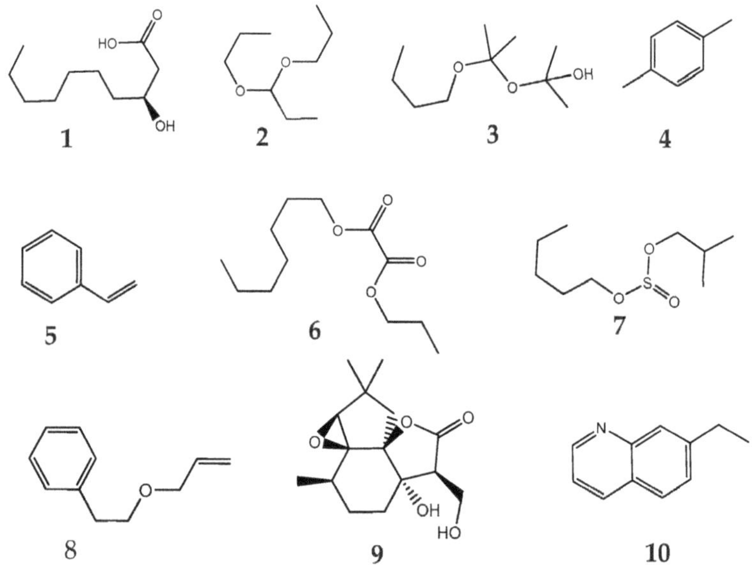

Figure 1. The volatile compounds detected from the methanol extract of ELC trunk bark via GCMS analysis.

Based on the application of UHPLC using commercial compounds as standards, one phenolic (symbolized as **11**) and ten flavonoid compounds (symbolized as **12–21**) were detected and identified (Figure 2), including chlorogenic acid (**11**) [39], epigallocatechin gallate (EGCG) (**12**) [40], epicatechin (**13**) [41], epicatechin gallate (**14**) [42], vitexin (**15**) [43], isovitexin (**16**) [44], rutin (**17**) [45], apigetrin (**18**) [46], myricetin (**19**) [47], quercetin (**20**) [48], and apigenin (**21**) [49]. The contents of these compounds were determined and are presented in Table 4. Among these compounds, a high yield of apigetrin (**18**) was found in the ELCTB extract at 2481.525 µg/g of dried extract. Chlorogenic acid (**11**), EGCG (**12**), epicatechin (**13**), and quercetin (**20**) were detected with a significant content of 395.808–576.809 µg/g of dried extract, while the content of other compounds, **14–17** and **21** was lower at 23.197–184.798 µg/g of dried extract. The UHPLC profiles of these identified compounds are presented in Figure A4.

Table 3. The volatile compounds in the MeOH extract of ELC trunk bark identified by GCMS analysis.

ID Compd	Compds	RT (min)	% Area
1	3-Hydroxydecanoic acid	3.40	48.63
2	Propane, 1,1-dipropoxy- (CAS)	4.77	2.91
3	(2-(2-butoxyisopropoxy)-2-isopropanol	4.88	2.62
4	*p*-Xylene	5.16	11.63
5	Styrene	5.56	8.53
6	Oxalic acid, heptyl propyl ester	7.45	4.62
7	Sulfurous acid, isobutyl pentyl ester	9.19	2.58
8	2-Phenylethyl allyl ether	10.28	0.38
9	12-Hydroxyalliacolide	12.76	8.04
10	7-ethyl-Quinoline	13.91	5.92

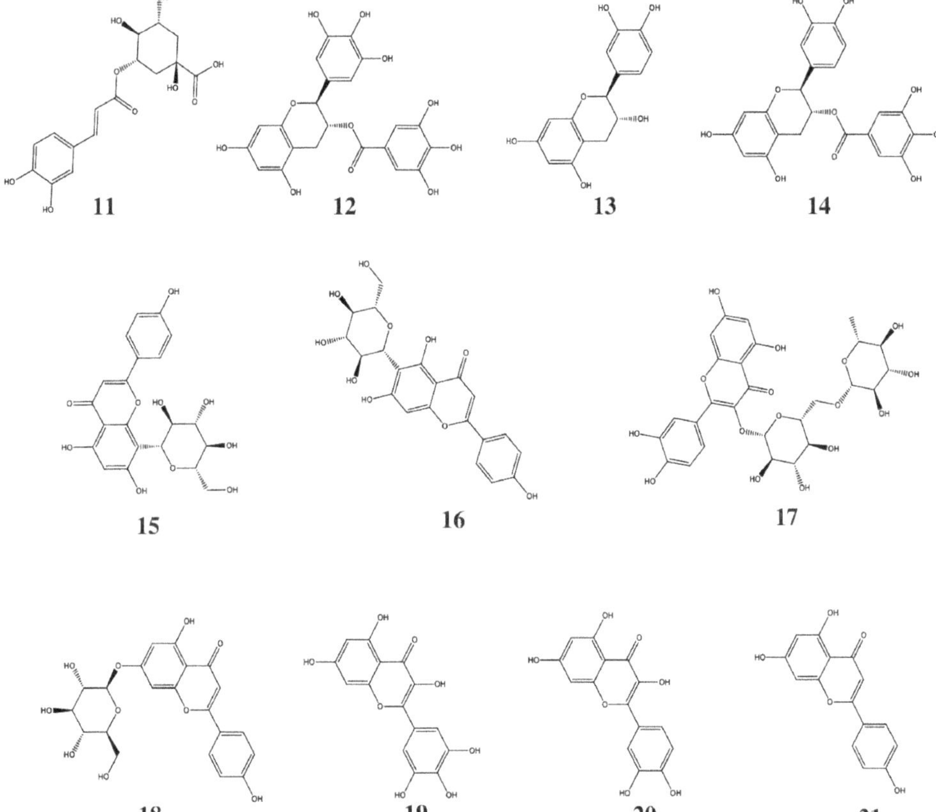

Figure 2. Flavonoid and phenolic compounds detected from the methanol extract of ELC trunk bark by UPLC.

Table 4. The content of phenolic compounds contained in the methanol extract of ELC trunk bark by UPLC.

ID Compd	Compds	RT (min)	Content µg/g of Dried Extract
11	Chlorogenic acid	13.700	395.808
12	EGCG	14.198	402.680
13	Epicatechin	14.760	576.809
14	Epicatechin gallate	16.230	79.513
15	Vitexin	17.785	23.197
16	Isovitexin	19.180	63.884
17	Rutin	19.603	61.581
18	Apigetrin	20.442	2481.525
19	Myricetin	20.882	221.843
20	Quercetin	22.830	487.600
21	Apigenin	24.772	184.798

Regarding the chemical profiles of ELC, a total of 36 compounds were previously purified and their chemical structures elucidated [13,19,21,22,27,50]. Of these, six compounds were first purified in ELC stems and leaves collected from Taiwan by Kuo et al., 2003 [50]. Recently, we isolated and identified chemical structures of 30 phenolics from the methanol extract of ELC trunk bark collected from the Central Highland of Vietnam [13,19,21,22,27]. However, there are no data on volatile compounds contained in the ELCTB identified through GCMS analysis so far. Thus, this work is the first to report the application of UHPLC in the detection and determination of major phenolic and flavonoid compounds contained in the extract of this herbal species. One phenolic compound, chlorogenic acid (**11**), and seven flavonoids including vitexin (**15**), isovitexin (**16**), rutin (**17**), apigetrin (**18**), myricetin (**19**), quercetin (**20**), and apigenin (**21**) in ELC extract are reported for the first time. Thus, the experimental data of this work contributed to enriching the chemical profiles of ELC.

3.3. Insight into the Interactions and Energy Binding Bioactive Compounds toward Targeting Enzyme—Acetylcholinesterase via Docking Study

The active compounds, the interaction and binding energy of the bioactive inhibitors toward AChE were predicted through the docking study using the MOE-2015.10 software. The data of AChE protein structure was obtained from the Worldwide Protein Data Bank. The most active site (on AChE for docking ligands) was found using the site finder function of the MOE software. Based on the out-put data of MOE, 25 binding sites on AChE were determined. The sizes, residues, and 3D structures of these binding sites were mapped and are presented in the Appendix A (Table A1). The ligands may bind to various sites on the enzyme; however, only the most active site was chosen to be presented and discussed in detail. Based on the site finder function of MOE, and the pre-screening results, binding site 1 is suggested as the most active binding site for further investigation. This active site was found (Figure 3a) to contain 39 residues. To determine whether binding site 1 was covered by the catalytic site of enzyme or not, the CASTp3.0 server was used in the prediction of the AChE catalytic site, which has the volume and the surface area of 904.278 Å^3 and 529.676 Å^2, respectively (Figure 3b, Table A2). Figure 3 indicated that binding site 1 was not located in the catalytic site. Thus, all these inhibitors show a high possibility of no binding to the catalytic site of AChE.

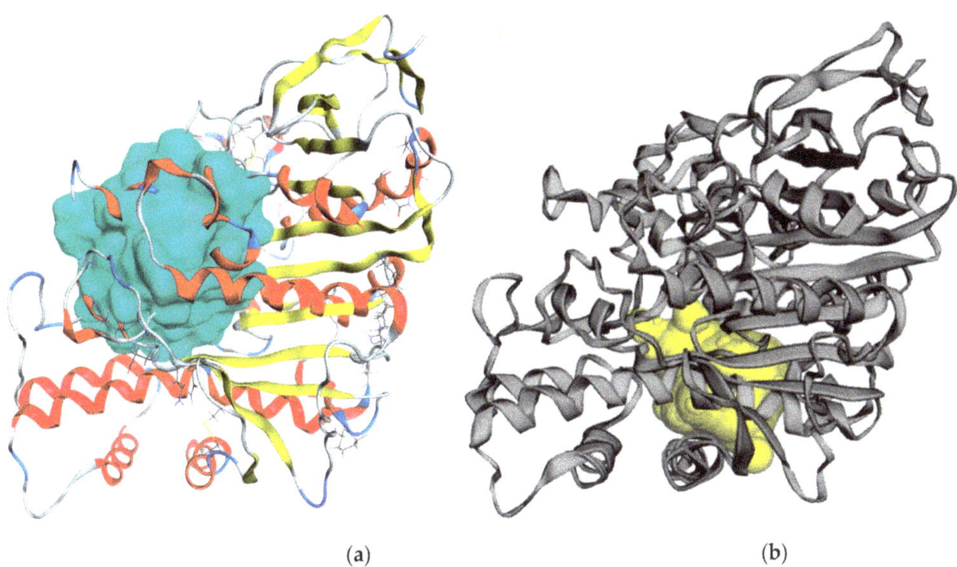

(a)   (b)

Figure 3. The 3D structure of the most active binding site 1 of acetylcholinesterase (**a**) was found using MOE-2015.10, and the 3D structure of the catalytic site of acetylcholinesterase (**b**) found by utilization of the CASTp3.0.

In the virtual study, RMSD and DS are important parameters to determine whether a compound (ligand/inhibitor) may bind to the target protein and inhibit its enzymatic activity [51,52]. For a successful binding, when a ligand interacts with a target protein with an RMSD value less than 2.0 Å, the binding is considered to be significant and widely accepted for further virtual analysis [51]. As shown in Table 5, all the identified compounds (**1–21**) interacted with AChE with low RMSD values in the range of 0.63–1.75 Å, while the interaction of the commercial inhibitor berberine chloride had an RMSD value of 1.65 Å. This result suggested the successful binding of all the ligands to the target protein with acceptable RMSD values. For virtual evaluation of effective inhibition, DS was commonly used. When a compound binds to the target enzyme with a DS value less than −3.20 kcal/mol, it is proposed as an enzyme inhibitor [52]. In comparison, the lower the DS value an inhibitor has, the greater is its inhibitory effect. As summarized in Table 5, all the compounds could bind to AChE with DS values greater than −6.3 kcal/mol, indicating that they may be possible AChE candidates. The commercial inhibitor berberine chloride showed high inhibition against AChE, with a DS value of −12.1 kcal/mol. Compared with the commercial inhibitor, eleven compounds (**1–10, 20**) exhibited weaker AChE inhibition (DS value in the range of −6.3 to −11.0 kcal/mol) than berberine chloride (**22**). Other compounds (**11–19** and **21**) demonstrated higher activity than berberine chloride (**22**) with low DS values of −12.3 to −14.4 kcal/mol. Overall, the order of inhibition of the AChE protein by active compounds was as follows: **17 > 13 > 18 > 16 > 14 > 11 = 19 > 15 > 12 > 21 > 22** (commercial inhibitor) **> 20 > 9 > 7 > 1 > 3 > 2 > 8 > 6 > 10 > 4 > 5**. The result of the in vitro test and virtual analysis could nearly fit and corroborate each other, indicating that the meOH extract of ELC trunk bark showed a high effectiveness against AChE, due to harboring various major active compounds (**11–19,21**).

Table 5. The RMSD and DS values of ligands (L) binding with AChE.

Compounds (Ligands)	Symbols of L-AChE	RMSD (Å)	DS (kcal/mol)
3-Hydroxydecanoic acid	1-AChE	0.82	−10.1
Propane, 1,1-dipropoxy-(CAS)	2-AChE	0.63	−9.6
(2-(2-butoxyisopropoxy)−2-isopropanol	3-AChE	1.72	−9.8
p-Xylene	4-AChE	1.57	−6.5
Styrene	5-AChE	0.93	−6.3
Oxalic acid, heptyl propyl ester	6-AChE	1.11	−8.8
Sulfurous acid, isobutyl pentyl ester	7-AChE	0.81	−10.3
2-Phenylethyl allyl ether	8-AChE	1.68	−8.9
12-Hydroxyalliacolide	9-AChE	1.40	−10.6
7-ethyl-Quinoline	10-AChE	0.82	−8.6
Chlorogenic acid	11-AChE	1.50	−13.1
EGCG	12-AChE	1.24	−12.5
Epicatechin	13-AChE	1.75	−14.3
Epicatechin gallate	14-AChE	1.52	−13.3
Vitexin	15-AChE	1.18	−12.8
Isovitexin	16-AChE	1.09	−13.8
Rutin	17-AChE	1.29	−14.4
Apigetrin	18-AChE	1.03	−13.9
Myricetin	19-AChE	0.77	−13.1
Quercetin	20-AChE	1.51	−11.0
Apigenin	21-AChE	1.26	−12.3
Berberine chloride	22-AChE	1.65	−12.1

To investigate the interaction of ligands and AChE, the detailed binding at the active site were recorded, as presented in Table 6 and Figure 4. The active inhibitor compounds (**11–19**, **21**) and commercial inhibitor (**22**) were examined. Rutin (**17**) demonstrated the highest inhibition against AChE by interacting with four amino acids, including His400, Glu199, Trp84, and Asp72 of this enzyme, resulting in the creation of seven linkages (4 H-donor, 2 H-pi, 1 pi-H). Among these, this ligand (**17**) was found to interact with His400 and Asp72 and formed one H-donor linkage and one pi-H linkage, respectively, while it interacted with Glu199 and Trp84 to create two linkages (H-donor), and three linkages (2 H-pi, 1 pi-H), respectively. This was followed by epicatechin (**13**), which showed efficient inhibition against AChE at a low DS value (−14.3 kcal/mol) by binding and interacting with this enzyme through four amino acids, Ser81, Asn85, Ser200, and Glu199, to generate four H-donor linkages. The next five compounds (**11**, **14**, **16**, **18**, and **19**) could also bind tightly to AChE with DS values lower than −13.1 kcal/mol by interacting with five, seven, two, two, and two amino acids to form five, seven, four, three, and four linkages, respectively. The binding of apigetrin (**18**) to AChE was through the least interaction (three linkages) but possessed the best binding energy (DS value of −13.9 kcal/mol), while other ligands had greater interaction with AChE (four to seven linkages) but showed weaker bind energies (DS values in the range of −13.1 to −13.8 kcal/mol). These results indicated that the DS values may be independent of the number of interactions between the ligand and the enzyme. The next three ligands (**12**, **15**, and **21**) also showed a slightly higher binding effect to AChE than the commercial inhibitor (**22**). These compounds, **12**, **15**, and **21**, could bind to AChE by interacting with two, four, and three amino acids and generated four,

four, and three linkages, respectively, while the commercial inhibitor (**22**) binds to AChE through only one amino acid, forming only one linkage. Among the prominent amino acids contained in the binding site, Glu199 had an important role in the interaction with all the above-mentioned ligands (**11–19, and 21**). This amino acid (Glu199) was bound to the ligands **11, 13, 14, 15, 20,** and **21** to generate one H-donor linkage (binding linkage energy in the range of −0.6 to −5.8 kcal/mol) with each ligand, while it interacted with ligands **12, 17,** and **18** to generate two H-donor linkages (binding linkage energy in the range of −0.5 to −3.8 kcal/mol) with each ligand. Especially, Glu199 was found to bind with ligands **16** and **19,** and for each ligand, up to three H-donor linkages (binding linkage energy in the range of −1.1 to −4.0 kcal/mol) were formed.

These potential molecules were further investigated in their frontier molecular orbitals. The data of the highest occupied molecular orbital (HOMO) and lowest unoccupied molecular orbital (LUMO) of these compounds are presented in Figure 5. These molecular structures showed their low E_{HOMO} values in the range of − 5.65 to − 8.65 eV. This indicates their significant electronic stability (commonly accepted under − 5 eV). In the previous investigation [53], theoretical complexes of the tetrylone family with E_{HOMO} values of − 3 to − 7 eV were also found as highly stable. It has been evidenced that all structures possess an insulation-to-semiconduction energy gap (3.2 eV < EG < 9 eV), showing good intermolecular binding capability toward protein structures [54], and this was further explained based on the super-exchange theory [55,56] and the electron hopping model [57]. In this study, the energy gap values of the inhibitors were found in the range of 3.77 to 5.61 eV; as such, they have potential intermolecular binding capability toward the targeting enzyme.

Based on RMSD, DS values and some data of frontier molecular orbitals recorded, some secondary metabolites (compounds **11–19, 21**) contained in the methanol extract of ELC trunk bark may be suggested as potential candidates of AChE inhibitors. However, some works summarized by Pagadala et al. [58] indicated unreliable binding affinity predictions using docking studies. Several reports also indicated some drawbacks of the docking study [59–61]. Thus, further works should be performed, including purifications of active compounds and testing activity via in vitro, in vivo, and clinical trials for development of these compounds to be drugs.

3.4. Lipinski's Rule of Five and ADMET-Based Pharmacokinetics and Pharmacology

Lipinski's rules have been applied to evaluate the drug-likeness of compounds; the five rules include "molecular mass must be less than 500 Da (rule 1), high lipophilicity with LogP value < 5 (rule 2), hydrogen bond donors < 5 (rule 3), hydrogen bond acceptors < 10 (rule 4), and the molar refractivity should be between 40–130 (rule 5)". A compound is considered to have drug-likeness properties and has a high possibility of being a drug when it satisfies at least 2/5 of Lipinski's rules. As presented in Table 7, all the identified compounds complied with five of Lipinski's rules, except compound **2,** which complied with four of Lipinski's rules. Thus, these compounds have a high probability of successfully being developed as a drug.

The ADMET properties of these compounds and commercial inhibitors were also compared, and the data are presented in Appendix A (Figures A3 and A4). In general, these compounds in the MeOH extract of ELCTB also showed good ADMET properties in the required allotted limitation. In addition, all the active inhibitor compounds (**11–19 and 21**) and the commercial inhibitor (**22**) were not toxic for human use. Furthermore, the evidence of safety on normal cells of some compounds was indicated via in vitro or in vivo tests. Chlorogenic acid was recorded to affect cancer cells without normal cell influence [62]. EGCG was found as toxic on HuCC-T1 cancer cells but safe on the viability of 293T normal cells [63]. In another report, EGCG from green tea at 40–200 microM caused a significant death for some tumor cell lines, but only 1% of the WI38 normal cells growth was affected in the same condition [64]. Epicatechin and Epicatechin gallate were discovered to cause death for various cancer cells but had no effect on normal cells in some reports [65–68]. Vitexin has a non-effect on bronchial epithelial 16HBE normal cells [69]. Rutin, Apigetrin, and Myricetin all are potential and safe adjuvant chemotherapeutic agents with trivial toxicity and side effects via in vivo and clinical trials [70–72]. The scientific proof also demonstrated that Quercetin and Apigenin have no or low effects on normal cells [71,73]. Almost phenolics in this study were showed safe for normal cells. However, the safe evidence for all compounds detected still needs to be performed continually in further research via in vitro, in vivo, and clinical trials.

Table 6. The docking results of ligands (L) binding with acetylcholinesterase (AChE).

L-AChE Complex	Linkages Number	Amino Acids Interacting with the Ligands [Distance (Å)/E (kcal/mol)/Linkage Type]
11-AChE	5 linkages (2 H-donor, 2 H-acceptor, 1 pi-H)	Tyr70 (2.75/−2.5/H-donor); Glu199 (3.42/−0.6/H-donor); Gly119 (2.78/1.6/H-acceptor); Ser200 (2.68/−1.2/H-acceptor); Asn85 (4.66/−0.6/pi-H)
12-AChE	4 linkages of H-donor	Ser81 (2.79/−3.7/H-donor); Ser81 (3.18/−2.6/H-donor); Glu199 (3.60/−0.5/H-donor); Glu199 (2.8/−5.0/H-donor)
13-AChE	4 linkages of H-donor	Ser81 (2.77/−2.8/H-donor); Asn85 (3.10/−1.0/H-donor); Ser200 (2.78/5.6/H-donor); Glu199 (2.99/−2.1/H-donor)
14-AChE	7 linkages (4 H-donor, 1 H-pi, 2 pi-pi)	His440 (2.59/−1.5/H-donor); Glu199 (2.87/−2.7/H-donor); Tyr70 (2.60/−2.5/H-donor); Asn85 (2.56/−1.9/H-donor); Phe330 (3.91/−1.1/H-pi); Trp84 (3.92/−0.0/pi-pi); Tyr334 (3.74/−0.0/pi-pi)
15-AChE	4 linkages (2 H-donor, 1 H-acceptor, 1 H-pi)	Asp72 (3.29/−1.0/H-donor); Glu199 (2.84/−4.3/H-donor); Ser200 (2.62/−1.1/H-acceptor); Phe330 (4.32/−1.3/H-pi)
16-AChE	4 linkages (3 H-donor, 1 H-pi)	Glu199 (3.17/−1.1/H-donor); Glu199 (2.61/−3.5/H-donor); Glu199 (2.89/−4.0/H-donor); Trp84 (3.55/−0.8/H-pi)
17-AChE	7 linkages (4 H-donor, 2 H-pi, 1 pi-H)	His400 (2.75/−2.2/H-donor); Glu199 (3.00/−2.0/H-donor); Glu199 (2.90/−3.0/H-donor); Trp84 (3.31/−0.9/H-donor); Trp84 (4.02/−0.9/H-pi); Trp84 (4.10/−0.6/H-pi); Asp72 (3.57/−0.8/pi-H)
18-AChE	3 linkages of H-donor	Glu199 (3.04/−2.7/H-donor); Glu199 (2.92/−3.8/H-donor); Asp72 (2.88/−1.5/H-donor)
19-AChE	4 linkages of H-donor	Tyr70 (2.98/−2.1/H-donor); Glu199 (2.81/−1.4/H-donor); Glu199 (2.80/−2.1/H-donor); Glu199 (2.84/−1.5/H-donor)
20-AChE	3 linkages (1 H-donor, 1 H-acceptor, 1 pi-H)	Glu199 (2.75/−5.8/H-donor); His440 (3.02/−1.5/H-acceptor); Gly118 (3.77/−0.7/pi-H)
21-AChE	3 linkages (1 H-donor, 1 H-acceptor, 1 pi-H)	Glu199 (2.76/−5.8/H-donor); His440 (3.04/−1.1/H-acceptor); Gly118 (3.74/−0.7/pi-H)
22-AChE	1 H-pi	Tyr121 (4.50/−0.7/H-pi)

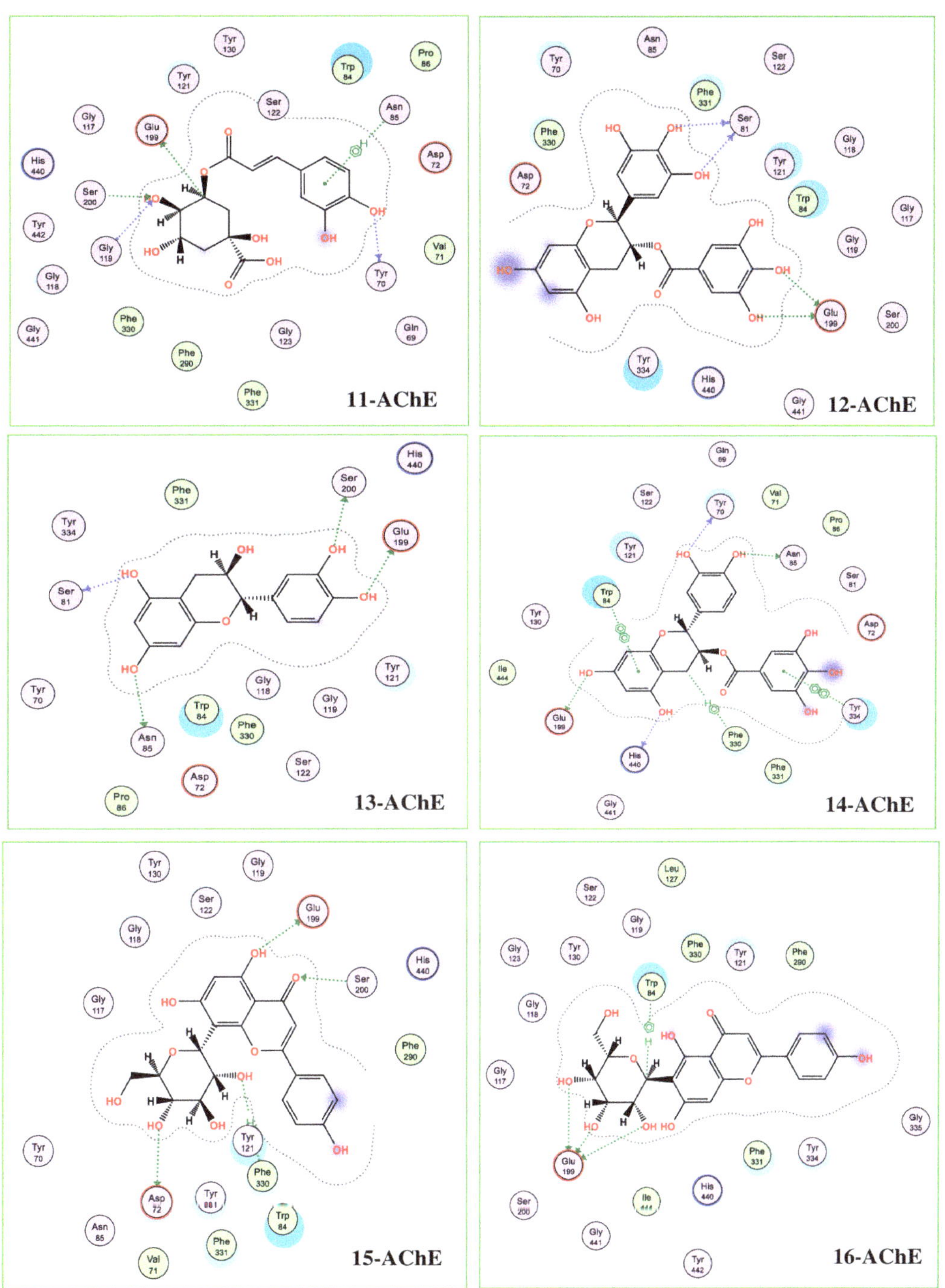

Figure 4. *Cont.*

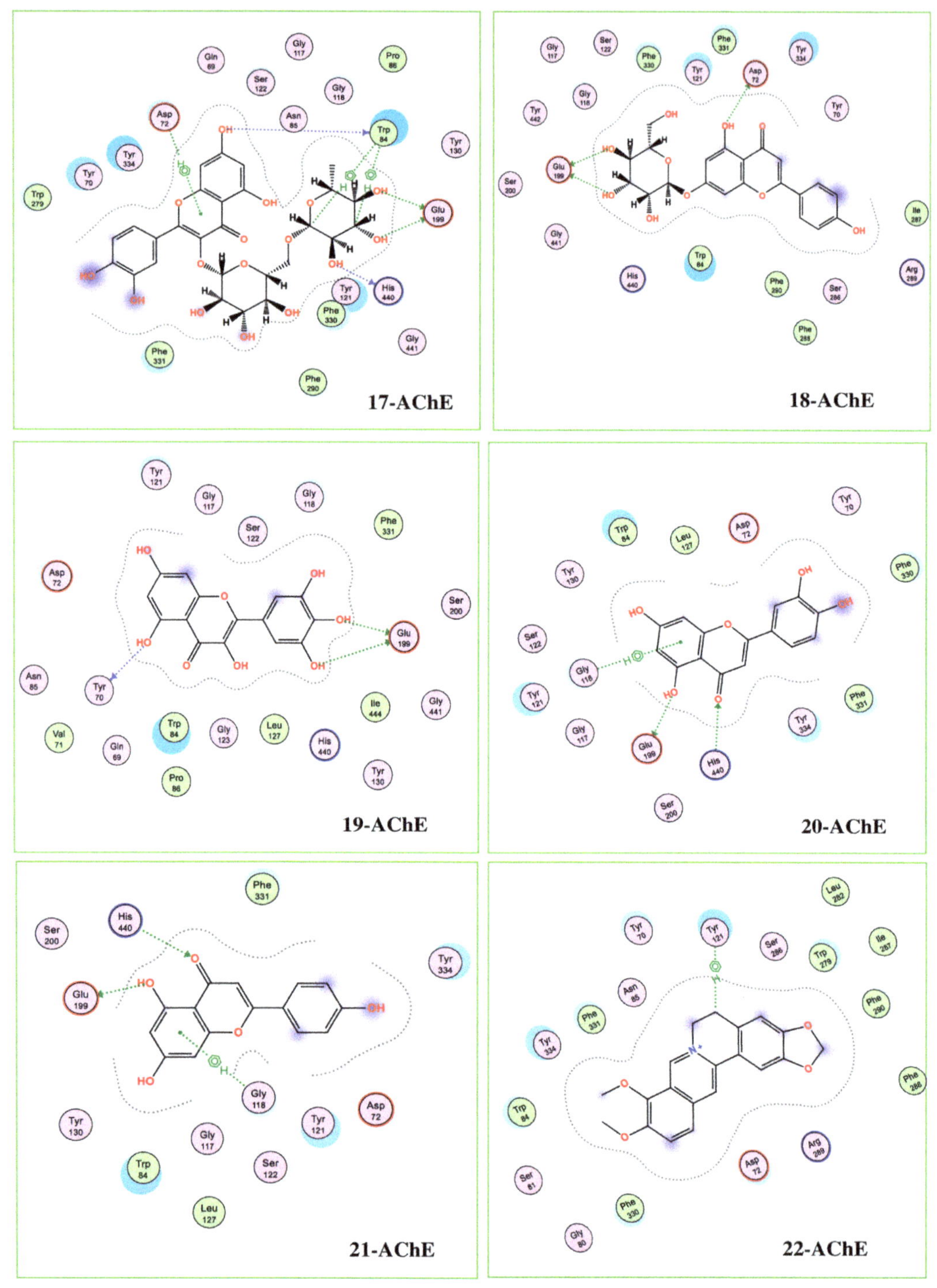

Figure 4. Insight into interactions of ligand–acetylcholinesterase (AChE) at the binding site on AChE (ligand: **11–22**).

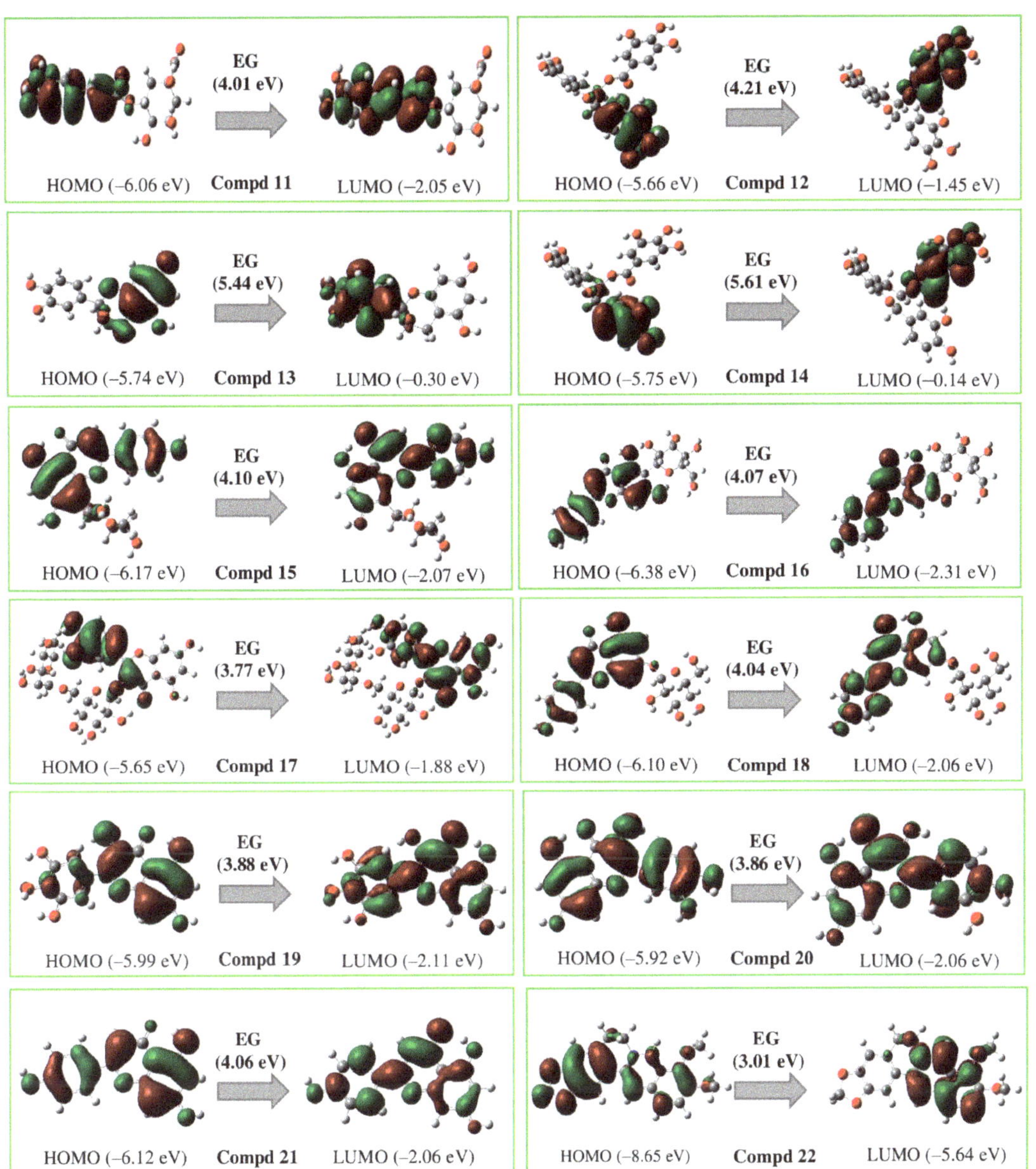

Figure 5. HOMO and LUMO of compounds **11–22** analyzed by DFT at level of theory B3LYP/6–31G.

Table 7. The result of Lipinski's Rule of Five of the compounds identified in the methanol extract of ELC trunk bark (**1–21**), and berberine chloride (**22**).

ID Compd	Mass (Dalton)	Hydrogen Bond Donor	Hydrogen Bond Acceptors	LogP	Molar Refractivity
1	187.0	1	3	0.848	49.01
2	160.0	0	2	22.576	46.56
3	190.0	1	3	2.284	52.32
4	106.0	0	0	2.303	35.92
5	104.0	0	0	2.33	36.53
6	230.0	0	4	2.45	60.97
7	209.0	1	3	2.787	56.26
8	162.0	0	1	2.432	51.16
9	187.0	1	3	0.848	49.01
10	157.0	0	1	2.797	51.210
11	157.0	0	1	2.797	51.121
12	353.0	5	9	−1.981	79.89
13	458.0	8	11	2.233	108.92
14	290.0	5	6	1.546	72.62
15	442.0	7	10	2.528	107.26
16	432.0	7	10	−0.066	103.53
17	432.0	7	10	−0.066	103.53
18	610.0	10	16	−1.879	137.50
19	432.0	6	10	−0.107	103.54
20	318.0	6	8	1.717	75.72
21	302.0	5	7	2.011	74.05
22	270.0	3	5	2.420	70.81
Lipinski's rules	≤500	≤5	≤10	≤5	40–130

4. Conclusions

This is the first report of AChE inhibition of the MeOH extract of ELCTB. From this extract, 21 secondary metabolites (**1–21**) were identified, including 10 volatile compounds (**1–10**), 1 phenolic (**11**), and 10 flavonoids (**12–21**). Of these, compounds **1–11, 15, 16, 17, 18, 19, 20,** and **21** were detected for the first time in this herbal extract. Compounds (**11–19** and **21**) exhibited greater effective inhibitory activity than the commercial inhibitor with good binding energy and acceptable RMSD values. The Lipinski rule of five and ADMET analyses indicated that the identified compounds possessed drug properties and were non-toxic for human use. The finding of this study indicates that ELC is a rich source of bioactive compounds that have the potential to be used as anti-Alzheimer drug candidates.

Author Contributions: Conceptualization, V.B.N.; methodology, V.B.N. and S.-L.W.; software, V.B.N. and T.Q.P.; validation, S.-L.W., M.D.D. and A.D.N.; formal analysis, V.B.N., A.D.N., T.Q.P., T.K.P.P., T.K.T.P., T.H.T.P. and T.K.P.P.; investigation, V.B.N., M.D.D. and T.Q.P.; resources, V.B.N.; data curation, V.B.N.; writing—original draft preparation, V.B.N.; writing—review and editing, S.-L.W., V.B.N. and A.D.N.; visualization, V.B.N.; supervision, V.B.N.; project administration, V.B.N. and S.-L.W. All authors have read and agreed to the published version of the manuscript.

Funding: This research was funded by grants from Tay Nguyen University, Vietnam (T2023-02CBTĐ); and supported in part by the Ministry of Science and Technology, Vietnam (NNĐT/TW/22/13); National Science and Technology Council, Taiwan (NSTC 111-2320-B-032-001; NSTC 111-2923-B-032-001).

Institutional Review Board Statement: Not applicable.

Informed Consent Statement: Not applicable.

Data Availability Statement: All the data for this article can be found in the article.

Conflicts of Interest: The authors declare no conflict of interest.

Appendix A

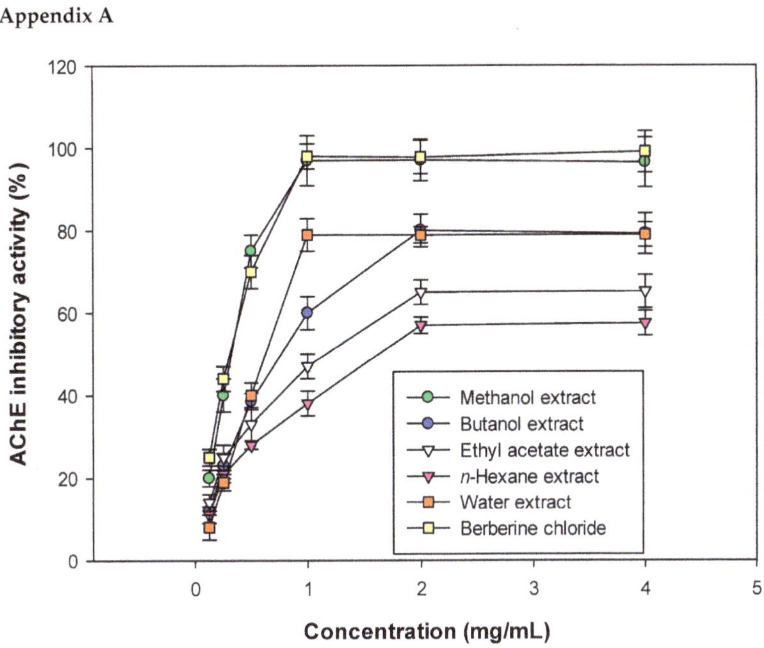

Figure A1. AChE inhibitory activity (%) of *Euonymus laxiflorus* Champ. trunk bark extracted by various.

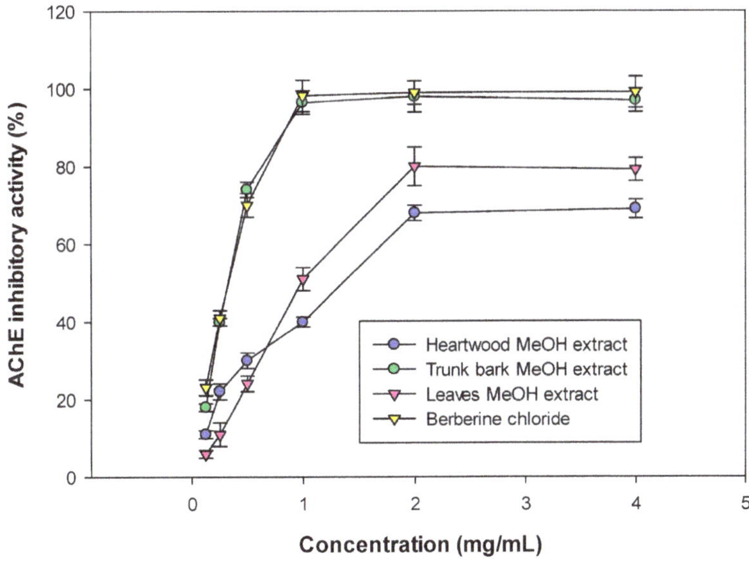

Figure A2. AChE inhibitory activity (%) of different parts of *Euonymus laxiflorus* Champ.

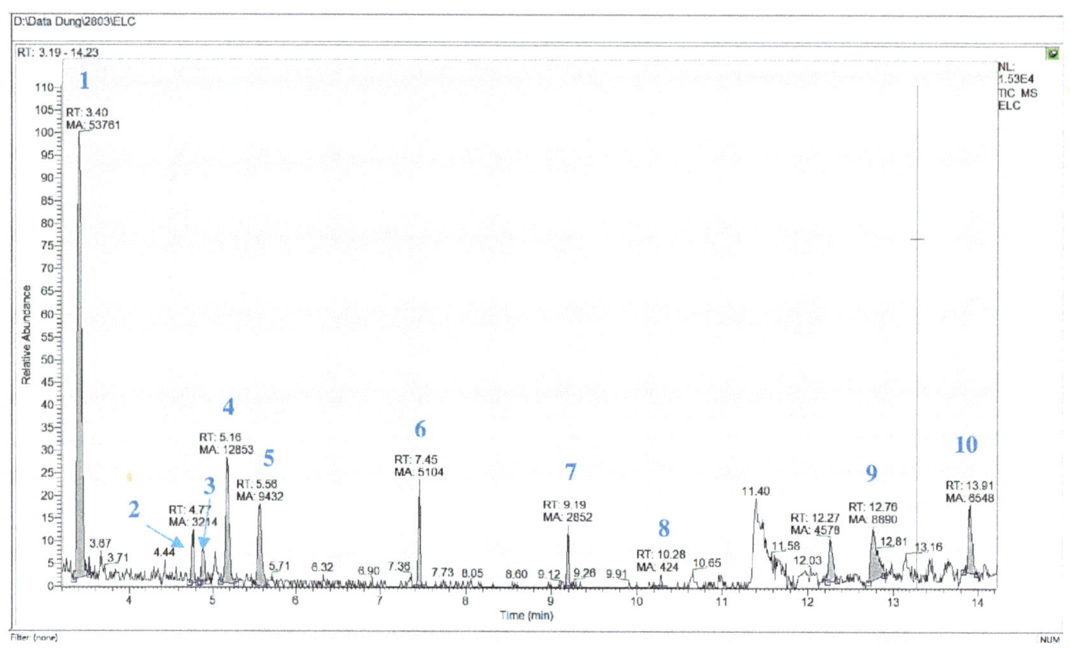

Figure A3. GC profile volatile compounds identified from the MeOH extract of *Euonymus laxiflorus* Champ. trunk bark.

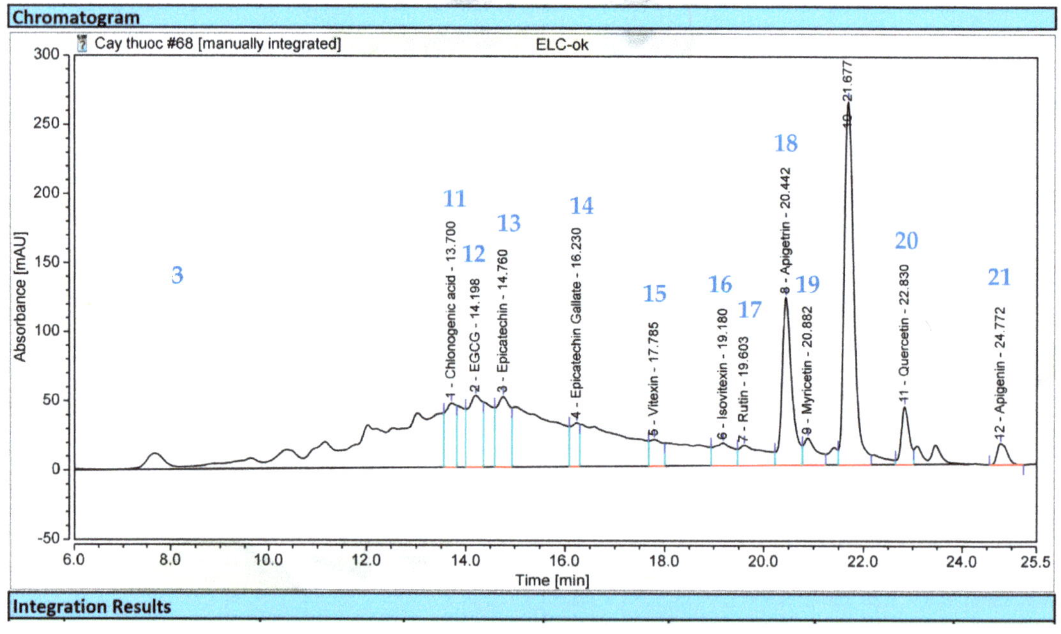

Figure A4. High-performance liquid chromatography finger printings of the MeOH extract of *Euonymus laxiflorus* Champ. trunk bark.

Table A1. The detail data of catalytic site prediction of extracellular domain of AChE found via using the CASTp3.0 server.

Name of Protein	Surface Area (SA) Å^2	Volume (SA) Å^3
Electrophorus electricus AChE (1EEA)	529.676	904.278

Amino acids located at the active site of 1EEA: PRO229 ASN230 CYS231 PRO232 TRP233 SER 235 VAL236 SER237 GLU240 ARG244 LEU282 PRO283 PHE284 SER286 ARG289 PHE290 VAL293 ILE296 SER304 LEU305 GLU306 PRO361 HIS362 HIS398 CYS402 PRO403 HIS406 TRP524 ASN525 GLN526 LEU 528 PRO529 LEU532 ASN533

Figure A5. Mapping the binding sites on AChE. The 3D structure of binding sites from the front surface (**A**) and back surface (**B**) of AChE. The 3D structure of 25 binding sites on AChE (named **1–25**).

Table A2. The detail size and residues of 25 binding sites of acetylcholinesterase was found using MOE-2015.10.

SITE	SIZE	Residues
1	202	1:(GLN69 TYR70 VAL71 ASP72 GLN74 SER81 TRP84 ASN85 PRO86 TYR116 GLY117 GLY118 GLY119 TYR121 SER122 GLY123 SER124 LEU127 TYR130 GLU199 SER200 TRP233 TRP279 LEU282 PHE284 ASP285 SER286 ILE287 PHE288 ARG289 PHE290 PHE330 PHE331 TYR334 GLY335 HIS440 GLY441 TYR442 ILE444)
2	39	1:(ASN230 PRO232 GLU306 ASP397 HIS398 CYS402 PRO403 HIS406 TRP524 ASN525 PRO529)
3	32	1:(LYS325 ASP326 ARG388 ASP389 ASP392 ASP393 ILE401 PHE422 GLU434 TRP435 ARG517)
4	20	1:(PRO232 GLU240 ARG244 LEU282 PRO283 PHE284 ASP285 SER286 ARG289 PRO361 HIS362)
5	20	1:(ARG349 PHE352 TYR375 THR376 ASP377 ASP380 ASP381 LYS386 ASN387 GLY390 LEU391)
6	27	1:(CYS402 MET405 HIS406 ASN409 LYS410 GLN500 ARG515 LEU516 ARG517 VAL518 CYS521 VAL522 ASN525 GLN526)
7	19	1:(ALA36 GLU37 PRO38 PRO39 MET43 ARG46 ARG47 PRO48 GLU49 LEU95 ARG149)
8	17	1:(ARG468 TYR472 SER487 GLU489 SER490 GLU508 PRO509 MET510)
9	22	1:(ARG47 GLY166 ASN167 LEU171 ARG174 SER212 PRO213 GLY214 ASP297 GLU299 PHE300)
10	19	1:(THR412 GLY415 ASN416 GLY417 THR418 LEU494 PHE495 THR496 THR497)
11	16	1:(GLU461 ALA464 LEU465 ARG468 ASN506 THR507 GLU508 PRO509)
12	24	1:(PRO39 ARG44 CYS67 GLN68 SER91 GLU92 CYS94 TYR148 ARG149 VAL150 PHE153)
13	12	1:(GLN68 GLN69 TYR70 TYR121 VAL150 GLY151 ALA152 PHE153 LEU274 ILE275 GLU278)
14	18	1:(GLU82 MET83 ASN85 ASN87 LEU127 ASP128 VAL129)
15	17	1:(GLN69 TYR70 VAL71 GLN272 ILE275 ASP276)
16	19	1:(LEU450 PRO451 LEU452 VAL453 LYS454 LEU456 ASN457 TYR458 THR459 ALA460 GLU463)
17	23	1:(ASP128 VAL129 ASN131 LYS133 TYR134 LEU450 VAL453 GLU455 LEU456)
18	20	1:(TYR375 THR376 ASP377 LYS386 ASP389 GLY390 ASP393 ARG517 MET520)
19	12	1:(LEU31 GLY32 TRP58 ASN59 ALA60 SER61 THR62 TYR63 PRO64)
20	21	1:(GLN318 ASN416 GLY417 TYR419 THR479 GLY480 ASN481 LEU494)
21	41	1:(MET83 VAL129 ASN429 LEU430 VAL431 TYR442 GLU445 LEU450 LEU456 TYR458)
22	7	1:(LYS11 SER12 LYS51 TRP179 ASP182 ASN183)
23	12	1:(GLN374 GLN519 MET520 VAL522 PHE523 PHE527)
24	17	1:(ASN42 GLU163 GLU260 ILE263 HIS264 ARG267)
25	16	1:(ARG47 PRO48 LEU171 ARG174 MET175 GLN178 LEU218)

Table A3. The absorption, distribution, metabolism, excretion, and toxicity (ADMET)-based pharmacokinetics and pharmacology of identified compounds contained in the methanol extract of *Euonymus laxiflorus* Champ. trunk bark (**1–11**).

Property \ ID Compd	1	2	3	4	5	6	7	8	9	10	11	Unit
Absorption												
Water solubility	−2.327	−2.051	−1.454	−2.522	−2.739	−2.3	−2.3	−2.633	−2.777	−2.445	−2.449	(1)
Caco2 permeability	1.398	1.653	1.726	1.547	1.544	1.45	1.755	1.563	0.336	1.619	−0.84	(2)
Intestinal absorption (human)	93.346	95.839	93.25	95.713	95.902	95.192	93.991	97.018	78.829	97.318	36.377	(3)
Skin Permeability	−2.713	−2.144	−3.472	−1.236	−1.104	−2.082	−2.018	−1.505	−3.257	−1.642	−2.735	(4)
P-glycoprotein substrate	No	No	No	No	No	No	No	No	No	Yes	Yes	(5)
P-glycoprotein I inhibitor	No	No	No	No	No	No	No	No	No	No	No	(5)
P-glycoprotein II inhibitor	No	No	No	No	No	No	No	No	No	No	No	(5)
Distribution												
VDss (human)	−0.799	0.054	−0.075	0.325	0.403	−0.034	−0.046	0.414	0.206	0.184	0.581	(6)
Fraction unbound (human)	0.454	0.55	0.603	0.362	0.331	0.454	0.489	0.284	0.622	0.27	0.658	(6)
BBB permeability	0.252	0.63	0.005	0.409	0.459	0.555	0.525	0.688	−0.008	0.396	−1.407	(7)
CNS permeability	−2.92	−2.69	−3.083	−1.677	−1.577	−2.849	−2.331	−1.884	−3.367	−1.92	−3.856	(8)
Metabolism												
CYP2D6 substrate	No	No	No	No	No	No	No	No	No	No	No	(5)
CYP3A4 substrate	No	No	No	No	No	No	No	No	No	No	No	(5)
CYP1A2 inhibitor	No	No	No	No	Yes	No	No	Yes	No	Yes	No	(5)
CYP2C19 inhibitor	No	No	No	No	No	No	No	No	No	Yes	No	(5)
CYP2C9 inhibitor	No	No	No	No	No	No	No	No	No	No	No	(5)
CYP2D6 inhibitor	No	No	No	No	No	No	No	No	No	No	No	(5)
CYP3A4 inhibitor	No	No	No	No	No	No	No	No	No	No	No	(5)
Excretion												
Total Clearance	1.598	1.639	1.35	0.254	0.265	1.774	0.535	0.35	0.803	0.296	0.307	(9)
Renal OCT2 substrate	No	No	No	No	No	No	No	No	No	No	No	(5)
Toxicity												
AMES toxicity	No	No	No	No	No	No	No	No	Yes	Yes	No	(5)
Max. tolerated dose (human)	0.084	0.796	0.848	0.921	0.943	0.771	0.805	0.999	0.529	0.331	−0.134	(10)
hERG I inhibitor	No	No	No	No	No	No	No	No	No	No	No	(5)
hERG II inhibitor	No	No	No	No	No	No	No	No	No	No	No	(5)
Oral Rat Acute Toxicity (LD50)	1.342	2.004	1.949	1.841	1.83	1.947	2.189	1.828	3.097	2.023	1.973	(11)
Oral Rat Chronic Toxicity	2.552	2.269	2.105	2.168	2.225	2.087	2.043	1.946	1.901	2.11	2.982	(12)
Hepatotoxicity	No	No	No	No	No	No	No	No	No	No	No	(5)
Skin Sensitization	No	Yes	Yes	No	No	Yes	Yes	Yes	No	No	No	(5)
T.Pyriformis toxicity	0.38	0.596	0.181	−0.022	−0.018	0.163	0.833	0.96	0.31	0.453	0.285	(13)
Minnow toxicity	0.711	1.336	1.956	1.31	1.136	0.477	0.854	0.67	2.574	0.643	5.741	(14)

Note: [1] $\log \mathrm{mol \cdot L^{-1}}$; [2] $\log \mathrm{Papp}\ (10^{-6}\ \mathrm{cm \cdot s^{-1}})$; [3] %; [4] $\log \mathrm{Kp}$; [5] Yes/No; [6] $\log \mathrm{L \cdot kg^{-1}}$; [7] $\log \mathrm{BB}$; [8] $\log \mathrm{PS}$; [9] $\log \mathrm{mL \cdot min^{-1} \cdot kg^{-1}}$; [10] $\log \mathrm{mg \cdot kg^{-1} \cdot day^{-1}}$; [11] $\mathrm{mol \cdot kg^{-1}}$; [12] $\log \mathrm{mg \cdot kg^{-1}\_bw \cdot day^{-1}}$; [13] $\log \mathrm{\mu g \cdot L^{-1}}$; [14] $\log \mathrm{mM}$.

Table A4. The absorption, distribution, metabolism, excretion, and toxicity (ADMET)-based pharmacokinetics and pharmacology of identified compounds contained in the methanol extract of *Euonymus laxiflorus* Champ. trunk bark (**12–21**) and berberine chloride (**22**).

Property	12	13	14	15	16	17	18	19	20	21	22	Unit
Absorption												
Water solubility	−2.894	−3.117	−2.911	−2.845	−2.812	−2.892	−2.559	−2.915	−2.925	−3.329	−6.715	(1)
Caco2 permeability	−1.521	−0.283	−1.264	−0.956	−0.618	−0.949	0.33	0.095	−0.229	1.007	1.212	(2)
Intestinal absorption (human)	47.395	68.829	62.096	46.695	64.729	23.446	37.609	65.93	77.207	93.25	94.642	(3)
Skin Permeability	−2.735	−2.735	−2.735	−2.735	−2.735	−2.735	−2.735	−2.735	−2.735	−2.735	−2.781	(4)
P-glycoprotein substrate	Yes	Yes	Yes	Yes	Yes	Yes	Yes	Yes	Yes	Yes	No	(5)
P-glycoprotein I inhibitor	No	No	No	No	No	No	No	No	No	No	Yes	(5)
P-glycoprotein II inhibitor	Yes	No	Yes	No	No	No	No	No	No	No	Yes	(5)
Distribution												
VDss (human)	0.806	1.027	0.664	1.071	1.239	1.663	0.342	1.317	1.559	0.822	0.179	(6)
Fraction unbound (human)	0.215	0.235	0.158	0.242	0.21	0.187	0.218	0.238	0.206	0.147	0	(6)
BBB permeability	−2.184	−1.054	−1.847	−1.449	−1.375	−1.899	−1.391	−1.493	−1.098	−0.734	0.764	(7)
CNS permeability	−3.96	−3.298	−3.743	−3.834	−3.754	−5.178	−3.746	−3.709	−3.065	−2.061	−1.652	(8)
Metabolism												
CYP2D6 substrate	No	No	No	No	No	No	No	No	No	No	No	(5)
CYP3A4 substrate	No	No	No	No	No	No	No	No	No	No	Yes	(5)
CYP1A2 inhibitor	No	No	No	No	No	No	No	Yes	Yes	Yes	No	(5)
CYP2C19 inhibitor	No	No	No	No	No	No	No	No	No	Yes	No	(5)
CYP2C9 inhibitor	No	No	No	No	No	No	No	No	No	No	No	(5)
CYP2D6 inhibitor	No	No	No	No	No	No	No	No	No	No	No	(5)
CYP3A4 inhibitor	Yes	No	No	No	No	No	No	No	No	No	No	(5)
Excretion												
Total Clearance	0.292	0.183	−0.169	0.444	0.442	−0.369	0.547	0.422	0.407	0.566	0.619	(9)
Renal OCT2 substrate	No	No	No	No	No	No	No	No	No	No	No	(5)
Toxicity												
AMES toxicity	No	No	No	No	No	No	No	No	No	No	No	(5)
Max. tolerated dose (human)	0.441	0.438	0.449	0.577	0.649	0.452	0.515	0.51	0.499	0.328	−0.653	(10)
hERG I inhibitor	No	No	No	No	No	No	No	No	No	No	No	(5)
hERG II inhibitor	Yes	No	Yes	No	No	Yes	No	No	No	No	Yes	(5)
Oral Rat Acute Toxicity (LD50)	2.522	2.428	2.558	2.595	2.558	2.491	2.595	2.497	2.471	2.45	2.553	(11)
Oral Rat Chronic Toxicity	3.065	2.5	2.777	4.635	5.37	3.673	4.359	2.718	2.612	2.298	0.89	(12)
Hepatotoxicity	No	No	No	No	No	No	No	No	No	No	No	(5)
Skin Sensitization	No	No	No	No	No	No	No	No	No	No	No	(5)
T.Pyriformis toxicity	0.285	0.347	0.285	0.285	0.285	0.285	0.285	0.286	0.288	0.38	0.432	(13)
Minnow toxicity	7.713	3.585	6.146	4.897	5.18	7.677	5.507	5.023	3.721	2.432	−1.711	(14)

Note: (1) log mol·L^{-1}; (2) log Papp (10^{-6} cm·s^{-1}); (3) %; (4) log Kp; (5) Yes/No; (6) log L·kg^{-1}; (7) log BB; (8) log PS; (9) log mL·min^{-1}·kg^{-1}; (10) log mg·kg^{-1}·day^{-1}; (11) mol·kg^{-1}; (12) log mg·kg^{-1}_bw·day^{-1}; (13) log μg·L^{-1}; (14) log mM.

References

1. Patterson, C. *World Alzheimer Report 2018*; Alzheimer's Disease International: London, UK, 2018.
2. Breijyeh, Z.; Karaman, R. Comprehensive review on Alzheimer's disease: Causes and treatment. *Molecules* **2020**, *25*, 5789. [CrossRef]
3. Majid, H.; Silva, F.V.M. Inhibition of enzymes important for Alzheimer's disease by antioxidant extracts prepared from 15 New Zealand medicinal trees and bushes. *J. R. Soc. N. Z.* **2020**, *50*, 538–551. [CrossRef]
4. Chen, B.W.; Li, W.X.; Wang, G.H.; Li, G.H.; Liu, J.Q.; Zheng, J.J.; Wang, Q.; Li, H.J.; Dai, S.X.; Huang, J.F. A strategy to find novel candidate anti-Alzheimer's disease drugs by constructing interaction networks between drug targets and natural compounds in medical plants. *PeerJ* **2018**, *6*, e4756. [CrossRef] [PubMed]
5. Singhal, A.K.; Naithani, V.; Bangar, O.P. Medicinal plants with a potential to treat Alzheimer and associated symptoms. *Int. J. Nutr. Pharmacol. Neurol. Dis.* **2012**, *2*, 84–91. [CrossRef]
6. Tuzimski, T.; Petruczynik, A. Determination of anti-Alzheimer's disease activity of selected plant ingredients. *Molecules* **2022**, *27*, 3222. [CrossRef] [PubMed]
7. Mukherjee, P.K.; Kumar, V.; Houghton, P.J. Screening of Indian medicinal plants for acetylcholinesterase inhibitory activity. *Phytother. Res.* **2007**, *21*, 1142–1145. [CrossRef]
8. Elufioye, T.O.; Obuotor, E.M.; Sennuga, A.T.; Joseph, M.A.; Saburi, A.A. Acetylcholinesterase and butyrylcholinesterase inhibitory activity of some selected Nigerian medicinal plants. *Rev. Bras. Farmacogn.* **2010**, *20*, 472–477. [CrossRef]
9. Mack, M.; Ashwell, R.N.; Jeffrey, F.F.; Johannes, V.S. Phenolic composition. antioxidant and acetylcholinesterase inhibitory activities of *Sclerocarya* birrea and *Harpephyllum caffrum* (*Anacardiaceae*) extracts. *Food Chem.* **2010**, *123*, 69–76.
10. Llorent, M.E.; Ortega, B.P.; Zengin, G.; Mocan, A.; Simirgiotis, M.; Ceylan, R.; Uysal, S.; Aktumsek, A. Evaluation of antioxidant potential, enzyme inhibition activity and phenolic profile of *Lathyrus cicera* and *Lathyrus digitatus*: Potential sources of bioactive compounds for the food industry. *Food Chem. Toxicol.* **2017**, *107*, 609–619. [CrossRef]
11. Nguyen, D.N.V.; Nguyen, T. *An Overview of the Use of Plants and Animals in Traditional Medicine Systems in Viet Nam*; Traffic Southeast Asia: Petaling Jaya, Malaysia, 2008.
12. Nguyen, V.B.; Nguyen, Q.V.; Nguyen, A.D.; Wang, S.L. Screening and evaluation of α-glucosidase inhibitors from indigenous medicinal plants in Dak Lak Province, Vietnam. *Res. Chem. Intermed.* **2017**, *43*, 3599–3612. [CrossRef]
13. Nguyen, V.B.; Wang, S.L.; Nguyen, A.D.; Vo, T.P.K.; Zhang, L.J.; Nguyen, Q.V.; Kuo, Y.H. Isolation and identification of novel α-amylase inhibitors from *Euonymus laxiflorus* Champ. *Res. Chem. Intermed.* **2018**, *44*, 1411–1424. [CrossRef]
14. Nguyen, V.B.; Nguyen, Q.V.; Nguyen, A.D.; Wang, S.L. Porcine pancreatic α-amylase inhibitors from *Euonymus laxiflorus* Champ. *Res. Chem. Intermed.* **2017**, *43*, 259–269. [CrossRef]
15. Lam, T.M.L.; Nguyen, T.M.T.; Nguyen, X.H.; Dang, P.H.; Nguyen, N.T.; Tran, M.H.; Nguyen, T.H.; Nguyen, N.M.; Min, B.S.; Kim, J.A.; et al. Anti-cholinesterases and memory improving effects of Vietnamese Xylia xylo-carpa. *Chem. Cent. J.* **2016**, *10*, 48. [CrossRef] [PubMed]
16. Nguyen, N.C.; Nguyen, T.T.H.; Tran, M.H.; Tran, C.L. Anti-cholinesterase activity of lycopodium alkaloids from *Vietnamese Huperzia squarrosa* (Forst.) Trevis. *Molecules* **2014**, *19*, 19172–19179.
17. Dung, H.V.; Cuong, T.D.; Chinh, N.M.; Quyen, D.; Kim, J.A.; Byeon, J.S.; Woo, M.H.; Choi, J.S.; Min, B.S. Compounds from the aerial parts of *Piper bavinum* and their anti-cholinesterase activity. *Arch. Pharm. Res.* **2015**, *38*, 677–682. [CrossRef]
18. Flora of China. Available online: http://www.efloras.org/florataxon.aspx?flora_id=2&taxon_id=200012808 (accessed on 24 June 2022).
19. Nguyen, V.B.; Wang, S.L.; Nguyen, A.D.; Lin, Z.H.; Doan, C.T.; Tran, T.N.; Huang, H.T.; Kuo, Y.H. Bioactivity-guided purification of novel herbal antioxidant and anti-NO compounds from *Euonymus laxiflorus* Champ. *Molecules* **2019**, *24*, 120. [CrossRef] [PubMed]
20. Nguyen, Q.V.; Nguyen, N.H.; Wang, S.L.; Nguyen, V.B.; Nguyen, A.D. Free radical scavenging and antidiabetic activities of *Euonymus laxiflorus* Champ. extract. *Res. Chem. Intermed.* **2017**, *43*, 5615–5624. [CrossRef]
21. Nguyen, V.B.; Ton, T.Q.; Nguyen, D.N.; Nguyen, T.T.; Nguyen, T.N.; Nguyen, T.H.; Doan, C.T.; Tran, T.N.; Nguyen, M.T.; Ho, N.D.; et al. Reclamation of beneficial bioactivities of herbal antioxidant condensed tannin extracted from *Euonymus laxiflorus*. *Res. Chem. Intermed.* **2020**, *46*, 4751–4766. [CrossRef]
22. Nguyen, V.B.; Wang, S.L.; Nguyen, T.H.; Nguyen, M.T.; Doan, C.T.; Tran, T.N.; Lin, Z.H.; Nguyen, Q.V.; Kuo, Y.H.; Nguyen, A.D. Novel potent hypoglycemic compounds from *Euonymus laxiflorus* Champ. and their effect on reducing plasma glucose in an ICR mouse model. *Molecules* **2018**, *23*, 1928. [CrossRef]
23. Ellman, G.L.; Courtney, K.D.; Andres, V.J.; Feather, S.R.M. A new and rapid colorimetric determination of acetylcholinesterase activity. *Biochem. Pharmacol.* **1961**, *7*, 88–95. [CrossRef]
24. Nguyen, T.H.; Wang, S.L.; Nguyen, A.D.; Doan, M.D.; Tran, T.N.; Doan, C.T.; Nguyen, V.B. Novel α-amylase inhibitor hemi-pyocyanin produced by microbial conversion of chitinous discards. *Mar. Drugs* **2022**, *20*, 283. [CrossRef] [PubMed]
25. Nguyen, V.B.; Wang, S.L.; Nguyen, A.D.; Phan, T.Q.; Techato, K.; Pradit, S. Bioproduction of prodigiosin from fishery processing waste shrimp heads and evaluation of its potential bioactivities. *Fishes* **2021**, *6*, 30. [CrossRef]
26. Pires, D.E.; Blundell, T.L.; Ascher, D.B. PkCSM: Predicting smallmolecule pharmacokinetic and toxicity properties using graphbased signatures. *J. Med. Chem.* **2015**, *58*, 4066–4072. [CrossRef] [PubMed]

27. Saxena, M.; Dubey, R. Target enzyme in Alzheimer's disease: Acetylcholinesterase inhibitors. *Curr. Top. Med. Chem.* **2019**, *19*, 264–275. [CrossRef] [PubMed]

28. Tian, J.; Shi, J.; Zhang, X. Herbal therapy: A new pathway for the treatment of Alzheimer's disease. *Alzheimer's Res. Therapy* **2010**, *2*, 30. [CrossRef] [PubMed]

29. Birks, J.; Grimley, E.J. *Ginkgo biloba* for cognitive impairment and dementia. *Cochrane Database Syst. Rev.* **2009**, *1*, CD003120. [CrossRef]

30. Ihl, R.; Bachinskaya, N.; Korczyn, A.D.; Vakhapova, V.; Tribanek, M.; Hoerr, R. Efficacy and safety of a oncedaily formulation of *Ginkgo biloba* extract EGb 761 in dementia with neuropsychiatric features: A randomized controlled trial. *Int. J. Geriatr. Psychiatry* **2010**, *9*, 123.

31. Fu, L.M.; Li, J.T. A systematic review of single Chinese herbs for Alzheimer's disease treatment. *Evid. Based Complement. Altern. Med.* **2009**, *2011*, 640284. [CrossRef]

32. Akhondzadeh, S.; Noroozian, M.; Mohammadi, M.; Ohadinia, S.; Jamshidi, A.H.; Khani, M. *Salvia officinalis* extract in the treatment of patients with mild to moderate Alzheimer's disease: A double blind, randomized and placebo-controlled trial. *J. Clin. Pharm. Ther.* **2003**, *28*, 53–59. [CrossRef]

33. Huijberts, G.N.; Eggink, G.; Waard, P.D.; Huisman, G.W.; Witholt, B. *Pseudomonas putida* KT2442 cultivated on glucose accumulates poly(3-hydroxyalkanoates) consisting of saturated and unsaturated monomers. *Appl. Environ. Microbiol.* **1992**, *58*, 536–544. [CrossRef]

34. Ekundayo, O.; Ajani, F.; Seppänen-Laakso, T.; Laakso, I. Volatile constituents of *Psidium Guajava* l. (guava) fruits. *Flavour Fragr. J.* **1991**, *6*, 233–236. [CrossRef]

35. Ronald, C.W.; Judy, C.J.; George, L.K.H. Volatile constituents of the muscadine grape (*Vitis rotundifolia*). *Agric. Food Chem.* **1982**, *30*, 681–684.

36. Ramadan, N.S.; Wessjohann, L.A.; Mocan, A.; Vodnar, C.; Abdou, M.D.; Abdel, A.Z.; Ehrlich, A. Nutrient and sensory metabolites profiling of Averrhoa Carambola L. (Starfruit) in the context of its origin and ripening stage by GC/MS and chemometric analysis. *Molecules* **2020**, *25*, 2423. [CrossRef]

37. Isaac, J.U.; Kerenhappuch, I.U.; Hauwa, A.U. Phytochemical screening, isolation, characterization of bioactive and biological activity of bungkang, (*Syzygium polyanthum*) root-bark essential oil. *Korean J. Food Health Converg.* **2020**, *6*, 5–21.

38. Farrell, I.W.; Halsall, T.G.; Thaller, V.; Bradshaw, A.P.W.; Hanson, J.R. Structures of some new sesquiterpenoid metabolites of *Marasmius alliaceus*. *J. Chem. Soc. Perkin Trans.* **1981**, *1*, 1790–1793. [CrossRef]

39. Awasthi, Y.C.; Mitra, C.R. Constituents of *Melia indica* leaves. *Phytochemistry* **1971**, *10*, 2842. [CrossRef]

40. Trabelsi, N.; Oueslati, S.; Caroline, H.V.; Pierre, W.T.; Medini, F.; Mérillon, J.M.; Abdelly, C.; Ksouri, R. Phenolic contents and biological activities of *Limoniastrum guyonianum* fractions obtained by centrifugal partition chromatography. *Ind. Crops Prod.* **2013**, *49*, 740–746. [CrossRef]

41. Margarita, H.P.; Teresa, H.; Carmen, G.C.; Estrella, I.; Rabanal, R.M. Phenolic composition of the "Mocán" *Visnea mocanera* L.f.). *J. Agric. Food Chem.* **1996**, *44*, 3512–3515.

42. Ohashi, K.; Winarno, H.; Mukai, M.; Inoue, M.; Prana, M.S.; Simanjuntak, P.; Shibuya, H. Indonesian medicinal plants. XXV. cancer cell invasion inhibitory effects of chemical constituents in the parasitic plant *Scurrula atropurpurea* (Loranthaceae). *Chem. Pharm. Bull.* **2003**, *51*, 343–345. [CrossRef] [PubMed]

43. Bruce, A.B.; Gavin, C.; Bhat, U.G. Flavonoids and the relationship of Itea to the saxifragaceae. *Phytochemistry* **1988**, *27*, 2651–2653.

44. Robert, H.J.; James, W.W.J. Taxonomic implications of the flavonoids of *Cymophyllus fraseri* (Cyperaceae). *Biochem. Systermatics Ecol.* **1988**, *16*, 521–523.

45. Backheet, E.Y.; Ahmed, A.S.; Sayed, H.M. Phytochemical study of the constituents of the leaves of *Ficus infectoria* (Roxb.). *Bull. Pharm. Sci.* **2001**, *24*, 21–27.

46. Barberán, F.A.T.; Gil, M.I.; Tomás, F.; Ferreres, F.; Arques, A. Flavonoid aglycones and glycosides from *Teucrium gnaphalodes*. *J. Nat. Prod.* **1985**, *48*, 859. [CrossRef]

47. Fishawy, A.A.; Rawia, Z.S.A. Phytochemical and pharmacological studies of *Ficus auriculata* Lour. (Moraceae) cultivated in Egypt. *Planta Med.* **2011**, *77*, 184–195.

48. Fischer, C.; Speth, V.; Fleig, E.S.; Neuhaus, G. Induction of zygotic polyembryos in wheat: Influence of auxin polar transport. *Plant. Cell* **1997**, *9*, 1767–1780. [CrossRef]

49. Venigalla, M.; Gyengesi, E.; Münch, G. Curcumin and Apigenin—Novel and promising therapeutics against chronic neuroinflammation in Alzheimer's disease. *Neural Regen. Res.* **2015**, *10*, 1181–1185.

50. Kuo, Y.H.; Huang, H.C.; Chiou, W.F.; Shi, L.S.; Wu, T.S.; Wu, Y.C. A novel NO-production-inhibiting triterpene and cytotoxicity of known alkaloids from *Euonymus laxiflorus*. *J. Nat. Prod.* **2003**, *66*, 554–557. [CrossRef]

51. Ding, Y.; Fang, Y.; Moreno, J.; Ramanujam, J.; Jarrell, M.; Brylinski, M. Assessing the similarity of ligand binding conformations with the contact mode score. *Comput. Biol. Chem.* **2016**, *64*, 403–413. [CrossRef]

52. Chandra, B.T.M.; Rajesh, S.S.; Bhaskar, B.V.; Devi, S.; Rammohan, A.; Sivaraman, T.; Rajendra, W. Molecular docking, molecular dynamics simulation, biological evaluation and 2D QSAR analysis of flavonoids from *Syzygium alternifolium* as potent anti-Helicobacter pylori agents. *RSC Adv.* **2017**, *7*, 18277–18292. [CrossRef]

53. Loan, H.T.P.; Bui, T.Q.; My, T.T.A.; Hai, N.T.T.; Quang, D.T.; Tat, P.V.; Hiep, D.T.; Trung, N.T.; Quy, P.T.; Nhung, N.T.A. In-Depth Investigation of a Donor-Acceptor Interaction on the Heavy-Group-14@Group-13-Diyls in transition-metal tetrylone complexes: Structure, bonding, and property. *ACS Omega* **2020**, *5*, 21271–21287. [CrossRef]
54. Rosenberg, B. Electrical conductivity of proteins. *Nature* **1962**, *193*, 364–365. [CrossRef] [PubMed]
55. Kharkyanen, V.N.; Petrov, E.G.; Ukrainskii, I.I. Donor-acceptor model of electron transfer through proteins. *J. Theor. Biol.* **1978**, *73*, 29–50. [CrossRef] [PubMed]
56. Cordes, M.; Giese, B. Electron transfer in peptides and proteins. *Chem. Soc. Rev.* **2009**, *38*, 892–901. [CrossRef] [PubMed]
57. Thao, T.T.P.; Bui, T.Q.; Hai, N.T.T.; Huynh, L.K.; Quy, P.T.; Bao, N.C.; Nhung, N.T.A. Newly synthesised oxime and lactone derivatives from *Dipterocarpus alatus* dipterocarpol as anti-diabetic inhibitors: Experimental bioassay-based evidence and theoretical computation-based prediction. *RSC Adv.* **2021**, *11*, 35765–35782. [CrossRef] [PubMed]
58. Pagadala, N.; Syed, K.; Tuszynski, J. Software for molecular docking: A review. *Biophys. Rev.* **2017**, *9*, 91–102. [CrossRef]
59. Pantsar, T.; Poso, A. Binding Affinity via Docking: Fact and Fiction. *Molecules* **2018**, *23*, 1899. [CrossRef]
60. Warren, G.L.; Andrews, C.W.; Capelli, A.M.; Clarke, B.; LaLonde, J.; Lambert, M.H.; Lindvall, M.; Nevins, N.; Semus, S.F.; Senger, S.; et al. A critical assessment of docking programs and scoring functions. *J. Med. Chem.* **2006**, *49*, 5912–5931. [CrossRef]
61. Marko, B.; Yunhui, G.; Joseph, B.; Hans, B.; David, H.; Clara, C.; Jérémie, M.; David, M.; Katharina, M. Prioritizing small sets of molecules for synthesis through in-silico tools: A comparison of common ranking methods. *Chem. Med. Chem.* **2023**, *18*, e202200425.
62. Huang, S.; Wang, L.L.; Xue, N.N.; Li, C.; Guo, H.H.; Ren, T.K.; Zhan, Y.; Li, W.B.; Zhang, J.; Chen, X.G.; et al. Chlorogenic acid effectively treats cancers through induction of cancer cell differentiation. *Theranostics* **2019**, *9*, 6745–6763. [CrossRef]
63. Kwak, T.W.; Park, S.B.; Kim, H.J.; Jeong, Y.I.; Kang, D.H. Anticancer activities of epigallocatechin-3-gallate against cholangiocarcinoma cells. *Onco Targets Ther.* **2017**, *10*, 137–144. [CrossRef]
64. Zong, P.C.; John, B.S.; Ho, C.T.; Chen, K.Y. Green tea epigallocatechin gallate shows a pronounced growth inhibitory effect on cancerous cells but not on their normal counterparts. *Cancer Lett.* **1998**, *129*, 173–179.
65. Babich, H.; Krupka, M.E.; Nissim, H.A.; Zuckerbraun, H.L. Differential in vitro cytotoxicity of (−)-epicatechin gallate (ECG) to cancer and normal cells from the human oral cavity. *Toxicol. Vitr.* **2005**, *19*, 231–242. [CrossRef]
66. Li, Z.; Feng, C.; Dong, H.; Jin, W.; Zhang, W.; Zhan, J.; Wang, S. Health promoting activities and corresponding mechanism of (−)-epicatechin-3-gallate. *Food Sci. Hum. Wellness* **2022**, *11*, 568–578. [CrossRef]
67. Hosseinimehr, S.J.; Rostamnezad, M.; Ghafarirad, V. Epicatechin enhances anti-proliferative effect of bleomycin in ovarian cancer cell. *Res. Mol. Med.* **2013**, *1*, 25–28. [CrossRef]
68. Elbaz, H.A.; Lee, I.; Antwih, D.A.; Liu, J.; Hüttemann, M.; Zielske, S.P. Epicatechin stimulates mitochondrial activity and selectively sensitizes cancer cells to radiation. *PLoS ONE* **2014**, *9*, e88322. [CrossRef]
69. Liu, X.; Jiang, Q.; Liu, H. Vitexin induces apoptosis through mitochondrial pathway and PI3K/Akt/mTOR signaling in human non-small cell lung cancer A549 cells. *Biol. Res.* **2019**, *52*, 7. [CrossRef] [PubMed]
70. Satari, A.; Ghasemi, S.; Habtemariam, S.; Asgharian, S.; Lorigooini, Z. Rutin: A Flavonoid as an Effective Sensitizer for Anticancer Therapy; Insights into Multifaceted Mechanisms and Applicability for Combination Therapy. *Evid.-Based Complement. Altern. Med. ECAM* **2021**, *2021*, 9913179. [CrossRef] [PubMed]
71. Yan, X.; Qi, M.; Li, P.; Zhan, Y.; Shao, H. Apigenin in cancer therapy: Anti-cancer effects and mechanisms of action. *Cell Biosci.* **2017**, *7*, 50. [CrossRef]
72. Afroze, N.; Pramodh, S.; Hussain, A.; Waleed, M.; Vakharia, K. A review on myricetin as a potential therapeutic candidate for cancer prevention. *Biotech* **2020**, *10*, 211. [CrossRef]
73. Srivastava, S.; Somasagara, R.R.; Hegde, M.; Nishana, M.; Tadi, S.K.; Srivastava, M.; Choudhary, B.; Raghavan, S.C. Quercetin, a natural flavonoid interacts with DNA, arrests cell cycle and causes tumor regression by activating mitochondrial pathway of apoptosis. *Sci. Rep.* **2016**, *6*, 24049. [CrossRef]

Article

Extraction, HPTLC Analysis and Antiobesity Activity of *Jatropha tanjorensis* and *Fraxinus micrantha* on High-Fat Diet Model in Rats

Swati Srivastava [1], Tarun Virmani [1,*], Md. Rafiul Haque [2], Abdulsalam Alhalmi [3], Omkulthom Al Kamaly [4], Samar Zuhair Alshawwa [4] and Fahd A. Nasr [5]

[1] School of Pharmaceutical Sciences, MVN University, Palwal 121105, India; swatisrivastva186@gmail.com
[2] School of Pharmacy, Al-Karim University, Katihar 854106, India; hrafiul@gmail.com
[3] Department of Pharmaceutical Sciences, College of Pharmacy, Aden University, Aden 6312, Yemen
[4] Department of Pharmaceutical Sciences, College of Pharmacy, Princess Nourah bint Abdulrahman University, P.O. Box 84428, Riyadh 11671, Saudi Arabia; omalkmali@pnu.edu.sa (O.A.K.); szalshawwa@pnu.edu.sa (S.Z.A.)
[5] Department of Pharmacognosy, College of Pharmacy, King Saud University, Riyadh 11451, Saudi Arabia; fnasr@ksu.edu.sa
* Correspondence: tarun.virmani@mvn.edu.in

Citation: Srivastava, S.; Virmani, T.; Haque, M.R.; Alhalmi, A.; Al Kamaly, O.; Alshawwa, S.Z.; Nasr, F.A. Extraction, HPTLC Analysis and Antiobesity Activity of *Jatropha tanjorensis* and *Fraxinus micrantha* on High-Fat Diet Model in Rats. *Life* **2023**, *13*, 1248. https://doi.org/10.3390/life13061248

Academic Editors: Efstathia Papada, Charalampia Amerikanou and Marisa Colone

Received: 27 March 2023
Revised: 8 May 2023
Accepted: 21 May 2023
Published: 25 May 2023

Abstract: The accumulation of body fat due to an imbalance between calorie intake and energy expenditure is called obesity. Metabolic syndrome increases the risk of heart disease, type 2 diabetes, and stroke. The purpose of this study was to determine the effect of *Jatropha tanjorensis* (J.T.) and *Fraxinus micrantha* (F.M.) leaf extracts on high-fat diet-induced obesity in rats. Normal control, high-fat diet (HFD) control, orlistat standard, and test groups were created using male Albino Wistar rats (n = 6 per group) weighing 190 ± 15 g. Except for the control group, all regimens were administered orally and continued for 6 weeks while on HFD. Evaluation criteria included body weight, food intake, blood glucose, lipid profile, oxidative stress, and liver histology. High-Performance Thin Layer Chromatography (HPTLC) analysis was performed using a solvent system (7:3 hexane: ethyl acetate for sitosterol solution and *Jatropha tanjorensis* extracts and 6:4 hexane: ethyl acetate: 1 drop of acetic acid for esculetin and *Fraxinus micrantha* extracts). There were no deaths during the 14 days before the acute toxicity test, indicating that aqueous and ethanolic extracts of both J.T. and F.M. did not produce acute toxicity at any dose (5, 50, 300, and 2000 mg/kg). The ethanolic and aqueous extracts of J.T. and F.M. leaves at 200 and 400 mg/kg/orally showed a reduction in weight gain, feed intake, and significant decreases in serum glucose and lipid profile. As compared to inducer HFD animals, co-treatment of aqueous and ethanolic extract of both J.T. and F.M. and orlistat increased the levels of antioxidant enzymes and decreased lipid peroxidation. The liver's histological findings showed that the sample had some degree of protection. These results indicate that ethanolic samples of J.T. have antidiabetic potential in diabetic rats fed an HFD. The strong antioxidant potential and restoration of serum lipid levels may be related to this. Co-treatment of samples JTE, JTAQ, FME, FMAQ and orlistat resulted in an increase in antioxidant enzymes and reduction in lipid peroxidation as compared to inducer HFD animals. We report, for the first time, on using these leaves to combat obesity.

Keywords: *Jatropha tanjorensis*; *Fraxinus micrantha*; HPTLC; anti-obesity activity; high-fat diet model; histological

1. Introduction

With more than 312 million clinically obese people worldwide, obesity is one of the most serious public health problems of the 21st century [1]. One of the hallmarks of obesity is the accumulation of fat in adipose tissue and other internal organs [2–4]. Many chronic diseases can be caused by obesity, including type 2 diabetes, heart disease, hyperlipidemia, atherosclerosis, certain cancers, and osteoarthritis [5]. Obesity, along with high social and health costs, increases morbidity and mortality worldwide. As a result, several attempts are being made to find anti-obesity drugs worldwide [6–8].

Consumers are drawn to naturally derived anti-obesity products due to the widespread belief that anything natural should be more effective and safer than traditional treatments. Sibutramine, orlistat, and rimonabant in particular have been associated with numerous serious side effects, including gastrointestinal problems and serious cardiovascular side effects [9]. Therefore, the current search for new anti-obesity drugs derived from natural sources due to their favorable pharmacological profile and low side effects has become the current strategy [10]. In wet forest areas of West Africa, *Jatropha tanjorensis* J. L. Ellis & Saroja (Euphorbiaceae) are typical weeds found in crops, shrub shoots, roadsides, and disturbed areas [11]. As a hybrid, it is a perennial herb with an intermediate phenotype between *Jatropha curcas* and *Jatropha gossypifolia* [12,13]. Catholic Vegetable, Jatropha, Hospital Too Far, and Jana Ifaya are just some of the common names [14,15]. In veterinary medicine, as well as in traditional and folk medicine, all parts of the plant, including the seeds, leaves, and bark, are used regardless of whether they are used fresh or in decoction form. According to early research, the antioxidant minerals phosphorus, selenium, zinc, and vitamins C and E are found in abundance in the *Jatropha tanjorensis* plant [16]. Phytochemical analysis of *Jatropha tanjorensis* leaves has revealed the presence of biologically active constituents such as alkaloids, flavonoids, tannins, cardiac glycosides, anthraquinones, and saponins [17,18]. The leaves of the plant have been widely used in tonics and soups originally from Nigeria and are known to increase blood volume. An infusion of *Jatropha tanjorensis* leaves is taken orally in southwestern Nigeria to treat symptoms of diabetes. The antidiabetic activities of the J.T. ethanolic extracts (JTE), namely JTE chloroform, JTE ethylacetate, JTE aqueous, have been evaluated [19]. The antibacterial and anti-inflammatory effects of leaf extracts in hexane, chloroform, and methanol are different [20].

The leaves and stems of *Jatropha curcas* and *Jatropha tanjorensis* have been studied from morphological and anatomical points of view [18,21]. Acute and subacute toxicity and microscopic examination of *Jatropha tanjorensis* leaves have been described [22], as well as the antibacterial activity of aqueous leaf extracts [23]. A versatile temperate deciduous tree species of the Himalayas of outstanding medicinal potential and ethnobotanical importance is *Fraxinus micrantha*, Lingelsh (also known by its native name Angu, English name Ash and surname Oleaceae). *Fraxinus micrantha* is one of the ashes found in Asia, mainly in India and Nepal. It can be found in the Indian states of Himachal Pradesh and Uttar Pradesh [24].

Since ancient times, *Fraxinus micrantha* has been researched for both its medicinal and economic benefits. The inner bark infusion is used locally by Dharchula, Himalayan residents to cure liver enlargement, jaundice, and other liver ailments [25]. Due to the presence of numerous glycosides, including fraxin, and an active diuretic agent called coumarin glycoside, Fraxinus species have been employed in folk medicine for their purgative and diuretic effects. Additionally, the leaves and the bark are used to cure cystitis, rheumatoid arthritis, constipation, and itchy scalps [26]. The discovery of the secoiridoid glucosides, which are significant metabolites in the genus of the family Oleaceae, has led to an increase in interest in the phytochemistry of Fraxinus in recent years [27].

Because of their similarity to human obesity and associated metabolic effects, the animal model of diet-induced obesity is one of the most widely used and reliable models for obesity research. The result is increased food intake, weight gain, body fat accumulation, impaired lipid profile, lack of antioxidant stability, and increased insulin resistance parameters [28,29]. Therefore, this study aimed to evaluate the effects of alcoholic and aqueous extracts of *Jatropha tanjorensis* and *Fraxinus micrantha* leaves on high-fat diet-induced obesity in rats.

2. Materials and Methods

2.1. Plant Material

Fresh leaves of *Jatropha tanjorensis* were collected from near Coimbatore, India, while the leaves of *Fraxinus micrantha* were collected in May from Nainital, Kumaun, Himalayas, India. Both drugs were identified from Vital Herbs, Uttam Nagar, Delhi. The leaves were dried at a constant weight in the air at room temperature, and then the leaves were crushed.

2.2. Chemical Reagents

All chemicals used in this study were obtained from the Hi Media Laboratories Pvt. Ltd. (Mumbai, India), Sigma-Aldrich Chemical Co. (Milwaukee, WI, USA), SD Fine-Chem. Ltd. (Mumbai, India), and SRL Pvt. Ltd. Chlorpheniramine maleate and clonidine were obtained from Unichem, Ltd. (Alchem, Mumbai). Only analytical grade compounds were used in the study.

2.3. Extract

2.3.1. Ethanol Extract

A total of 150 gm of powdered leaves of F.M. and 50 gm of dry powder leaves of J.T. were placed in a Soxhlet device with a thimble. An organic solvent was used for extraction, i.e., ethanol, for 8–10 h and the temperature of the mantle heater was adjusted to 40–60 °C. After the extraction process, the sample extract was filtered and concentrated to dryness. Extracts were collected in sealed containers [30]. Yields of all extracts were calculated.

2.3.2. Water Extract

Coarse powder of leaves (150 g of J.T. and 50 g of F.M.) was boiled in distilled water for 15 min. After leaving this at room temperature for 15 min, it was filtered through a muslin cloth. The resulting solution was boiled again to obtain a thick concentrated extract. After drying, the extract was collected in a sealed container [31]. Extraction yields of all extracts were calculated.

2.4. High-Performance Thin Layer Chromatography (HPTLC)

HPTLC was performed on silica gel 60 F_{254} 100 × 100 mm plates (Merck) with hexane: ethyl acetate (7:3 v/v) as mobile phase for the standard (β-sitosterol) solution (2–10 μL) and J.T. (0.2–0.3 μL); and hexane: ethyl acetate: 1 drop of acetic acid (6:4 v/v) as mobile phase for the standard (esculetin) solution (1–4 μL) and F.M. (1–3 μL). These were applied to the plate as 8 mm bands. Application of the sample was performed with CAMAG-Linomat 5 Automated spray on a band applicator equipped with a 100 μL syringe and operated with the settings: band length 8 mm, application rate 150 nL /s, application volume 0.20 μL, distance between track 14.4 mm, distance from the plate side edge 15.0 mm and solvent front position 70 mm. CAMAG TLC visualizes 2 and was used densitometrically to scan the bands. The scanner operating parameters were set to mode absorption/reflection at an optimized wavelength of 254, 366 nm, and in the visible range. Integration parameters were set to gauss (legacy) with sensitivity 0.1, separation 1, and threshold 0.1.

2.5. HFD-Induced Obesity

Albino Wistar rats weighing 190 ± 15 g obtained from the breeding farm of the Pinnacle Biomedical Research Institute (PBRI), Bhopal, were selected. They were then divided into groups of 6 with controlled temperature and humidity (25 ± 2 °C, 55–65%). Rats were provided regular rodent chow and unlimited water. Adult male Wistar rats were acclimated to laboratory conditions for 2 weeks.

All studies were conducted indoors without background noise. Each study set used a different group of rats (n = 6). The Institutional Animal Ethics Committee (IAEC) of the Pinnacle Biomedical Research Institute (PBRI) approved animal studies by Bhopal, India (Reg. No. 1824/PO/ERe/S/15/CPCSEA). The protocol approval reference number is PBRI/IAEC/10-09-22/012. A normal control group (NC) of 6 rats was randomly divided and fed a typical diet. Afterwards, animals in different experimental groups were given sufficient food and water along with a high-fat diet for 6 weeks. The high-fat content was obtained by mixing coconut oil with vanaspati Indian ghee in a ratio of 3:1 (*v/v*). Rats were fed daily at a dose of 3 mL per kg of body weight.

2.6. Acute Toxicity Studies

J.T. and F.M.'s acute toxicity investigation was completed according to the Organisation for Economic Cooperation and Development (OECD) guidelines [32]. Four treatment groups with dosages of 5 mg/kg, 50 mg/kg, 300 mg/kg, and 2000 mg/kg body weight were included in the test groups. Using an appropriate intubation canula or a specifically developed oral needle, the test drug was gavaged in a single dose. Before dosing, animals were fasted for three hours (only food was withheld for 3 h but not water). The animals were closely monitored for behavioral changes, mortality, and appearance starting in the first four hours, then occasionally over the next twenty-four hours, and finally daily for a period of two weeks up to fourteen days [33].

2.7. Experimental Design

Wistar rats were divided into control and preventive groups with at least six animals each.

Group I: normal control; rats were administered saline/vehicle.

Group II: HFD Control; animals were fed a standard granulated diet with an HFD of 3 mL daily for 6 weeks.

Group III: Standard Medication Treatment; animals were fed a standard granular diet with 3 mL of HFD daily for 6 weeks. After 3 weeks of the study, 30 mg/kg/day of orlistat was administered after 3 weeks of HFD, and this continued for the remaining 3 weeks; this was regarded as the standard group.

Group IV: Animals were fed a standard granular diet with an HFD of 3 mL daily for 6 weeks. After 3 weeks of the study, HFD was fed for 3 weeks, then JTAQ 200 mg/kg/day was administered and continued until the remaining 3 weeks; this was considered a treatment group.

Group V: Animals were fed a standard granulated diet with an HFD of 3 mL daily for 6 weeks. After 3 weeks of the study, HFD was fed for 3 weeks, then JTAQ 400 mg/kg/day was administered and continued until the remaining 3 weeks; this was considered a treatment group.

Group VI: Animals were fed a standard granular diet with 3 mL of HFD daily for 6 weeks. After 3 weeks of the study, 200 mg/kg/day of JTE was administered after 3 weeks of HFD and continued for the remaining 3 weeks; this was considered a treatment group.

Group VII: Animals were fed a standard granulated diet with an HFD of 3 mL daily for 6 weeks. After 3 weeks of study, 400 mg/kg/day of JTE was administered after 3 weeks of HFD and continued for the remaining 3 weeks; this was considered a treatment group.

Group VIII: Animals were fed a standard granulated diet with an HFD of 3 mL daily for 6 weeks.

After 3 weeks of the study, 200 mg/kg/day of FMAQ was administered after 3 weeks of high-fat diet food, and continued until the remaining 3 weeks; this was considered a treatment group.

Group IX: Animals were fed a standard granulated diet with an HFD of 3 mL daily for 6 weeks. After 3 weeks of the study, HFD was fed for 3 weeks, and FMAQ 400 mg/kg/day was administered, and continued until the remaining 3 weeks; this was considered a treatment group.

Group X: Animals were fed a standard granulated diet with an HFD of 3 mL daily for 6 weeks. After 3 weeks of the study, HFD was fed for 3 weeks, and FME 200 mg/kg/day was administered, and continued until the remaining 3 weeks; this was considered a treatment group.

Group XI: Animals were fed a standard granular diet containing 3 mL of an HFD for 6 weeks. After 3 weeks of the study, HFD was fed for 3 weeks, then PME 400 mg/kg/day was administered and continued until the remaining 3 weeks; this was considered a treatment group.

2.8. Study Parameters

Body weight, blood glucose, lipid profile, oxidative stress, and liver histology were investigated as variables. After animals were sacrificed at the end of the study, blood parameters and oxidative stress markers were examined. Then, the liver was washed thoroughly with ice-cold saline and Tris-HCl buffer (0.1 M, pH 7.4) for antioxidant assay. Total cholesterol (TC) and triglyceride (TG) levels were assessed using a commercial assay kit (SPAN Diagnostics Ltd., Surat, India) according to the manufacturer's instructions. High-density lipoprotein (HDL) levels were measured using the HDL test kit (Reckon Diagnostics Pvt. Ltd.,. Baroda, India). The Friedewald equation was used to calculate the concentration of LDL [34].

2.8.1. LDL Level

Serum LDL level (mg/dl) = total cholesterol − (HDL level + VLDL level)

2.8.2. Oxidation Markers

Superoxide Dismutase (SOD)

The red formazan dye reduction process produces superoxide radicals, which are detected by the superoxide dismutase assay kit using a tetrazolium salt. One unit (U) of SOD activity is the amount of enzyme required to demonstrate 50% superoxide radical dismutation. In a nutshell, 0.1 mL of sample was introduced together with 1.2 mL of buffer containing 0.052 M sodium pyrophosphate, 186 μM phenazine methosulphate, 300 μM nitroblutetrazolium, and 0.2 μM NADH, with 90 s of incubation time at 30 °C. Glacial acetic acid, 0.1 mL, was added and then we added 4.0 mL of n-butanol and stirred. After 10 min of standing time, the butanol layer and centrifuge were separated. At 560 nm, the reduction was measured, and the percentage of SOD inhibition in comparison to the control was calculated. Unit U/mg tissue was used to determine and express one unit of SOD [35].

Malondialdehyde (MDA)

Through the measurement of thiobarbituric acid reactive material, liver peroxidation was discovered (TBARS). In a nutshell, 0.2 mL of an aliquot had 0.2 mL 8.1% SDS, 1.5 mL 20% acetic acid, and 1.5 mL 8% TBA added to it (made up to a volume of 4 mL with distilled water). Using a glass ball as the condenser, this was heated in a water bath for 60 min, then cooled and made up to a volume of 5 mL. Then, 5 mL of butanol: pyridine was added (15:1). The mixture was centrifuged at 3000 rpm for 10 min after being vortexed for 2 min. Then we removed the top layer of the mixture, and then used a UV-Vis Spectrophotometer to measure the pink product's absorbance at 532 and 600 nm wavelengths (Systronic-2202). The difference in absorbance was measured and contrasted with that of standard solutions of varying concentrations of malonaldehyde tetramethyl acetal. Nmol MDAm/g tissue was used to express MDA activity [36].

Reduced Glutathione (GSH)

Through a linked reaction with GR, the glutathione peroxidase test kit indirectly assesses GPx activity. Glutathione becomes oxidized when hydroperoxide is reduced by GPx; this oxidized glutathione is then recycled to its reduced state by GR and NADPH. The absorbance falls as NADPH is converted to NADP+ through oxidation. A total of 10% by weight of tissue homogenate in a pH 7.4 phosphate buffer solution, 1.5 mL of 20% TCA and 1.5 mL of 1 mM EDTA were combined with 0.2 mL of homogenate and left to sit for 15 min. Then it received 10 min of centrifuging at 2000 rpm. The supernatant (400 L) was collected, and the supernatant was transferred to a fresh tube containing 1.8 mL of Elman's reagent (0.1 mM 5,5′-dithiol bis-2-nitrobenzoic acid produced in 0.3 M phosphate buffer, pH 7 with 1% sodium citrate solution; keep at 0–4 °C in the dark). Distilled water was used to maintain volumes of up to 2 mL. At 412 nm, the absorbance was measured [37].

2.8.3. Histopathology

Liver samples were immersed in a 10% formalin solution for histological analysis. These tissues were prepared, dehydrated with different concentrations of alcohol, washed with toluene, and immersed in molten paraffin wax for specified times. Freshly melted paraffin wax was used to pour the treated fabric and then the wax was allowed to harden. To demonstrate the general structure of the tissue, sections were produced at 3 μm thickness, dried on a hot plate for 15 min, and then stained with hematoxylin and 1% eosin aqueous solution. Stained glass slides were dehydrated in alcohol of varying strength, then purified in xylene and embedded in Canadian balsam. Sections were examined up close using a ×10 objective.

2.9. Statistical Analysis

Data are presented as mean standard deviation (n = 6). Results were statistically tested using a one-way analysis of variance (ANOVA) followed by Bonferroni's t-test. When comparing the two groups, the significance level was $p < 0.05$.

3. Results

To achieve the real yield of extraction, the crude extracts produced after each successive Soxhlet extraction and decoction technique were concentrated in a water bath by entirely evaporating the solvents. The percentage yields of the resultant extracts were determined to be 2.20 and 0.83% for J.T. and 10.0 and 12.6% for F.M. in various solvents such as ethanol and water. At 254 and 366 nm, the J.T. ethanolic extract HPTLC chromatogram data were examined. Six spots were seen in the extract, and their Rf values were 0.45, 0.38, 0.28, 0.22, 0.19, and 0.01. Figures 1 and 2 show that the standard (-sitosterol) Rf value was 0.33. Analysis of the F.M. extract HPTLC chromatogram results was carried out at 254 and 366 nm. The chromatogram showed nine spots for the extract at 0.8, 0.76, 0.65, 0.55, 0.40, 0.24, 0.08, 0.04, and 0.06 and one spot for the standard at 0.08 (esculetin). The outcomes demonstrated that esculetin was present in the F.M. extract (Figures 3 and 4). No deaths occurred throughout the 2 weeks of the acute toxicity research, proving that the administration of Sample-JT and FM did not, at any dose, produce toxicity (5, 50, 300 and 2000 mg/kg). None of the treatment groups experienced any discernible weight loss throughout the 14-day observation period. The rats showed no indication of any abnormalities. Test samples JT and FM did not significantly alter the parameters tested (5, 50, 300, and 2000) such as urine, convulsions, tremors, changes in skin color, etc.

Therefore, based on the most recent findings from an acute toxicity study, the final doses chosen for further research were 1/10th and 1/5th of 2000 mg/kg bw. At the beginning of the trial, the mean body weights of the seven experimental groups were comparable. The initial body weights (i.e., initial body weights) of the healthy control rats in Group I were normal, and they continued to be so for the next six weeks of the trial. After the trial, HFD-treated rats (Group II) showed a significantly higher body weight than the healthy control group (Group I) ($p < 0.05$). After receiving treatment with test samples JTAQ, JTE, FMAQ, and FME for six weeks, the changes in body weight to normal body weight were significantly ($p < 0.05$) maintained. However, compared to the HFD control group, therapy with the common medicine orlistat (30 mg/kg, p.o.) once daily for six weeks dramatically reduced body weight and feed intake. When compared to the HFD control group, once-daily treatment for six weeks with JTAQ, JTE, FMAQ, and FME significantly reduced body weight and feed intake (Tables 1 and 2).

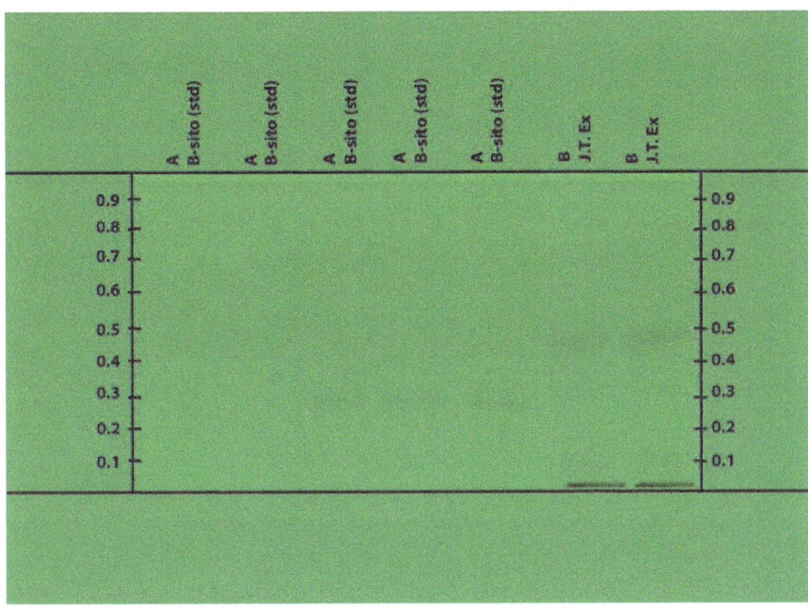

Figure 1. Chromatogram obtained from separation of J.T extract and visualized under UV light of wavelength 254 nm.

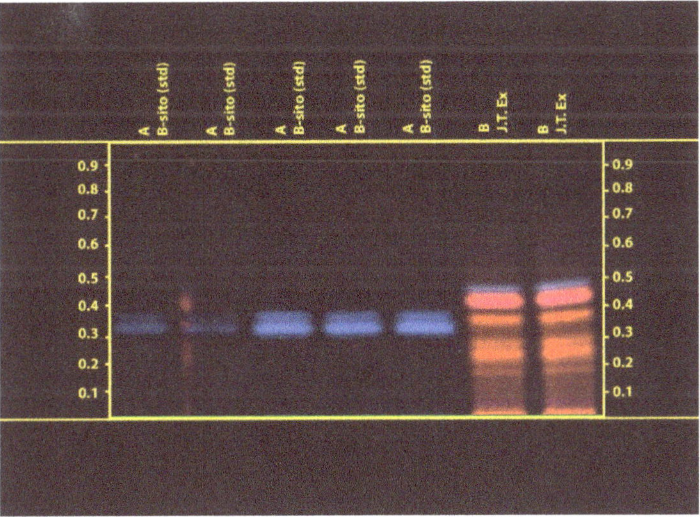

Figure 2. Chromatogram obtained from the separation of J.T extract and visualized under UV light of wavelength 366 nm.

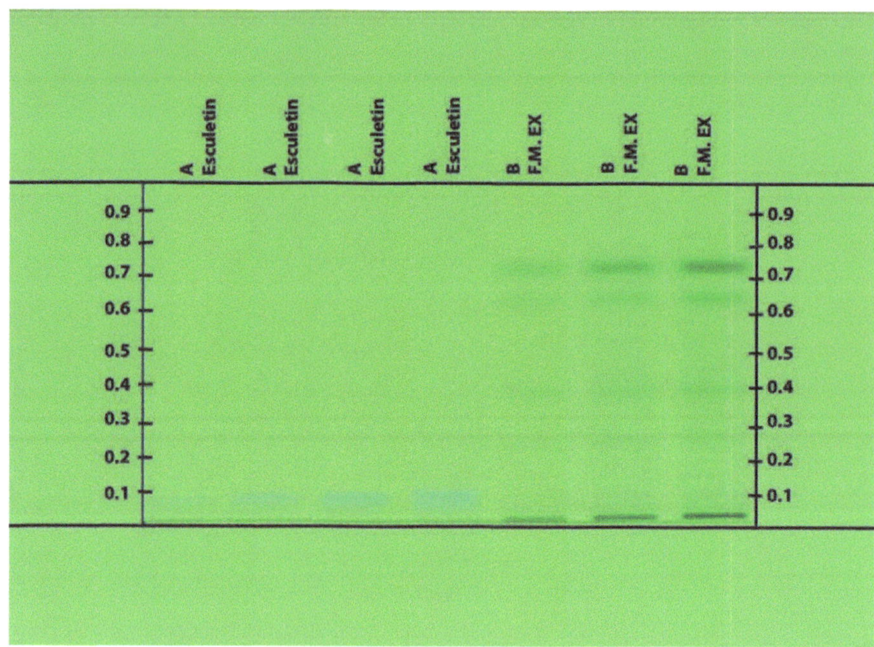

Figure 3. Chromatogram obtained from the separation of F.M extract and visualized under UV light of wavelength 254 nm.

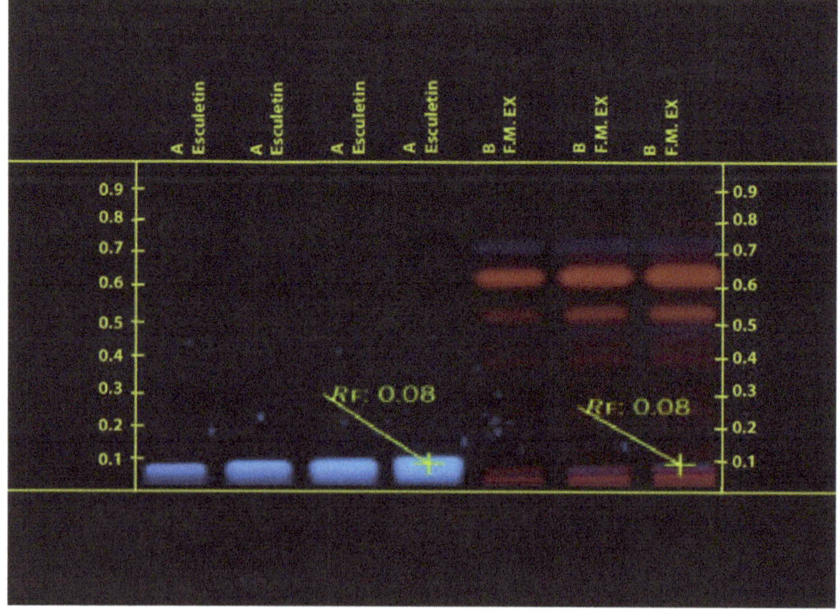

Figure 4. Chromatogram obtained from the separation of F.M extract and visualized under UV light of wavelength 366 nm.

Table 1. Food intake in normal control, inducer, standard, and test samples JTAQ, JTE, FMAQ, and FME (200 and 400 mg/kg) treated groups.

Group No.	Treatment	0 Week	3 Week	6 Week
1.	Normal Control (Vehicle treated)	12.16 ± 0.820	13.33 ± 0.982	15.14 ± 0.864
2.	HFD induced only	13.52 ± 1.306	17.70 ± 1.170	23.06 ± 1.582
3.	HFD + Standard Orlistat (30 mg/kg)	12.33 ± 1.034 [NS]	14.10 ± 1.119 [**]	16.07 ± 1.169 [**]
4.	HFD + JTAQ (200 mg/kg)	12.33 ± 0.977 [NS]	14.71 ± 0.872 [*]	18.25 ± 1.249 [**]
5.	HFD + JTAQ (400 mg/kg)	13.37 ± 1.338 [NS]	15.85 ± 1.611 [NS]	18.50 ± 1.149 [**]
6.	HFD + JTE (200 mg/kg)	11.54 ± 1.288 [NS]	13.47 ± 1.459 [**]	15.57 ± 1.688 [**]
7.	HFD + JTE (400 mg/kg)	12.10 ± 1.229 [NS]	14.08 ± 1.314 [**]	16.28 ± 1.854 [**]
8.	HFD + FMAQ (200 mg/kg)	11.89 ± 1.092 [NS]	15.35 ± 1.317 [NS]	19.59 ± 1.166 [**]
9.	HFD + FMAQ (400 mg/kg)	12.45 ± 0.932 [NS]	15.53 ± 0.911 [NS]	18.72 ± 1.948 [**]
10.	HFD + FME (200 mg/kg)	13.06 ± 1.721 [NS]	15.50 ± 1.377 [NS]	18.52 ± 1.205 [**]
11.	HFD + FME (400 mg/kg)	10.99 ± 0.976 [*]	12.82 ± 0.641 [**]	14.09 ± 0.748 [**]

One-way ANOVA followed by the Bonferroni test, with values reported as MEAN $\pm$ SD at n = 6, [*] $p < 0.050$, [**] $p < 0.001$, and [NS] $p > 0.001$ compared to group HFD treated.

Table 2. Variations in body weight in normal control, inducer, standard, and test samples JTAQ, JTE, FMAQ, and FME (200 and 400 mg/kg).

Group No.	Treatment	0 Week	3 Week	6 Week
I.	Normal Control (Vehicle treated)	202.58 ± 5.218	207.33 ± 5.354	214.50 ± 4.505
II.	HFD induced only	207.37 ± 5.845	215.03 ± 6.267	223.74 ± 7.090
III.	HFD + Standard Orlistat (30 mg/kg)	195.53 ± 4.398 [*]	199.64 ± 4.052 [*]	206.60 ± 3.706 [**]
IV.	HFD + JTAQ (200 mg/kg)	199.26 ± 4.707 [NS]	206.63 ± 2.763 [NS]	$213.68 + 2.633$ [*]
V.	HFD + JTAQ (400 mg/kg)	195.98 ± 4.796 [*]	201.33 ± 4.134 [*]	208.72 ± 3.824 [**]
VI.	HFD + JTE (200 mg/kg)	202.94 ± 2.464 [NS]	208.60 ± 2.643 [NS]	215.16 ± 2.639 [NS]
VII.	HFD + JTE (400 mg/kg)	208.97 ± 3.941 [NS]	214.45 ± 4.110 [NS]	221.66 ± 4.718 [NS]
VIII.	HFD + FMAQ (200 mg/kg)	194.25 ± 4.502 [*]	200.67 ± 4.110 [*]	210.82 ± 4.944 [*]
IX.	HFD + FMAQ (400 mg/kg)	206.46 ± 1.965 [NS]	213.00 ± 1.965 [NS]	220.50 ± 2.664 [NS]
X.	HFD + FME (200 mg/kg)	206.00 ± 6.693 [NS]	212.03 ± 5.748 [NS]	219.74 ± 6.053 [NS]
XI.	HFD + FME (400 mg/kg)	199.21 ± 8.699 [NS]	205.53 ± 7.311 [*]	213.00 ± 6.870 [*]

One-way ANOVA followed by the Bonferroni test, with values reported as MEAN $\pm$ SD at n = 6, [*] $p < 0.050$, [**] $p < 0.001$, and [NS] $p > 0.001$ compared to group HFD treated.

Throughout the trial, the serum glucose levels of the HFD control group significantly increased. The HFD control group (Group II) demonstrated increased blood glucose, indicative of impaired glucose tolerance, whereas HFD animals receiving treatment with orlistat, JTE, and FME (Group III) demonstrated decreased blood glucose levels in comparison to the HFD-treated group, except JTAQ, FMAQ treated animals, which displayed less effective outcomes (Table 3). When compared to normal control rats, the HFD inducer group (Group II) showed a two-fold rise in the level of TGs (Group I). When compared to the HFD inducer group (Group II), the serum TG concentrations of the rats in Group IV-XI treated for 6 weeks with JTAQ, JTE, FMAQ, and FME decreased, respectively ($p < 0.05$). As compared to the HFD inducer group (Group II), TC levels were also affected in Group IV-XI treated rats (JTAQ, JTE, FMAQ, and FME) and decreased dose-dependently ($p < 0.05$). Additionally, JTE and FME treatment for six weeks in HFD rats (Groups VII and XI) resulted in significant ($p < 0.05$) reductions in LDL (44.53 ± 6.396 and 58.06 ± 7.510, respectively), in comparison to the HFD inducer group (99.65 ± 3.879) (Group II). When compared to control rats, the HFD inducer group (Group II) showed a drop (15.69 ± 0.667) in HDL (Group I). Contrarily, administration of JTAQ, JTE, FMAQ, and FME at 200 and 400 mg/kg raised HDL levels in HFD rats (Table 4). As compared to the normal control group, there was a significant drop in reduced glutathione (GSH), superoxide dismutase (SOD), and an increase in malondialdehyde (MDA) in the HFD control group. When compared to the HFD control group, the once-daily oral dose of orlistat for six weeks combined with

HFD significantly enhanced the levels of GSH and SOD with a decrease in MDA. Additionally, compared to the HFD control group and comparable to standard medication (orlistat) treatment, the once-daily treatment with JTAQ, JTE, FMAQ, and FME (200 and 400 mg/kg, p.o.) for six weeks dramatically reduced the levels of GSH and SOD with a drop in MDA. Distribution of the test sample JTAQ, JTE, FMAQ, and FME inactivity demonstrates that JTAQ, JTE, FMAQ, and FME at 200 and 400 mg/kg had free radical scavenging activity, which may operate favorably against pathological changes brought on by the presence of O2– and OH– (Table 5). Histology of the liver sections from normal control animals showed normal hepatic cells with well-preserved cytoplasm, a prominent nucleus and nucleolus, and a well-brought-out central vein. The liver sections of Group 2 animals showed hepatic cells with severe toxicity, degeneration, edema and necrosis of hepatocytes with hemorrhage (h) and destruction of hepatic sinusoids. Induced rats treated with JTAQ, JTE, FMAQ, and FME had livers that were intensely affected by fatty changes, sinusoidal feathery degeneration and necrosis, congestion in the central vein with fibrous tissue proliferation, ballooning, and severe hepatocyte degeneration, whereas rats treated with JTE had livers that more closely resembled normal hepatic structure (Figure 5).

Table 3. Variations in blood glucose level after 0, 3, and 6 weeks of the treatment period.

Group No.	Treatment	0 Week	3 Week	6 Week
I.	Normal Control (Vehicle treated)	91.16 ± 6.911	92.83 ± 7.083	94.66 ± 5.610
II.	HFD induced only	193.33 ± 11.003	248.16 ± 12.922	286.83 ± 9.174
III.	HFD + Standard Orlistat (30 mg/kg)	188.50 ± 16.706 [NS]	159.83 ± 8.886 **	118.00 ± 4.690 **
IV.	HFD + JTAQ (200 mg/kg)	211.00 ± 5.831 [NS]	267.66 ± 9.026 [NS]	200.66 ± 5.715 **
V.	HFD + JTAQ (400 mg/kg)	200.50 ± 7.232 [NS]	233.50 ± 9.138 [NS]	161.00 ± 13.161 **
VI.	HFD + JTE (200 mg/kg)	205.50 ± 13.576 [NS]	228.16 ± 9.239 [NS]	169.33 ± 11.911 **
VII.	HFD + JTE (400 mg/kg)	209.16 ± 14.634 [NS]	189.16 ± 6.014 **	123.16 ± 4.309 **
VIII.	HFD + FMAQ (200 mg/kg)	207.08 ± 7.619 [NS]	253.03 ± 9.826 [NS]	211.03 ± 7.593 **
IX.	HFD + FMAQ (400 mg/kg)	192.66 ± 12.628 [NS]	226.00 ± 14.615 *	161.33 ± 10.033 **
X.	HFD + FME (200 mg/kg)	209.83 ± 8.183 [NS]	250.33 ± 17.294 [NS]	193.00 ± 14.993 **
XI.	HFD + FME (400 mg/kg)	192.83 ± 16.167 [NS]	209.83 ± 10.028 **	156.83 ± 5.776 **

One-way ANOVA followed by the Bonferroni test, with values reported as MEAN $\pm$ SD at n = 6, * $p < 0.050$, ** $p < 0.001$, and [NS] $p > 0.001$ compared to group HFD treated.

Table 4. Variations in lipid profile in normal control, inducer, standard and test samples JTAQ, JTE, FMAQ, and FME (200 and 400 mg/kg).

Group No.	Treatment	TC	TG	HDL	LDL
I	Normal Control (Vehicle treated)	100.22 ± 2.751	86.21 ± 0.967	52.81 ± 4.489	30.16 ± 6.473
II	HFD induced only	153.66 ± 2.934	191.57 ± 8.099	15.69 ± 0.667	99.65 ± 3.879
III	HFD + Standard Orlistat (30 mg/kg)	93.14 ± 2.631 **	90.35 ± 1.723 **	46.66 ± 7.242 **	28.41 ± 8.784 **
IV	HFD + JTAQ (200 mg/kg)	130.74 ± 2.673 **	111.22 ± 12.938 **	37.87 ± 3.529 **	70.61 ± 3.597 **
V	HFD + JTAQ (400 mg/kg)	112.92 ± 2.000 **	102.14 ± 2.177 **	40.90 ± 7.519 **	51.58 ± 7.774 **
VI	HFD + JTE (200 mg/kg)	119.48 ± 3.752 **	94.42 ± 2.529 **	35.75 ± 5.100 **	64.84 ± 7.082 **
VII	HFD + JTE (400 mg/kg)	107.14 ± 1.718 **	100.94 ± 2.298 **	42.42 ± 6.867 **	44.53 ± 6.396 **
VIII	HFD + FMAQ (200 mg/kg)	111.74 ± 1.706 **	137.54 ± 17.086 **	23.03 ± 2.969 [NS]	61.20 ± 4.256 **
IX	HFD + FMAQ (400 mg/kg)	135.81 ± 0.947 **	120.35 ± 9.360 **	26.66 ± 4.833 [NS]	85.07 ± 6.331 *
X	HFD + FME (200 mg/kg)	140.25 ± 2.105 **	122.49 ± 3.921 **	28.18 ± 7.862 *	87.57 ± 7.555 [NS]
XI	HFD + FME (400 mg/kg)	111.74 ± 1.706 **	115.36 ± 2.530 **	30.60 ± 8.246 *	58.06 ± 7.510 **

One-way ANOVA followed by the Bonferroni test, with values reported as MEAN $\pm$ SD at n = 6, * $p < 0.050$, ** $p < 0.001$, and [NS] $p > 0.001$ compared to group HFD treated.

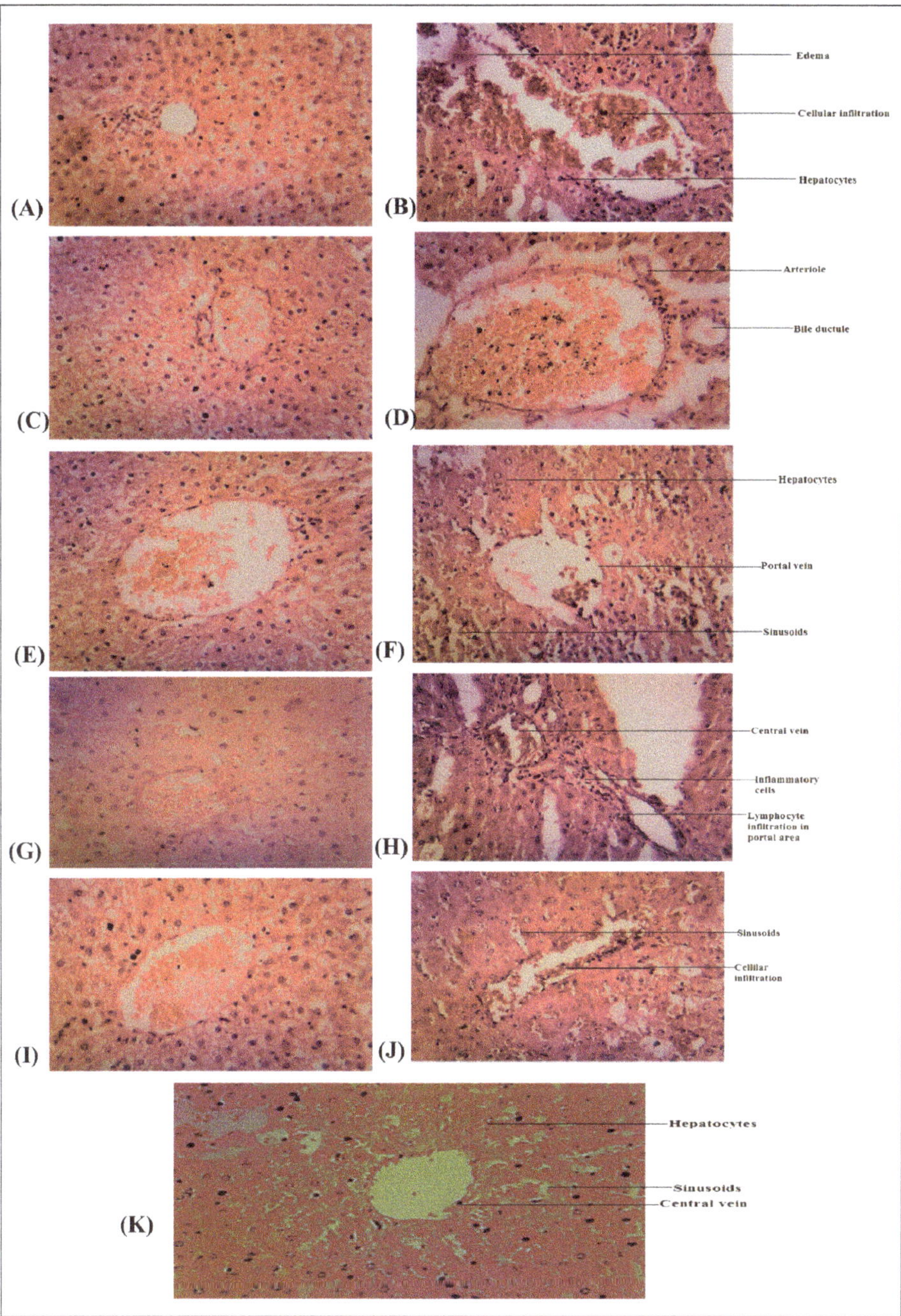

Figure 5. (**A**) Group I: Normal control (vehicle-treated), (**B**) Group II: HFD induced, (**C**) Group III: Standard orlistat treated, (**D**) Group IV: JTAQ 200 mg/kg, (**E**) Group V: JTAQ 400 mg/kg, (**F**) Group VI: JTE 200 mg/kg, (**G**) Group VII: JTE 400 mg/kg, (**H**) Group VIII: FMAQ 200 mg/kg, (**I**) Group IX: FMAQ 400 mg/kg, (**J**) Group X. FME 200 mg/kg, (**K**) Group XI: FME 400 mg/kg.

Table 5. Variations in oxidative markers (SOD, LPO and GSH).

Group No.	Treatment	SOD (Unit/mg Tissue)	LPO (nmol MDA/mg Tissue)	GSH (nmol/mg Tissue)
I	Normal Control (Vehicle treated)	78.85 ± 10.362	12.06 ± 0.703	6.67 ± 0.475
II	HFD induced only	13.46 ± 5.990	39.74 ± 0.825	0.85 ± 0.024
III	HFD + Standard Orlistat (30 mg/kg)	81.93 ± 10.578 **	17.35 ± 0.241 **	6.61 ± 0.245 **
IV	HFD + JTAQ (200 mg/kg)	66.29 ± 4.608 **	20.86 ± 0.948 **	4.93 ± 0.107 **
V	HFD + JTAQ (400 mg/kg)	77.95 ± 8.697 **	18.25 ± 0.675 **	4.32 ± 0.227 **
VI	HFD + JTE (200 mg/kg)	84.44 ± 10.905 **	18.70 ± 0.452 **	4.92 ± 0.010 **
VII	HFD + JTE (400 mg/kg)	106.76 ± 4.690 **	14.83 ± 0.475 **	5.02 ± 0.045 **
VIII	HFD + FMAQ (200 mg/kg)	48.05 ± 9.998 **	23.22 ± 1.045 **	3.05 ± 0.052 **
IX	HFD + FMAQ (400 mg/kg)	51.44 ± 2.728 **	22.71 ± 0.747 **	4.05 ± 0.627 **
X	HFD + FME (200 mg/kg)	56.22 ± 7.715 **	24.59 ± 0.492 **	3.93 ± 0.979 **
XI	HFD + FME (400 mg/kg)	60.41 ± 8.517 **	21.88 ± 0.825 **	4.01 ± 0.599 **

One-way ANOVA followed by the Bonferroni test, with values reported as MEAN ± SD at n = 6, ** $p < 0.001$, compared to group HFD treated.

The present study reveals that co-treatment of samples JTE, JTAQ, FME, FMAQ and orlistat resulted in an increase in the antioxidant enzymes and reduction in lipid peroxidation as compared to inducer HFD animals. The fact that the extracts were seen to significantly lower serum total lipids, total cholesterol, and LDL cholesterol suggests that it can be used in hyperlipidemia. As compared to other extracts, the ethanolic extract of *Jatropha tanjorensis* (JTE) leaves exhibits a promising role in the control of high-fat-induced obesity and explains the traditional use of Jatropha tanjorensis leaves to treat cardiac diseases [38]. In future research, we are planning to conduct GC-MS analysis, isolation and purification of various active ingredients present in the plants responsible for various kinds of pharmacological activities.

4. Conclusions

Significant progress has been made in the literature linking crude extracts and bioactive scaffolds from edible and medicinal plants to obesity. Up to this point, there have been many reports on the anti-obesity properties of various extracts and constituents of *Jatropha tanjorensis* and *Fraxinus micrantha*. In this work, we investigated the anti-obesity effect of ethanolic and aqueous extracts of *Jatropha tanjorensis* and *Fraxinus micrantha* leaves as we continue to focus on the anti-obesity potential of these plants. The present study claims that, compared to other extracts, the ethanolic extract of *Jatropha tanjorensis* (JTE) leaves exhibits a promising role in the control of high-fat-induced obesity. The present study provides scientific evidence and support for the traditional use of *Jatropha tanjorensis* leaf extract for the treatment of obesity.

Author Contributions: Conceptualization, S.S. and T.V.; methodology, S.S., T.V. and M.R.H.; formal analysis, S.S., T.V. and M.R.H.; investigation, S.S., T.V.; resources, S.S., T.V.; data curation, S.S., T.V.; writing—original draft preparation, S.S., T.V.; writing—review and editing, A.A., S.Z.A. and F.A.N. and O.A.K.; funding acquisition, O.A.K., S.Z.A. and F.A.N.; supervision, T.V. All authors have read and agreed to the published version of the manuscript.

Funding: This research was funded by Princess Nourah bint Abdulrahman University Researchers Supporting Project number (PNURSP2023R165), Princess Nourah bint Abdulrahman University, Riyadh, Saudi Arabia.

Institutional Review Board Statement: The Institutional Animal Ethics Committee (IAEC) of the Pinnacle Biomedical Research Institute (PBRI) approved animal studies by Bhopal (Reg. No. 1824/PO/ERe/S/15/CPCSEA). The protocol approval reference number is PBRI/IAEC/10-09-22/012.

Informed Consent Statement: Not applicable.

Data Availability Statement: Not applicable.

Acknowledgments: The authors extend their appreciation to Princess Nourah bint Abdulrahman University Researchers Supporting Project number (PNURSP2023R165), Princess Nourah bint Abdulrahman University, Riyadh, Saudi Arabia. Thankful also to Researchers Supporting Project number (RSPD2023R732), King Saud University, Riyadh, Saudi Arabia.

Conflicts of Interest: The authors declare no conflict of interest.

References

1. Murray-Davis, B.; Darling, E.K.; Berger, H.; Melamed, N.; Li, J.; Guarna, G.; Syed, M.; Barrett, J.; Geary, M.; Mawjee, K.; et al. Midwives Perceptions of Managing Pregnancies Complicated by Obesity: A Mixed Methods Study. *Midwifery* **2022**, *105*, 103225. [CrossRef] [PubMed]
2. Davanzo, G.G.; Castro, G.; Monteiro, L.D.B.; Castelucci, B.G.; Jaccomo, V.H.; da Silva, F.C.; Marques, A.M.; Francelin, C.; de Campos, B.B.; de Aguiar, C.F.; et al. Obesity Increases Blood-Brain Barrier Permeability and Aggravates the Mouse Model of Multiple Sclerosis. *Mult. Scler. Relat. Disord.* **2023**, *72*, 104605. [CrossRef] [PubMed]
3. Bays, H.E.; Bindlish, S.; Clayton, T.L. Obesity, Diabetes Mellitus, and Cardiometabolic Risk: An Obesity Medicine Association (OMA) Clinical Practice Statement (CPS) 2023. *Obes. Pillars* **2023**, *5*, 100056. [CrossRef]
4. Ahima, R.S. Obesity Epidemic in Need of Answers. *Gastroenterology* **2006**, *131*, 991. [CrossRef]
5. Jin, X.; Qiu, T.; Li, L.; Yu, R.; Chen, X.; Li, C.; Proud, C.G.; Jiang, T. Pathophysiology of Obesity and Its Associated Diseases. *Acta Pharm. Sin. B*, 2023; in press. [CrossRef]
6. Malandrino, N.; Bhat, S.Z.; Alfaraidhy, M.; Grewal, R.S.; Kalyani, R.R. Obesity and Aging. *Endocrinol. Metab. Clin. N. Am.* **2023**, *52*, 317–339. [CrossRef]
7. Wilson, R.; Aminian, A. Obesity-Associated Cancer Risk Reduction after Metabolic Surgery: Insights from the SPLENDID Study and the Path Forward. *Surg. Obes. Relat. Dis.* 2023; in press. [CrossRef]
8. Lee, S.H.T.; Ng, J.J.; MTL Choong, A. The Effect of Obesity on Outcomes after Arteriovenous Fistulae Creation: A Systematic Review. *Ann. Vasc. Surg.* **2023**, *92*, 304–312. [CrossRef]
9. Padwal, R.S.; Majumdar, S.R. Drug Treatments for Obesity: Orlistat, Sibutramine, and Rimonabant. *Lancet* **2007**, *369*, 71–77. [CrossRef]
10. Agarwal, A.; Remedies, N. Review Article: Herbal Approach for Obesity Management Review Article: Herbal Approach for Obesity Management. *Am. J. Plant Sci.* **2012**, *3*, 1003–1014. [CrossRef]
11. Idu, M.; Igbafe, G.; Erhabor, J.O. Anti-Anaemic Activity of *Jatropha tanjorensis* Ellis & Saroja in Rabbits. *J. Med. Plants Stud.* **2014**, *2*, 64–72.
12. Prabakaran, A.J.; Sujatha, M. *Jatropha tanjorensis* Ellis & Saroja, a Natural Interspecific Hybrid Occurring in Tamil Nadu, India. *Genet. Resour. Crop Evol.* **1999**, *46*, 213–218. [CrossRef]
13. Omoregie, E.S.; Osagie, A.U. Phytochemical Screening and Anti-Anaemic Effect of *Jatropha tanjorensis* Leaf in Protein Malnourished Rats. *Plant Arch.* **2007**, *7*, 509–516.
14. Omoregie, E.S.; Osagie, A.U. Effect of *Jatropha tanjorensis* Leaves Supplement on The Activities of Some Antioxidant Enzymes, Vitamins and Lipid Peroxida-Tion in Rats. *J. Food Biochem.* **2011**, *35*, 409–424. [CrossRef]
15. Oyewole, O.I.; Oladipupo, O.T.; Atoyebi, B.V. Assessment of Renal and Hepatic Functions in Rats Administered Methanolic Leaf Extract of *Jatropha tanjorensis* Scholars Research Library. *Ann. Biol. Res.* **2016**, *3*, 837–841.
16. Omobuwajo, O.R.; Alade, G.O.; Akanmu, M.A.; Obuotor, E.; Osasan, S.A. Microscopic and Toxicity Studies on the Leaves of *Jatropha tanjorensis*. *Afr. J. Pharm. Pharmacol.* **2011**, *5*, 12–17.
17. Mishra, S.B.; Mukerjee, A.; Vijayakumar, M.; Ehhq, K.D.V.; Iru, V.; Qxpehu, H.E.D.; Wd, R.I. Pharmacognostical and Phytochemical Evaluation of Leaves Extract of *Jatropha curcas* Linn. *Pharmacogn. J.* **2010**, *2*, 9–14. [CrossRef]
18. Byrappa, M.; Doss, J.; Ananthi, J.; Sathish, P. Fitoterapia Antimicrobial Activity of Bioactive Compounds and Leaf Extracts in *Jatropha tanjorensis*. *Fitoterapia* **2012**, *83*, 1153–1159. [CrossRef]
19. Adebajo, A.C.; Olayiwola, G.; Omobuwajo, O.R. The Antidiabetic Potential Of *Jatropha tanjorensis* Leaves. *Niger. J. Nat. Prod. Med.* **2004**, *8*, 55–58. [CrossRef]
20. Viswanathan, M.B.; Jeyaananthi, J. Antimicrobial and Antiinflammatory Activities of Various Extracts of the Leaves of Jatropha Tanjorenesis. *Biosci. Biotechnol. Res. Asia* **2009**, *6*, 297–300.
21. Abdelgadir, H.A.; Van Staden, J. Ethnobotany, Ethnopharmacology and Toxicity of *Jatropha curcas* L. (Euphorbiaceae): A Review. *S. Afr. J. Bot.* **2013**, *88*, 204–218. [CrossRef]
22. Danborno, A.M.; Tarfa, F.; Toryila, J.E.; Awheela, E.U.; Shekarau, V.T. The Effects of *Jatropha tanjorensis* Aqueous Leaf Extract on Haematological Parameters in Wistar Rats. *J. Afr. Assoc. Physiol. Sci.* **2019**, *7*, 133–137.
23. Ntak, O.; Ekpo, E.; State, A.I.; State, C.R. Nutritional Studies and Antimicrobial Activities of *Jatropha tanjorensis* Leaves Extracts against Escherichia Coli Isolates. *Int. J. Innov. Sci. Res. Technol.* **2019**, *4*, 945–955.
24. Sharma, J.; Gairola, S.; Gaur, R.D.; Painuli, R.M. The Treatment of Jaundice with Medicinal Plants in Indigenous Communities of the Sub-Himalayan Region of Uttarakhand, India. *J. Ethnopharmacol.* **2012**, *143*, 262–291. [CrossRef] [PubMed]
25. Garbyal, S.S.; Grover, A.; Aggarwal, K.K.; Babu, C.R. Traditional Phytomedicinal Knowledge of Bhotias of Dharchula in Pithoragarh. *Indian J. Tradit. Knowl.* **2007**, *4*, 199–207.

26. Sarfraz, I.; Rasul, A.; Jabeen, F.; Younis, T.; Zahoor, M.K.; Arshad, M.; Ali, M. Fraxinus: A Plant with Versatile Pharmacological and Biological Activities. *Evid.-Based Complement. Altern. Med.* **2017**, *2017*, 4269868. [CrossRef] [PubMed]
27. Kostova, I.; Iossifova, T. Chemical Components of Fraxinus Species. *Fitoterapia* **2007**, *78*, 85–106. [CrossRef]
28. Buettner, R.; Schölmerich, J.; Bollheimer, L.C. High-Fat Diets: Modeling the Metabolic Disorders of Human Obesity in Rodents. *Obesity* **2007**, *15*, 798–808. [CrossRef] [PubMed]
29. Hariri, N.; Thibault, L. High-Fat Diet-Induced Obesity in Animal Models. *Nutr. Res. Rev.* **2010**, *23*, 270–299. [CrossRef]
30. Alara, O.R.; Abdurahman, N.H.; Ukaegbu, C.I.; Kabbashi, N.A. Extraction and Characterization of Bioactive Compounds in Vernonia Amygdalina Leaf Ethanolic Extract Comparing Soxhlet and Microwave-Assisted Extraction Techniques. *J. Taibah Univ. Sci.* **2019**, *13*, 414–422. [CrossRef]
31. Martins, N.; Barros, L.; Santos-Buelga, C.; Silva, S.; Henriques, M.; Ferreira, I.C.F.R. Decoction, Infusion and Hydroalcoholic Extract of Cultivated Thyme: Antioxidant and Antibacterial Activities, and Phenolic Characterisation. *Food Chem.* **2015**, *167*, 131–137. [CrossRef]
32. Alhalmi, A.; Amin, S.; Khan, Z.; Beg, S.; Al Kamaly, O.; Saleh, A.; Kohli, K. Nanostructured Lipid Carrier-Based Codelivery of Raloxifene and Naringin: Formulation, Optimization, In Vitro, Ex Vivo, In Vivo Assessment, and Acute Toxicity Studies. *Pharmaceutics* **2022**, *14*, 1771. [CrossRef]
33. Jonsson, M.; Jestoi, M.; Nathanail, A.V.; Kokkonen, U.-M.; Anttila, M.; Koivisto, P.; Karhunen, P.; Peltonen, K. Application of OECD Guideline 423 in Assessing the Acute Oral Toxicity of Moniliformin. *Food Chem. Toxicol. Int. J. Publ. Br. Ind. Biol. Res. Assoc.* **2013**, *53*, 27–32. [CrossRef]
34. Friedewald, W.T.; Levy, R.I.; Fredrickson, D.S. Estimation of the Concentration of Low-Density Lipoprotein Cholesterol in Plasma, without Use of the Preparative Ultracentrifuge. *Clin. Chem.* **1972**, *18*, 499–502. [CrossRef]
35. Zaidun, N.H.; Thent, Z.C.; Latiff, A.A. Combating Oxidative Stress Disorders with Citrus Flavonoid: Naringenin. *Life Sci.* **2018**, *208*, 111–122. [CrossRef]
36. Treml, J.; Šmejkal, K. Flavonoids as Potent Scavengers of Hydroxyl Radicals. *Compr. Rev. Food Sci. Food Saf.* **2016**, *15*, 720–738. [CrossRef] [PubMed]
37. Jaishree, V.; Badami, S.; Krishnamurthy, P.T. Antioxidant and Hepatoprotective Effect of the Ethyl Acetate Extract of Enicostem-maAxillare (Lam). Raynal against CCL4-Induced Liver Injury in Rats. *Indian J. Exp. Biol.* **2010**, *48*, 896–904. [PubMed]
38. Amaechi, D.; Yisa, B.N.; Ekpe, I.P.; Nwawuba, P.I.; Rabbi, A. Phytochemical Screening, Anti-obesity and Hepatoprotective Activities of Ethanol Leaf Extract of Jatropha tanjorensis in Wistar Rats. *Asian J. Appl. Chem. Res.* **2022**, *12*, 20–26. [CrossRef]

Article

Anti-Inflammatory, Antinociceptive, Antipyretic, and Gastroprotective Effects of *Eurycoma longifolia* Jack Ethanolic Extract

Subhawat Subhawa [1,2], Warangkana Arpornchayanon [2], Kanjana Jaijoy [3], Sunee Chansakaow [4], Noppamas Soonthornchareonnon [5] and Seewaboon Sireeratawong [1,2,6,*]

[1] Clinical Research Center for Food and Herbal Product Trials and Development (CR-FAH), Faculty of Medicine, Chiang Mai University, Chiang Mai 50200, Thailand; subhawat.s@cmu.ac.th
[2] Department of Pharmacology, Faculty of Medicine, Chiang Mai University, Chiang Mai 50200, Thailand; warangkana@gmail.com
[3] McCormick Faculty of Nursing, Payap University, Chiang Mai 50000, Thailand; joi.kanjana@gmail.com
[4] Department of Pharmaceutical Sciences, Faculty of Pharmacy, Chiang Mai University, Chiang Mai 50200, Thailand; chsunee@gmail.com
[5] Department of Pharmacognosy, Faculty of Pharmacy, Mahidol University, Bangkok 10400, Thailand; noppamas.sup@mahidol.ac.th
[6] Department of Preclinical Science, Division of Pharmacology, Faculty of Medicine, Rungsit Campus, Thammasat University, Pathum Thani 12120, Thailand
* Correspondence: seewaboon.s@cmu.ac.th or seewaboon@gmail.com; Tel.: +66-53-934591

Citation: Subhawa, S.; Arpornchayanon, W.; Jaijoy, K.; Chansakaow, S.; Soonthornchareonnon, N.; Sireeratawong, S. Anti-Inflammatory, Antinociceptive, Antipyretic, and Gastroprotective Effects of *Eurycoma longifolia* Jack Ethanolic Extract. *Life* **2023**, *13*, 1465. https://doi.org/10.3390/life13071465

Academic Editors: Marisa Colone, Efstathia Papada and Charalampia Amerikanou

Received: 11 May 2023
Revised: 25 June 2023
Accepted: 26 June 2023
Published: 28 June 2023

Abstract: Tongkat ali (*Eurycoma longifolia* Jack) (ELJ) is a plant in the Simaroubaceae family. Its roots are used in traditional Thai medicine to treat inflammation, pain, and fever; however, the antiulcer abilities of its ethanolic extract have not been studied. This study examined the anti-inflammatory, antinociceptive, antipyretic, and gastroprotective effects of ethanolic ELJ extract in animal models and found that ELJ effectively reduced EPP-induced ear edema in a dose-dependent manner and that a high dose of ELJ inhibited carrageenan-induced hind paw edema formation. In cotton-pellet-induced granuloma formation, a high dose of ELJ suppressed the increases in wet granuloma weight but not dry or transudative weight. In the formalin-induced nociception study, ELJ had a significant dose-dependent inhibitory impact. Additionally, the study found that yeast-induced hyperthermia could be significantly reduced by antipyretic action at the highest dose of ELJ. In all the gastric ulcer models induced by chemical substances or physical activity, ELJ extracts at 150, 300, and 600 mg/kg also effectively prevented gastric ulcer formation. In the pyloric ligation model, however, the effects of ELJ extract on gastric volume, gastric pH, and total acidity were statistically insignificant. These findings support the current widespread use of *Eurycoma longifolia* Jack in traditional medicine, suggest the plant's medicinal potential for development of phytomedicines with anti-inflammatory, antinociceptive, and antipyretic properties, and support its use in the treatment of gastric ulcers due to its gastroprotective properties.

Keywords: *Eurycoma longifolia* Jack; extract; anti-inflammatory; antinociceptive effect; antipyretic effects; gastroprotective activity; gastric ulcer; Thai traditional medicine

1. Introduction

Inflammation is a complicated response of living tissues to injury, one which can significantly affect health and lifestyle [1]. For decades, inflammation has been treated with anti-inflammatory medications, primarily non-steroidal anti-inflammatory drugs (NSAIDs), although several side effects of those medications have been reported, including nausea, vomiting, dyspepsia, stomach pain, ulcers, and bleeding [2]. One anti-inflammatory drug, cyclooxygenase-2 (COX-2) selective inhibitors, has been linked to cardiovascular side effects [3], prompting the search for suitable alternatives to these often-prescribed drugs in foods and herbs [4,5]. In addition to reducing the risk of adverse effects of long-term exposure to anti-inflammatory drugs, another desirable objective is to identify natural ingredients that not only do not cause gastric ulcers, but which can also reduce their incidence.

Eurycoma longifolia Jack (ELJ), also known as tongkat ali, is an herbaceous plant belonging to the Simaroubaceae family which possesses several therapeutic properties. The therapeutic pharmacological effects of ELJ are attributed to the presence of numerous bioactive chemicals, including quassinoids, tirucallanes of the triterpene type, squalene derivatives, eurycomalactone, and bioactive steroids [6]. Various biologically active compounds found in ELJ roots, stems, and leaves, and even in bark, are linked to various pharmacological effects [7]. In the Thai medical tradition, the roots of ELJ have antipyretic properties extracted by boiling 1 handful of dried roots (weighing 8–15 g) with drinking water before breakfast and dinner twice daily [8,9]. In addition, ELJ root extracts have been used as analgesic and anti-inflammatory treatments [10,11]. Furthermore, root preparations of ELJ have been used to treat various illnesses, including malaria, fever, impotence, and loss of sexual desire [12–14]. In previous studies, the analgesic activity of ELJ was evaluated using the hot plate and acetic acid tests in mice, and its anti-inflammatory effect was demonstrated in carrageenan-induced paw edema in mice [15]. Notably, it is a component in the Chantaleela recipe, which in Thai traditional folk medicine is prescribed for fever relief and as an anti-inflammatory [14,16]. Apart from antipyretic effects and anti-inflammatory uses, it has been shown that "Radix" herbal remedies made from *Althaea officinalis* L., including ELJ, can successfully protect gastric mucosa against ethanol-induced gastric lesions [17]. Although ELJ roots have been used in Thai traditional medicine, particularly in the Chantaleela formula, which has exhibited efficacy as an antipyretic treatment [16], there have been limited animal investigations on the anti-inflammatory, antipyretic, or gastroprotective effects of ELJ.

To investigate the therapeutic characteristics traditional medicine has attributed to this medicinal plant, we studied the anti-inflammatory and antinociceptive effects of the ethanolic root extract of *Eurycoma longifolia* Jack (ELJ) in vivo on experimental animals. As various non-steroidal anti-inflammatory medicines (NSAIDs) frequently produce mucosal lesions in human stomachs, we also investigated the gastroprotective effect of ELJ extracts using models of gastric ulcers.

2. Materials and Methods

2.1. Chemicals and Reagents

Aspirin, phenylbutazone, morphine, and prednisolone were purchased from Schering (Bangkok) Ltd., Bangkok, Thailand. Absolute ethanol, carrageenan, ethyl phenylpropiolate (EPP), scopoletin, umbelliferone, cimetidine, and indomethacin were obtained from Sigma Chemical Company (St. Louis, MO, USA). All other chemicals were of analytical grade.

2.2. Extract Preparation

Eurycoma longifolia Jack (ELJ) was obtained from Vejpong-Osot Drug Store in Bangkok, Thailand. Associate Professor Dr. Noppamas Soonthornchareonnon conducted the species identification, including comparing the ELJ preparation to an authentic sample of the crude drug at the Faculty of Pharmacy Plant Museum, Mahidol University, Bangkok, Thailand, and evaluating its properties in accordance with Thai Herbal Pharmacopoeia methods, e.g., organoleptic examination, extractive values, % loss on drying, total ash, and acid insoluble ash [18]. To prepare an ethanolic extract of ELJ, 5 kg of dried ELJ was macerated in 18 L of 95% ethanol for 5 days, then filtered through Whatman No. 1 filter paper (Sigma-Aldrich, St. Louis, MO, USA) and concentrated at 40 °C using a rotary evaporator (EYELA, Tokyo, Japan). The root extract of *Eurycoma longifolia* Jack yielded a weight-to-weight ratio of 1.80%.

2.3. Phytochemical Constituents of E. longifolia as Determined by TLC

The chemical constituents of the ethanolic ELJ extracts were investigated using a modified Tung et al. approach, utilizing thin-layer chromatography (TLC) [19]. The ELJ extract was spread on aluminum plates precoated with silica gel 60 GF (254) (Merck, Darmstadt, Germany), and dichloromethane:methanol (95:5) was used for the mobile phase. The reference standards were scopoletin and eurycomalectone. After development, the TLC plates were placed in a drying chamber. Components were detected under 254 nm and 366 nm UV light, and the plate was then sprayed with 10% KOH, anisaldehyde-sulfuric acid, anisaldehyde-sulfuric acid, and a natural product spraying reagent. The migration distances of unidentified points were compared to standard compounds using Rf values, which were determined as follows: Rf = spot migration distance/solvent migration distance.

2.4. Identification of Phytochemicals by High-Pressure Liquid Chromatography (HPLC)

As in previous studies, each sample was analyzed using high-performance liquid chromatography (HPLC) [20,21]. Briefly, a Spheri-5 RP18 column, 220 mm × 4.6 mm i.d. (Perkin-Elmer®, Waltham, MA, USA) HPLC column, was used. Mobile phases consisted of methanol and water (content 0.02% phosphoric acid) (23:77) with isocratic systems at a 1.2 mL/min flow rate. A quantum of 20 µL of each sample was injected into the column, with a flow rate of 1.0 mL/min, and monitored with a diode array detector at 345 nm; the column temperature was kept at 23–25 °C.

2.5. Animals

Male Sprague-Dawley rats weighing 180 and 200 g, bred by the National Laboratory Animal Center, Mahidol University, Nakhon Pathom, Thailand, were used in this study. Standard animal care was maintained throughout the study period in accordance with the laws and regulations governing animal care and use. The room was kept at 25 ± 1 °C and 60% humidity, with a 12-h day/night interval. Food and water were provided ad libitum throughout the study. After transfer to the animal room, all animals received at least 1 week of care before the experiment began. All animal studies were approved by the Research Ethics Committee for Animal Studies (Study code: 0003/2008), Faculty of Medicine, Thammasat University, Pathum Thani, Thailand.

2.6. Test Substance Administration

In each animal experiment, the control group received the same route and volume of vehicles as the test group. In the ear edema model, 20 µL/ear of ELJ extracts and a reference drug (ibuprofen) were applied topically to the ears of rats. A total of 5 mL/kg body weight of ELJ extracts and reference drugs (aspirin, morphine, and prednisolone) were administered by oral gavage to the other models.

2.7. Anti-Inflammatory Activity

2.7.1. Ethyl Phenylpropiolate (EPP)-Induced Ear Edema

To investigate the topical anti-inflammatory impact of ELJ extracts, a modified version of the method reported by Brattsand et al. (1982) [22] was used. Thirty male Sprague-Dawley rats weighing 40–60 g were randomly assigned into five groups of six rats each. EPP was dissolved in acetone and topically administered to the inner and outer surfaces of both ears at a dose of 1 mg/ear. The acetone (vehicle control), ELJ extracts (1, 2, and 4 mg/ear), or ibuprofen (1 mg/ear) were applied to the ear immediately before EPP administration. The ear thickness was measured with digital vernier calipers beforehand, and at 15, 30, 60, and 120 min after edema induction. The test substances' ear edema inhibition was then determined. The percentage of inhibition was calculated by comparing the increase in ear thickness of each test group to that of its control group.

2.7.2. Carrageenan-Induced Hind Paw Edema in Rats

Thirty-six rats weighing 100–120 g were divided randomly into six groups (6 rats per group). On the plantar side of the right hind paw of the rats, carrageenan (0.05 mL, 1% w/v in NSS) was injected intradermally. Aspirin (300 mg/kg, p.o.) or ELJ extracts (300, 600, and 1200 mg/kg) were administered 1 h before carrageenan administration. The edema of the right hind paw was measured beforehand, and at 1, 3, and 5 h after carrageenan injection using a plethysmometer (model 7150, Ugo Basile, Italy) [23]. The degree of decrease in the difference in paw edema volume between the left (without carrageenan) and right (with carrageenan) hind paws were used to determine the anti-inflammatory activity of the ELJ extracts (mL). The percentage of inhibition was estimated by comparing each test group's decreased paw thickness to that of the corresponding control group.

2.7.3. Cotton-pellet-induced Granuloma Formation in Rats

Twenty-four rats weighing 200 to 250 g were divided randomly into four groups of six each. Two sterile cotton pellets (19–21 mg) were implanted subcutaneously, one on each side of the rat's abdomen, while the rats were under anesthesia. Each group of rats then received distilled water (0.5 mL/100 g, p.o.), aspirin (300 mg/kg/day, p.o.), prednisolone (5 mg/kg/day, p.o.), or ELJ extract (1200 mg/kg/day, p.o.) daily for 7 days. On the eighth day, the rats were anesthetized, and the implanted pellets were dissected. The pellets' wet and dry weights (dried at 60 °C for 18 h) were determined. The granuloma inhibition and transudative weight were calculated. The thymus gland dry weight and the animals' body weight were also measured [24].

2.8. Formalin-Induced Nociception Model

The formalin test was conducted using Swiss albino mice following the procedures described above [25]. The behavior of the mice after receiving an intraplanar injection of 20 mL 2.7% formalin (1% formaldehyde) in saline into the ventral surface of the right hind paw was observed. The licking reaction time during the first five minutes following the injection was recorded as the neurogenic or initial response; the licking reaction during the period 20–30 min after injection was recorded as the late or inflammatory pain response. One hour before the formalin injection, the mice were administered either ELJ extracts (300, 600, or 1200 mg/kg) or saline solution (10 mL/kg) orally (by gavage). As an indicator of nociception, the duration of the animals' licking of the injected paw was measured with a chronometer after formalin injection and placement in 20-cm-diameter glass cylinders. The data were analyzed using the following equation to determine the mean percent inhibition of licking response (PIL).

$$\text{PIL} = \frac{\text{Licking response time (control)} - \text{Licking response time (treated)}}{\text{Licking response time (control)}} \times 100$$

2.9. Antipyretic Activity

Thirty male rats weighing 180–200 g, randomly divided into five groups of six rats each, were subjected to the yeast-induced hyperthermia model, following Mizui et al. [26]. A 12-channel electronic thermometer (LETICA, model TMP 812 RS, Panlab S.L., Cornellà de Llobregat, Spain) was used to determine rectal temperatures at baseline. To induce hyperthermia, yeast was injected subcutaneously (1 mL/100 g body weight, 25% brewers' yeast w/v in NSS). After 18 h, rectal temperatures were measured again. Dosages of 5% Tween 80, aspirin (300 mg/kg), or ELJ extracts (75, 150, or 300 mg/kg) were administered orally to rats with a temperature increase of greater than 1 °C. Rectal temperatures were measured at 30, 60, 90, 120, and 180 min after treatment, and again after 18 h.

2.10. Gastric Ulcer Models

All gastric ulcer experiments were performed in 12 h fasted rats. Water ad libitum was allowed up to 1 h before drug administration. For each experiment, 30 rats were randomly allocated into 5 groups of 6 rats each as follows: group 1 received sterile water (control), group 2 received standard treatment (100 mg/kg of cimetidine), group 3 received 150 mg/kg of ELJ extract, group 4 received 300 mg/kg of ELJ extract, and group 5 received 600 mg/kg of ELJ extract. The rats were monitored for symptoms and any adverse effects. The rats were also given a high dose of ELJ extract (600 mg/kg) daily for 14 days to investigate whether ELJ could cause a gastric ulcer.

2.10.1. Ethanol/Hydrochloric Acid (EtOH/HCl)-Induced Gastric Lesions

The gastric ulceration ethanol evaluation was conducted using an acidified solution, as modified from a published methodology [27], using thirty rats (6 rats per group). One hour after oral administration of the test drugs, 0.1 mL of EtOH/HCl solution (consisting of 60 mL of absolute ethanol, plus 12.5 mL of HCl and 27.5 mL of water) was orally administered to the rats in all groups. After one hour, the rats were sacrificed, using thiopental sodium induction, before evaluating gastric lesions. An incision was made along the greater curvature of the stomach, exposing the gastric mucosa. The size of the gastric ulcers was measured in millimeters (mm) under a 10-x microscope. The size of the lesions was converted to an ulcer index (UI), using the following formula:

$$\mathrm{UlcerIndex(UI)} = \frac{\text{Sum of the total length of lesions in each group}}{\text{Number of rats in that group}} \tag{1}$$

Then, the percentage of gastric ulcer inhibition (% inhibition) of each test drug was estimated using the following formula:

$$\%\mathrm{Inhibition} = \frac{\mathrm{UIc} - \mathrm{UIt}}{\mathrm{UIc}} \times 100 \tag{2}$$

where UIc is the "ulcer index of the control group" and UIt is the "ulcer index of the test group".

2.10.2. Indomethacin-Induced Gastric Lesions

Indomethacin is known to cause significant adverse reactions, including petechial hemorrhage, inflammatory lesions, and erosions of the stomach mucosa [28–30]. A dose of 30 mg/kg indomethacin suspension in 5% Tween 80 was administered intraperitoneally to each of the animals in all groups. Five hours later, the rats were sacrificed to determine gastric lesions. Ulcer indexes and % inhibition of gastric ulcers were calculated.

2.10.3. Restraint Water Immersion Stress-Induced Gastric Lesions

Water immersion and stress-induced gastric lesions in rats, a combination of physical and psychological stressors, resulted in lesions that mimicked those caused by sepsis, trauma, or surgery [31,32]. After the rats had been fasted for 48 h without food and 1 h without water, they were randomly divided into 5 groups of six each, and the test drugs were administered orally. One hour after drug administration, the rats were placed in stainless steel cages that fit only one animal per cage, with the head upright for 5 h, and the cages immersed in a cold-water tank (20 ± 2 °C). The water level was maintained at the rats' chest level. The rats were then sacrificed, and the gastric lesions were measured. The ulcer indexes and % inhibition were compared between groups.

2.11. Pylorus Ligation

One hour after drug administration, general anesthesia was administered to each rat. Laparotomy and pyloric ligation were performed, followed by skin closure [33]. Five hours later, the animals were sacrificed. Their stomach and gastric contents were removed. After being centrifuged at 2500 rpm for 5 min, the gastric juice was tested for total volume and pH, as well as total acidity in the supernatant. Quantifying the total acidity of gastric juice was performed by titration with 0.1 N NaOH to an endpoint of pH 7.4 using phenolphthalein as the indicator to determine the amount in ml and μEq per 100 g body weight of the rat each hour. Gastric ulcers were measured at their greatest length (mm), and then UI and percentage of ulcer inhibition were calculated.

2.12. Statistical Analysis

One-way analysis of variance (ANOVA) and the post hoc least significant difference (LSD) test were used to compare the data between groups using GraphPad Prism 9.0 software (GraphPad Software, Inc., San Diego, CA, USA). Data are shown as mean $\pm$ standard error of the mean (SEM). p-values less than 0.05 were considered statistically significant.

3. Results

3.1. Specification of E. longifolia Jack by TLC and HPLC Determinations

Screening for phytochemicals was performed on the root ELJ in accordance with the protocol for the standard method. The saponin, the terpenoids, and an unidentified blue-colored compound were found in the tested sample. The results of their respective quality tests, which were conducted according to the 2018 Thai Herbal Pharmacopoeia, are shown in Table 1.

Table 1. Physical and chemical properties of the root of *E. longifolia* Jack.

Test	Result
Foreign matter (%w/w)	Not found
Hexane extractive (%w/w)	0.66
Dichloromethane extractive (%w/w)	0.94
Ethanol extractive content (%w/w)	1.51
Water extractive content (%w/w)	10.57
Loss on drying (%v/w)	9.58
Total ash (%w/w)	2.54
Acid-insoluble ash (%w/w)	0.90
Chemical composition	saponin, terpenoids

Scopoletin and eurycomalectone were selected as reference standards, and their Rf values were measured, as represented in Figure 1. Scopoletin was seen as spots under 254 and 366 nm UV light in 10%KOH, an anisaldehyde–sulfuric-acid reagent with UV at 366 nm, and a natural product spraying reagent with Rf = 0.40. While under 254 nm UV light and with an anisaldehyde–sulfuric-acid reagent, eurycomalectone appeared as a spot with Rf = 0.50. The ethanolic extract of ELJ resulted in a blue band with Rf = 0.40 under 254 and 366 nm UV light, 10%KOH, an anisaldehyde–sulfuric-acid reagent with UV at 366 nm, and natural product spraying, suggesting the presence of scopoletin. Unlike the standard eurycomalectone, the ethanolic extract ELJ exhibited bands under 366 nm UV light, 10% KOH, and a natural product spraying reagent with Rf = 0.52. Consistent with the previous study, it was reported that scopoletin and eurycomalectone are components of an ethanolic extract of *E. longifolia* Jack [34,35].

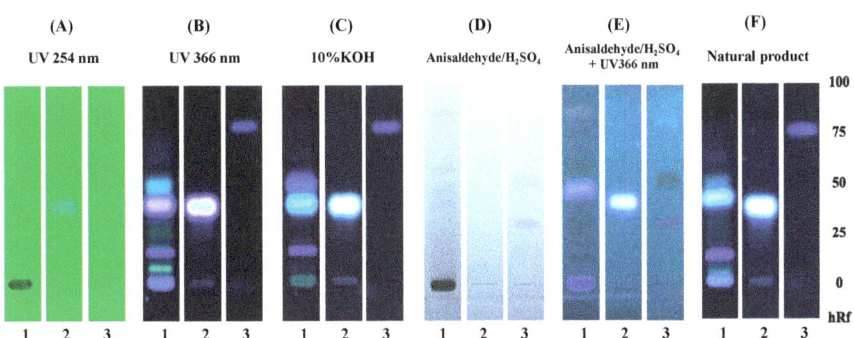

Figure 1. TLC chromatogram of 95% ethanolic extract of *Eurycoma longifolia* Jack: TLC chromatogram of ethanolic extract of *Eurycoma longifolia* Jack (1), scopoletin (2), and eurycomalectone (3), observed under (**A**) UV at 254 nm, (**B**) UV at 366 nm, (**C**) 10%KOH, (**D**) anisaldehyde–sulfuric-acid reagent, (**E**) anisaldehyde–sulfuric-acid reagent with UV at 366 nm, and (**F**) natural product spraying reagent.

Apart from TLC, scopoletin was also seen as a clear peak in HPLC, and the amount of it in the ELJ extract was then quantitatively compared to the standard (Figure 2A) and umbelliferone with a well-separated retention time from scopoletin was used as an internal standard to correct for volume errors (Figure 2B). According to HPLC analysis, scopoletin accounted for 12.590 µg/mL of the ELJ extract, which was determined to be 0.252% *w/w* of crude extract. The result implies that scopoletin in TLC results were consistent with HPLC analyses. Therefore, the presence of scopoletin and eurycomalectone in our study on the analyzed ELJ extracts was consistent with the previous research. In the future, the TLC technology could be paired with mass spectrometry for the identification of chemical components that have been separated.

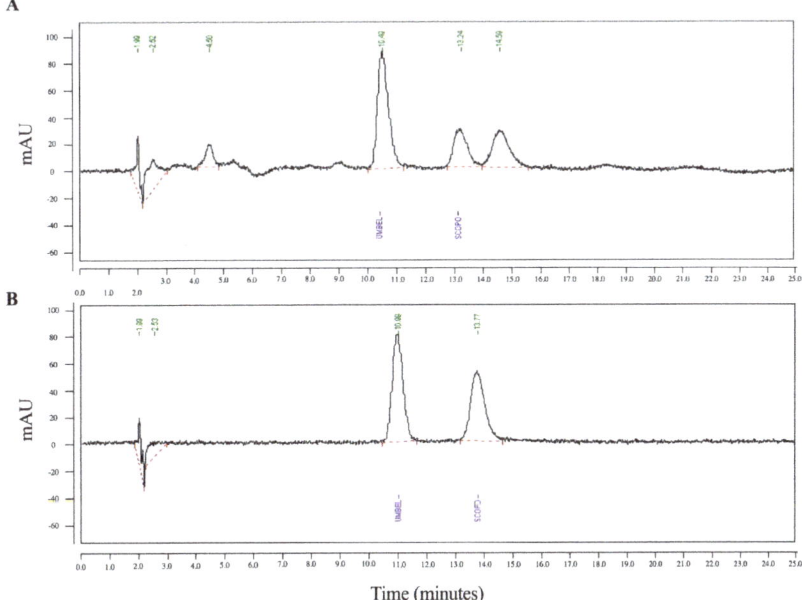

Figure 2. The HPLC of scopoletin of ethanolic ELJ extract (**A**) compared to standard (**B**). Umbelliferone used as an internal standard, with a retention time well-separated from the scopoletin, was mixed with scopoletin to correct for volume errors.

3.2. EPP-Induced Ear Edema

As indicated in Figure 3, EPP applied topically to the ears of rats caused ear edema after 15 min. The maximal effect lasted one hour and then progressively diminished. All treatments, including ELJ extracts and phenylbutazone, significantly decreased ear edema formation with equivalent inhibition percentages at all evaluation time points.

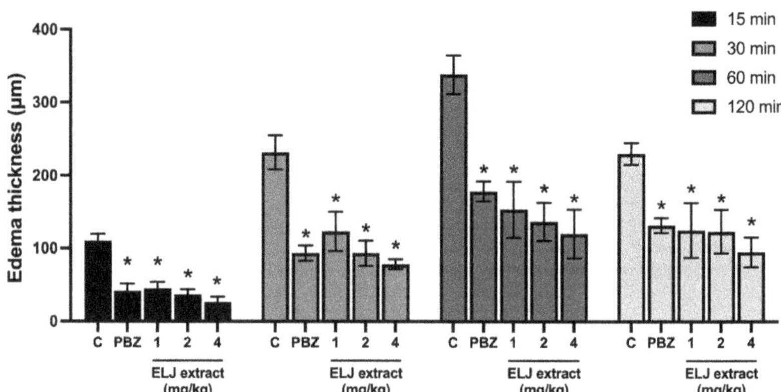

Figure 3. Effects of ELJ extract and phenylbutazone on EPP-induced ear edema in rats. Values are expressed as mean $\pm$ SEM (n = 6). Control: treated with acetone. * Significantly different from its control group, $p > 0.05$. C, Control. PBZ, phenylbutazone (1 mg/kg).

3.3. Carrageenan-Induced Hind Paw Edema

Figure 4 shows the inhibitory effect of oral treatment of ELJ extracts on carrageenan-induced paw edema in rats. The injection of carrageenan caused edema of the hind paw within 1 h, with the maximum effect occurring after 3 h. Only ELJ extract at a high dose (1200 mg/kg) and aspirin (300 mg/kg) effectively prevented hind paw edema formation at all assessment time points, with the greatest inhibition at 1 h after carrageenan administration. This inhibitory impact of ELJ extracts appeared to be dose dependent.

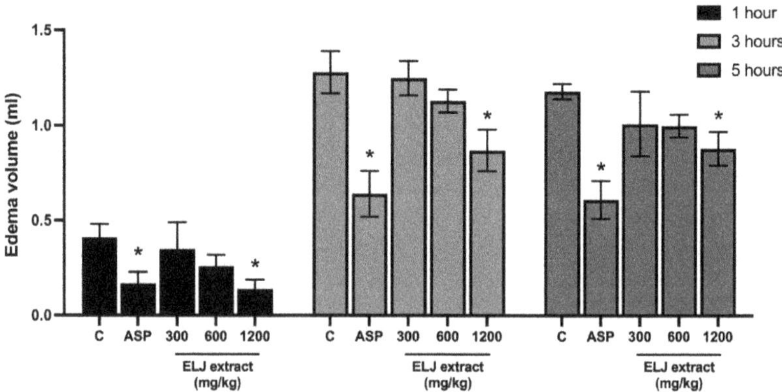

Figure 4. Effects of ELJ extracts and aspirin on carrageenan-induced hind paw edema in rats. Values are expressed as mean $\pm$ SEM (n = 6). Control: treated with acetone. * Significantly different from its control group, $p > 0.05$. C, Control. ASP, Aspirin (300 mg/kg).

3.4. Cotton-Pellet-Induced Granuloma Formation

Aspirin (300 mg/kg/day) and ELJ extract (1200 mg/kg/day) tended to reduce transudative weights and granuloma weights, as shown by their granuloma inhibition of 4% and 25%, respectively, except for prednisolone (5 mg/kg/day), which significantly reduced from those of control groups, as shown by their granuloma inhibition of 37%. (Table 2). As shown in Table 3, the body weight gain and dry thymus weight did not differ substantially between the prednisolone, aspirin, and ELJ extract groups.

Table 2. Effects of ELJ extract, prednisolone, and aspirin on transudative weight, granuloma weight, and percentage of granuloma inhibition on cotton-pellet-induced granuloma formation in rats.

Groups	Dose (mg/kg)	Granuloma Wet Weight (mg)	Granuloma Dry Weight (mg)	Transudative Weight (mg)	Granuloma Weight (mg/mg Cotton)	GI (%)
Control	-	450.0 ± 27.0	81.8 ± 4.4	368.2 ± 23.4	3.1 ± 0.2	-
Prednisolone	5	291.0 ± 12.4 *	58.6 ± 2.4 *	232.5 ± 10.9 *	1.9 ± 0.1 *	37
Aspirin	300	415.8 ± 33.1	79.5 ± 4.9	336.3 ± 30.1	3.0 ± 0.2	4
ELJ extract	1200	347.4 ± 25.6 *	66.2 ± 3.0 *	281.2 ± 21.7	2.3 ± 0.3	25

Values are expressed as mean ± SEM (n = 6). Control: treated with acetone. * Significantly different from its control group, $p > 0.05$.

Table 3. Effects of ELJ extract, prednisolone, and aspirin on dry thymus weight with reference to cotton-pellet-induced granuloma formation in rats.

Groups	Dose (mg/kg)	Body Weight (g)			Dry Thymus Weight (mg/100 g)
		Initial	Final	Gain	
Control	-	387.50 ± 13.34	394.17 ± 11.06	6.67 ± 3.07	21.91 ± 1.97
Prednisolone	5	365.00 ± 9.83	353.33 ± 10.14 *	−11.67 ± 6.54 *	17.24 ± 1.79
Aspirin	300	360.83 ± 12.74	361.67 ± 11.38 *	0.83 ± 4.36	16.96 ± 1.27
ELJ extract	1200	383.33 ± 3.80	380.00 ± 5.16	−3.33 ± 4.22	16.99 ± 1.82

Values are expressed as mean ± SEM (n = 6). Control: treated with acetone. * Significantly different from its control group, $p > 0.05$.

3.5. Formalin-Induced Nociception Model

In the formalin test, the control group's average licking duration was 67.8 s in the early phase and increased to 93.8 s in the late phase (Table 4). All treatments (aspirin, morphine, and ELJ extracts) significantly inhibited the paw licking in both phases in a dose-dependent manner. In the early phase, the percentage inhibition of paw licking by ELJ extracts at 300, 600, and 1200 mg/kg were almost the same, accounting for 34%, 38%, and 43%, respectively. However, in the late phase, maximum inhibition was recorded for aspirin (100%), followed by morphine (98%) and ELJ extracts ranging from 88–97%, which was comparable to standard.

Table 4. Effects of ELJ extract, prednisolone, and aspirin on dry thymus weight with reference to cotton-pellet-induced granuloma formation in rats.

Groups	Dose (mg/kg)	Early Phase		Late Phase	
		Licking Time (s)	% Inhibition of Licking Response	Licking Time (s)	% Inhibition of Licking Response
Control	-	67.8 ± 4.5	-	93.8 ± 7.1	-
Aspirin	300	45.0 ± 5.3 *	34	0.0 ± 0.0 *	100
Morphine	10	0.0 ± 0.0 *	100	1.5 ± 1.5 *	98
ELJ extract	300	44.8 ± 5.0 *	34	11.2 ± 9.5 *	88
	600	41.8 ± 3.3 *	38	4.8 ± 3.4 *	95
	1200	38.7 ± 3.2 *	43	3.2 ± 3.2 *	97

Values are expressed as mean ± SEM (n = 6). Control: treated with acetone. * Significantly different from its control group, $p > 0.05$.

3.6. Antipyretic Activity

As shown in Table 5, the rectal temperature of all mice was increased 18 h after the baker's yeast injection. Aspirin at 300 mg/kg and ELJ extract at 1200 mg/kg dose significantly reduced hyperthermia at all time points (Table 5).

Table 5. Effects of ELJ extracts and aspirin on yeast-induced hyperthermia in rats.

| Groups | Dose (mg/kg) | Rectal Temperature (°C) | | | | | |
| | | Baseline | 18 h after Yeast Injection | Time after Drug Administration (min) | | | |
				30 min	60 min	90 min	120 min
Control	-	37.98 ± 0.22	39.20 ± 0.23	39.17 ± 0.30	39.03 ±0.23	39.08 ± 0.21	39.03 ± 0.22
Aspirin	300	38.03 ± 0.21	39.07 ± 0.20	38.30 ± 0.25 *	37.98 ± 0.22 *	37.87 ± 0.18 *	37.75 ± 0.20 *
ELJ extract	300	38.03 ± 0.23	39.23 ± 0.09	39.02 ± 0.15	38.85 ± 0.17	38.85 ± 0.17	38.83 ± 0.20
	600	37.88 ± 0.16	39.23 ± 0.10	38.83 ± 0.20	38.60 ± 0.13	38.65 ± 0.10	38.73 ± 0.07
	1200	38.07 ± 0.15	39.00 ± 0.13	38.30 ± 0.27 *	38.18 ± 0.25 *	38.27 ± 0.27 *	38.08 ± 0.31 *

Values are expressed as mean ± SEM (n = 6). Control: treated with acetone. * Significantly different from its control group, $p > 0.05$.

3.7. Anti-Ulcerogenic Activities of ELJ Extract in Gastric Ulcer Models

Administration of ELJ to rats at the highest dose (600 mg/kg) for 14 days did not result in the development of stomach ulcers (Figure 5). The anti-ulcerogenic effect of ELJ extract was investigated using three gastric ulcer models. The EtOH/HCl-acid-induced gastric lesions model results in the most severe gastric mucosal injury in several hemorrhagic areas in the glandular part of the stomach. Administration of both ELJ extracts at 150, 300, and 600 mg/kg and cimetidine at 100 mg/kg appeared to reduce stomach lesions compared to the induction group. In comparison to the EtOH/HCl-acid-induced gastric lesions model, the indomethacin and restraint water immersion stress models produced fewer spots and less damage to the stomach glandular mucosa. In both models, stomach mucosal damage was reduced after treatment with either ELJ extract or cimetidine. In addition, there was no evidence of harm to the stomach in the group that received only ELJ extract.

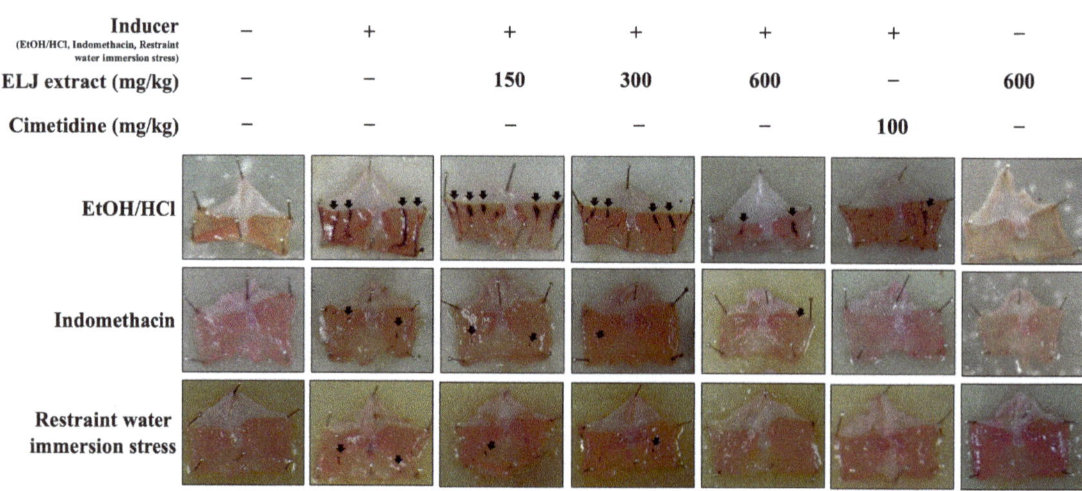

Figure 5. Evaluation of the effect of *Eurycoma longifolia* Jack (ELJ) extract at 150, 300, and 600 mg/kg and cimetidine at 100 mg/kg on the gastric surface in rats, comparing three different models of gastric ulcers to the control group. The black arrows represent the typical necrotic bands or spots and small erosions that can develop into gastric ulcers.

3.7.1. Effect of ELJ on the EtOH/HCl-Induced Gastric Ulcer in Rats

In the EtOH/HCl-acid-induced gastric lesions model, ELJ extract at 150, 300, and 600 mg/kg demonstrated gastric ulcer inhibition levels of 55, 60, and 62%, respectively (Supplementary Table S3). The ulcer indexes were dramatically reduced compared to the control group after treatment with 150, 300, and 600 mg/kg of ELJ extract (Figure 6). Furthermore, ELJ extract had identical ulcer indexes and inhibition percentages at higher doses of 300 and 600 mg/kg. Additionally, cimetidine was found to have a more pronounced effect on reducing gastric ulcers than did ELJ extract, suggesting that treatments with ELJ extract may help reduce and prevent gastric lesions induced by EtOH/HCl.

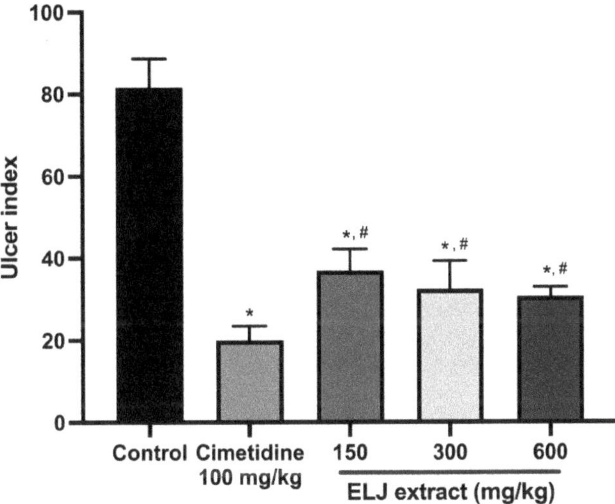

Figure 6. Effects of ELJ extract, control, and cimetidine in a rat model of EtOH/HCl-acid-induced gastric lesions. Ulcer indexes are mean $\pm$ S.E.M. ($n = 6$). * Significantly different from the control group, $p < 0.05$; # Significantly different from the cimetidine group, $p < 0.05$.

3.7.2. Effect of ELJ and Cimetidine on Indomethacin-Induced Gastric Ulcer in Rats

With an ulcer index of 9.30 ± 0.59 mm, the indomethacin-induced damage to the gastric glandular mucosa of the control group was evident (Supplementary Table S4). To the contrary, at all doses, rats treated with either cimetidine or ELJ exhibited dramatically reduced gastric lesions. At a dose of 600 mg/kg, ELJ inhibited gastric ulcers by a maximum of 93%, whereas cimetidine inhibited ulcers by 95% (Figure 7).

3.7.3. Effect of ELJ on Restrained Water Immersion Stress-Induced Gastric Ulcer in Rats

ELJ extract at doses of 150, 300, and 600 mg/kg and cimetidine at a dose of 100 mg/kg significantly reduced gastric ulcer formation induced by restrained water immersion stress in the rats (Figure 8). The percentage of inhibition from cimetidine at a dose of 100 mg/kg and ELJ extract at doses of 150, 300, and 600 mg/kg were 95, 71, 89, and 89% in the restrained water immersion stress-induced gastric lesions model (Supplementary Table S5). These findings suggest that ELJ extract might have the ability to reduce the severity of chemically and physiologically induced gastric ulcers.

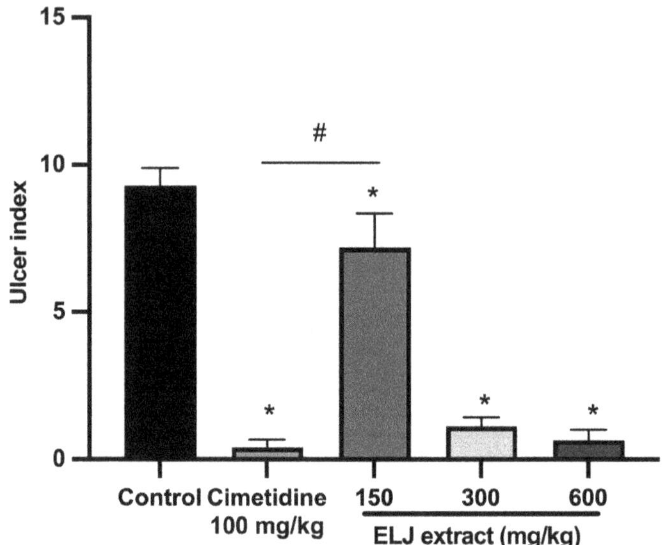

Figure 7. Effect of ELJ and cimetidine on indomethacin-induced gastric ulcers in rats. Ulcer indexes are mean $\pm$ S.E.M. (n = 6). * Significantly different from the control group, $p < 0.05$; # Significantly different from the cimetidine group, $p < 0.05$.

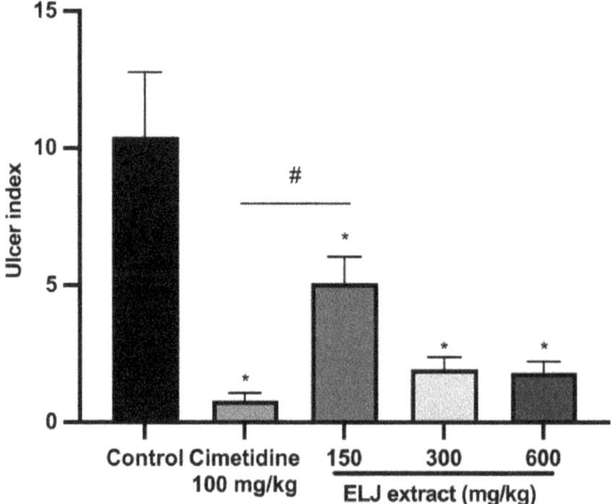

Figure 8. Effects of ELJ extract, control, and cimetidine in a rat model of the restrained water immersion stress-induced gastric lesions. Ulcer indexes are mean $\pm$ S.E.M. (n = 6). * Significantly different from the control group, $p < 0.05$; # Significantly different from the cimetidine group, $p < 0.05$.

3.8. Pylorus Ligation Model

None of the doses of ELJ extract reduced the gastric acid's secretory rate or total acidity, or increased the intragastric pH in the pylorus ligation rat model (Table 6). Conversely, the secretory rate and total acidity in rats receiving 100 mg/kg of cimetidine decreased significantly compared to the controls.

Table 6. Effects of ELJ extract, control, and cimetidine in the pylorus ligation rat model.

Group	Gastric Volume (mL)	Secretory Rate (mL/100 g/5 h)	pH	Total Acidity (mEq/100 g/5 h)
Control	3.25 ± 0.35	1.54 ± 0.19	1.95 ± 0.32	29.20 ± 2.65
Cimetidine 100 mg/kg	2.37 ± 0.16	0.87 ± 0.05 *	5.06 ± 0.79 *	13.30 ± 2.35 *
ELJ extract 150 mg/kg	3.43 ± 0.39	1.18 ± 0.17	1.87 ± 0.10	27.02 ± 4.83
ELJ extract 300 mg/kg	3.05 ± 0.43	1.07 ± 0.13	2.31 ± 0.28	27.42 ± 3.25
ELJ extract 600 mg/kg	3.38 ± 0.24	1.26 ± 0.15	2.03 ± 0.42	34.76 ± 5.61

All values are mean ± S.E.M. (n = 6). * Significantly different from the control group, $p < 0.05$.

4. Discussion

Presently, medical herbs are commonly used in the context of food additives for health maintenance, and plant products have also been increasingly found in the market, because herbs contain several natural materials, including phenolic acids and flavonoids, that are used in the treatment of diseases [36,37]. *E. longifolia* leaf decoctions are used to wash away itches, while its root bark is used to treat gastric ulcers, diarrhea, and fever. Additionally, its roots are utilized as an appetite stimulant and nutritional supplement [38]. Numerous investigations on the bioactivity of this plant have been conducted; however, the anti-inflammatory, antinociceptive, and antipyretic activities of ethanolic *E. longifolia* Jack (ELJ) and its protective efficacy against gastric ulcers have not been investigated. The present study focuses on ELJ's antiulcer potential in animal models.

Apart from extraction, ethanol is commonly used to screen for phytoconstituents in plant extracts because of the ease with which it penetrates the cellular membrane, thus gaining access to intracellular targets [39]. Additionally, plants used in ethanolic extraction contain phytochemicals such as phenolic acids and flavonoids, which may be directly comparable to other solvents (e.g., hexane, ethyl acetate, or acetone), and ethanol can be used effectively to treat industrial waste [40]. Extraction methods increase composite amounts depending on the solvent used and the extraction process [41]. In 2010, Huang and Chang reported that ethanol may offer a higher yield [42]; in this study, we therefore chose to use ethanol for extraction.

From the chemical pattern of the effect of ELJ extract on TLC and HPLC, which was found in this study, we now qualify the efficacy of herbs using a control model approach following the Thai Herbal Pharmacopoeia (THP) formula [18] to generate the monograph of this plant. TLC studies have demonstrated that our ELJ extract contains scopoletin, while phytochemical screening revealed the presence of phenolics, tannins, flavonoids, and terpenes [43]. In addition to promoting wound healing, these compounds have been demonstrated as displaying antioxidant properties. In fact, the antioxidant activity of ELJ extract is positively correlated with flavonoid contents [44]. TLC fingerprints of the ELJ extract (Figure 1) were compared with the reference markers, such as scopoletin. Compared to eurycomalectone, the chromatographic pattern of the ELJ extract was noticeably weaker. Thus, eurycomalectone in ELJ extract contained fewer chemical elements than did the other samples, indicating that it was less effective. This discrepancy could have been caused by differences in the environmental conditions under which the plants were grown, or by the age of the plants when they were picked, among other factors [45]. The scopoletin TLC pattern resembles the ELJ extract pattern which we have observed in this study (Figure 1). Along with the first two components of ELJ (scopoletin and eurycomalectone), quassinoids are another crucial component that has been shown to protect against gastric ulcers in rats with pylorus ligation and indomethacin-induced gastric lesions [46]. The fractionation approach should be investigated further to study phytochemicals and active substances. In addition, other quassinoids, such as eurycolactone A, B, D, E, and F, process anti-inflammatory properties, while both eurycolactone and eurycomanol regulate signaling pathways involved in cell development, cell death, and inflammation [47,48].

In a previous study of an aqueous extract of *Eurycoma longifolia* Jack (ELJ) on reproductive functions in the rat [49], the effective dose range for ELJ was 200–800 mg/kg of body weight. This range of ELJ extracted by methanol or water has been used in various investigations [50,51]. The gastroprotective efficacy of ELJ extract administered at a dose of 100 mg/kg, compared to that of the standard drug cimetidine (100 mg/kg), for the prevention of development of acute gastric ulcers was investigated in restrained rats [52]. Based on that study, a dose range of 150–600 mg/kg body weight was selected as a safe and efficacious concentration for the ELJ administration to rats in this study. Due to its safety, the European Food Safety Authority (EFSA) classifies ELJ as a new food in Europe, and a 12-month chronic toxicity effect in rats at 250, 500, and 1000 mg/kg bodyweight per day revealed no toxicity for ELJ extract [53]. Moreover, Thailand Traditional Medicine has used it for years to treat malaria [14,16]. These results indicate that ELJ extract is safe for both animals and humans. Using various animal models, this study could demonstrate, for the first time, the anti-antinociceptive of *Eurycoma longifolia* Jack, but not the anti-inflammatory and analgesic activities.

ELJ has been a procedure widely used to screen and assess the anti-inflammatory activity of test compounds using the EPP-induced ear edema mode [22]. EPP triggers the release of pro-inflammatory mediators, including histamine, serotonin, kinins, and prostaglandins, resulting in an acute inflammatory response that causes vascular alterations, including vasodilation and an increase in vascular permeability, resulting in the creation of ear edema [54,55]. This study used phenylbutazone as a reference drug to inhibit cyclooxygenase (COX) [56]. Those pro-inflammatory mediators, as well as both phenylbutazone and ELJ extracts, effectively reduced EPP-induced ear edema. This indicates that the potential mechanism of action of ELJ extracts may involve the suppression of the production of these pro-inflammatory mediators.

Injecting carrageenan into the plantar surface of the hind paw induces an initial inflammatory reaction that results in a biphasic phase of paw edema, and it is a well-established model for testing the anti-inflammatory efficacy of various treatments [57,58]. During the first phase following carrageenan injection (0–2.5 h), kinins, serotonin, and histamine are released. The second phase (2.5–6 h) is characterized by the release of bradykinin, NO, and PGs, which are associated with the elevation of COX-2 and nitric oxide synthase (iNOS) activities, as well as oxygen-derived free radicals, in addition to neutrophil infiltration and activation in the injured areas [58,59]. For our current study, only an oral treatment with a high dose of either ELJ (1200 mg/kg) or aspirin (300 mg/kg) significantly reduced the production of carrageenan-induced rat paw edema during all measurement periods, demonstrating the recipe's ability to modify the intensity of inflammation.

Cotton-pellet-induced granuloma formation, whose responses can be classified into three stages, is commonly used to examine the anti-inflammatory effects of test compounds on chronic inflammation [57]. The initial or transudative phase, which occurs 0–3 h following cotton pellet implantation, is characterized by fluid leaking from blood vessels due to increased vascular permeability. The second or exudative phase occurs 3–72 h after cotton pellet implantation, and is characterized by protein leaking from the bloodstream around the granuloma due to the significant increase in vascular permeability. The final or proliferative phase, lasting 3–6 days, is characterized by the development of granulomatous tissues resulting from the continual release of pro-inflammatory mediators [24,60]. According to Swingle and Shideman (1972), the weight of the fluid in the granuloma is correlated with the transudative component. In contrast, the weight of the dry pellet is associated with the amount of granuloma tissue deposited. In terms of therapy, steroids can decrease both the transudative and proliferative stages significantly, whereas NSAIDs, such as aspirin, can reduce these effects only moderately [61]. The present study showed that, although ELJ extract was unable to reduce transudative weight or suppress granuloma formation effectively, these effects were more pronounced when compared to aspirin but less pronounced when compared to prednisolone. Furthermore, the thymus' weight and overall weight gain were unaffected by the ELJ extract. Thus, it appears improbable that ELJ extract has a steroid-like action. Taken together, these data suggest that ELJ extract can reduce chronic inflammation with an effect superior to that of aspirin but inferior to that of prednisolone. However, the mechanical effects of the ELJ extract need to be further explored.

The formalin test is a suitable and exhaustive model for testing the antinociceptive efficacy of pharmaceuticals. The early neurogenic pain phase is caused by the direct stimulation of nociceptors, whereas the late inflammatory pain phase is caused by the release of pro-inflammatory mediators [62]. In this investigation, all ELJ extract treatments significantly inhibited paw licking responses in both phases (Table 4). This finding indicated that ELJ extracts have analgesic efficacy through reducing early neurogenic pain and, in particular, late inflammatory pain. The percentage inhibition of licking response in the late phase was higher than in the early phase for all treatments, with comparable effects to aspirin, possibly due to decreased production or preferred inhibition of pro-inflammatory mediators such as prostaglandin, adenosine, and histamine in peripheral tissue.

It is generally believed that yeast-induced hyperthermia is caused by the activation of endogenous pyrogen and the consequent synthesis of pro-inflammatory mediators, which then act on the hypothalamus and drive prostaglandin E2 (PGE2) synthesis in the preoptic area of the anterior hypothalamus thermoregulatory centers [63,64]. The inhibition of prostaglandin synthesis could be a mechanism of antipyretic action comparable to that of acetaminophen and aspirin [65], and the inhibition of prostaglandin can be achieved by decreasing the activity of the cyclooxygenase enzyme. Several mediators contribute to fever, and their suppression is responsible for the antipyretic effect [66]. Only a high dose of ELJ extract (1200 mg/kg) effectively reduced the rectal temperature of yeast-infected rats. Thus, it can be hypothesized that the ELJ extract contained pharmacologically active components that could inhibit prostaglandin release. This result led to the hypothesis that ELJ extract could exert central antipyretic effects.

Since the undesirable gastrointestinal side effects of existing anti-inflammatory medications are major downsides, patients and physicians will welcome new anti-inflammatory drugs that could help avoid these problems, such as ELJ, an extract derived from a flowering plant *Eurycoma longifolia* Jack (*Tongkat ali*) of the family Simaroubaceae, which is native to Indonesia, Myanmar, Laos, and Thailand [6]. Most NSAIDs used to treat inflammation and pain also result in stomach ulcers [2]. We administered ELJ extract to rats for fourteen days, after first investigating its impact on gastric ulcers in rats. It was found that EJL extract could not be the cause of gastric ulcers. Subsequently, rats administered 150 to 600 mg/kg of ELJ extract were examined to determine whether it could prevent gastrointestinal tract ulcers.

Acidified ethanol can be administered orally, and is known to cause intracellular oxidative stress in the gastric mucosa, as well as disrupt gastric mucus, which can perturb superficial epithelial mucosa, cause cellular necrosis, and cause a gastric ulcer [67]. Moreover, the EtOH/HCl model showed that hydrochloric acid induces gastric ulcer formation via direct irritation of the gastric mucosa, while ethanol decreases gastric mucous production, endogenous glutathione, mucosal blood flow, and prostaglandins [68,69]. The results of this study show that rats given ELJ extract at doses of 150, 300, and 600 mg/kg had considerably lower ulcer indexes than did the controls. According to previous research, an aqueous extract of *E. longifolia* includes significant amounts of superoxide dismutase (SOD) and hydroxyl radical scavenging capabilities. SOD is an antioxidant that plays a crucial role in preventing oxidative stress. Thus, it is probable that the antioxidative impact of ELJ extract contributes to the reduction of acidified-alcohol-induced gastric ulcer formation. These findings are comparable with studies of the cimetidine group, indicating that ELJ extract exhibits gastroprotective activity. However, the mechanism of ELJ extract's antioxidant activity in relation to the EtOH/HCl-induced gastric ulcer model should be explored for further.

The use of NSAIDs such as indomethacin typically causes gastric ulcers by disrupting the natural gastric mucosal barrier of hydrophobic mucous and bicarbonates. Inhibition of the cyclooxygenase-1 (COX-1) enzyme disturbs cytoprotective prostaglandin synthesis [70]. Additionally, indomethacin can cause direct damage to the mucosa by uncoupling the mitochondrial oxidative phosphorylation, which leads to an increase in the formation of reactive oxygen species (ROS) [71]. The anti-ulcerogenic effects were confirmed in the indomethacin-induced gastric lesion model. At different dosages of ELJ extract, the ulcer indexes were statistically significantly lower than those of the controls. At a high dose of ELJ extract (600 mg/kg), the ulcer index of treated rats was non-inferior to the 100 mg/kg group of cimetidine-treated rats.

When experimental rats are exposed to cold water, stress can induce vagal overactivity, thereby increasing gastric acid secretion, while decreasing gastric mucosal blood flow, leading to gastric ulcer formation [72]. As demonstrated in the restrained water immersion stress-induced gastric lesion model, the ulcer indexes of the ELJ-treated groups were significantly lower than that of the non-ELJ-treated group and were almost the same as that of the cimetidine-treated group.

In addition to the EtOH/HCl and indomethacin-induced peptic ulcer models, macroscopic and microscopic characteristics are critical for elucidating cellular mechanisms. Both models give total scores generated from similar partial evaluation scores at the macroscopic level (size, number, and site of hemorrhagic lesions). When anti-ulcerative treatments were studied macroscopically, gastric lesions were decreased, most notably with cimetidine and ELJ extract at a high doses (600 mg/kg). In both animal models, anti-ulcerative medicines significantly reduced the depth of mucosal erosion [73,74]. However, the microscopic evaluation score, which includes erosion depth, the severity of hemorrhage, inflammation, and apoptosis, should be investigated further to identify the possible mechanisms of action of the ELJ extract.

Pyloric ligation is a surgical model used to determine anti-secretory activity. Ligating the pylorus causes intraluminal HCl accumulation, which aggressively damages the gastric mucosa [75]. In this study, gastric acid output was examined in all groups, but only the animals in the cimetidine group showed a significant reduction of gastric acid secretory rates and total acidity. ELJ extract did not decrease the volume of gastric acid or increase the intragastric pH. Scopoletin, a purified compound found in ELJ extract, has been shown to have a preventative effect against the development of esophagitis [76]. In the present study, however, it is probable that the concentration of scopoletin in ELJ extract at 12.590 g/mL was insufficient to suppress gastric acid secretion and total acidity (Table 6). Moreover, obtaining enough scopoletin to use in each experiment was problematic, because the amount of ELJ extract available was limited. Cimetidine is an antagonist of histamine-2 receptors (HH2R) that controls gastric acid secretion by parietal cells. Thus, the results indicate that the anti-ulcerogenic activity of ELJ extract is not related to its anti-secretory activity.

A previous study of a product containing ELJ as a primary constituent on gastroprotective effects and gastrointestinal antioxidants discovered that this product could prevent gastric ulcers in the EtOH-induced gastric lesion model, but could not reduce MDA (malondialdehyde) levels or gastric acid [17]. As previous research on ELJ has been inconclusive, additional study of its protective efficacy against gastric ulcers is needed. The present study investigated the gastroprotective effects of ELJ using three types of gastric ulcers induced in rats by means of acidified ethanol, indomethacin, and restraint water immersion, respectively. In all gastric ulcer models, ELJ extract at concentrations of 150, 300, and 600 mg/kg effectively prevented gastric ulcer formation. The efficacy of ELJ extract at 600 mg/kg was roughly comparable to that of cimetidine at 100 mg/kg, accounting for 62%, 93%, and 82% in the EtOH/HCl-induced, indomethacin-induced, and restrained water immersion stress-induced gastric lesions models, respectively. In contrast, the efficacy in the cimetidine-treated groups was superior to that of ELJ at 75%, 95%, and 92%. In the pyloric ligation model, however, the effects of ELJ extract on gastric volume, gastric pH, and total acidity were not statistically significant, indicating that ELJ extract has potent anti-gastric properties, but has no effect on gastric pH or acidity.

This non-clinical investigation was conducted to examine the gastroprotective effects (i.e., inflammation, fever, and pain) of ELJ extract in vivo using animal models. Some clinical studies have shown that ELJ extract is safe to use, and may even be effective in the induction of immunomodulation by ingestion of capsules containing 200 mg of a standardized *E. longifolia* water-soluble TA (Tongkat Ali) root extract, without inducing toxicity in blood parameters or biochemical analyses [77]. Even though there have been only a few clinical studies on ELJ extract, it is anticipated that the results of this study could lead to future clinical trials to investigate the action of ELJ extract and establish possible mechanisms for its gastric-ulcer protection ability.

5. Conclusions

The present study demonstrates that ethanolic extract of *Eurycoma longifolia* Jack extract (ELJ) exerts dose-dependent anti-inflammatory and antinociceptive action against formalin-induced nociception, whereas only a high dose of ELJ extract has antipyretic activity. This study provides scientific evidence, in the preclinical phase, of the pharmacological activity of *E. longifolia* extract in its anti-inflammatory, antinociceptive, antipyretic, and gastroprotective activities and supports its use, without causing gastric ulcers, in gastric ulcer treatment. For these reasons, ELJ extract should be regarded as having advantages over some NSAIDS. Prior to promotion of routine clinical use, additional preclinical studies of its attributes, e.g., short- and long-term pharmacological activities and toxicity, are needed to evaluate the safety ELJ extract.

Supplementary Materials: The following supporting information can be downloaded at: https://www.mdpi.com/article/10.3390/life13071465/s1, Table S1: Effects of ELJ extract and phenylbutazone on EPP-induced ear edema in rats.; Table S2: Effects of ELJ extracts and aspirin on carrageenan-induced hind paw edema in rats.; Table S3: Effects of ELJ extract, control, and cimetidine in a rat model of EtOH/HCl-acid-induced gastric lesions.; Table S4: Effect of ELJ and cimetidine on indomethacin-induced gastric ulcer in rat.; Table S5: Effects of ELJ extract, control, and cimetidine in a rat model of the restrained water immersion stress-induced gastric lesions.

Author Contributions: Conceptualization, S.S. (Seewaboon Sireeratawong); methodology, S.S. (Subhawat Subhawa), K.J., and S.C.; resources and validation, N.S.; investigations, S.S. (Subhawat Subhawa), W.A., and K.J.; writing—original draft preparation, S.S. (Subhawat Subhawa) and K.J.; writing—review and editing, S.S. (Seewaboon Sireeratawong), S.S. (Subhawat Subhawa), W.A., K.J., and S.C.; funding acquisition, S.S. (Seewaboon Sireeratawong). All authors have read and agreed to the published version of the manuscript.

Funding: The research was sponsored by the National Research Council of Thailand (NRCT) (grant number 016/2552).

Institutional Review Board Statement: The animal study protocol was approved by the Research Ethics Committee for Animal Studies (Study code: 0003/2008), Faculty of Medicine, Thammasat University, Pathum Thani, Thailand.

Informed Consent Statement: Not applicable.

Data Availability Statement: Not applicable.

Acknowledgments: The authors wish to express our sincere appreciation to Lamar Robert for his assistance in proofreading the manuscript.

Conflicts of Interest: The authors declare no conflict of interest.

References

1. Schmid-Schönbein, G.W. Analysis of inflammation. *Annu. Rev. Biomed. Eng.* **2006**, *8*, 93–131. [CrossRef]
2. Sostres, C.; Gargallo, C.J.; Arroyo, M.T.; Lanas, A. Adverse effects of non-steroidal anti-inflammatory drugs (NSAIDs, aspirin and coxibs) on upper gastrointestinal tract. *Best Pract. Res. Clin. Gastroenterol.* **2010**, *24*, 121–132. [CrossRef]
3. Cui, J.; Jia, J. Natural COX-2 Inhibitors as Promising Anti-inflammatory Agents: An Update. *Curr. Med. Chem.* **2021**, *28*, 3622–3646. [CrossRef]
4. Marinaccio, L.; Zengin, G.; Pieretti, S.; Minosi, P.; Szucs, E.; Benyhe, S.; Novellino, E.; Masci, D.; Stefanucci, A.; Mollica, A. Food-inspired peptides from spinach Rubisco endowed with antioxidant, antinociceptive and anti-inflammatory properties. *Food Chem.* **2023**, *18*, 100640. [CrossRef]

5. Mollica, A.; Zengin, G.; Sinan, K.I.; Marletta, M.; Pieretti, S.; Stefanucci, A.; Etienne, O.K.; Jekő, J.; Cziáky, Z.; Bahadori, M.B.; et al. A Study on Chemical Characterization and Biological Abilities of Alstonia boonei Extracts Obtained by Different Techniques. *Antioxidants* **2022**, *11*, 2171. [CrossRef]
6. Rehman, S.U.; Choe, K.; Yoo, H.H. Review on a Traditional Herbal Medicine, Eurycoma longifolia Jack (Tongkat Ali): Its Traditional Uses, Chemistry, Evidence-Based Pharmacology and Toxicology. *Molecules* **2016**, *21*, 331. [CrossRef]
7. Kuo, P.-C.; Damu, A.G.; Lee, K.-H.; Wu, T.-S. Cytotoxic and antimalarial constituents from the roots of Eurycoma longifolia. *Bioorg. Med. Chem.* **2004**, *12*, 537–544. [CrossRef]
8. Herbal Medicine Reference Textbook Preparation Subcommittee. Monograph of selected Thai materia medica: PLA LAI PUEAK-RAK. *J. Thai Trad. Alt. Med.* **2018**, *16*, 158–162.
9. Office of Resource and Information Technology Kamphaeng Phet Rajabhat University. PLA LAI PUEAK. Available online: https://arit.kpru.ac.th/ap2/local/?nu=pages&page_id=1660&code_db=610010&code_type=01 (accessed on 29 December 2022).
10. Bedir, E.; Abou-Gazar, H.; Ngwendson, J.N.; Khan, I.A. Eurycomaoside: A new quassinoid-type glycoside from the roots of Eurycoma longifolia. *Chem. Pharm. Bull.* **2003**, *51*, 1301–1303. [CrossRef]
11. Guo, Z.; Vangapandu, S.; Sindelar, R.W.; Walker, L.A.; Sindelar, R.D. Biologically active quassinoids and their chemistry: Potential leads for drug design. *Curr. Med. Chem.* **2005**, *12*, 173–190. [CrossRef]
12. Ang, H.H.; Cheang, H.S.; Yusof, A.P.M. Effects of Eurycoma longifolia Jack (Tongkat Ali) on the initiation of sexual performance of inexperienced castrated male rats. *Exp. Anim.* **2000**, *49*, 35–38. [CrossRef]
13. Low, B.S.; Choi, S.B.; Abdul Wahab, H.; Das, P.K.; Chan, K.L. Eurycomanone, the major quassinoid in Eurycoma longifolia root extract increases spermatogenesis by inhibiting the activity of phosphodiesterase and aromatase in steroidogenesis. *J Ethnopharmacol.* **2013**, *149*, 201–207. [CrossRef]
14. Bhat, R.; Karim, A.A. Tongkat Ali (Eurycoma longifolia Jack): A review on its ethnobotany and pharmacological importance. *Fitoterapia* **2010**, *81*, 669–679. [CrossRef]
15. Han, Y.M.; Woo, S.U.; Choi, M.S.; Park, Y.N.; Kim, S.H.; Yim, H.; Yoo, H.H. Antiinflammatory and analgesic effects of Eurycoma longifolia extracts. *Arch. Pharm. Res.* **2016**, *39*, 421–428. [CrossRef]
16. Sireeratawong, S.; Khonsung, P.; Piyabhan, P.; Nanna, U.; Soonthornchareonnon, N.; Jaijoy, K. Anti-inflammatory and anti-ulcerogenic activities of Chantaleela recipe. *Afr. J. Tradit. Complement. Altern. Med.* **2012**, *9*, 485–494. [CrossRef]
17. Qodriyah, H.; Asmadi, A. Eurycoma longifolia in Radix for the treatment of ethanol-induced gastric lesion in rats. *Pak. J. Biol. Sci.* **2013**, *16*, 1815–1818. [CrossRef]
18. Bureau of Drug and Narcotic, Department of Medicine Sciences, Ministry of Public Health. *Thai Herbal Pharmacopoeia Volume IV*; Department of Medical Sciences: Bangkok, Thailand, 2014.
19. Tung, N.H.; Uto, T.; Hai, N.T.; Li, G.; Shoyama, Y. Quassinoids from the root of Eurycoma longifolia and their antiproliferative activity on human cancer cell lines. *Pharmacogn. Mag.* **2017**, *13*, 459.
20. Bao, Y.; Ji, W.H.; Ma, Y.H.; Ji, L.J. Simultaneous determination of six main constituents in Swertia of Qinghai Province and Sichuan Province by HPLC. *Zhongguo Zhong Yao Za Zhi* **2006**, *31*, 2036–2038.
21. Nie, Y.; Lin, P. Determination of five active constituents in aerial part of Tibetan medicine Gentiana straminea by HPLC. *Zhongguo Zhong Yao Za Zhi* **2010**, *35*, 1276–1279. [CrossRef]
22. Brattsand, R.; Thalén, A.; Roempke, K.; Källström, L.; Gruvstad, E. Influence of 16 alpha, 17 alpha-acetal substitution and steroid nucleus fluorination on the topical to systemic activity ratio of glucocorticoids. *J. Steroid. Biochem.* **1982**, *16*, 779–786. [CrossRef]
23. Winter, C.A.; Risley, E.A.; Nuss, G.W. Carrageenin-induced edema in hind paw of the rat as an assay for antiiflammatory drugs. *Proc. Soc. Exp. Biol. Med.* **1962**, *111*, 544–547. [CrossRef]
24. Swingle, K.F.; Shideman, F.E. Phases of the inflammatory response to subcutaneous implantation of a cotton pellet and their modification by certain anti-inflammatory agents. *J. Pharmacol. Exp. Ther.* **1972**, *183*, 226–234.
25. Hunskaar, S.; Hole, K. The formalin test in mice: Dissociation between inflammatory and non-inflammatory pain. *Pain* **1987**, *30*, 103–114. [CrossRef]
26. Teotino, U.M.; Friz, L.P.; Gandini, A.; Dellabella, D. Thio Derivatives of 2,3-Dihydro-4h-1,3-Benzoxazin-4-One. Synthesis And Pharmacological Properties. *J. Med. Chem.* **1963**, *6*, 248–250. [CrossRef]
27. Mizui, T.; Doteuchi, M. Effect of polyamines on acidified ethanol-induced gastric lesions in rats. *Jpn. J. Clin. Pharmacol. Ther.* **1983**, *33*, 939–945. [CrossRef]
28. Djahanguiri, B. The production of acute gastric ulceration by indomethacin in the rats. *Scand J. Gastroenterol.* **1969**, *4*, 265–267.
29. Hayden, L.J.; Thomas, G.; West, G. Inhibitors of gastric lesions in the rat. *J. Pharm. Pharmacol.* **1978**, *30*, 244–246. [CrossRef]
30. Kim, J.-H.; Kim, B.-W.; Kwon, H.-J.; Nam, S.-W. Curative effect of selenium against indomethacin-induced gastric ulcers in rats. *J. Microbiol. Biotechnol.* **2011**, *21*, 400–404. [CrossRef]
31. Takagi, K.; Kasuya, Y.; Watanabe, K. Studies on the Drugs for Peptic Ulcer. A Reliable Method for Producing Stress Ulcer in Rats. *Chem. Pharm. Bull.* **1964**, *12*, 465–472. [CrossRef]
32. Lu, C.-L.; Li, Z.-P.; Zhu, J.-P.; Zhao, D.-Q.; Ai, H.-B. Studies on functional connections between the supraoptic nucleus and the stomach in rats. *J. Physiol. Sci.* **2011**, *61*, 191–199. [CrossRef]
33. SHAY, H. A simple method for the uniform production of gastric ulceration in the rats. *Gastroenterology* **1945**, *5*, 143–149.

34. Chaingam, J.; Juengwatanatrakul, T.; Yusakul, G.; Kanchanapoom, T.; Putalun, W. HPLC-UV-Based Simultaneous Determination of Canthin-6-One Alkaloids, Quassinoids, and Scopoletin: The Active Ingredients in Eurycoma Longifolia Jack and Eurycoma Harmandiana Pierre, and Their Anti-Inflammatory Activities. *J. AOAC Int.* **2021**, *104*, 802–810. [CrossRef]

35. Miyake, K.; Tezuka, Y.; Awale, S.; Li, F.; Kadota, S. Quassinoids from Eurycoma longifolia. *J. Nat. Prod.* **2009**, *72*, 2135–2140. [CrossRef]

36. George, V.C.; Dellaire, G.; Rupasinghe, H.V. Plant flavonoids in cancer chemoprevention: Role in genome stability. *J. Nutr. Biochem.* **2017**, *45*, 1–14. [CrossRef]

37. Cicero, A.F.; Allkanjari, O.; Busetto, G.M.; Cai, T.; Larganà, G.; Magri, V.; Perletti, G.; Della Cuna, F.S.R.; Russo, G.I.; Stamatiou, K. Nutraceutical treatment and prevention of benign prostatic hyperplasia and prostate cancer. *Arch. Ital. Urol. Androl.* **2019**, *91*. [CrossRef]

38. Majidi Wizneh, F.; Zaini Asmawi, M. Eurycoma longifolia Jack (*Simarubaceae*); Advances in Its Medicinal Potentials. *Pharmacogn. J.* **2014**, *6*, 1–9. [CrossRef]

39. Pandey, A.; Tripathi, S. Concept of standardization, extraction and pre phytochemical screening strategies for herbal drug. *J Pharmacogn. Phytochem.* **2014**, *2*, 115–119.

40. Dhawan, D.; Gupta, J. Research article comparison of different solvents for phytochemical extraction potential from datura metel plant leaves. *Int. J. Biol. Chem.* **2017**, *11*, 17–22. [CrossRef]

41. Truong, D.H.; Nguyen, D.H.; Ta, N.T.A.; Bui, A.V.; Do, T.H.; Nguyen, H.C. Evaluation of the Use of Different Solvents for Phytochemical Constituents, Antioxidants, and In Vitro Anti-Inflammatory Activities of *Severinia buxifolia*. *J. Food Qual.* **2019**, *2019*, 9. [CrossRef]

42. Huang, Y.-P.; Chang, J.I. Biodiesel production from residual oils recovered from spent bleaching earth. *Rene. Energ.* **2010**, *35*, 269–274. [CrossRef]

43. Hou, W.; Xiao, X.; Guo, W.; Zhang, T. Advances in Studies on Chemistry, Pharmacological Effect, and Pharmacokinetics of Eurycoma longifolia. *Chin. Herb. Med.* **2011**, *3*, 186–195.

44. Varghese, C.P.; Ambrose, C.; Jin, S.; Lim, Y.; Keisaban, T. Antioxidant and anti-inflammatory activity of Eurycoma longifolia Jack, a traditional medicinal plant in Malaysia. *Int. J. Pharm. Sci. Nanotech.* **2013**, *5*, 1875–1878. [CrossRef]

45. Kunle, O.F.; Egharevba, H.O.; Ahmadu, P.O. Standardization of herbal medicines-A review. *Int. J. Biodivers. Conserv.* **2012**, *4*, 101–112. [CrossRef]

46. Tada, H.; Yasuda, F.; Otani, K.; Doteuchi, M.; Ishihara, Y.; Shiro, M. New antiulcer quassinoids from Eurycoma longifolia. *Eur. J. Med. Chem.* **1991**, *26*, 345–349. [CrossRef]

47. Hajjouli, S.; Chateauvieux, S.; Teiten, M.H.; Orlikova, B.; Schumacher, M.; Dicato, M.; Choo, C.Y.; Diederich, M. Eurycomanone and eurycomanol from Eurycoma longifolia Jack as regulators of signaling pathways involved in proliferation, cell death and inflammation. *Molecules* **2014**, *19*, 14649–14666. [CrossRef]

48. Tran, T.V.A.; Malainer, C.; Schwaiger, S.; Atanasov, A.G.; Heiss, E.H.; Dirsch, V.M.; Stuppner, H. NF-κB Inhibitors from Eurycoma longifolia. *J. Nat. Prod.* **2014**, *77*, 483–488. [CrossRef]

49. Solomon, M.; Erasmus, N.; Henkel, R. In vivo effects of Eurycoma longifolia Jack (Tongkat Ali) extract on reproductive functions in the rat. *Andrologia* **2014**, *46*, 339–348. [CrossRef]

50. Ang, H.H.; Cheang, H.S. Effects of Eurycoma longifolia jack on laevator ani muscle in both uncastrated and testosterone-stimulated castrated intact male rats. *Arch. Pharm. Res.* **2001**, *24*, 437–440. [CrossRef]

51. Ang, H.H.; Sim, M.K. Eurycoma longifolia JACK and orientation activities in sexually experienced male rats. *Biol. Pharm. Bull.* **1998**, *21*, 153–155. [CrossRef]

52. Lee, S.P.; Tasman-Jones, C. Prevention of acute gastric ulceration in the rat by cimetidine, a histamine H2-receptor antagonist. *Clin. Exp. Pharmacol. Physiol.* **1978**, *5*, 61–66. [CrossRef]

53. Turck, D.; Bohn, T.; Castenmiller, J.; De Henauw, S.; Hirsch-Ernst, K.I.; Maciuk, A.; Mangelsdorf, I.; McArdle, H.J.; Naska, A.; Pelaez, C.; et al. Safety of Eurycoma longifolia (Tongkat Ali) root extract as a novel food pursuant to Regulation (EU) 2015/2283. *EFSA J.* **2021**, *19*, e06937. [CrossRef]

54. Carlson, R.P.; O'Neill-Davis, L.; Chang, J.; Lewis, A.J. Modulation of mouse ear edema by cyclooxygenase and lipoxygenase inhibitors and other pharmacologic agents. *Agents Actions* **1985**, *17*, 197–204. [CrossRef] [PubMed]

55. Rubin, R.; Strayer, D.S.; Rubin, E. *Rubin's Pathology: Clinicopathologic Foundations of Medicine*; Lippincott Williams & Wilkins: Philadelphia, PA, USA, 2008.

56. Borges, R.S.; Palheta, I.C.; Ota, S.S.B.; Morais, R.B.; Barros, V.A.; Ramos, R.S.; Silva, R.C.; Costa, J.D.S.; Silva, C.; Campos, J.M.; et al. Toward of Safer Phenylbutazone Derivatives by Exploration of Toxicity Mechanism. *Molecules* **2019**, *24*, 143. [CrossRef]

57. Patil, K.R.; Mahajan, U.D.; Unger, B.S.; Goyal, S.N.; Belemkar, S.; Surana, S.J.; Ojha, S.; Patil, C.R. Animal models of inflammation for screening of anti-inflammatory drugs: Implications for the discovery and development of phytopharmaceuticals. *Int. J. Mol. Sci.* **2019**, *20*, 4367. [CrossRef]

58. Di Rosa, M. Biological properties of carrageenan. *J. Pharm. Pharmacol.* **1972**, *24*, 89–102. [CrossRef]

59. Rosa, S.G.; Brüning, C.A.; Pesarico, A.P.; de Souza, A.C.G.; Nogueira, C.W. Anti-inflammatory and antinociceptive effects of 2, 2-dipyridyl diselenide through reduction of inducible nitric oxide synthase, nuclear factor-kappa B and c-Jun N-terminal kinase phosphorylation levels in the mouse spinal cord. *J. Trace. Elem. Med. Biol.* **2018**, *48*, 38–45. [CrossRef] [PubMed]

60. Sarraf, P.; Sneller, M.C. Pathogenesis of Wegener's granulomatosis: Current concepts. *Expert Rev. Mol. Med.* **2005**, *7*, 1–19. [CrossRef] [PubMed]
61. Katzung, B.G.; Trevor, A. Chapter 36: Nonsteroidal Anti-Inflammatory Drugs, Disease-Modifying Antirheumatic Drugs, Nonopioid Analgesics, & Drugs Used in Gout. In *Basic & Clinical Pharmacology*, 9th ed.; McGraw-Hill: New York, NY, USA, 2007.
62. Tjølsen, A.; Berge, O.G.; Hunskaar, S.; Rosland, J.H.; Hole, K. The formalin test: An evaluation of the method. *Pain* **1992**, *51*, 5–17. [CrossRef] [PubMed]
63. Devi, B.P.; Boominathan, R.; Mandal, S.C. Evaluation of antipyretic potential of Cleome viscosa Linn. (Capparidaceae) extract in rats. *J. Ethnopharmacol.* **2003**, *87*, 11–13. [CrossRef]
64. Moltz, H. Fever: Causes and consequences. *Neurosci. Biobehav. Rev.* **1993**, *17*, 237–269. [CrossRef]
65. Lovejoy, F.H., Jr. Aspirin and acetaminophen: A comparative view of their antipyretic and analgesic activity. *Pediatrics* **1978**, *62*, 904–909.
66. Rawlins, M.; Postgrad, R. Mechanism of salicylate-induced antipyresis. In Proceedings of the Pharmacology Thermoregulatory Proceeding Satellite Symposium, San Francisco, CA, USA, 23–28 July 1973; pp. 311–324.
67. Szelenyi, I.; Brune, K. Possible role of oxygen free radicals in ethanol-induced gastric mucosal damage in rats. *Dig. Dis. Sci.* **1988**, *33*, 865–871. [CrossRef] [PubMed]
68. Guslandi, M. Effects of ethanol on the gastric mucosa. *Dig. Dis.* **1987**, *5*, 21–32. [CrossRef] [PubMed]
69. Cho, C.H.; Ogle, C.W. The pharmacological differences and similarities between stress-and ethanol-induced gastric mucosal damage. *Life Sci.* **1992**, *51*, 1833–1842. [CrossRef] [PubMed]
70. Drini, M. Peptic ulcer disease and non-steroidal anti-inflammatory drugs. *Aust. Prescr.* **2017**, *40*, 91. [CrossRef]
71. Wallace, J.L.; McKnight, W.; Reuter, B.K.; Vergnolle, N. NSAID-induced gastric damage in rats: Requirement for inhibition of both cyclooxygenase 1 and 2. *Gastroenterol* **2000**, *119*, 706–714. [CrossRef]
72. Guo, S.; Gao, Q.; Jiao, Q.; Hao, W.; Gao, X.; Cao, J.-M. Gastric mucosal damage in water immersion stress: Mechanism and prevention with GHRP-6. *World. J. Gastroenterol.* **2012**, *18*, 3145. [CrossRef]
73. Simões, S.; Lopes, R.; Campos, M.C.D.; Marruz, M.J.; da Cruz, M.E.M.; Corvo, L. Animal models of acute gastric mucosal injury: Macroscopic and microscopic evaluation. *AMEM* **2019**, *2*, 121–126. [CrossRef]
74. Adinortey, M.B.; Ansah, C.; Galyuon, I.; Nyarko, A. In vivo models used for evaluation of potential antigastroduodenal ulcer agents. *Ulcers* **2013**, *2013*, 796405. [CrossRef]
75. Jayachitra, C.; Jamuna, S.; Ali, M.A.; Paulsamy, S.; Al-Hemaid, F.M. Evaluation of traditional medicinal plant, Cissus setosa Roxb.(*Vitaceae*) for antiulcer property. *Saudi J. Biol. Sci.* **2018**, *25*, 293–297. [CrossRef]
76. Mahattanadul, S.; Ridtitid, W.; Nima, S.; Phdoongsombut, N.; Ratanasuwon, P.; Kasiwong, S. Effects of Morinda citrifolia aqueous fruit extract and its biomarker scopoletin on reflux esophagitis and gastric ulcer in rats. *J. Ethnopharmacol.* **2011**, *134*, 243–250. [CrossRef] [PubMed]
77. George, A.; Suzuki, N.; Abas, A.B.; Mohri, K.; Utsuyama, M.; Hirokawa, K.; Takara, T. Immunomodulation in Middle-Aged Humans Via the Ingestion of Physta® Standardized Root Water Extract of Eurycoma longifolia Jack—A Randomized, Double-Blind, Placebo-Controlled, Parallel Study. *Phytother. Res.* **2016**, *30*, 627–635. [CrossRef] [PubMed]

 *life*

Article

Ocimum sanctum Alters the Lipid Landscape of the Brain Cortex and Plasma to Ameliorate the Effect of Photothrombotic Stroke in a Mouse Model

Inderjeet Yadav [1,2,†], Nupur Sharma [3,†], Rema Velayudhan [1], Zeeshan Fatima [2,4,*] and Jaswinder Singh Maras [3,*]

[1] National Brain Research Centre, Gurugram 122052, India; inderjeet@nbrc.ac.in (I.Y.); rema.velayudhan@gmail.com (R.V.)
[2] Department of Medical Laboratory Sciences, College of Applied Medical Sciences, University of Bisha, Bisha 61922, Saudi Arabia
[3] Department of Molecular and Cellular Medicine, Institute of Liver and Biliary Sciences, New Delhi 110070, India; nupush1995@gmail.com
[4] Amity Institute of Biotechnology, Amity University Haryana, Gurugram 122413, India
*   Correspondence: drzeeshanfatima@gmail.com (Z.F.); jassi2param@gmail.com (J.S.M.)
†   These authors contributed equally to this work.

Citation: Yadav, I.; Sharma, N.; Velayudhan, R.; Fatima, Z.; Maras, J.S. *Ocimum sanctum* Alters the Lipid Landscape of the Brain Cortex and Plasma to Ameliorate the Effect of Photothrombotic Stroke in a Mouse Model. *Life* **2023**, *13*, 1877. https://doi.org/10.3390/life13091877

Academic Editors: Marisa Colone, Balazs Barna, Charalampia Amerikanou and Efstathia Papada

Received: 31 May 2023
Revised: 25 July 2023
Accepted: 14 August 2023
Published: 7 September 2023

Abstract: Stroke-like injuries in the brain result in not only cell death at the site of the injury but also other detrimental structural and molecular changes in regions around the stroke. A stroke-induced alteration in the lipid profile interferes with neuronal functions such as neurotransmission. Preventing these unfavorable changes is important for recovery. *Ocimum sanctum* (Tulsi extract) is known to have anti-inflammatory and neuroprotective properties. It is possible that Tulsi imparts a neuroprotective effect through the lipophilic transfer of active ingredients into the brain. Hence, we examined alterations in the lipid profile in the cerebral cortex as well as the plasma of mice with a photothrombotic-ischemic-stroke-like injury following the administration of a Tulsi extract. It is also possible that the lipids present in the Tulsi extract could contribute to the lipophilic transfer of active ingredients into the brain. Therefore, to identify the major lipid species in the Tulsi extract, we performed metabolomic and untargeted lipidomic analyses on the Tulsi extract. The presence of 39 molecular lipid species was detected in the Tulsi extract. We then examined the effect of a treatment using the Tulsi extract on the untargeted lipidomic profile of the brain and plasma following photothrombotic ischemic stroke in a mouse model. Mice of the C57Bl/6j strain, aged 2–3 months, were randomly divided into four groups: (i) Sham, (ii) Lesion, (iii) Lesion plus Tulsi, and (iv) Lesion plus Ibuprofen. The cerebral cortex of the lesioned hemisphere of the brain and plasma samples were collected for untargeted lipidomic profiling using a Q-Exactive Mass Spectrometer. Our results documented significant alterations in major lipid groups, including PE, PC, neutral glycerolipids, PS, and P-glycerol, in the brain and plasma samples from the photothrombotic stroke mice following their treatment with Tulsi. Upon further comparison between the different study groups of mice, levels of MGDG (36:4), which may assist in recovery, were found to be increased in the brain cortexes of the mice treated with Tulsi when compared to the other groups ($p < 0.05$). Lipid species such as PS, PE, LPG, and PI were commonly altered in the Sham and Lesion plus Tulsi groups. The brain samples from the Sham group were specifically enriched in many species of glycerol lipids and had reduced PE species, while their plasma samples showed altered PE and PS species when compared to the Lesion group. LPC (16:1) was found in the Tulsi extract and was significantly increased in the brains of the PTL-plus-Tulsi-treated group. Our results suggest that the neuroprotective effect of Tulsi on cerebral ischemia may be partially associated with its ability to regulate brain and plasma lipids, and these results may help provide critical insights into therapeutic options for cerebral ischemia or brain lesions.

Keywords: lipidomic profile; Tulsi; *Ocimum sanctum*; ischemic stroke; photothrombotic ischemia; medicinal plant

1. Introduction

Cerebral ischemia is one of the leading causes of all human strokes [1]. It results in neuronal death, primarily in the ischemic region, and neuronal degeneration in the penumbral region [2]. It is the first leading cause of disability and the second leading cause of mortality worldwide, accounting for about 6,000,000 deaths annually [3]. The World Health Organization (WHO) speculates that there will be over 7.8 million stroke-linked deaths per year by 2030 [4]. A stroke occurs due to an inadequate supply of oxygen to parts of the brain, either due to a disrupted supply of blood (ischemic stroke) to some parts of the brain or sometimes due to a sudden breach in a blood vessel in the brain (hemorrhagic stroke). So, if the blood supply in part of the brain is interrupted, it results in neuronal death and neuronal degeneration, due to which symptoms of cerebral ischemia can be seen in the body.

Many published studies have reported a link between dyslipidemia and stroke [5–7]. Dyslipidemia, mostly hyperlipidemia, is an abnormal amount of blood lipids. An increase in blood lipid levels could be a predisposing factor for the pathogenesis of stroke, mainly ischemic stroke. Previous reports in the literature have suggested a strong association between stroke and lipid metabolism [8–11]. Lipid profile assessments in brain anomalies could be helpful as lipids are considered the main constituents of the brain. Lipids comprise 60% of its dry weight and represent the second-most abundant tissue in the brain after adipose tissue [12]. They are essentially required for maintaining the physiological functions of the neurons and also aid in their structural development. Thus, understanding the dynamics of lipid molecules through lipidomics could help interpret altered brain functions in stroke. Although significant advances have been made in understanding stroke pathogenesis in recent years, studies focusing on brain and plasma lipidomics have been limited [13–15].

Ayurvedic medicine is one of the ancient systems of medicine [16] in which many natural and herbal compounds are used to cure human diseases because of their medicinal properties. One such plant, *Ocimum sanctum* (Tulsi), has been very well known as a medicinal herb since ancient times. It has many medicinal properties and has been reported to possess minimal to zero adverse side effects [17,18]. *Ocimum sanctum*, commonly known as holy basil or "Tulsi", is called the "Elixir of Life" for its healing powers and is frequently used as a medicinal agent in the Ayurvedic and Siddha medical systems to ameliorate numerous body ailments [19–21]. Tulsi has been used for decades for its potential to treat a number of diseases, including anxiety, cough, asthma, diarrhea, fever, dysentery, arthritis, eye diseases, otalgia, indigestion, hiccups, vomiting, gastric, cardiac, and genitourinary disorders, back pain, skin diseases, ringworm, insect, snake, and scorpion bites, and malaria [22–25], due to its broad spectrum of pharmacological activities. It has been found to have diverse protective effects, including hepato-protective, immuno-modulatory, anti-ulcer, anti-diabetic, anti-hypercholesterolemic, nerve tonic, chemo-protective, nootropic, antitussive, anti-inflammatory, wound healing, anti-tumorigenesis, anti-convulsant, anthelmintic, anti-bacterial, anti-anxiety, and anti-stress activities [26–30].

However, thus far, the effects of Tulsi treatments on cerebral ischemia are limited [31]. The oral administration of an aqueous extract of Tulsi for 15 days before MCAO demonstrated a marked reduction in infarct size, reduced neurological deficits, and suppressed neuronal loss in MCAO rats [31]. Pretreatment with a methanolic extract of Tulsi for 7 days significantly prevented cerebral-hypoperfusion-induced functional and structural disturbances and was useful in the treatment of cerebral reperfusion injury and cerebrovascular insufficiency states [27]. Tulsi has demonstrated anti-inflammatory effects in animal models of acute and chronic inflammation [32]. Additionally, nanostructured lipid carriers of a Tulsi leaf extract were shown to inhibit both the Cox-1 and Cox-2 enzyme pathways, highlighting the potent anti-inflammatory potential of Tulsi and its compounds [33]. Tulsi was shown to be useful for the management of experimentally-induced cognitive dysfunctions in rats [32]. The effect of Tulsi treatment following ischemia on alterations of the lipid profile of the brain and plasma is not known. Also, there is a lack of information on whether there are changes in the lipid species in the brain with the progression of stroke.

In the current study, lipidomic profiles of the brain and plasma samples of animals that had undergone photothrombotic stroke were evaluated following treatment with Tulsi. Lipid changes in the brain cortex and plasma of mice were analyzed using liquid chromatography coupled with mass spectrometry (LC-MS). Lipid changes associated with a change in stroke pathophysiology due to the healing effect of Tulsi were identified and reported. In addition, the change in brain lipid composition was correlated to that seen in the plasma samples of lesioned animals that were treated with Tulsi. The identified lipids could be used as a clinical indicator of brain recovery and could also be exploited as therapeutic targets.

2. Material and Methods

2.1. Animals

For this study, 24 male mice of the C57Bl/6j strain aged 2–3 months were procured from the animal house of the National Brain Research Centre, Gurgaon, India. All the experimental procedures were duly approved by the Institutional Animal Ethics Committee (IAEC) of the National Brain Research Center. Animals were housed in standard cages with dimensions of 44 × 29 × 16 cm (LXWXH). They were maintained under controlled environmental conditions with a temperature maintained at 22 ± 1 °C, relative humidity between 45 and 55%, a 12:12 h light-dark cycle, and 12–15 air changes per hour as specified in CPCSEA (Committee for the Purpose of Control and Supervision of Experiments on Animals, GOI) guidelines. Pelleted feed procured from Altromin, Germany, and autoclaved water were provided *ad libitum* to all the animals.

2.2. Experimental Groups

The animals were randomly divided into four groups: (i) Sham; (ii) PTL only; (iii) PTL plus Tulsi; (iv) PTL plus Ibu.

Sham animals ($n = 8$) were the controls for the experiment. They underwent all surgical procedures similar to all other animals but were not subjected to Photothrombotic Lesion (PTL). They were anesthetized, and the skin above the skull was incised. The skull was exposed, and subsequently, the skin was sutured. An antibiotic cream, Neosporin, was applied to the wound. Brains and plasma from four animals in this group were used for lipidomic analyses. The brain sections of four animals were used as controls for determining the extent of PTL lesions using cytochrome oxidase staining.

PTL-only animals ($n = 8$) were subjected to PTL. Animals in this group had focal unilateral photothrombotic lesions made by exposing the cortex to laser for 10 min after injecting Rose Bengal red dye (I.V). Four animals from this group were used for the brain and plasma lipidome analyses. Brain sections from four animals were analyzed for cytochrome oxidase to determine the extent of the lesion.

PTL plus Tulsi ($n = 4$) animals were subjected to PTL, and for seven days post-PTL, Tulsi leaf extract was orally administered to them. Tulsi ethanolic extract (Batch Number U/1443/17-18) was purchased from M/S Saiba Industries PVT. Ltd., Mumbai, India. The extract was dissolved in sterile water at a concentration of 60 mg/mL. Each animal was given oral gavage of 500 mg/kg body weight daily. We selected the dosage based on studies showing beneficial effects at this concentration [34–36]. Studies examining the toxicity of *Ocimum sanctum* [37,38] have shown no adverse effects at this concentration.

PTL plus Ibuprofen animals ($n = 4$) were subjected to PTL, and subsequently, for seven days, they were given a 100 mg/kg oral dose of Ibuprofen (Cipla) oral syrup [39,40].

2.3. Photothrombotic Lesion (PTL)

The surgical area was sterilized using 70% alcohol and betadine, and all surgical instruments were sterilized by autoclaving before the surgery. Animals were anesthetized with a mixture of ketamine (100 mg/kg body weight) and xylazine (10 mg/kg body weight) given through the I/P route. The photothrombotic cortical lesion was made in mice's brains as described by Watson et al. (1985) [41] with few modifications. The head of the animal was firmly secured on the stereotaxic apparatus by inserting the ear bars carefully into the external meatus to avoid any damage to the eardrum, and a midline incision was made from eye level down to the neck using a scalpel. Skin retractors were applied to keep the skull exposed. The Bregma and lambda were the landmarks for stereotaxic coordinates. The somatosensory cortex, 0.7 mm posterior from the bregma and 2.8 mm lateral to the midline (2.5 mm diameter), was marked with a marker pen. An area of about 2.5 mm in diameter, which includes a large part of the somatosensory cortex according to the mouse brain atlas by Franklin and Paxinos [42] on the left hemisphere, was marked. A sterilized black plastic paper with a hole of 2.5 mm diameter was placed on the skull such that the hole was above the marking of the somatosensory cortical region while the other regions were covered. The body temperature of the animals was maintained at 37 ± 0.5 °C during the surgery with the help of a heating pad.

Rose Bengal solution 20 mg/kg was given as a slow intravenous injection through the tail vein. All sources of light in the room were turned off. A green laser (532 nm laser irradiation, 50 mW/cm^2) was switched on for 10 min. Following laser exposure, the skin of the head region was sutured. The mouse was removed from the stereotaxic apparatus and placed on a heating pad (pre-warmed) until it became fully awake. It was then returned to its home cage.

2.4. Cytochrome Oxidase Reaction

The four animals with sham control and the four animals with PTL were given saline for seven days post-lesion. On the eighth day, the animals were perfused transcardially with PBS (NaCl (80 gm), KCl (2 gm), Na$_2$HPO$_4$ (11.4 gm), and KH$_2$PO$_4$ (23.2 g) in 900 mL of water and then forming a volume of 1 L, pH-7.4) to clear the blood and then with 4% paraformaldehyde in PBS (PFA) to fix the tissues. The brains were removed and postfixed for 24 h in 10% sucrose in a PFA solution at 4 °C. The brains were cryoprotected by sequentially allowing them to sink in 20% sucrose in PBS, followed by 30% sucrose in PBS at 4 °C. The forebrain was cut into 30 μm thick sections on a sliding microtome. Every sixth section was reactive for cytochrome oxidase. The sections were washed three times with PBS at room temperature and then immersed in freshly prepared cytochrome oxidase staining solution (in 30 mL of 0.1 M phosphate buffer, sucrose (5 g), cytochrome C (Sigma, St. Louis, MO, USA; 25 mg), and DAB (Sigma; 20 mg) were dissolved, and the volume of solution was made up to 50 mL with phosphate buffer). The sections were incubated for 5 h at 37 °C. The sections were then washed twice with PBS, mounted on gelatin-coated glass slides, air dried, and coverslipped with DPX mounting medium. The sections were observed under a light microscope and imaged. Figure 1 shows sections from the brains of sham control animals (Figure 1A) and PTL animals (Figure 1B). In Figure 1B, the cortical region where the photothrombotic lesion was done appears pale, suggesting that cells in the area are dead, as indicated by the absence of cytochrome oxidase activity.

2.5. Sample Collection and Preparation

Twenty-four hours following the last treatment dose, samples were collected for lipidome analyses. Blood (600 μL) from the heart of each animal was collected after cervical dislocation and kept on ice. The animals were then decapitated, and the lesioned hemisphere was rapidly removed and flash-frozen in liquid nitrogen. Blood samples were centrifuged at 3000 rpm for 15 min at 4 °C, and plasma (the supernatant) was collected. Both brain and plasma samples were stored at −80 °C until lipid extraction. Figure 2A shows the schematic diagram of the procedure for untargeted lipidomic analyses of the brain and plasma.

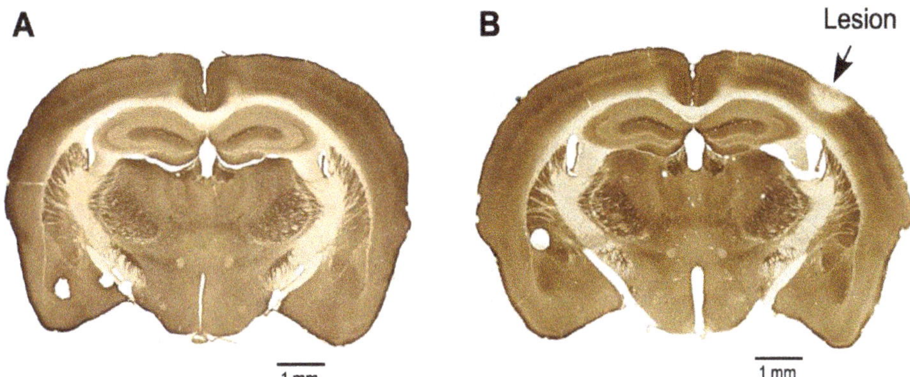

Figure 1. Photomicrograph showing brain sections that were reacted for cytochrome oxidase from (**A**) a sham control mouse and (**B**) a PTL mouse. The lesioned area, indicated by the arrow, shows a region with inactive cells.

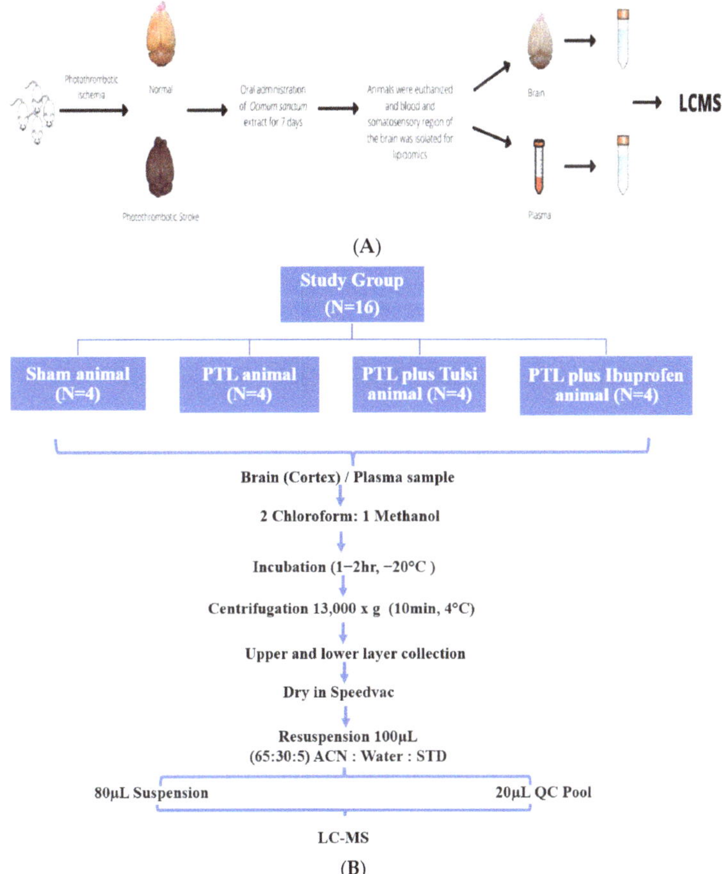

Figure 2. (**A**) The overall scheme for the untargeted lipidomic analysis of brain and plasma. (**B**) Workflow to perform lipidomic analysis.

Brain (cortex) and plasma samples were prepared for lipidomic analysis as described in Sharma N et al., 2022 [43]. Briefly, 100 µL brain/plasma samples were added to a chloroform: methanol (2:1) mixture and homogenized. After that, the samples were centrifuged at 13,000 rpm for 10 min to extract the dissolved lipids (supernatant). Each sample was then vacuum-dried. These dried samples were reconstituted in 100 µL of a 65:30:5:5 standard solution consisting of acetonitrile (65%): isopropanol (30%): water (5%): internal and external standards (5%). Using an ultra-high-performance liquid chromatographic system, they were subjected to reverse-phase chromatography in the C18 column (Thermo Scientific™25003102130: 3 µm, 2.1 mm, 100 mm). Mobile phase A was 0.1% formic acid, and mobile phase B was 100% acetonitrile. The sample processing flow chart is shown in Figure 2B.

2.6. Mass Spectrometry

For performing mass spectroscopy, a 10 µL sample was passed through the column, which was directed into the heated electrospray ionization (HESI) source of a Q-Exactive mass spectrometer (Thermo Scientific, San Jose, CA, USA), and analysis was carried out in both positive and negative ionization modes in two independent runs. The HESI source parameters were as follows: The spray voltage was set to 3.7 kV in positive ionization mode and −3.1 kV in negative ionization mode. The heated capillary temperature was maintained at 360 °C, and the sheath and auxiliary gas flow were set to 15 and 10 (arbitrary units), respectively. All the samples were pooled together and spiked with internal standards (Dinose b: 1 mg/mL; MCPA: 1 mg/mL; Dimetrazole: 1 mg/mL) and external standards (Cholesterol: 0.1 mg/mL; Colchicine: 4 mg/mL; Impramine: 2 mg/mL; Roxithromycin: 2 mg/mL; Amiloride: 1 mg/mL; Atropine: 2 mg/mL; 2-aminoanthroceae: 370.5 mg/mL; Prednisolone: 2 mg/mL) to make dilutions of 1:1, 1:2, 1:4 and 1:8 which were operated as QC for the higher-energy collisional dissociation (HCD) MS/MS experiment. This study used metabolite internal and external standards to attain analytical sensitivity for the MS/MS experiments. In MS/MS mode, the isolation width was set to m/z 1.5, the normalized collision energy was 32%, and the mass resolution was set at 17,500 FWHM at m/z 200.

Mass Spectrometry for Metabolomics of Tulsi Extract

Mass spectrometry for metabolomics of Tulsi extract was performed using a 100 µL sample of plant extract and chilled methanol in a 1:4 ratio (400 µL methanol). The sample was incubated at −20 °C for 10 h or overnight. After incubation, the sample was centrifuged at 13,000 rpm for 10 min, and the supernatant was taken and discarded in the pellet. The supernantant was freeze dried completely, and the dried sample was reconstituted in 100 µL of 90:5:5 reconstitution buffer (90% water, 5% acetonitrile, 5% external standard and internal standard). The sample was run for mass spectrometry as described by Sharma et al. [43].

Composition of internal standards (Dinose b: 5 µg/mL; MCPA (2 methyl-4-chlorophenoxy acetic acid): 5 µg/mL; Dimetrazole: 5 µg/mL, AMPA: 5 µg/mL) and external standards (Dihydrostreptomycin: 20 µg/mL; Colchicine: 0.5 µg/mL; Impramine: 0.5 µg/mL; Roxithromycin: 10 µg/mL; Amiloride: 10 µg/mL; Atropine: 1 µg/mL; 2-aminoanthroceae: 1 µg/mL; Prednisolone: 1 µg/mL; Metformin: 1 µg/mL; Ethylmalonic acid: 3 µg/mL).

2.7. Software Analysis

Lipid features were identified using LipidSearch 4.0 software (Thermo Scientific, San Jose, CA, USA). The feature identification and quantitation parameters used are mentioned in Supplementary File S2.

2.8. Statistical Analysis

Annotated lipid features were subjected to different statistical software platforms. First, missing value imputation was applied to data in which half of the minimum positive value was estimated for lipids that were undetected in the samples. Subsequently, data were filtered based on non-parametric relative standard deviation (MAD/median) and subjected to log normalization and Pareto-scaled using the Metaboanalyst 5.0 server "http://metaboanalyst.ca" (17 October 2022) [44]. Unpaired (two-tail) Student's *t*-test and the Mann–Whitney U test were performed for comparison of two groups. For more than two groups, one-way ANOVA (analysis of variance), and the Kruskal–Wallis test were performed. PCA, PLS-DA, heat map, random-forest analysis, and other statistical analyses were performed. Venn diagram analysis was utilized to understand the correlation between brain and plasma lipid profiles, and *p*-values of <0.05 using Benjamini-Hochberg correction were considered statistically significant.

3. Results

This study examined the effect of oral administration of Tulsi for seven days on the lipid profile of the cerebral cortex of the brain and the plasma of mice with ischemic lesions induced by photothrombosis. Untargeted lipidomic analysis was performed on the cerebral cortex of the brain and also on the blood plasma. We have analyzed changes associated with lipid molecules in the lesioned brain cortex of mice following oral administration of Tulsi for seven days. One of the study's objectives was to determine whether the changes induced by Tulsi were similar to those induced by a known anti-inflammatory drug, Ibuprofen, which was also administered orally for seven days. A comparison of lipid profiles between animal groups with sham, PTL, PTL plus Tulsi, and PTL plus Ibuprofen-treated animals was performed to determine whether treatment with Tulsi could restore the normal lipid profile in lesioned animals.

3.1. Tulsi Modulates Lipidomic Signature in the Lesioned Cortical Hemisphere of Mice with Photothrombotic Ischemic Stroke-like Lesion

Untargeted lipidomic analysis was performed on cerebral cortex samples. The cerebral cortex of the left hemisphere of all the experimental animals was used for lipidomic profiling. Figure 3A illustrates the different types of comparisons done on the lipids of the cerebral cortex of experimental and sham-lesioned animals. Figure 3B demonstrates the results of PLS-DA analysis, highlighting the similarities and differences of the brain lipidome amongst the four study groups, i.e., sham, PTL, PTL plus Tulsi, and PTL plus Ibu.

The score plot with PC1 (29.8%) and PC2 (9.1%) clearly distinguished each group. Compact and distinct clustering was seen for Sham, PTL, PTL plus Ibu, and PTL plus Tulsi. The PTL plus Tulsi-treated mice group was distinctly positioned as compared to other groups; this observation suggested a critical role for Tulsi in the modulation of brain lipid profiles. Hierarchical clustering analysis was performed to better illustrate the differences in the lipid profile of the brain amongst the four groups (Figure 3C). A visual comparison of the lipids of the different groups in the heat map suggested that the PTL plus Tulsi animals have similarities in many of the upregulated as well as downregulated lipid species with PTL plus Ibu.

Hierarchical clustering analysis also showed that the lipidomic profile of the mice's brains in the PTL plus Tulsi group was strikingly similar to that of the sham-operated group. These results again reconfirm that treatment with Tulsi normalizes the brain lipidomic profile. Multi-group random forest analysis of different lipid groups from the cortex of the four groups of animals identified 15 lipid species ranked by mean decrease in accuracy (Supplementary Figure S1).

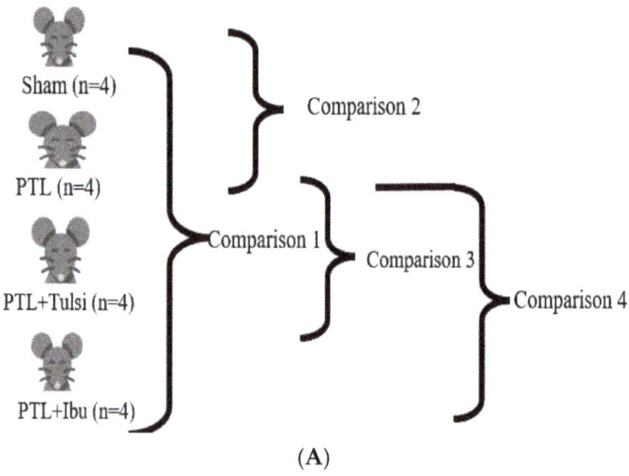

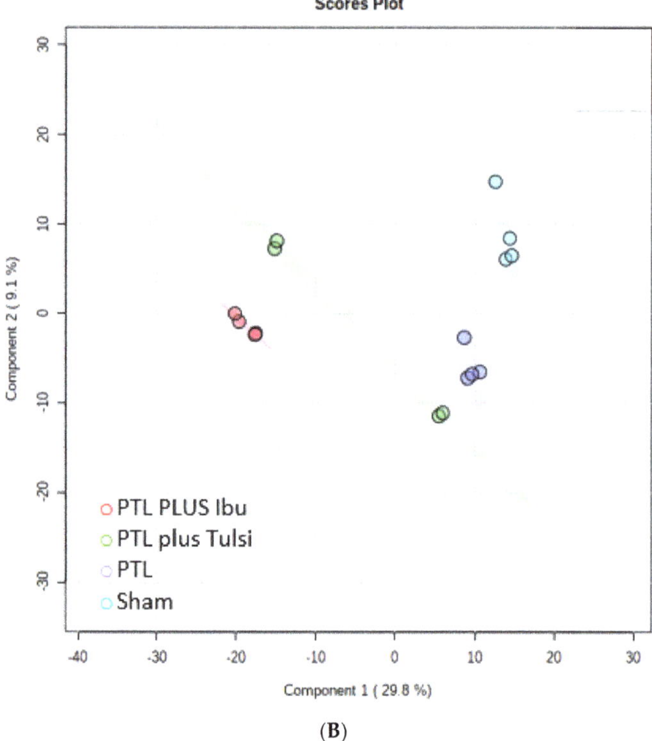

Figure 3. *Cont.*

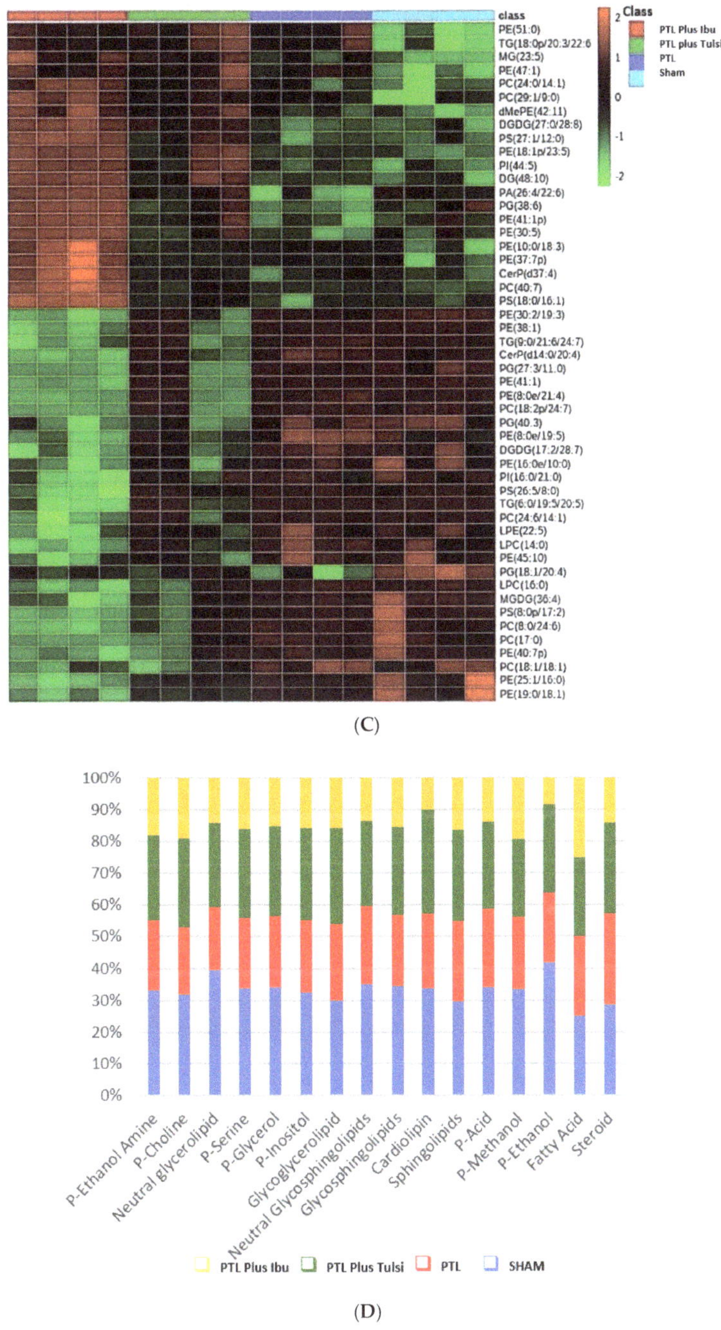

Figure 3. *Cont.*

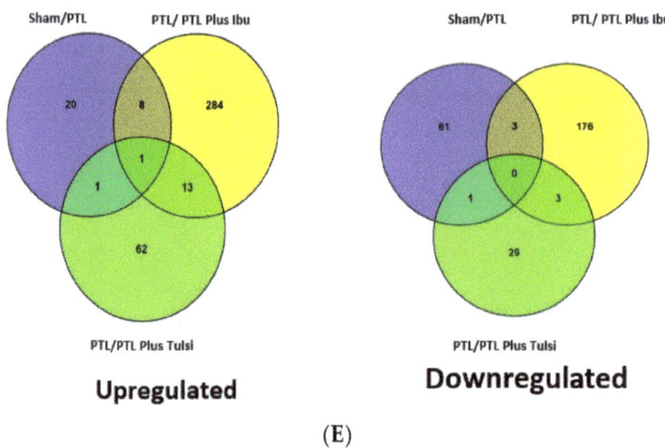

(E)

Figure 3. (**A**) The schematic representation of different comparisons (sham, PTL, PTL plus Tulsi, and PTL plus Ibu) for the brain and plasma is denoted by comparisons 1, 2, 3, and 4. (**B**) PLS−DA score plot of lipid species in the ipsilateral hemisphere of the brain cortex of the sham, PTL, PTL plus Tulsi, and PTL plus Ibu. (**C**) A heatmap showing the expression patterns of identified differential lipids in the brain from different comparison groups. Upregulated lipids are shown in red, while downregulated lipids are shown in green. (**D**) Relative abundance of different lipid groups in the ipsilateral hemisphere of the brain cortex of the Sham, PTL, PTL plus Tulsi, and PTL plus Ibu. (**E**) The common and unique upregulated and downregulated lipids are identified through the Venn diagram in the ipsilateral hemisphere of the brain cortex.

Additional comparisons of alterations in the lipid profiles of the brain were performed between Sham vs. PTL, PTL vs. PTL plus Tulsi and PTL vs. PTL plus Ibu (Supplementary Figures S2–S10). A volcano map comparison (Supplementary Figure S2) of the lipids of brain samples from the sham vs. PTL group revealed 32 upregulated lipid molecules and 65 downregulated lipid molecules ($\log_2$ FC = 1.5 and $p < 0.05$). The PLS-DA analysis with a component variance of 17.8% (component 1) and 19.2% (component 2) separated the lipids from Sham and PTL into two distinct clusters (Supplementary Figure S3A). The VIP score from PLS-DA of lipid species in the brain of Sham vs. PTL has identified 35 lipid species (Supplementary Figure S3B). The heat map showed the differential upregulation and downregulation of lipids in the brains of Sham and PTL mice (Supplementary Figure S4A). Random forest modeling between lipid groups in the brain of Sham vs. PTL has identified 15 lipid species ranked by mean decrease in accuracy (Supplementary Figure S4B).

Treatment with Tulsi induced changes in the lipid composition in the brain of lesioned mice (PTL plus Tulsi) compared with PTL mice. Volcano map (Supplementary Figure S5) shows significant alteration of 110 lipid species (77 upregulated and 33 downregulated; $\log_2$ FC = 1.5 and $p < 0.05$). Segregation and distinct clustering of the PTL plus Tulsi group from the PTL groups were evident from PLS-DA analyses based on the component variance of 36.1% (PC1) and 14.3% (PC2), indicating significant differences in lipid composition between the two groups (Supplementary Figure S6A). The VIP score plot showed 35 lipid species that were highly modulated between PTL plus Tulsi and PTL groups (Supplementary Figure S6B). The upregulated and downregulated lipids in the brain of PTL plus Tulsi vs. PTL mice displayed in the heat map also showed that the lipids were differentially altered (Supplementary Figure S7A). Random forest modeling identified 15 species of lipids ranked by mean decrease in accuracy that differentially modulated brain lipid composition between PTL plus Tulsi and PTL (Supplementary Figure S7B).

We determined whether treatment with the non-steroidal anti-inflammatory drug Ibuprofen can affect the lipidome profile in the brains of mice with PTL by comparing PTL plus Ibu vs. PTL using the Volcano Map (Supplementary Figure S8). There were significant changes in 488 lipid species (306 upregulated and 182 downregulated; $\log_2$

FC = 1.5 and *p*-value < 0.05). PLS-DA revealed that the lipids from the brain of PTL plus Ibu mice form a distinct cluster well separated from the PTL cluster with a component variance of 47.1% (PC1) and 12.5% (PC2) (Supplementary Figure S9A). The plot presenting the VIP score showed the 35 most modified lipid species (Supplementary Figure S9B). A heat map of the lipids in the brains of PTL plus Ibu and PTL mice showed that there was differential regulation of lipid species such that the lipids that were downregulated in the PTL condition were upregulated in the PTL plus Ibu brains (Supplementary Figure S10A). Random forest modeling showed 15 species of lipids ranked by mean decrease in accuracy that differentially modulated brain lipid composition between PTL plus Ibu and PTL (Supplementary Figure S10B).

We also determined the relative abundance of lipid groups in the brains of the four groups of mice, i.e., Sham, PTL, PTL plus Tulsi, and PTL plus Ibu. (Figure 3D; Supplementary Table S1). When we examined the lipid profile of the cerebral cortex, we found that 16 groups were abundant in the brains of the four groups of mice. The relative abundance of these 16 lipid groups was compared among the brains of four different groups of mice to determine the effects of different treatments on lipid metabolism in the brain.

The sham group exhibited the highest number of lipid species compared to the P-ethanolamine group. The relative abundance of phosphatidyl-choline, neutral glycerolipid, phosphatidyl-serine, phosphatidyl-glycerol, and phosphatidyl-inositol was also high in the brains of the four groups of animals. The results showed that the highest representation of 13 lipid groups was in the brains of sham group animals. This suggested that the normal brain lipid profile was different from that of lesioned animals. When comparing the brains of the PTL mice to the sham mice, it was found that the relative abundance of all 13 lipid groups was lower in the PTL group. This indicated that the PTL significantly affected lipid metabolism in the brain.

When comparing the brains of the PTL plus Ibu mice to the PTL mice, it was found that the relative abundance of 13 lipid groups was increased in the PTL plus Ibu mice. However, the relative abundance of most lipid groups was still lower than that of the sham mice. This suggests that ibuprofen treatment was ineffective in fully restoring the normal lipid profile in the brains of lesioned animals. On the other hand, when comparing the brains of the PTL plus Tulsi mice to the PTL mice, it was found that the relative abundance of all 13 lipid groups was higher in the PTL plus Tulsi mice. In fact, all 13 lipid groups that were highest in the brains of sham mice were also higher in the brains of PTL plus Tulsi mice compared to PTL or PTL plus Ibu mice. This indicated that treatment with Tulsi effectively restored the normal lipid profile in the brains of lesioned animals. The significance of these observations was that Tulsi might possess the ability to reverse the deleterious effect of a lesion on the lipidome profile in the ischemic brain.

3.2. Comparison of Lipidome in Photothrombotic Ischemia-Induced Brain of Mice Treated with Tulsi vs. Ibuprofen

In order to find out the crucial lipids that were altered in photothrombotic lesions by Tulsi or Ibuprofen treatment, we used a Venn diagram (Figure 3E) to find out the lipid species that were commonly modulated in the brains of sham, PTL, PTL plus Tulsi, and PTL plus Ibu groups.

Venn analysis of upregulated lipids between sham/PTL, PTL/PTL plus Tulsi, and PTL/PTL plus Ibu showed only one lipid species, monogalactosyldiacylglycerol (36:4), to be commonly upregulated between these groups (Figure 3E: Supplementary Table S3). Its presence in response to the lesion and treatment with Tulsi or Ibuprofen indicated that a significant increase in MGDG (36:4) lipid might play a role in the transition from pathological to physiological conditions in brain samples. This lipid could be an important marker to determine the injury-induced reaction in the brain.

When the 77 lipid species that were upregulated in brains of PTL /PTL plus Tulsi were compared with the 306 lipid species upregulated in PTL/PTL plus Ibu there were 13 common upregulated lipid species, dimethyl-phosphatidylethanolamine (18:1/14:0), phosphatidyl-ethanolamine (16:0/22:4), sphingomyelin (d44:7), phosphatidylcholine (17:0), phosphatidyl-ethanolamine (43:10), monogalactosyl-monoacylglycerol (14:2), di-galactosyl-diacylglycerol (42:9), phosphatidylinositol (28:5), phosphatidylserine (8:0e/21:6), lyso-phosphatidylcholine (16:1), phosphatidyl-ethanolamine (39:6), di-galactosyl-diacylglycerol (45:9) and diglyceride (42:5e) (Figure 3E: Supplementary Table S3).

When downregulated lipid species were compared between PTL/PTL plus Tulsi (33 lipid species) and PTL/PTL plus Ibu (182 lipid species), we found three common lipid species, diglyceride (4:0/20:5), phosphatidyl-ethanolamine (35:0p), and di-galactosyl-diacylglycerol (42:14), were downregulated (Figure 3E: Supplementary Table S4). These common upregulated and downregulated lipid species in the brains of Tulsi as well as Ibuprofen-treated mice suggested that both Tulsi and Ibuprofen might have similar effects on altering the lipidome of the ischemic brain. These results were also indicative that perhaps Tulsi acts as an anti-inflammatory agent similar to Ibuprofen.

3.3. Effect of Tulsi in Modulating Lipidome Signature of Plasma in Mice with Photothrombotic Ischemic Lesion of the Cerebral Cortex

Untargeted lipidomic analysis was performed on plasma samples to identify markers of ischemic lesion and recovery, and other different types of comparisons were done on the plasma lipids as described for the brain (Figure 3A). PLS-DA analysis was performed to determine whether there were any significant differences in the plasma lipidomic profiles of different groups of mice. The results showed that there was significant variance in the plasma lipidomic profile of different comparison groups, as indicated by the component variances of PC1 (20.3%) and PC2 (9.9%) (Figure 4A).

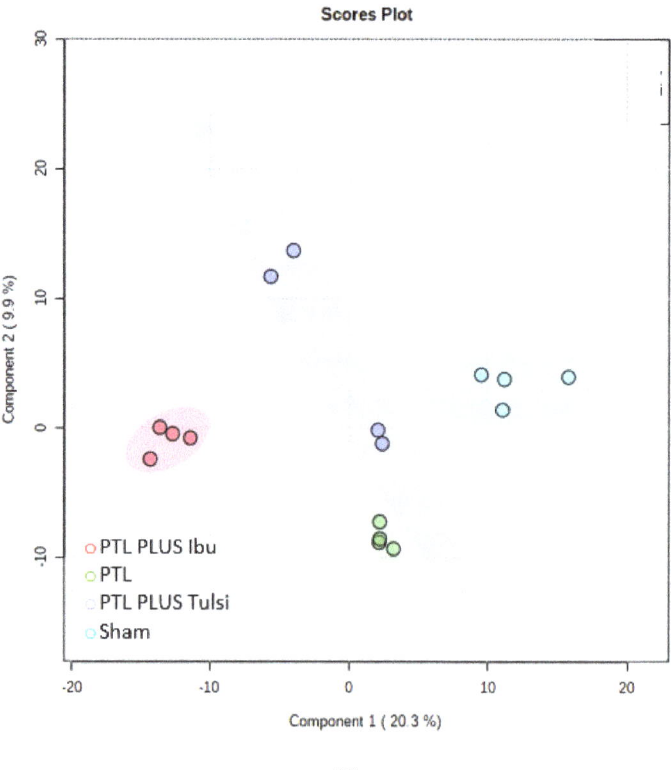

(A)

Figure 4. *Cont.*

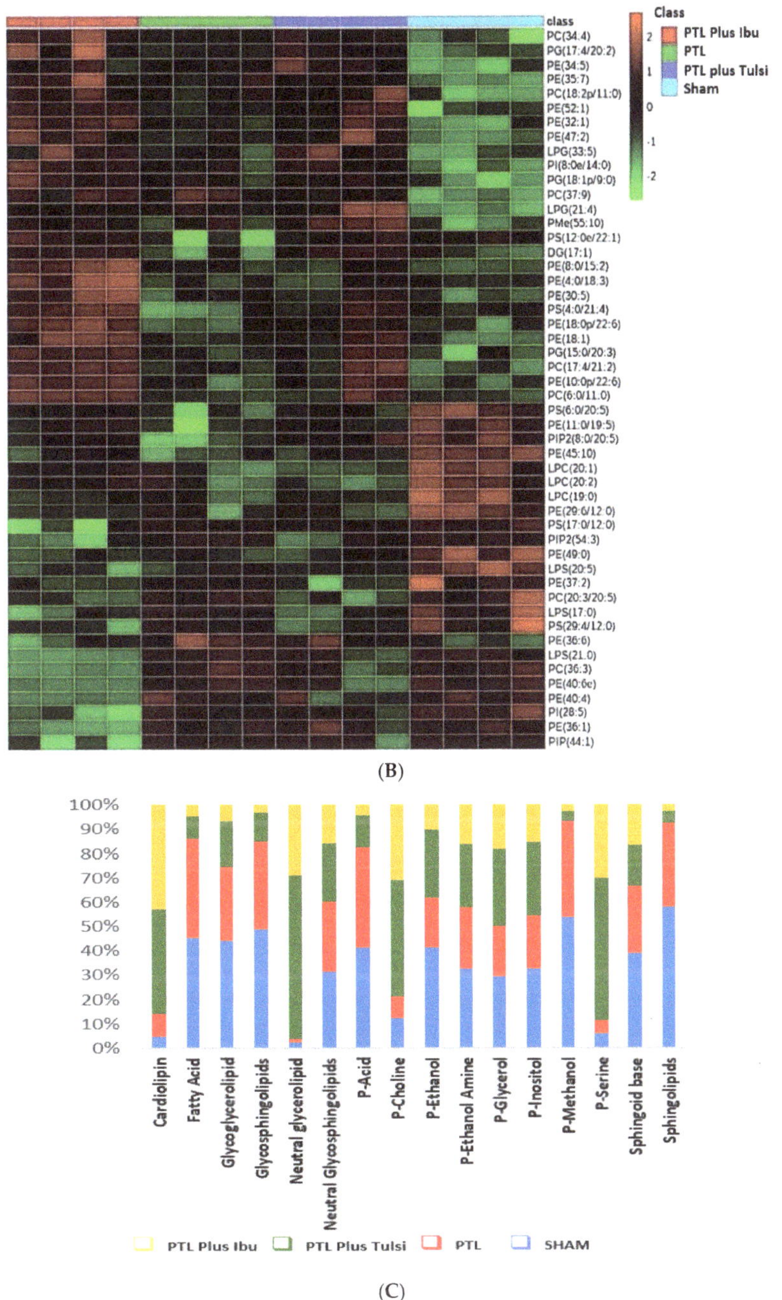

(B)

(C)

Figure 4. *Cont.*

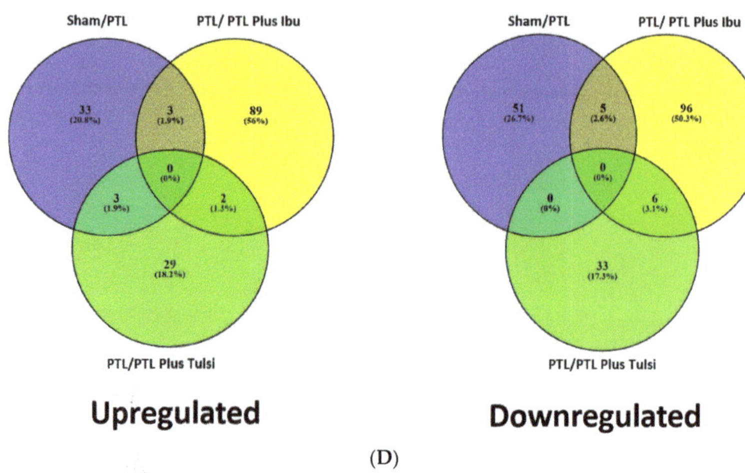

Upregulated   Downregulated

(D)

Figure 4. (**A**) PLS–DA score plot of lipid species in the plasma lipid samples of the sham, PTL, PTL plus Tulsi, and PTL plus Ibuprofen. (**B**) A heatmap showing the expression patterns of identified differential lipids in the plasma from different comparison groups. Upregulated lipids are shown in red while downregulated lipids are shown in green. (**C**) Relative abundance of different lipid groups in the plasma lipid samples of the sham, PTL, PTL plus Tulsi, and PTL plus Ibuprofen. (**D**) The common and unique upregulated and downregulated lipids were identified through the Venn diagram in the plasma samples of mice.

This suggested that the different treatments significantly affected plasma lipid metabolism. Furthermore, the plasma lipidomics of the PTL plus Tulsi group was found to be similar to the sham group, which was consistent with the findings from the brain groups. This indicates that treatment with Tulsi was effective in restoring normal lipid metabolism not only in the brain but also in the plasma of lesioned animals. We employed hierarchical clustering analysis to explore the potential relationship between the lipids in the different groups.

The heat map showed apparent differences in the top 50 lipid species (Figure 4B). The visual comparison suggests that lipid species in both PTL plus Tulsi and PTL plus Ibu have similar profiles. Volcano map of plasma samples ($\log_2$ FC= 1.5 and p-value < 0.05) of PTL group mice vs. sham group mice showed significant upregulation of 39 lipid molecules and downregulation of 56 lipid molecules (Supplementary Figure S11). Comparison of plasma lipids between PTL vs. PTL plus Tulsi using the volcano map analysis ($\log_2$ FC = 1.5 and p < 0.05) showed significant upregulation of 34 lipid species and downregulation of 39 lipid species (Supplementary Figure S14). While the volcano map analysis ($\log_2$ FC = 1.5 and p < 0.05) between PTL vs. PTL plus Ibu showed significant alterations in 201 lipid species (94 up- and 107 down-regulated; $\log_2$ FC = 1.5 and p < 0.05; Supplementary Figure S17).

PLS-DA score plots for PTL vs. Sham (component 1 = 20.8%, component 2 = 18%; Supplementary Figure S12A), PTL vs. PTL plus Tulsi (component 1 = 27.4%, component 2 = 18.5%; Supplementary Figure S15A) and PTL vs. PTL plus Ibu (component 1 = 34.7%, component 2 = 17.7%; Supplementary Figure S18A) showed that these groups were distinct from each other. The VIP score of the relative concentration of lipid species in plasma of Sham vs. PTL (Supplementary Figure S12B), PTL vs. PTL plus Tulsi (Supplementary Figure S15B), and PTL vs. PTL plus Ibu (Supplementary Figure S18B) showed that several lipids were differentially regulated in the comparison groups. These differential alterations were further seen when heat map generated from hierarchical clustering analysis was examined between the PTL vs. Sham (Supplementary Figure S13A), PTL vs. PTL plus Tulsi (Supplementary Figure S16A), PTL vs. PTL plus Ibu (Supplementary Figure S19A). Using Random Forests analysis, the lipid species with significant alterations and ranked by the mean decrease in classification accuracy have been identified between PTL vs. Sham (Supplementary Figure S13B), PTL vs. PTL plus Tulsi (Supplementary Figure S16B), PTL vs. PTL plus Ibu (Supplementary Figure S19B).

Relative abundance of lipids in the plasma of the Sham, PTL, PTL plus Tulsi, and PTL plus Ibu mice by arranging lipids into different groups (Figure 4C; Supplementary Table S2). Similar to the brain, a total of 16 lipid groups were found with high relative abundance in the plasma. Among the 16 groups, P-ethanol amine was highest in plasma samples of sham, PTL, PTL plus Tulsi, and PTL plus Ibu mice. P-inositol was also present in moderately high abundance in the plasma of all four groups of mice.

The abundance of these two lipid groups was found to be highest In the sham group and reduced in the plasma of the PTL group. Interestingly, treatment with Tulsi was found to increase the abundance of P-ethanol amine and P-inositol groups, as well as neutral glycerolipids, P-serine, P-glycerol, and P-choline. Similarly, treatment with Ibuprofen also increased the abundance of these lipid groups, except P-glycerol. However, the relative abundance was found to be higher after treatment with Tulsi as compared to Ibuprofen. Furthermore, compared to the sham and lesion groups, there was a very high abundance of neutral glycerolipids, P-serine, cardiolipin, and P-choline in the plasma of Tulsi as well as Ibuprofen-treated mice. This suggested that treatment with Tulsi or Ibuprofen may have a significant impact on plasma lipid composition, which may have important implications for various physiological processes.

Venn analysis (Figure 4D) showed the upregulation of 39 lipid species and the downregulation of 56 lipid species in sham vs. PTL, and 95 upregulated and 107 downregulated lipid species in the plasma of PTL vs. PTL plus Ibu mice. In the plasma of PTL vs. PTL plus Tulsi, there were 34 upregulated and 39 downregulated lipid species (Figure 4D; Supplementary Tables S7 and S12). There are two common lipid species, PS (46:4) and PC (38:4), that are upregulated, and six common lipid species, PI (33:3/23:2), PE (44:11), PC (27:2), PE (4:0/19:5), PG (32:4), and PE (42:3), in the plasma of PTL vs. PTL plus Tusli and PTL vs. PTL plus Ibuprofen, which suggested that both Tulsi and Ibuprofen could be having a similar effect on these lipid species (Figure 4D; Supplementary Tables S5 and S6).

3.4. Integration of Brain and Plasma Lipidomic Analysis

Although there were lipid species that were commonly altered in the plasma of both Tulsi- and Ibuprofen-treated animals, it may be possible that this could be indicative of a generalized reaction to either Tulsi or Ibuprofen. Hence, a comparative analysis of upregulated and downregulated lipid species in the brain and plasma of lesioned animals treated with Tulsi vs. Ibuprofen was performed. When sham vs. PTL conditions were compared between brain and plasma, there were 28 unique lipid species in the brain and 37 in the plasma, as well as two commonly upregulated lipid species, TG (8:0/24:6/24:7) and PS (8:0p/8:0) (Figure 5A; Supplementary Table S7). Analysis between PTL plus Tulsi and the PTL group showed only PE (10:0p/9:0) was commonly upregulated in the brain and plasma, though the brain and plasma showed 76 and 33 uniquely expressed lipid species, respectively (Figure 5B; Supplementary Table S9). A comparison of PTL plus Ibu and PTL only groups for the brain and plasma lipidome showed 11 lipid species commonly up-regulated. We also found 295 lipid species that were up-regulated and specific to the brain and 83 lipid species that were up-regulated and specific to the plasma (Figure 5C; Supplementary Table S11).

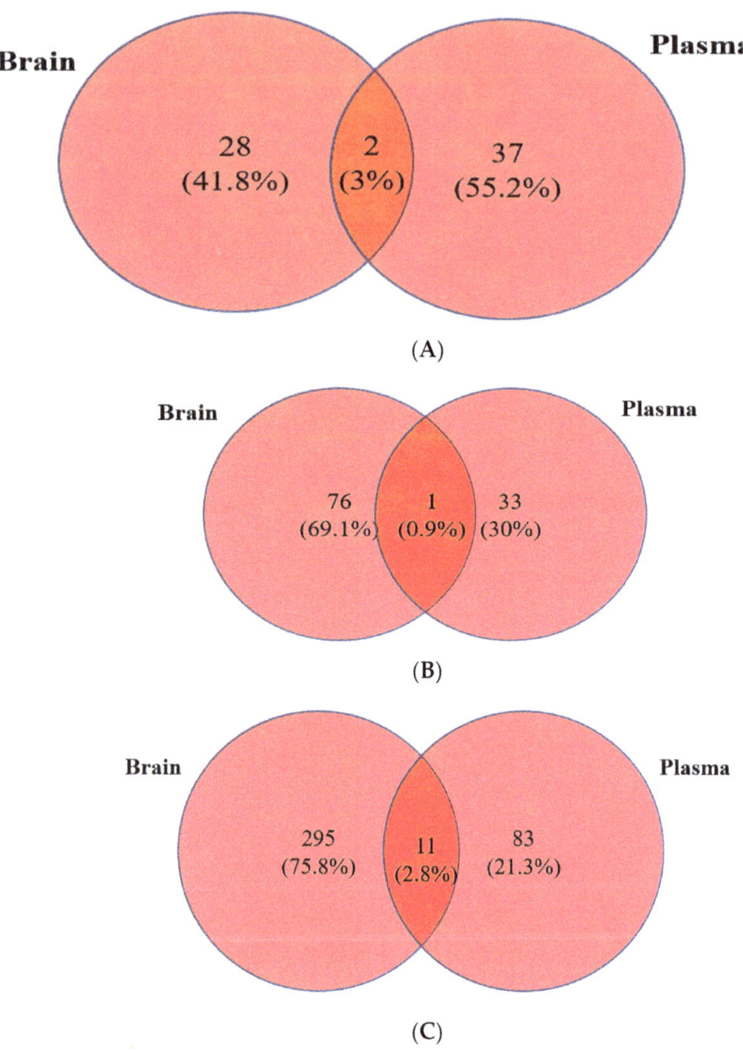

Figure 5. (**A**) Venny of upregulated lipid species in the brain and plasma between the Sham group and the PTL group. (**B**) Venny of upregulated lipid species in the brain and plasma between the PTL group and the PTL plus Tulsi group. (**C**) Venny of upregulated lipid species in the brain and plasma between the PTL group and the PTL plus Ibu group.

We have also analyzed the lipid species that were down-regulated in different groups. PTL as compared to sham showed significant down-regulation of lipid species: 63 in brains and 54 in plasma. The commonly downregulated lipid species were PE (45:10) and PE (42:3p) (Figure 6A, Supplementary Table S8). Interestingly, there were no commonly downregulated lipid species between PTL plus Tulsi and the PTL group. However, in lesioned animals treated with Tulsi, a significant reduction was seen for 33 lipid species in the brain and 39 lipid species in plasma (Figure 6B; Supplementary Table S10).

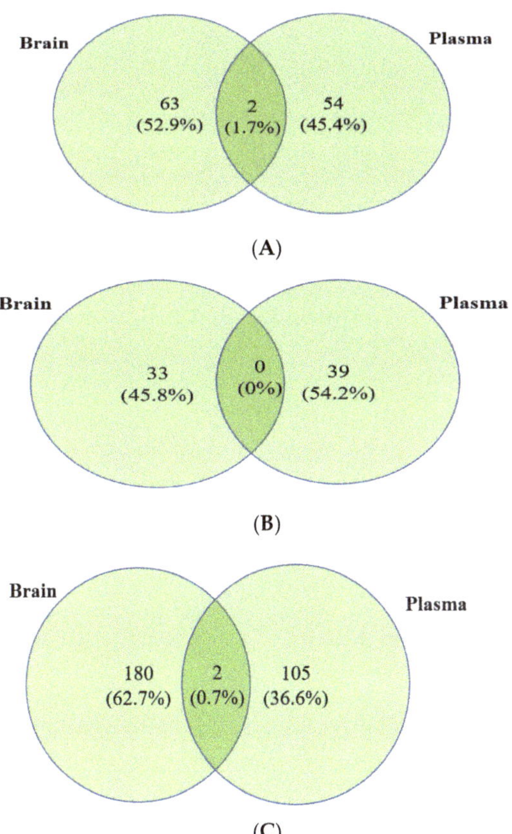

Figure 6. (**A**) Venny of downregulated lipid species in the brain and plasma between the Sham group and the PTL group. (**B**) Venny of downregulated lipid species of the brain and plasma between the PTL group and the PTL plus Tulsi group. (**C**) Venny of downregulated lipid species of the brain and plasma between the PTL group and the PTL plus Ibu group.

Finally, a comparison of PTL plus Ibuprofen vs. PTL groups showed LPE (18:0) and PE (39:2) were commonly downregulated. A total of 180 lipid species were downregulated in the brain and 105 lipid species in plasma samples (Figure 6C; Supplementary Table S12).

3.5. Untargeted Lipidomic and Metabolomics of Ocimum sanctum

The extract of Tulsi leaves contains a diverse array of constituents that are known to possess potential biological activity. These bioactive compounds found in Tulsi extract are known for their various pharmacological properties and may contribute to the plant's medicinal benefits. In the present study, we performed phytochemical screening of Tulsi extract. The standard protocol for phytochemical screening was employed to confirm the presence of metabolites and active compounds in the extract of Tulsi. Analysis of the ethanolic extract revealed the major presence of ursolic acid, alpha-curcumene, aesculin (esculin), coumarin, and p-cymene (Figure 7A), along with various other phytoconstituents in the extract of Tulsi, as listed in Supplementary Table S13.

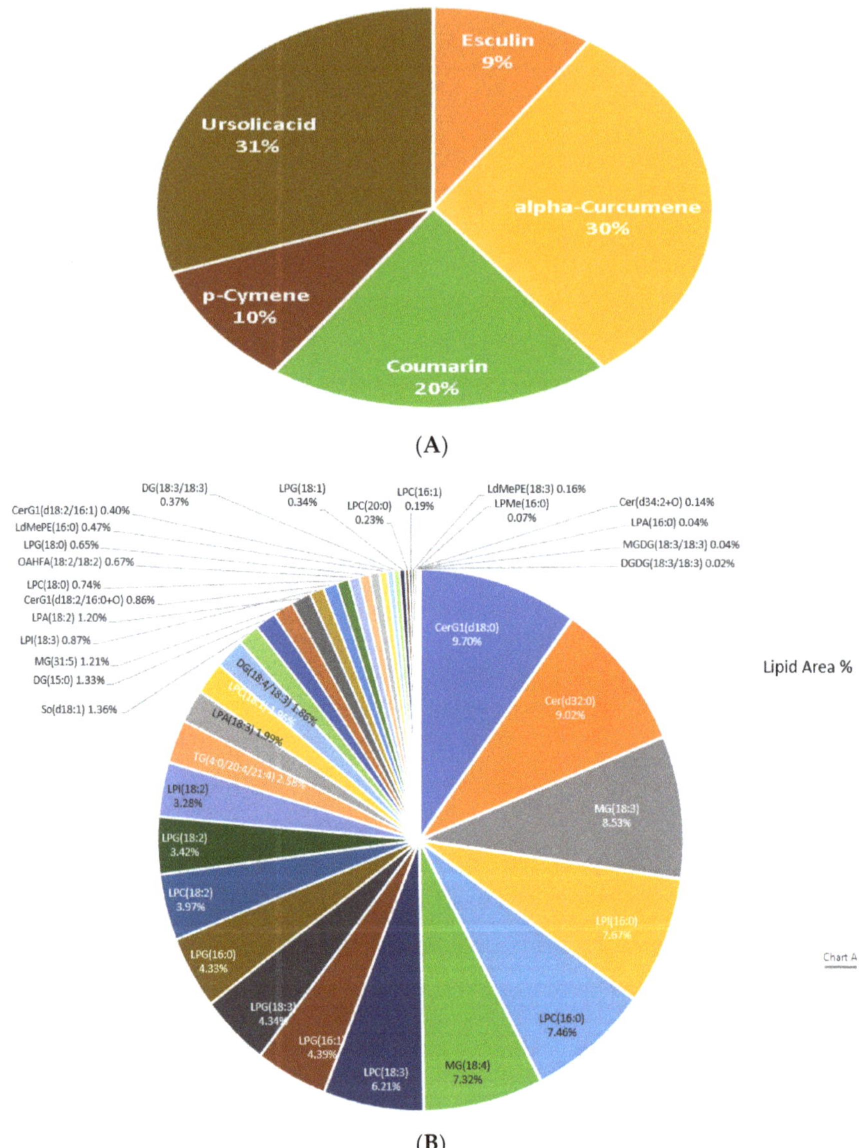

Figure 7. *Cont.*

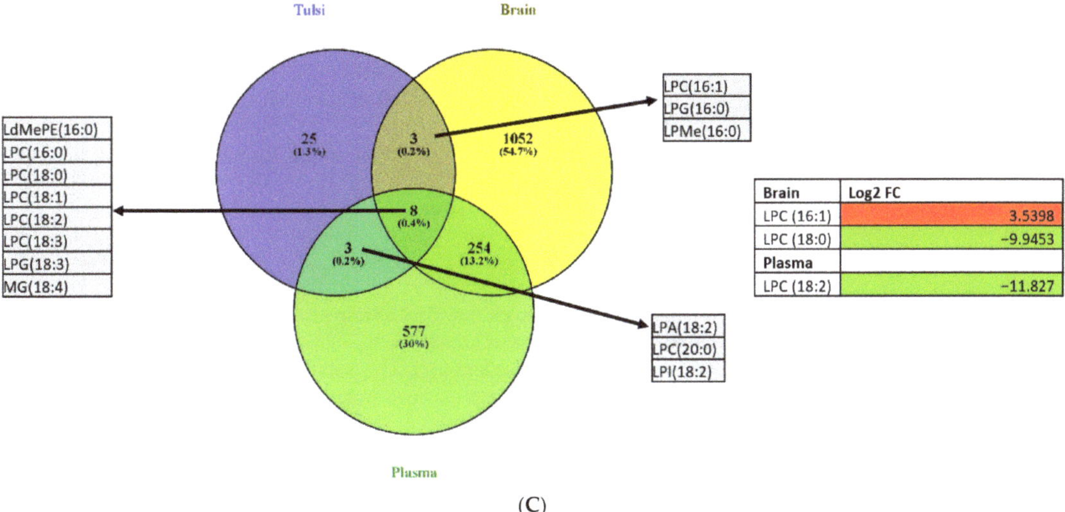

Brain	Log2 FC
LPC (16:1)	3.5398
LPC (18:0)	−9.9453
Plasma	
LPC (18:2)	−11.827

(C)

Figure 7. (**A**) Phytoconstituents in Tulsi extract. (**B**) A sunburst of different lipid molecules obtained in the Tulsi extract. (**C**) The common and unique lipids were identified through the Venn diagram in the Tulsi extract, brain cortex, and plasma samples of mice.

Untargeted lipidomic analysis was also performed on the Tulsi extract to correlate the effect of Tulsi treatment on the lipidomic profiling of the brain and plasma after its treatment. We have obtained a total of 39 lipid molecules from the Tulsi extract (Figure 7B, Supplementary Table S14).

Further, to find out lipids commonly found in tulsi extract, brain tissue, and plasma of the study group, we used a Venn diagram (Figure 7C) to find out lipid species that were modulated by Tulsi in the brain and plasma of mice from the PTL plus Tulsi group. Venn analysis showed eight lipid species were common with PTL plus the Tulsi group of the brain and plasma. These eight common lipid species were lyso-di-methyl phosphatidyl ethanolamine (16:0), lyso-phosphatidyl choline (16:0), lyso-phosphatidyl choline (18:0), lyso-phosphatidyl choline (18:1), lyso-phosphatidyl choline (18:2), lyso-phosphatidyl choline (18:3), lyso-phosphatidyl glycerol (18:3), monoglyceride (18:4). Venn analysis between lipid species of Tulsi and lipid species in the brain of PTL plus Tulsi has found three common lipids: lyso-phosphatidyl choline (16:1), lyso-phosphatidyl glycerol (16:0), and lyso-phosphatidyl methanol (16:0). Whereas Venn analysis of Tulsi lipids with plasma lipids of PTL plus Tulsi also found three common lipids: lyso-phosphatidic acid (18:2), lyso-phosphatidyl choline (20:0), and lyso-phosphatidyl inositol (18:2).

LPC (16:1) was found in Tulsi extract and found to be significantly increased in the brains of the PTL plus Tulsi-treated group. This finding provides valuable insights into the potential therapeutic effects of Tulsi extract in promoting healing and repair processes in the brain.

Our results show that lipidomic profiles in mouse models of ischemic injury in the brain and plasma were altered and distinct. These results also suggest that the alterations in some of the lipids seen with treatment with Tulsi or Ibuprofen were present in both the brain and plasma. These lipids might play a significant role in the recovery processes following ischemic injury. Importantly, the presence of these lipids in the circulatory system could help in the prognosis and evaluation of recovery.

4. Discussion

Tulsi is well known for its neuroprotective action; in addition, the effect of ibuprofen on amelioration of brain lesions is also well known [45–48]. In the current pilot study, the lipidomic landscape of photothrombotic ischemic lesions in a mouse model was studied. We also examined the effect of the Tulsi treatment following a photothrombotic ischemic lesion on the modulation of the lipidomic profile of the brain and plasma in the mouse model. The alterations in lipidome profile with Tulsi treatment were compared to treatment with a non-steroid anti-inflammatory drug, Ibuprofen.

Our objective was to see whether changes in the brain and plasma lipidomic profiles could be attributed to the changes in the lesion status in the mouse model of ischemic stroke following treatment with Tulsi.

It has been shown in the past that oral administration of Tulsi to rats with MCAO (middle cerebral artery occlusion) markedly reduced the infarct size, reduced neurological deficits, and suppressed neuronal loss [31]. The potential of Tulsi in the management of experimentally induced cognitive dysfunctions in rats has been determined [49]. However, there was no data available about the effect of Tulsi on brain and plasma lipidomic profiles in ischemic stroke. This study represents the first untargeted lipidomic profiling of the brain and plasma to obtain information about the treatment effects of Tulsi after the manifestation of ischemia in the cerebral cortex of the brain.

Previous findings revealed that there is a strong relationship between lipid profiles and ischemic stroke [10]. Our comparison of sham and PTL mice in brain and plasma also showed a concordant effect with changes in the lipidomic profile, suggesting the effect of stroke upon the lipid species. We found the distinct lipidomic profile of the mouse model with the lesion, which, on treatment with either Tulsi or ibuprofen, showed apparent changes in the lipidomic profile that were systemic and evident even in the plasma samples.

Ischemic brain injury releases free radicals in the form of hydroxyl radicals, radical superoxide anion, and nitric oxide (NO). During ischemic-stroke injury, R.O.S. are primarily produced in the mitochondria [50,51]. While excessive ROS formation secondary to reperfusion injury was attenuated by Tulsi [27]. Several active compounds present in Tulsi contribute to its anti-inflammatory activity [52–56]. Our study showed that after treatment with Tulsi, there was a change in the brain and plasma lipidomic profiles, which could possibly be one of the reasons for the decrease in lesion size seen post-treatment with Tulsi in previous studies [31].

Next, we classified the lipid species into groups to broadly understand the lipid groups involved in the treatment of photothrombotic ischemia with Tulsi. In the brain, lipid groups such as P-ethanol amine, P-choline, neutral glycerolipids, glycoglycerolipids, and others were significantly reduced in the lesion but surged after Tulsi treatment. On the other hand, in plasma, P-choline, neutral glycerolipids, P-serine, and P-glycerol were significantly increased compared to sham or lesion groups, which could be due to the systemic effect of Tulsi, which was administered orally to the mice. Increases in P-ethanol amine, P-choline, or P-glycerol were already known to have anti-inflammatory functions [57–63].

Further analysis between different comparison groups of the brain identified monoglactosyl diacylglycerol (MGDG) species that assist in the recovery process [64], which was found to be increased with the treatment of Tulsi or ibuprofen. The increased levels of MGDG (36:4) may be attributed to oral treatment with Tulsi or ibuprofen. Previous studies have found that MGDG has strong anti-inflammatory [65–68] and anti-proliferative activities [69]. It has been shown earlier that MGDG also has an inhibitory effect on cancer cells [70]. This was in concurrence with our results, which showed that the overall level of MGDG (36:4) was increased in the animals after the treatment with Tulsi and ibuprofen, though these observations require further study.

Lipid species of phosphatidyl serine (PS), phosphatidyl ethanolamine (PE), lysyl-phosphatidylglycerol (LPG), and phosphatidyl inositol (PI) have shown common expression in sham and lesion plus Tulsi, which indicated the association of Tulsi with an increase in these lipids in the brain. These lipid groups were reported to perform several metabolic pathways that were found altered in disease conditions [71,72].

Further, we integrated the brain and plasma lipidomes in different comparison groups to attain site-specific or common alteration. The brain of Sham was found to be enriched explicitly in many species of glycerol lipids and to have reduced P-ethanol amine species, while in plasma, most of the altered lipid species belong to P-ethanol amine and P-serine compared to the lesion. These results were concordant with the literature [13,14,73,74]. The significant lipids that changed following treatment with Tulsi belong primarily to P-ethanolamine, while downregulated brain lipids were mainly composed of phosphatidyl serine. Phosphatidylserine is required for healthy nerve cell membranes, myelin, and cognitive functions [75,76]. Phosphatidylethanolamine was also found to be significantly altered in an ibuprofen-treated photothrombotic ischemia animal model. All this suggests that changes in phosphatidyl ethanolamine and phosphatidyl serine were induced due to treatment with Tulsi or ibuprofen and may be associated with lesion regression. It can also be considered for systemic representation of site-specific injury or improvement.

Metabolomic analysis of Tulsi leaf extract has provided robust evidence confirming the presence of several important phytoconstituents. Notably, ursolic acid, alpha-curcumene, aesculin (Esculin), coumarin, and p-Cymene have been identified as significant components in the Tulsi extract, as supported by existing literature. [77,78]. The identification and confirmation of these phytoconstituents, alongside other metabolites in Tulsi extract through metabolomic analysis, provide scientific support for the traditional uses and health-promoting effects associated with Tulsi. These findings expand our understanding of the chemical composition of Tulsi and pave the way for further exploration of its therapeutic applications. It was reported in the literature that the leaf extract of Tulsi contains a higher concentration of ursolic acid [79]. Ursolic acid has several medicinal properties such as analgesic, anti-inflammatory, anti-atherosclerotic, anti-cancer, anti-diabetic, anti-epileptic, hepato-protective, anti-hyperlipidemic, anti-fertility, anti-platelet aggregation, anti-tuberculosis, and anti-HIV activities [80–85]. Ursolic acid, derived from *Ocimum sanctum*, has been recognized for its potent anti-inflammatory properties. The administration of Ursolic acid has been shown to reduce brain edema and neurological insufficiencies after TBI induction in a murine model [86]. Ursolic acid has significantly helped in reducing intercellular adhesion molecule-1 (ICAM-1), toll-like receptor 4 (TLR4), nuclear factor-κB (NF-κB) P65, interleukin-1β (IL-1β), tumor necrosis factor-α (TNF-α), interleukin-6 (IL-6), inducible nitric oxide synthase (iNOS), and matrix metalloproteinase-(MMP)-9 in a sub-arachnoid hemorrhage brain injury model in rats [87,88]. In the MCAO model in rats, administration of ursolic acid helped decrease neurological deficits along with reduced levels of proinflammatory cytokine concentrations (IL-1β, TNF-α, and IL-6), TLR4, and inactivated NF-κB [89]. It also suppressed the activity of cyclooxygenase-2 (COX-2), an enzyme involved in the synthesis of inflammatory prostaglandins [89]. Ursolic acid, being the active constituent of Tulsi, may be responsible for its anti-inflammatory action, as numerous studies have highlighted the anti-inflammatory effects of ursolic acid, making it a promising therapeutic agent for various inflammatory conditions.

Untargeted lipidomic analysis of the Tulsi extract has shown the presence of 39 important lipid molecules, which may help in regulating the perturbed lipidomic of the brain and plasma and might be responsible for the recovery process in various disease ailments. LPC (16:1), an important lipid in Tulsi extract, was found to be upregulated in the brains of PTL after treatment with Tulsi. Lysophosphatidylcholine (LPC) was considered an important membrane constituent implicated in signaling and immune regulation [90]. It was reported in the literature that the level of LPC was altered in the brain following both focal and global cerebral ischemia in rats and mice [13,91–94]. LPC is secreted from apoptotic cells, which play a role in the inflammatory reaction mediated by microglia [95]. The observed elevation of LPC (16:1) suggested a potential involvement of Tulsi lipids in the recovery process following brain lesions. Therefore, targeting LPC (16:1) might be a potential therapeutic method for brain ischemia. However, further studies are needed to fully elucidate the mechanisms underlying the role of Tulsi lipids, including LPC (16:1), in brain recovery. The current pilot study aimed to generate preliminary data and valuable insights that would guide future research. The findings from this study provided valuable insights into the potential therapeutic effects of Tulsi extract in promoting healing and repair processes in the brain and plasma. However, considering the intricate nature of stroke and its multifaceted pathophysiology, it is crucial to go beyond the identification of lipidomic changes. Establishing a clear correlation between lipidomic changes and the overall improvement or deterioration of stroke outcomes is essential. Hence, a future larger-scale study with a more substantial sample size along with examining and assessing the correlation between inflammatory markers, neurological function, infarct size measurement, functional recovery, anti-inflammatory markers, neurotransmitters, and the mechanisms of action of treatments should be planned to gain comprehensive insights into stroke management and treatment efficacy.

5. Conclusions

Nowadays, lipidomic-based studies have become an important tool for obtaining lipidomic snapshots. In the present study, we have performed metabolomics and untargeted lipidomic analysis of the Tulsi extract to see the presence of lipid molecules and various other metabolites in the extract and further correlated the effect of treatment of the Tulsi extract after a photothrombotic lesion on the modulation of lipidomic profiling in the brain and plasma. This approach will provide insights into the roles of specific lipids and help establish a therapeutic solution for human stroke and related disorders. Our study found the deregulation of various lipid species as a characteristic feature of the mouse model of an ischemic stroke lesion. We also reported that brain and plasma lipids were altered in animals with stroke-like lesions following treatment with Tulsi extract. Specifically, lipid species such as Phosphatidyl Serine (PS), Phosphatidyl Ethanolamine (PE), Lysyl-phosphatidylglycerol (LPG), and Phosphatidyl Inositol (PI) were increased after treatment with Tulsi. Notably, the cortex of mice treated with Tulsi showed an upregulation of Monogalactosyldiacylglycerol (36:4). One intriguing finding was the significant increase of LPC (16:1), present in Tulsi extract, in the brains of the PTL plus Tulsi-treated group. Our study suggested that significant changes in the lipidomic profile in the brain and plasma caused by either Tulsi or Ibuprofen may help ameliorate brain injury. The change in the brain and plasma lipidomic profile induced by Tulsi could be relevant to the reduction of lesion seen after Tulsi treatment in previous stroke studies; however, further validation with various stroke-related measures is needed to draw more robust conclusions.

Supplementary Materials: The following supporting information can be downloaded at: https://www.mdpi.com/article/10.3390/life13091877/s1, Figure S1: Random Forest from mice brains of different groups identified 15 different lipid features and ranked by the mean decrease in classification accuracy. Figure S2: The volcano map showed the differential lipids expression level in the mice brain of sham vs PTL groups (log2 FC = 1.5 and *p*-value < 0.05 considered significant). Upregulated lipids are shown by red color dots while downregulated lipids are shown by blue colored dots. Figure S3: (A): Partial Least Squares-Discriminant Analysis (PLS-DA) plot of lipid species in mice brain of sham vs PTL groups revealing the component variance 19.2% and 17.8%. (B): VIP score of lipid species in PLS-DA in the mice brain of sham vs. PTL groups. The colored boxes on the right side represent the relative concentration of the corresponding lipid species. Red indicates high and blue indicates low. Figure S4: (A): Heatmap of the expression of identified differential lipids in the mice brain between sham vs PTL groups. The colored boxes on the right side represent the relative concentration of the corresponding lipid species. Red indicates high and green indicates low. (B): Random Forest from mice brains of sham vs PTL groups identified 15 different lipid features and ranked by the mean decrease in classification accuracy. Figure S5: The volcano map showed the differential lipids expression level in the mice brain of PTL vs. PTL plus Tulsi groups (log2 FC = 1.5 and *p*-value < 0.05 considered significant). Upregulated lipids are shown by red color dots while downregulated lipids are shown by blue colored dots. Figure S6: (A): Partial Least Squares - Discriminant Analysis (PLS-DA) plot of lipid species in mice brain of PTL vs. PTL plus Tulsi groups revealing the component variance 36.1% and 14.3%. (B): VIP score of lipid species in PLS-DA in the mice brain of PTL vs. PTL plus Tulsi groups. The colored boxes on the right side represent the relative concentration of the corresponding lipid species. Red indicates high and blue color indicates low. Figure S7: (A): Heatmap of the expression of identified differential lipids in the mice brain between PTL vs. PTL plus Tulsi groups. The colored boxes on the right side represent the relative concentration of the corresponding lipid species. Red indicates high and green indicates low. (B): Random Forest from mice brains of PTL vs. PTL plus Tulsi groups identified 15 different lipid features and ranked by the mean decrease in classification accuracy. Figure S8: The volcano map showed the differential lipids expression level in the mice brain of PTL vs. PTL plus Ibu groups (log2 FC = 1.5 and p-value < 0.05 considered significant). Upregulated lipids are shown by red color dots while down-regulated lipids are shown by blue colored dots. Figure S9: (A): Partial Least Squares - Discriminant Analysis (PLS-DA) plot of lipid species in mice brain of PTL vs. PTL plus Ibu groups revealing the component variance 47.1% and 12.5%. (B): VIP score of lipid species in PLS-DA in the mice brain of PTL vs. PTL plus Ibu groups. The colored boxes on the right side represent the relative concentration of the corresponding lipid species. Red color indicates high and blue indicates low. Figure S10: (A): Heatmap of the expression of identified differential lipids in the mice brain between PTL and PTL plus Ibu groups. The colored boxes on the right side represent the relative concentration of the corresponding lipid species. Red indicates high and green indicates low. (B): Random Forest from mice brains of PTL vs. PTL plus Ibu groups identified 15 different lipid features and ranked by the mean decrease in classification accuracy. Figure S11: The volcano map showed the differential lipids expression level in the mice plasma of sham vs PTL groups (log2 FC = 1.5 and p-value < 0.05 considered significant). Up-regulated lipids are shown by red color dots while downregulated lipids are shown by blue colored dots. Figure S12: (A): Partial Least Squares - Discriminant Analysis (PLS-DA) plot of lipid species in mice plasma of sham vs. PTL groups revealing the component variance 20.8% and 18%. (B): VIP score of lipid species in PLS-DA in the mice plasma of sham vs PTL groups. The colored boxes on the right side represent the relative concentration of the corresponding lipid species. Red color indicates high and blue color indicates low. Figure S13: (A): Heatmap of the expression of identified differential lipids in the mice plasma between sham vs PTL groups. The colored boxes on the right side represent the relative concentration of the corresponding lipid species. Red indicates high and green indicates low. (B): Random Forest from mice plasma of sham vs. PTL groups identified 15 different lipid features and ranked by the mean decrease in classification accuracy. Figure S14: The volcano map showed the differential lipids expression level in the mice plasma of PTL vs PTL plus Tulsi groups (log2 FC = 1.5 and p-value < 0.05 considered significant). Upregulated lipids are shown by red color dots while downregulated lipids are shown by blue colored dots. Figure S15: (A): Partial Least Squares-Discriminant Analysis (PLS-DA) plot of lipid species in mice plasma of PTL vs PTL plus Tulsi groups revealing the component variance 27.4% and 18.5%. (B): VIP score of lipid species in PLS-DA in the mice plasma of PTL vs PTL plus Tulsi groups. The colored boxes on the right side represent the relative concentration of the corresponding lipid species. Red indicates high and blue indicates low. Figure S16: (A): Heatmap of the expression of identified differential lipids in the mice plasma between PTL vs. PTL plus Tulsi groups. The colored boxes on the right side represent the relative concentration of the corresponding lipid species. Red indicates high and green indicates low. (B): Random Forest from mice plasma of PTL vs. PTL plus Tulsi groups identified 15 different lipid features and ranked by the mean decrease in classification accuracy. Figure S17: The volcano map showed the differential lipids expression level in the mice plasma of PTL vs PTL plus Ibu groups (log2 FC = 1.5 and p-value < 0.05 considered significant). Upregulated lipids are shown by red color dots while downregulated lipids are shown by blue colored dots. Figure S18: (A): Partial Least Squares-Discriminant Analysis (PLS-DA) plot of lipid species in mice plasma of PTL vs. PTL plus Ibu groups revealing the component variance 34.7% and 17.7%. (B): VIP score of lipid species in PLS-DA in the mice plasma of PTL vs. PTL plus Ibu groups. The colored boxes on the right side represent the relative concentration of the corresponding lipid species. Red indicates high and blue indicates low. Figure S19: (A): Heatmap of the expression of identified differential lipids in the mice plasma between PTL vs. PTL plus Ibu groups. The colored boxes on the right side represent the relative concentration of the corresponding lipid species. Red indicates high and green indicates low. (B): Random Forest from mice plasma of PTL vs. PTL plus Ibu groups identified 15 different lipid features and ranked by the mean decrease in classification accuracy. Table S1: Table showing relative abundance of lipid groups in the mice brain among the Sham group, PTL group, PTL plus Tulsi group and PTL plus Ibu groups. Table S2: Table showing relative abundance of lipid groups in the mice plasma among the Sham group, PTL group, PTL plus Tulsi group and PTL plus Ibu groups. Table S3: Venny brain upregulated lipid species among the Sham group, PTL group, PTL plus Tulsi group and PTL plus Ibu groups. Table S4: Venny brain downregulated lipid species among the Sham group, PTL group, PTL plus Tulsi group and PTL plus Ibu groups. Table S5: Venny plasma upregulated lipid species among the Sham group, PTL group, PTL plus Tulsi group and PTL plus Ibu groups. Table S6: Venny plasma downregulated lipid species among the Sham group, PTL group, PTL plus Tulsi group and PTL plus Ibu groups. Table S7: Venny of upregulated lipid species of brain and plasma between the Sham group and PTL group. Table S8: Venny of downregulated lipid species of brain and plasma between the Sham group and PTL group. Table S9: Venny of upregulated lipid species of brain and plasma between PTL group and PTL plus Tulsi group. Table S10: Venny of downregulated lipid species of brain and plasma between PTL group and PTL plus Tulsi group. Table S11: Venny of upregulated lipid species of brain and plasma between PTL group and PTL plus Ibuprofen group. Table S12: Venny of downregulated lipid species of brain and plasma between PTL group and PTL plus Ibuprofen group. Table S13: Phytoconstituents in the Tulsi extract. Table S14: List of Lipid molecules obtained from Tulsi extract. Supplementary document File S2.

Author Contributions: Conceptualization, J.S.M., Z.F. and R.V.; methodology, I.Y., N.S., R.V., J.S.M. and Z.F.; software, I.Y. and N.S.; validation, R.V., J.S.M. and Z.F.; formal analysis, I.Y. and N.S.; investigation, I.Y. and N.S.; resources, R.V., J.S.M. and Z.F.; data curation, I.Y., N.S., J.S.M. and Z.F.; writing—original draft preparation, I.Y. and N.S.; writing—review and editing, J.S.M., Z.F. and R.V.; supervision, J.S.M., Z.F. and R.V. All authors have read and agreed to the published version of the manuscript.

Funding: This research received no external funding and the APC was funded by the National Brain Research Center, India.

Institutional Review Board Statement: The animal study protocol was approved by the Institutional Animal Ethics Committee (NBRC/IAEC/2018/140) of the National Brain Research Center, India.

Informed Consent Statement: Not applicable.

Data Availability Statement: Data are contained within the article.

Acknowledgments: We would like to thank the National Brain Research Centre (NBRC, India) and Institute of Liver and Biliary Sciences (ILBS, India) for providing all the necessary support. The author would like to thank Zeeshan Fatima, Amity Institute of Biotechnology, Amity University Haryana for her guidance and support.

Conflicts of Interest: The authors declare no conflict of interest.

Abbreviations

ANOVA—Analysis of Variance, FWHM—Full Width at Half Maximum, HCD—Higher-Energy Collisional Dissociation, HESI—Heated Electrospray Ionization, IV—Intravenous, LdMePE—lyso-dimethylphosphatidylethanolamine, LPA—lysophosphatidic acid, LPC—Lysyl-Phosphatidyl Choline, LPE—lysophosphatidylethanolamine, LPG—Lysyl-phosphatidylglycerol, LPI—lysophosphatidylinositol, LPMe—lysophosphatidylmethanol, MCAO—Middle Cerebral Artery Occlusion, MG—monoglyceride, MGDG—Monogalactosyldiacylglycerol, PCA—Principal Component Analysis, PC—Phosphatidyl Choline, PE—Phosphatidyl Ethanolamine, P-Glycerol—Phosphatidyl Glycerol, PI—Phosphatidyl Inositol, PLS-DA—Partial Least Squares-Discriminant Analysis, PS—Phosphatidyl Serine, PTL—Photothrombotic Lesion.

References

1. Mozaffarian, D.; Benjamin, E.J.; Go, A.S.; Arnett, D.K.; Blaha, M.J.; Cushman, M.; De Ferranti, S.; Després, J.P.; Fullerton, H.J.; Howard, V.J.; et al. Heart Disease and Stroke Statistics-2015 Update: A Report from the American Heart Association. *Circulation* **2015**, *131*, e29–e39. [CrossRef] [PubMed]
2. Barber, P.A.; Demchuk, A.M.; Hirt, L.; Buchan, A.M. Biochemistry of Ischemic Stroke. *Adv. Neurol.* **2003**, *92*, 151–164. [PubMed]
3. Jivad, N.; Rabiei, Z. A Review Study on Medicinal Plants Used in the Treatment of Learning and Memory Impairments. *Asian Pac. J. Trop. Biomed.* **2014**, *4*, 780–789. [CrossRef]
4. Mackay, J.; Mensah, G. *The Atlas of Heart Disease and Stroke*; World Health Organization: Geneva, Switzerland, 2004.

5. Chang, Y.; Eom, S.; Kim, M.; Song, T.J. Medical Management of Dyslipidemia for Secondary Stroke Prevention: Narrative Review. *Medicina* **2023**, *59*, 776. [CrossRef]
6. Chen, K.N.; He, L.; Zhong, L.M.; Ran, Y.Q.; Liu, Y. Meta-Analysis of Dyslipidemia Management for the Prevention of Ischemic Stroke Recurrence in China. *Front. Neurol.* **2020**, *11*, 1–8. [CrossRef]
7. Lee, J.S.; Chang, P.Y.; Zhang, Y.; Kizer, J.R.; Best, L.G.; Howard, B.V. Triglyceride and HDL-C Dyslipidemia and Risks of Coronary Heart Disease and Ischemic Stroke by Glycemic Dysregulation Status: The Strong Heart Study. *Diabetes Care* **2017**, *40*, 529–537. [CrossRef]
8. Yaghi, S.; Elkind, M.S.V. Lipids and Cerebrovascular Disease: Research and Practice. *Stroke* **2015**, *46*, 3322–3328. [CrossRef]
9. Gu, X.; Li, Y.; Chen, S.; Yang, X.; Liu, F.; Li, Y.; Li, J.; Cao, J.; Liu, X.; Chen, J.; et al. Association of Lipids with Ischemic and Hemorrhagic Stroke a Prospective Cohort Study among 267,500 Chinese. *Stroke* **2019**, *50*, 3376–3384. [CrossRef]
10. Kloska, A.; Malinowska, M.; Gabig-Cimińska, M.; Jakóbkiewicz-Banecka, J. Lipids and Lipid Mediators Associated with the Risk and Pathology of Ischemic Stroke. *Int. J. Mol. Sci.* **2020**, *21*, 3618. [CrossRef]
11. Liu, X.; Yan, L.; Xue, F. The Associations of Lipids and Lipid Ratios with Stroke: A Prospective Cohort Study. *J. Clin. Hypertens.* **2019**, *21*, 127–135. [CrossRef]
12. Hussain, G.; Wang, J.; Rasul, A.; Anwar, H.; Imran, A.; Qasim, M.; Zafar, S.; Kamran, S.K.S.; Razzaq, A.; Aziz, N.; et al. Role of Cholesterol and Sphingolipids in Brain Development and Neurological Diseases. *Lipids Health Dis.* **2019**, *18*, 26. [CrossRef] [PubMed]
13. Jia, Z.; Tie, C.; Wang, C.; Wu, C.; Zhang, J. Perturbed Lipidomic Profiles in Rats With Chronic Cerebral Ischemia Are Regulated by Xiao-Xu-Ming Decoction. *Front. Pharmacol.* **2019**, *10*, 264. [CrossRef] [PubMed]
14. Wang, R.; Liu, S.; Liu, T.; Wu, J.; Zhang, H.; Sun, Z.; Liu, Z. Mass Spectrometry-Based Serum Lipidomics Strategy to Explore the Mechanism of: *Eleutherococcus senticosus* (Rupr. & Maxim.) Maxim. Leaves in the Treatment of Ischemic Stroke. *Food Funct.* **2021**, *12*, 4519–4534. [CrossRef] [PubMed]
15. Liu, M.; Chen, M.; Luo, Y.; Wang, H.; Huang, H.; Peng, Z.; Li, M.; Fei, H.; Luo, W.; Yang, J. Lipidomic Profiling of Ipsilateral Brain and Plasma after Celastrol Post-Treatment in Transient Middle Cerebral Artery Occlusion Mice Model. *Molecules* **2021**, *26*, 4124. [CrossRef]
16. Jaiswal, Y.S.; Williams, L.L. A Glimpse of Ayurveda—The Forgotten History and Principles of Indian Traditional Medicine. *J. Tradit. Complement. Med.* **2017**, *7*, 50–53. [CrossRef]
17. Sharma, P.; Kumar, P.; Sharma, R.; Gupta, G.; Chaudhary, A. Immunomodulators: Role of Medicinal Plants in Immune System. *Natl. J. Physiol. Pharm. Pharmacol.* **2017**, *7*, 552–556. [CrossRef]
18. Almatroodi, S.A.; Alsahli, M.A.; Almatroudi, A.; Rahmani, A.H. Ocimum Sanctum: Role in Diseases Management through Modulating Various Biological Activity. *Pharmacogn. J.* **2020**, *12*, 1198–1205. [CrossRef]
19. Hanumanthaiah, P.; Panari, H.; Chebte, A.; Haile, A.; Belachew, G.T. Tulsi (*Ocimum sanctum*)—A Myriad Medicinal Plant, Secrets behind the Innumerable Benefits. *Arab. J. Med. Aromat. Plants* **2020**, *6*, 106–127. [CrossRef]
20. Jamshidi, N.; Cohen, M.M. The Clinical Efficacy and Safety of Tulsi in Humans: A Systematic Review of the Literature. *Evid.-Based Complement. Altern. Med.* **2017**, *2017*, 9217567. [CrossRef]
21. Cohen, M.M. Tulsi-Ocimum Sanctum: A Herb for All Reasons. *J. Ayurveda Integr. Med.* **2014**, *5*, 251–259. [CrossRef]
22. Singh, N.; Hoette, Y.; Miller, D.R. *Tulsi: The Mother Medicine of Nature*; International Institute of Herbal Medicine: Lucknow, India, 2002; ISBN 8188007005.
23. Mohan, L.; Amberkar, M.V.; Kumari, M. *Ocimum sanctum* Linn. (TULSI)—An Overview. *Int. J. Pharm. Sci. Rev.* **2011**, *7*, 51–53.
24. Pattanayak, P.; Behera, P.; Das, D.; Panda, S.K. *Ocimum sanctum* Linn. A Reservoir Plant for Therapeutic Applications: An Overview. *Pharmacogn. Rev.* **2010**, *4*, 95. [CrossRef] [PubMed]
25. Mondal, S.; Mirdha, B.R.; Mahapatra, S.C. The Science behind Sacredness of Tulsi (*Ocimum sanctum* Linn.). *Indian J. Physiol. Pharmacol.* **2009**, *53*, 291–306.
26. Jaggi, R.K.; Madaan, R.; Singh, B. Anticonvulsant Potential of Holy Basil, *Ocimum sanctum* Linn., and Its Cultures. *Indian J. Exp. Biol.* **2003**, *41*, 1329–1333. [PubMed]
27. Yanpallewar, S.U.; Rai, S.; Kumar, M.; Acharya, S.B. Evaluation of Antioxidant and Neuroprotective Effect of *Ocimum sanctum* on Transient Cerebral Ischemia and Long-Term Cerebral Hypoperfusion. *Pharmacol. Biochem. Behav.* **2004**, *79*, 155–164. [CrossRef]
28. De Almeida, I.; Alviano, D.S.; Vieira, D.P.; Alves, P.B.; Blank, A.F.; Lopes, A.H.C.S.; Alviano, C.S.; Rosa, M.D.S.S. Antigiardial Activity of *Ocimum basilicum* Essential Oil. *Parasitol. Res.* **2007**, *101*, 443–452. [CrossRef]
29. Bhattacharyya, D.; Sur, T.K.; Jana, U.; Debnath, P.K. Controlled Programmed Trial of *Ocimum sanctum* Leaf on Generalized Anxiety Disorders. *Nepal Med. Coll. J.* **2008**, *10*, 176–179.
30. Baliga, M.S.; Jimmy, R.; Thilakchand, K.R.; Sunitha, V.; Bhat, N.R.; Saldanha, E.; Rao, S.; Rao, P.; Arora, R.; Palatty, P.L. *Ocimum sanctum* L (Holy Basil or Tulsi) and Its Phytochemicals in the Prevention and Treatment of Cancer. *Nutr. Cancer* **2013**, *65*, 26–35. [CrossRef]
31. Ahmad, A.; Khan, M.M.; Raza, S.S.; Javed, H.; Ashafaq, M.; Islam, F.; Safhi, M.M.; Islam, F. *Ocimum sanctum* Attenuates Oxidative Damage and Neurological Deficits Following Focal Cerebral Ischemia/Reperfusion Injury in Rats. *Neurol. Sci.* **2012**, *33*, 1239–1247. [CrossRef]

32. Kothari, S.K.; Bhattacharya, A.K.; Ramesh, S.; Garg, S.N.; Khanuja, S.P.S. Volatile Constituents in Oil from Different Plant Parts of Methyl Eugenol-Rich *Ocimum tenuiflorum* L.F. (Syn. *O. Sanctum* L.) Grown in South India. *J. Essent. Oil Res.* **2005**, *17*, 656–658. [CrossRef]

33. Ahmad, A.; Abuzinadah, M.F.; Alkreathy, H.M.; Banaganapalli, B.; Mujeeb, M. Ursolic Acid Rich *Ocimum sanctum* L Leaf Extract Loaded Nanostructured Lipid Carriers Ameliorate Adjuvant Induced Arthritis in Rats by Inhibition of COX-1, COX-2, TNF-α and IL-1: Pharmacological and Docking Studies. *PLoS ONE* **2018**, *13*, e0193451. [CrossRef] [PubMed]

34. Godhwani, J.L.; Vyas, D.S. *Ocimum sanctuim*: An Experimental Study Evaluating Its Anti-Inflammatory, Analgesic Andantipyretic Activity in Animals. *J. Ethnopharmacol.* **1987**, *21*, 153–163. [CrossRef] [PubMed]

35. Hannan, J.M.A.; Das, B.K.; Uddin, A.; Bhattacharjee, R.; Das, B.; Chowdury, H.S.; Mosaddek, A.S.M. Analgesic and Anti-Inflammatory Effects of *Ocimum sanctum* (Linn) in Laboratory Animals. *Int. J. Pharm. Sci. Res.* **2011**, *2*, 2121–2125.

36. Yuniarti, W.M.; Krismaharani, N.; Ciptaningsih, P.; Celia, K.; Veteriananta, K.D.; Ma'ruf, A.; Lukiswanto, B.S. The Protective Effect of *Ocimum sanctum* Leaf Extract against Lead Acetate-Induced Nephrotoxicity and Hepatotoxicity in Mice (Mus Musculus). *Vet. World* **2021**, *14*, 250–258. [CrossRef] [PubMed]

37. Raina, P.; Chandrasekaran, C.V.; Deepak, M.; Agarwal, A.; Ruchika, K.G. Evaluation of Subacute Toxicity of Methanolic/Aqueous Preparation of Aerial Parts of *O. Sanctum* in Wistar Rats: Clinical, Haematological, Biochemical and Histopathological Studies. *J. Ethnopharmacol.* **2015**, *175*, 509–517. [CrossRef]

38. Gautam, M.K.; Goel, R.K. Toxicological Study of *Ocimum sanctum* Linn Leaves: Hematological, Biochemical, and Histopathological Studies. *J. Toxicol.* **2014**, *2014*, 135654. [CrossRef]

39. Das, S.; Das, S.; Das, M.K.; Basu, S.P. Evaluation of Anti-Inflammatory Effect of Calotropis Gigantea and Tridax Procumbens on Wistar Albino Rats. *J. Pharm. Sci. Res.* **2009**, *1*, 123–126.

40. Baghdadi, H.H.; El-Demerdash, F.M.; Hussein, S.; Radwan, E.H. The Protective Effect of *Coriandrum sativum* L. Oil against Liver Toxicity Induced by Ibuprofen in Rats. *J. Biosci. Appl. Res.* **2016**, *2*, 197–202. [CrossRef]

41. Watson, B.D.; Dietrich, W.D.; Busto, R.; Wachtel, M.S.; Ginsberg, M.D. Induction of Reproducible Brain Infarction by Photochemically Initiated Thrombosis. *Ann. Neurol.* **1985**, *17*, 497–504. [CrossRef]

42. Paxinos, G.; Franklin, K.B.J. *The Mouse Brain in Stereotaxic Coordinates*; Academic Press: Cambridge, MA, USA, 2001.

43. Sharma, N.; Yadav, M.; Tripathi, G.; Mathew, B.; Bindal, V.; Falari, S.; Pamecha, V.; Maras, J.S. Bile Multi-Omics Analysis Classifies Lipid Species and Microbial Peptides Predictive of Carcinoma of Gallbladder. *Hepatology* **2022**, *76*, 920–935. [CrossRef]

44. Pang, Z.; Chong, J.; Zhou, G.; de Lima Morais, D.A.; Chang, L.; Barrette, M.; Gauthier, C.; Jacques, P.-É.; Li, S.; Xia, J. MetaboAnalyst 5.0: Narrowing the Gap between Raw Spectra and Functional Insights. *Nucleic Acids Res.* **2021**, *49*, W388–W396. [CrossRef]

45. Chi, Y.; Ma, Q.; Ding, X.Q.; Qin, X.; Wang, C.; Zhang, J. Research on Protective Mechanism of Ibuprofen in Myocardial Ischemia-Reperfusion Injury in Rats through the PI3K/Akt/MTOR Signaling Pathway. *Eur. Rev. Med. Pharmacol. Sci.* **2019**, *23*, 4465–4473. [CrossRef] [PubMed]

46. Dokmeci, D.; Kanter, M.; Inan, M.; Aydogdu, N.; Basaran, U.N.; Yalcin, O.; Turan, F.N. Protective Effects of Ibuprofen on Testicular Torsion/Detorsion-Induced Ischemia/Reperfusion Injury in Rats. *Arch. Toxicol.* **2007**, *81*, 655–663. [CrossRef] [PubMed]

47. Iwata, Y.; Nicole, O.; Zurakowski, D.; Okamura, T.; Jonas, R.A. Ibuprofen for Neuroprotection after Cerebral Ischemia. *J. Thorac. Cardiovasc. Surg.* **2010**, *139*, 489–493. [CrossRef] [PubMed]

48. Park, E.M.; Cho, B.P.; Volpe, B.T.; Cruz, M.O.; Joh, T.H.; Cho, S. Ibuprofen Protects Ischemia-Induced Neuronal Injury via up-Regulating Interleukin-1 Receptor Antagonist Expression. *Neuroscience* **2005**, *132*, 625–631. [CrossRef]

49. Giridharan, V.V.; Thandavarayan, R.A.; Mani, V.; Ashok Dundapa, T.; Watanabe, K.; Konishi, T. *Ocimum sanctum* Linn. Leaf Extracts Inhibit Acetylcholinesterase and Improve Cognition in Rats with Experimentally Induced Dementia. *J. Med. Food* **2011**, *14*, 912–919. [CrossRef]

50. Crack, P.J.; Taylor, J.M. Reactive Oxygen Species and the Modulation of Stroke. *Free Radic. Biol. Med.* **2005**, *38*, 1433–1444. [CrossRef]

51. Saeed, S.A.; Shad, K.F.; Saleem, T.; Javed, F.; Khan, M.U. Some New Prospects in the Understanding of the Molecular Basis of the Pathogenesis of Stroke. *Exp. Brain Res.* **2007**, *182*, 1–10. [CrossRef]

52. Kothari, A.; Sharma, S. Evaluation of Anti-Inflammatory Effect of Fresh Tulsi Leaves (*Ocimum sanctum*) against Different Mediators of Inflammation in Albino Rats. *Int. J. Pharm. Sci. Rev. Res.* **2012**, *14*, 119–123.

53. Fernández, P.B.; Figueredo, Y.N.; Dominguez, C.C.; Hernández, I.C.; Sanabria, M.L.G.; González, R. Anti-Inflammatory Effect of Lyophilized Aqueous Extract of *Ocinum tenuiflorum* on Rats. *Acta Farm Bonaer.* **2004**, *23*, 92–97.

54. Thakur, K.; Chem, K.P.-R.J. Undefined Anti-Inflammatory Activity of Extracted Eugenol from *Ocimum sanctum* L. Leaves. *Citeseer* **2009**, *2*, 472–474.

55. Singh, S.; Majumdar, D.K. Evaluation of Antiinflammatory Activity of Fatty Acids of *Ocimum sanctum* Fixed Oil. *Indian J. Exp. Biol.* **1997**, *35*, 380–383. [PubMed]

56. Singh, S. Comparative Evaluation of Antiinflammatory Potential of Fixed Oil of Different Species of Ocimum and Its Possible Mechanism of Action. *Indian J. Exp. Biol.* **1998**, *36*, 1028–1031. [PubMed]

57. Klein, M.E.; Mauch, S.; Rieckmann, M.; Martínez, D.G.; Hause, G.; Noutsias, M.; Hofmann, U.; Lucas, H.; Meister, A.; Ramos, G.; et al. Phosphatidylserine (PS) and Phosphatidylglycerol (PG) Nanodispersions as Potential Anti-Inflammatory Therapeutics: Comparison of in Vitro Activity and Impact of Pegylation. *Nanomed. Nanotechnol. Biol. Med.* **2020**, *23*, 102096. [CrossRef] [PubMed]

58. Klein, M.E.; Rieckmann, M.; Sedding, D.; Hause, G.; Meister, A.; Mäder, K.; Lucas, H. Towards the Development of Long Circulating Phosphatidylserine (Ps)- and Phosphatidylglycerol (Pg)-enriched Anti-inflammatory Liposomes: Is Pegylation Effective? *Pharmaceutics* **2021**, *13*, 282. [CrossRef]
59. Klein, M.E.; Rieckmann, M.; Lucas, H.; Meister, A.; Loppnow, H.; Mäder, K. Phosphatidylserine (PS) and Phosphatidylglycerol (PG) Enriched Mixed Micelles (MM): A New Nano-Drug Delivery System with Anti-Inflammatory Potential? *Eur. J. Pharm. Sci.* **2020**, *152*, 105451. [CrossRef]
60. Treede, I.; Braun, A.; Sparla, R.; Kühnel, M.; Giese, T.; Turner, J.R.; Anes, E.; Kulaksiz, H.; Füllekrug, J.; Stremmel, W.; et al. Anti-Inflammatory Effects of Phosphatidylcholine. *J. Biol. Chem.* **2007**, *282*, 27155–27164. [CrossRef]
61. Erős, G.; Varga, G.; Váradi, R.; Czóbel, M.; Kaszaki, J.; Ghyczy, M.; Boros, M. Anti-Inflammatory Action of a Phosphatidylcholine, Phosphatidylethanolamine and N-Acylphosphatidylethanolamine-Enriched Diet in Carrageenan-Induced Pleurisy. *Eur. Surg. Res.* **2008**, *42*, 40–48. [CrossRef]
62. Chen, L.; Beppu, F.; Takatani, N.; Miyashita, K.; Hosokawa, M. N-3 Polyunsaturated Fatty Acid-Enriched Phosphatidylglycerol Suppresses Inflammation in RAW264.7 Cells through Nrf2 Activation via Alteration of Fatty Acids in Cellular Phospholipids. *Fish. Sci.* **2021**, *87*, 727–737. [CrossRef]
63. Ireland, R.; Schwarz, B.; Nardone, G.; Wehrly, T.D.; Broeckling, C.D.; Chiramel, A.I.; Best, S.M.; Bosio, C.M. Unique Francisella Phosphatidylethanolamine Acts as a Potent Anti-Inflammatory Lipid. *J. Innate Immun.* **2018**, *10*, 291–305. [CrossRef]
64. Liu, Z.; Xu, P.; Gong, F.; Tan, Y.; Han, J.; Tian, L.; Yan, J.; Li, K.; Xi, Z.; Liu, X. Altered Lipidomic Profiles in Lung and Serum of Rat after Sub-Chronic Exposure to Ozone. *Sci. Total Environ.* **2022**, *806*, 150630. [CrossRef]
65. Ulivi, V.; Lenti, M.; Gentili, C.; Marcolongo, G.; Cancedda, R.; Descalzi Cancedda, F. Anti-Inflammatory Activity of Monogalacto-syldiacylglycerol in Human Articular Cartilage in Vitro: Activation of an Anti-Inflammatory Cyclooxygenase-2 (COX-2) Pathway. *Arthritis Res. Ther.* **2011**, *13*, R92. [CrossRef] [PubMed]
66. Bruno, A.; Rossi, C.; Marcolongo, G.; Di Lena, A.; Venzo, A.; Berrie, C.P.; Corda, D. Selective in Vivo Anti-Inflammatory Action of the Galactolipid Monogalactosyldiacylglycerol. *Eur. J. Pharmacol.* **2005**, *524*, 159–168. [CrossRef] [PubMed]
67. Zi, Y.; Yao, M.; Lu, Z.; Lu, F.; Bie, X.; Zhang, C.; Zhao, H. Glycoglycerolipids from the Leaves of *Perilla frutescens* (L.) Britton (Labiatae) and Their Anti-Inflammatory Activities in Lipopolysaccharide-Stimulated RAW264.7 Cells. *Phytochemistry* **2021**, *184*, 112679. [CrossRef] [PubMed]
68. Leutou, A.S.; McCall, J.R.; York, R.; Govindapur, R.R.; Bourdelais, A.J. Anti-Inflammatory Activity of Glycolipids and a Polyunsaturated Fatty Acid Methyl Ester Isolated from the Marine Dinoflagellate Karenia Mikimotoi. *Mar. Drugs* **2020**, *18*, 138. [CrossRef]
69. Maeda, N.; Kokai, Y.; Hada, T.; Yoshida, H.; Mizushina, Y. Oral Administration of Monogalactosyl Diacylglycerol from Spinach Inhibits Colon Tumor Growth in Mice. *Exp. Ther. Med.* **2013**, *5*, 17–22. [CrossRef]
70. Murakami, C.; Kumagai, T.; Hada, T.; Kanekazu, U.; Nakazawa, S.; Kamisuki, S.; Maeda, N.; Xu, X.; Yoshida, H.; Sugawara, F.; et al. Effects of Glycolipids from Spinach on Mammalian DNA Polymerases. *Biochem. Pharmacol.* **2003**, *65*, 259–267. [CrossRef]
71. Calzada, E.; Onguka, O.; Claypool, S.M. Phosphatidylethanolamine Metabolism in Health and Disease. *Int. Rev. Cell Mol. Biol.* **2016**, *321*, 29–88. [CrossRef]
72. Akyol, S.; Ugur, Z.; Yilmaz, A.; Ustun, I.; Gorti, S.K.K.; Oh, K.; McGuinness, B.; Passmore, P.; Kehoe, P.G.; Maddens, M.E.; et al. Lipid Profiling of Alzheimer's Disease Brain Highlights Enrichment in Glycerol(Phospho)Lipid, and Sphingolipid Metabolism. *Cells* **2021**, *10*, 2591. [CrossRef]
73. Rao, A.M.; Hatcher, J.F.; Dempsey, R.J. Lipid Alterations in Transient Forebrain Ischemia: Possible New Mechanisms of CDP-Choline Neuroprotection. *J. Neurochem.* **2000**, *75*, 2528–2535. [CrossRef]
74. Yang, Y.; Zhong, Q.; Zhang, H.; Mo, C.; Yao, J.; Huang, T.; Zhou, T.; Tan, W. Lipidomics Study of the Protective Effects of Isosteviol Sodium on Stroke Rats Using Ultra High-Performance Supercritical Fluid Chromatography Coupling with Ion-Trap and Time-of-Flight Tandem Mass Spectrometry. *J. Pharm. Biomed. Anal.* **2018**, *157*, 145–155. [CrossRef] [PubMed]
75. Glade, M.J.; Smith, K. Phosphatidylserine and the Human Brain. *Nutrition* **2015**, *31*, 781–786. [CrossRef] [PubMed]
76. Kim, H.Y.; Huang, B.X.; Spector, A.A. Phosphatidylserine in the Brain: Metabolism and Function. *Prog. Lipid Res.* **2014**, *56*, 1–18. [CrossRef]
77. Dharsono, H.D.A.; Putri, S.A.; Kurnia, D.; Dudi, D.; Satari, M.H. Ocimum Species: A Review on Chemical Constituents and Antibacterial Activity. *Molecules* **2022**, *27*, 6350. [CrossRef] [PubMed]
78. Ramaiah, M.; Prathi, A.; Singam, B.; Tulluru, G.; Tummala, L. A Review on Ocimum Species: *Ocimum americanum* L., *Ocimum basilicum* L., *Ocimum gratissimum* L. and *Ocimum tenuiflorum* L. *Int. J. Res. Ayurveda Pharm.* **2019**, *10*, 41–48. [CrossRef]
79. Rahman, S.; Islam, R.; Kamruzzaman, M.; Alam, K.; Jamal, A.H.M. *Ocimum sanctum* L.: A Review of Phytochemical and Pharmacological Profile. *Am. J. Drug Discov. Dev.* **2011**, *1*, 1–15.
80. Liu, J. Pharmacology of Oleanolic Acid and Ursolic Acid. *J. Ethnopharmacol.* **1995**, *49*, 57–68. [CrossRef]
81. Kazmi, I.; Narooka, A.R.; Afzal, M.; Singh, R.; Al-Abbasi, F.A.; Ahmad, A.; Anwar, F. Anticancer Effect of Ursolic Acid Stearoyl Glucoside in Chemically Induced Hepatocellular Carcinoma. *J. Physiol. Biochem.* **2013**, *69*, 687–695. [CrossRef]
82. Kazmi, I.; Afzal, M.; Gupta, G.; Anwar, F. Antiepileptic Potential of Ursolic Acid Stearoyl Glucoside by Gaba Receptor Stimulation. *CNS Neurosci. Ther.* **2012**, *18*, 799–800. [CrossRef]

83. Kazmi, I.; Rahman, M.; Afzal, M.; Gupta, G.; Saleem, S.; Afzal, O.; Shaharyar, M.A.; Nautiyal, U.; Ahmed, S.; Anwar, F. Anti-Diabetic Potential of Ursolic Acid Stearoyl Glucoside: A New Triterpenic Gycosidic Ester from Lantana Camara. *Fitoterapia* **2012**, *83*, 142–146. [CrossRef]

84. Kim, S.H.; Hong, J.H.; Lee, Y.C. Ursolic Acid, a Potential PPARγ Agonist, Suppresses Ovalbumin-Induced Airway Inflammation and Penh by down-Regulating IL-5, IL-13, and IL-17 in a Mouse Model of Allergic Asthma. *Eur. J. Pharmacol.* **2013**, *701*, 131–143. [CrossRef] [PubMed]

85. Dhandayuthapani, S.; Azad, H.; Rathinavelu, A. Apoptosis Induction by *Ocimum sanctum* Extract in LNCaP Prostate Cancer Cells. *J. Med. Food* **2015**, *18*, 776–785. [CrossRef]

86. Ding, H.; Wang, H.; Zhu, L.; Wei, W. Ursolic Acid Ameliorates Early Brain Injury After Experimental Traumatic Brain Injury in Mice by Activating the Nrf2 Pathway. *Neurochem. Res.* **2016**, *42*, 337–346. [CrossRef] [PubMed]

87. Zhang, T.; Su, J.; Wang, K.; Zhu, T.; Li, X. Ursolic Acid Reduces Oxidative Stress to Alleviate Early Brain Injury Following Experimental Subarachnoid Hemorrhage. *Neurosci. Lett.* **2014**, *579*, 12–17. [CrossRef] [PubMed]

88. Zhang, T.; Su, J.; Guo, B.; Zhu, T.; Wang, K.; Li, X. Ursolic Acid Alleviates Early Brain Injury after Experimental Subarachnoid Hemorrhage by Suppressing {TLR}4-Mediated Inflammatory Pathway. *Int. Immunopharmacol.* **2014**, *23*, 585–591. [CrossRef] [PubMed]

89. Wang, Y.; Li, L.; Deng, S.; Liu, F.; He, Z. Ursolic Acid Ameliorates Inflammation in Cerebral Ischemia and Reperfusion Injury Possibly via High Mobility Group Box 1/Toll-like Receptor 4/NFκB Pathway. *Front. Neurol.* **2018**, *9*, 253. [CrossRef]

90. Papangelis, A.; Ulven, T. Synthesis of Lysophosphatidylcholine and Mixed Phosphatidylcholine. *J. Org. Chem.* **2022**, *87*, 8194–8197. [CrossRef]

91. Whitehead, S.N.; Chan, K.H.N.; Gangaraju, S.; Slinn, J.; Li, J.; Hou, S.T. Imaging Mass Spectrometry Detection of Gangliosides Species in the Mouse Brain Following Transient Focal Cerebral Ischemia and Long-Term Recovery. *PLoS ONE* **2011**, *6*, e20808. [CrossRef]

92. Jove, M.; Mauri-Capdevila, G.; Suarez, I.; Cambray, S.; Sanahuja, J.; Quilez, A.; Farre, J.; Benabdelhak, I.; Pamplona, R.; Portero-Otin, M.; et al. Metabolomics Predicts Stroke Recurrence after Transient Ischemic Attack. *Neurology* **2014**, *84*, 36–45. [CrossRef]

93. Koizumi, S.; Yamamoto, S.; Hayasaka, T.; Konishi, Y.; Yamaguchi-Okada, M.; Goto-Inoue, N.; Sugiura, Y.; Setou, M.; Namba, H. Imaging Mass Spectrometry Revealed the Production of Lyso-Phosphatidylcholine in the Injured Ischemic Rat Brain. *Neuroscience* **2010**, *168*, 219–225. [CrossRef]

94. Wang, H.Y.J.; Liu, C.B.; Wu, H.W.; Kuo, S. Direct Profiling of Phospholipids and Lysophospholipids in Rat Brain Sections after Ischemic Stroke. *Rapid Commun. Mass Spectrom.* **2010**, *24*, 2057–2064. [CrossRef] [PubMed]

95. Takenouchi, T.; Sato, M.; Kitani, H. Lysophosphatidylcholine Potentiates Ca^{2+} Influx, Pore Formation and P44/42 MAP Kinase Phosphorylation Mediated by P2X7 Receptor Activation in Mouse Microglial Cells. *J. Neurochem.* **2007**, *102*, 1518–1532. [CrossRef] [PubMed]

Article

Hesperetin and Capecitabine Abate 1,2 Dimethylhydrazine-Induced Colon Carcinogenesis in Wistar Rats via Suppressing Oxidative Stress and Enhancing Antioxidant, Anti-Inflammatory and Apoptotic Actions

Asmaa K. Hassan [1], Asmaa M. El-Kalaawy [2], Sanaa M. Abd El-Twab [1], Mohamed A. Alblihed [3] and Osama M. Ahmed [1,*]

[1] Physiology Division, Zoology Department, Faculty of Science, Beni-Suef University, Beni-Suef 62521, Egypt
[2] Pharmacology Department, Faculty of Medicine, Beni-Suef University, Beni-Suef 62521, Egypt
[3] Department of Microbiology, College of Medicine, Taif University, Taif 21944, Saudi Arabia
* Correspondence: osamamoha@yahoo.com or osama.ahmed@science.bsu.edu.eg

Abstract: Colon cancer is a major cause of cancer-related death, with significantly increasing rates of incidence worldwide. The current study was designed to evaluate the anti-carcinogenic effects of hesperetin (HES) alone and in combination with capecitabine (CAP) on 1,2 dimethylhydrazine (DMH)-induced colon carcinogenesis in Wistar rats. The rats were given DMH at 20 mg/kg body weight (b.w.)/week for 12 weeks and were orally treated with HES (25 mg/kg b.w.) and/or CAP (200 mg/kg b.w.) every other day for 8 weeks. The DMH-administered rats exhibited colon-mucosal hyperplastic polyps, the formation of new glandular units and cancerous epithelial cells. These histological changes were associated with the significant upregulation of colon Ki67 expression and the elevation of the tumor marker, carcinoembryonic antigen (CEA), in the sera. The treatment of the DMH-administered rats with HES and/or CAP prevented these histological cancerous changes concomitantly with the decrease in colon-Ki67 expression and serum-CEA levels. The results also indicated that the treatments with HES and/or CAP showed a significant reduction in the serum levels of lipid peroxides, an elevation in the serum levels of reduced glutathione, and the enhancement of the activities of colon-tissue superoxide dismutase, glutathione reductase and glutathione-S-transferase. Additionally, the results showed an increase in the mRNA expressions of the anti-inflammatory cytokine, IL-4, as well as the proapoptotic protein, p53, in the colon tissues of the DMH-administered rats treated with HES and/or CAP. The TGF-β1 decreased significantly in the DMH-administered rats and this effect was counteracted by the treatments with HES and/or CAP. Based on these findings, it can be suggested that both HES and CAP, singly or in combination, have the potential to exert chemopreventive effects against DMH-induced colon carcinogenesis via the suppression of oxidative stress, the stimulation of the antioxidant defense system, the attenuation of inflammatory effects, the reduction in cell proliferation and the enhancement of apoptosis.

Keywords: colon carcinogenesis; 1,2 dimethylhydrazine; hesperetin; capecitabine

Citation: Hassan, A.K.; El-Kalaawy, A.M.; Abd El-Twab, S.M.; Alblihed, M.A.; Ahmed, O.M. Hesperetin and Capecitabine Abate 1,2 Dimethylhydrazine-Induced Colon Carcinogenesis in Wistar Rats via Suppressing Oxidative Stress and Enhancing Antioxidant, Anti-Inflammatory and Apoptotic Actions. *Life* **2023**, *13*, 984. https://doi.org/10.3390/life13040984

Academic Editors: Marisa Colone, Charalampia Amerikanou and Efstathia Papada

Received: 17 February 2023
Revised: 7 April 2023
Accepted: 8 April 2023
Published: 11 April 2023

1. Introduction

Colorectal cancer (CRC) is the third and second most prevalent cancer in males and females, respectively, around the world. It accounts for 10% of all malignancies and is thought to be the cause of approximately 600,000 deaths each year [1,2]. There are numerous risk factors linked to the development of CRC, including exogenous risk factors such as obesity, lack of physical exercise, nicotine use, moderate-to-excessive alcohol consumption, hypertension, increased blood lipids and colonization by *Streptococcus gallolyticus*, as well as endogenous risk factors, such as a personal or family history of colon polyps and hereditary CRC, inflammatory bowel illness, type 2 diabetes, hereditary nonpolyposis colon cancer (CC) and Cowden's disease [3,4]. Although CC is frequently discovered in the late stages when the symptoms become clear, the early detection of cancer can save the lives of patients [5].

The most commonly used CC animal model is the 1-dimethylhydrazine (DMH)-induced animal model [6]. Colon tumors produced by DMH, a powerful colon carcinogen, resemble human CC in many ways, including how they react to several promotion- and prevention-related drugs [7]. A number of pathogenic alterations, including the creation of aberrant cryptic foci, occur as a result of DMH-induced CC in a multi-step process [8].

Oxidative stress due to the excessive production of reactive oxygen species (ROS) is a cellular state that overrides the antioxidant defense mechanisms of cells. Many studies have demonstrated a substantial correlation between oxidative stress and the development or advancement of a number of human diseases, including cancer [9–11]. Chronic oxidative stress has been connected to cancer in epidemiological studies [12], proving its role in the development of cancer. The function of ROS in tumor genesis, development and progression is supported by a large body of experimental evidence [13–15]. Reactive oxygen species are created during typical cellular metabolism. Although ROS generation is essential for healthy cell-signaling pathways, excessive ROS can harm mitochondrial and genomic deoxyribonucleic acid (DNA), causing mutations in molecules, as well as DNA damage [16].

Apoptosis is a tightly controlled physiological process of cell death that eliminates unneeded, severely damaged, mutant, ageing and/or unrepairable cells while maintaining the integrity of the remaining cells and the organism as a whole [17,18]. Apoptosis imbalance, which can involve levels of apoptosis that are either excessively high or excessively low, may contribute to the pathogenesis of a variety of illnesses, including cancer, ischemia, neurodegeneration and autoimmunity [19]. Apoptosis is triggered by toxic carcinogens or mutagenic substances, viral infections and UV light. Extracellular or intracellular cues can commit cells to undergoing apoptosis, which involves activating the caspase family through intrinsic and extrinsic mechanisms [20,21].

The versatile cytokine known as transforming growth factor-beta (TGF-β) was shown to have both physiological and pathological uses. The three main TGF-family isoforms, TGF-β1, TGF-β2 and TGF-β3, exhibit various biological functions [22,23]. Interestingly, only the promoter region of TGF-β1 can be activated directly by reactive oxygen species (ROS), involving different trans-activating proteins, such as plasmin; due to its multiple regulatory sites [24], this highlights its pleiotropic nature in carcinogenesis, fibrogenesis, immunomodulation, cell proliferation and cell differentiation [25,26].

Many biological processes, including death, differentiation and proliferation, have been intensively examined in relation to the molecular pathways of TGF-β signaling [27].

Particularly in cancer, TGF-βsignaling results in many downstream effects in a context-dependent manner. It has two functions: one as a tumor suppressor in pre-malignant cells and the other as a tumor promoter in cancer cells [28]. Through acquired mutations, cancer cells are able to deactivate the tumor-suppressive elements of TGF-β/Smad signaling, whilst tumor-suppressive effects can selectively apply pressure on pre-malignant cells [29].

Surgery and chemotherapy are the key components of the current clinical treatment for CC. Nonetheless, finding new and more potent medications for the treatment of CC is urgently needed due to the development of side effects and the emergence of drug resistance [30].

Capecitabine (CAP) is a fluoropyrimidine-based chemotherapeutic drug used to treat a variety of malignancies, including colon, colorectal and breast cancer [31]. As an antimetabolite, it causes cell-cycle arrest and apoptosis by blocking DNA polymerase. It has estrogenic properties, cytotoxicity, toxicity and teratogenic properties [32,33]. Additionally, CAP and its metabolites have inter-individual variability in their pharmacokinetic characteristics; this is most likely due to variations in the activity of enzymes involved in CAP metabolism [34,35]. Cytotoxic medications not only kill cancer cells but also harm healthy cells. This toxic reaction has led to concern regarding drug dose and has become a factor influencing patients' quality of life. Bone-marrow suppression, gastrointestinal problems and hair loss are among the most prevalent adverse effects [36].

Citrus aurantium L. (Rutaceae) fruit peel contains a naturally occurring flavone called hesperetin (HES), a phytoestrogen with anti-tumor effects [37]. Hesperetin has been shown to apply a cytotoxic mechanism against a variety of cancer cells, including those from breast cancer [38], pancreatic cancer [39], prostate cancer [40], glioblastoma [41], liver cancer [42], kidney cancer [43], colon cancer [44], lung cancer [45], oral cancer [46], esophageal cancer [47], osteosarcoma [48], ovarian cancer [49], thyroid [50], leukemia [51] and others.

The aim of this study was to investigate the potential of the promising effects of HES on colon carcinogenesis, both alone and in combination with CAP, on DMH-induced colon carcinogenesis in rats.

2. Materials and Methods

2.1. Drugs and Chemicals

The HES (3′,5,7-trihydroxy-4′-methoxy flavanone) and DMH were purchased from Sigma-Aldrich (St. Louis, MO, USA) and stored at 2–4 °C. The CAP was obtained from the Roche Company and stored at 20–25 °C. Carcinoembryonic antigen (CEA)-enzyme-linked immunosorbent assay (ELISA) kit was supplied by R&D Systems (Minneapolis, MN, USA). The primary antibody for TGF-β1 was obtained from ABclonal Technology (Wuhan, China). All other chemicals used in the experimental procedures and assays were of analytical grade.

2.2. Animals and Treatment

Fifty adult male Wistar rats with body weight (b.w.) of approximately 100 ± 20 g were obtained from the National Research Center, Doki and Giza, Egypt. They were kept under observation for two weeks prior to the experiment to exclude any with infections at the time at which the study began. The chosen animals were housed in polystyrene-well aerated cages at normal atmospheric temperature (25 ± 5 °C) and humidity ($55 \pm 5\%$) and under a 12-h light/dark cycle. During the study period, the rats were provided with water and a normal basal diet. All animal procedures were in accordance with the guidelines and recommendations of the Experimental Animal Ethics Committee for Use and Care of Animals, Faculty of Science, Beni-Suef University, Egypt (ethical approval number BSU/FS/2018/17).

The experimental animals were randomly allocated into five groups (with ten animals in each), as follows: Group 1 served as a normal control, in which rats were orally administered equivalent volumes of saline (0.9% NaCl) each week for 12 weeks and 1% carboxymethylcellulose (CMC) every other day during the last 8 weeks; the rats in Group 2, the DMH-administered group, were orally given DMH (20 mg/kg b.w.) [52] dissolved in saline (0.9% NaCl) each week for 12 weeks and the equivalent volume of 1% CMC every other day during the last 8 weeks; the rats in Group 3, the DMH-administrated group treated with HES, were orally given DMH as described for Group 2 and orally treated with HES (25 mg/kg b.w.) [53] dissolved in 0.1% CMC every other day for 8 weeks, starting from the 5th week of the DMH administration; the rats in Group 4, the DMH-administered group treated with CAP, were orally given DMH as described for Group 2 and orally treated with CAP (200 mg/kg b.w.) [54] dissolved in 0.1% CMC every other day for 8 weeks, starting from the 5th week of the DMH administration; and the rats in Group 5, the DMH-administered group treated with HES and CAP combination, were orally given DMH as described for Group 2 and orally treated with HES (25 mg/kg b.w.) and CAP (200 mg/kg b.w.) dissolved in 0.1% CMC every other day for 8 weeks, starting from the 5th week of the DMH administration (Figure 1).

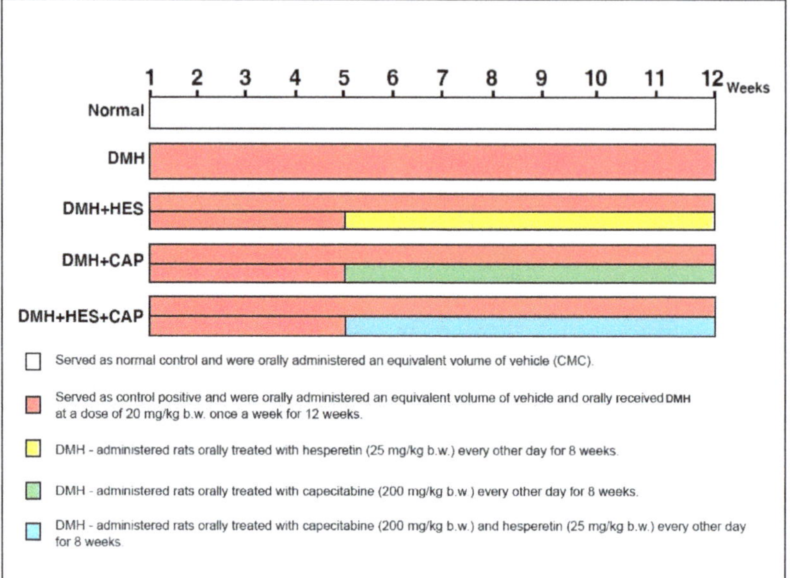

Figure 1. Experimental design.

2.3. Blood and Colon Sampling

After 12 weeks, the animals were given inhalation anesthesia, blood samples from the jugular vein were taken and colon-tissue samples were removed for biochemical, histological and molecular investigations. The animals were then decapitated and dissected. After allowing the blood samples to clot, the sera were separated using centrifugation at 3000 r.p.m. for 15 min. The obtained sera were collected into sterilized tubes and stored at -30 °C. Half gram of each frozen colon was homogenized in 10 mL 0.9% NaCl to yield 1% homogenate (w/v), and then centrifuged at 3000 r.p.m. for 15 min at 4 °C; the supernatant was separated and kept at -30 °C until it was used for the determinations of oxidative-stress and antioxidant-defense parameters. Other pieces from the colon of each rat were gathered on 10% neutral buffered formalin for histological evaluation and others were stored at -70 °C in sterilized Eppendorf tubes for RNA isolation and real-time PCR (RT-PCR) analysis.

2.4. Biochemical Investigations

The serum levels of CEA were estimated using ELISA kits (R&D Systems, Minneapolis, MN, USA), as per the manufacturer's instructions.

Serum levels of lipid peroxides (LPO) were estimated according to the method described by Preuss et al. [55]. In brief, the proteins were precipitated by adding 0.15 mL 76% trichloroacetic acid (TCA) to 1 mL serum. In order to develop the color of the isolated supernatant, 0.35 mL of thiobarbituric acid (TBA) was added. After 30 min of incubation in an 80 °C water bath, a faint pink color developed and was detected at 532 nm. Malondialdehyde (MDA; 1,1,3,3-tetramethoxypropane) was used as standard. Serum level of reduced glutathione (GSH) content was estimated according to the method described by Beutler [56] by adding 0.5 mL 5,5′-dithiobis(2-nitrobenzoic acid), known as Ellman's reagent (a color-developing agent) and phosphate-buffer solution (pH 7) to the serum after protein precipitation. The yellow color developed in samples and GSH standard was measured at 412 nm against blank.

The activities of glutathione reductase (GR), glutathione-S-transferase (GST) and superoxide dismutase (SOD) were determined in colon homogenates using the methods presented by Goldberg [57], Mannervik and Guthenberg [58] and Marklund and Marklund [59], respectively. The colon-GR activity was determined by mixing 40 µL of colon-homogenate supernatant with 1 mL substrate (2.2 mmol/L oxidized glutathione) dissolved in buffer (250 mmol/L potassium phosphate; pH 7.3). A volume of 200 µL 0.17 mmol/L NADPH (nicotinamide adenine dinucleotide reduced form) was added and the mixture was incubated in an incubator at 37 °C. The GR activity was calculated from the formula: activity (U/L) = 4983 × ΔA nm/min. To determine colon GST activity, 250 µL mM 1-chloro-2,4-dinitrobenzene (CDNB) was added to a Wasserman tube that contained 250 µL sample, 250 µL GSH solution (4 mM) and 250 µL phosphate buffer (pH 7.3). The developed color was measured after 10 min of incubation at 25 °C at 430 nm. Colon-SOD activity was determined based on the inhibition of auto-oxidation of pyrogallol by the enzyme. The process was dependent on the presence of superoxide ions. The amount of enzyme that caused a 50% inhibition in the extinction changes in 1 min compared to the control was regarded as one unit of the enzyme. Briefly, 50 µL of pyrogallol (10 mM) was added to 1 mL of the colon-homogenate supernatant in the presence of Tris buffer (pH 8). The initial absorbance was measured after adding pyrogallol and at 10 min. The inhibition of the yellow color at 430 nm and the enzyme activity were calculated.

2.5. Ribonucleic Acid (RNA) Isolation and Reverse Transcriptase–Polymerase Chain Reaction (RT-PCR) Analysis

The total RNA was separated from the colon tissues based on the method described by Chomczynski and Sacchi [60], using a Qiagen tissue-extraction kit (USA). The isolated RNA was quantified at 260 nm and transcribed into cDNA using My Taq One-Step RT-PCR Kit (Bioline, Meridian Bioscience, Memphis, TN, USA) in the presence of specific primers (LGC Biosearch Technologies, Petaluma, CA, USA) of proliferator marker (Ki67), interleukin-14 (IL-4), proapoptotic protein 53 (p53) and β-actin (Table 1). The resultant PCR products were analyzed following electrophoresis in 1× Tris-Borate-EDTA buffer (pH 8.3–8.5) on 1.5% agarose gel stained with ethidium bromide. A gel-documentation system was used to visualize the electrophoretic pattern. The relative values of gene expression were normalized to that of β-actin.

Table 1. Primer sequences used in qRT-PCR analysis.

Gene	Sequence (5′–3′)	References
Ki67	F: 5d CTTTGCGCCATGCTGAAACT3′ R: 5d ATGACGACCTGGAACATCGG3′	Yanai et al. [61]
IL-4	F: 5d GGAACACCACGGAGAACG3′ R: 5d GCACGGAGGTACATCACG3′	Zhou et al. [62]
p53	F: 5d CAGCGTGATGATGGTAAGGA3′ R: 5d GCGTTGCTCTGATGGTGA3′	Ahmed et al. [63]
β-actin	F: 5d TCACCCTGAAGTACCCCATGGAG3′ R: 5d TTGGCCTTGGGGTTCAGGGGG3′	Ahmed et al. [63]

2.6. Histopathological Studies

Colon pieces of each rat were fixed in 10% neutral buffered formalin for 24 h before dehydration in an ascending series of alcohol concentrations, clearing in xylene and embedding in paraffin wax. The paraffin-wax blocks with the tissues were prepared by cutting 5 µm sections. Next, the tissue sections were processed for staining using hematoxylin and eosin (H & E) [64] and the examination was conducted using an electric-light microscope.

2.7. Immunohistochemistry

For the immunohistochemical investigations, colon sections (4 µm thick) were mounted onto positive-charged slides (Thermo Fisher Scientific, Pittsburgh, PA, USA) and immunostaining was conducted according to the methods described by Ahmed and Ahmed [65].

Briefly, the sections were incubated in 3% H_2O_2 solution for 15 min following deparaffinization, rehydration, antigen retrieval and sealing. Next, they were blocked and incubated with TGF-β1 antibody (Santa Cruz Biotechnology, Santa Cruz, CA, USA) (1:200 dilution) at 4 °C overnight. After washing with phosphate-buffered saline, the sections with the peroxidase-labeled secondary antibody (1:200 dilution) were incubated for 30 min. The bound antibody complex was visualized by the reaction of 3,3-diaminobenzidine (DAB) substrate and counterstaining with hematoxylin. This method was applied according to the instructions of ABclonal Inc. Company, Wuhan, China. The immunohistochemically stained sections were examined by a light microscope at high power ($\times 400$). The positive reaction appeared brown in color. The integrated intensities of the TGF-β1 response were measured using the ImageJ program.

2.8. Statistical Analysis

The results were expressed as mean ± standard error (SE), which equals SD/$\sqrt{n}$ (n represents the number of animals). All statistical comparisons were made by one-way ANOVA test followed by Duncan's method for post hoc analysis using Statistical Package for the Social Sciences (SPSS) version 22 for Windows (New York, NY, USA) [66]. Symbols a, b, c and d were used to indicate significance between groups for each parameter. The means, which had different symbols, were statistically significant at $p < 0.05$.

3. Results

3.1. Effect of HES and CAP on Serum CEA Level

The oral intake of DMH induced a significant ($p < 0.05$) elevation in the serum levels of CEA when compared to the normal control rats. BY contrast, the treatment of the DMH-administered rats with HES and CAP, both individually and in combination, produced a significant improvement ($p < 0.05$) in the serum levels of CEA in comparison with the DMH-administered control (Table 2); the combinatory effect seemed to be the most potent.

Table 2. Effects of HES and CAP on serum CEA levels in DMH-administered rats.

Groups	CEA (ng/mL)
Normal control	1.90 ± 0.07 [a]
DMH control	12.83 ± 0.65 [d]
DMH + HES	5.01 ± 0.29 [bc]
DMH + CAP	6.01 ± 0.35 [c]
DMH + HES + CAP	4.17 ± 0.15 [b]

Data are presented as the mean ± SE ($n = 6$). Means with different superscript symbols ([a–d]) are significantly different at $p < 0.05$.

3.2. Effect on Oxidative-Stress and Antioxidant-Defense Markers

3.2.1. Effects of HES and CAP on Serum Levels of LPO and GSH

The DMH-administered rats exhibited a significant ($p < 0.05$) increase in their serum LPO levels compared to the normal control rats. The oral supplementation of HES and CAP, both individually and in combination, significantly ($p < 0.05$) and successfully prevented the LPO elevation when compared to the DMH-administered control group (Table 3).

By contrast, the serum level of GSH was significantly ($p < 0.05$) decreased in the DMH-administered rats compared to the normal control rats. The supplementation of HES alone and/or in combination with CAP to the DMH-administered rats significantly ($p < 0.05$) prevented the depletion of the serum GSH level when compared to the DMH-administered control (Table 3).

Table 3. Effects of HES and CAP on serum LPO and GSH levels in DMH-administered rats.

Groups	LPO (nmol/mL)	GSH (µmol/L)
Normal control	6.04 ± 0.8 [a]	4.30 ± 0.23 [d]
DMH control	18.10 ± 0.4 [c]	1.23 ± 0.15 [a]
DMH + HES	13.65 ± 0.62 [b]	3.83 ± 0.26 [cd]
DMH + CAP	15.00 ± 1.07 [b]	2.71 ± 0.10 [b]
DMH + HES + CAP	15.55 ±1.15 [b]	3.69 ± 0.06 [c]

Data are presented as the mean ± SE (n = 6). Within the same column, means with different superscript symbols ([a–d]) are significantly different at $p < 0.05$.

3.2.2. Effects of HES and CAP on Colon SOD, GR and GST Activities in DMH-Administered Rats

The data presented in Table 4 exhibit a significant ($p < 0.05$) decrease in the colon-homogenate activities of the SOD, GR and GST in the DMH-administered group compared with those of the normal control. By contrast, supplementation with the CAP and HES both alone and in combination prevented the depletion of SOD, GR and GST activities ($p < 0.05$). The effect of HES and CAP in combination on the colon GR and GST activities seemed to be the most potent.

Table 4. Effects of HES and CAP on colon SOD, GST and GR activities in DMH-administered rats.

Groups	SOD (U/g)	GR (U/g)	GST (U/g)
Normal control	19.8 ± 0.82 [c]	90.21 ± 4.25 [b]	632.11 ± 4.71 [bc]
DMH control	3.77 ± 0.21 [a]	36.38 ± 5.08 [a]	254.26 ± 28.17 [a]
DMH + HES	10.71 ± 0.24 [b]	134.54 ± 13.73 [c]	608.95 ± 10.40 [b]
DMH + CAP	11.44 ± 0.20 [b]	166.18 ± 13.49 [cd]	651.09 ± 7.82 [bc]
DMH + HES + CAP	11.46 ± 0.07 [b]	185.36 ± 15.94 [d]	658.58 ± 7.24 [c]

Data are presented as the mean ± SE (n = 6). Within the same column, means with different superscript symbols ([a–d]) are significantly different at $p < 0.05$.

3.3. Effects of HES and CAP on the mRNA Expressions of Ki67, IL-4 and p53

The DMH-supplemented rats exhibited a significant ($p < 0.05$) increase in the mRNA expressions of colon Ki67 in comparison with the normal control rats. The treatment with HES alone and in combination with CAP resulted in a significant ($p < 0.05$) decrease in the mRNA expression of ki67 (Figure 2); the effects in the three treated groups were more or less similar.

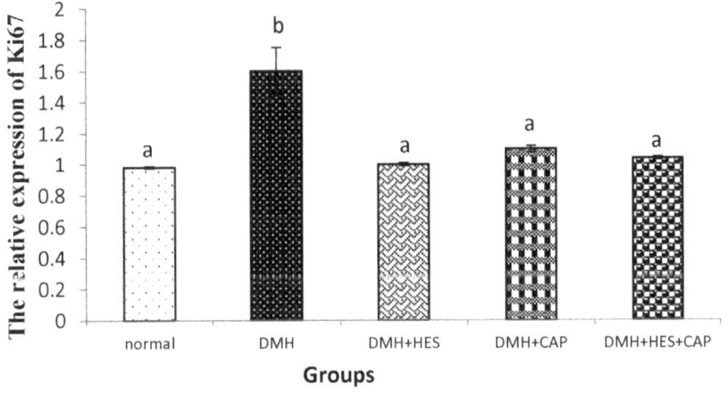

Figure 2. Effects of HES and CAP on Ki76-mRNA expressions in colon tissues of rats given DMH. Data are presented as mean values ± SE with results from 3 independent biological repeats. Means with different symbols (a,b) are significantly different at $p < 0.05$.

As illustrated in Figure 3, the administration of DMH significantly ($p < 0.05$) down-regulated the mRNA expression of IL-4 in comparison with the normal control rats. By contrast, the treatment with HES alone and in combination with CAP suppressed the expression ($p < 0.05$) of IL-4, but the effect was not significant ($p > 0.05$) with CAP alone when compared with the DMH-administered group.

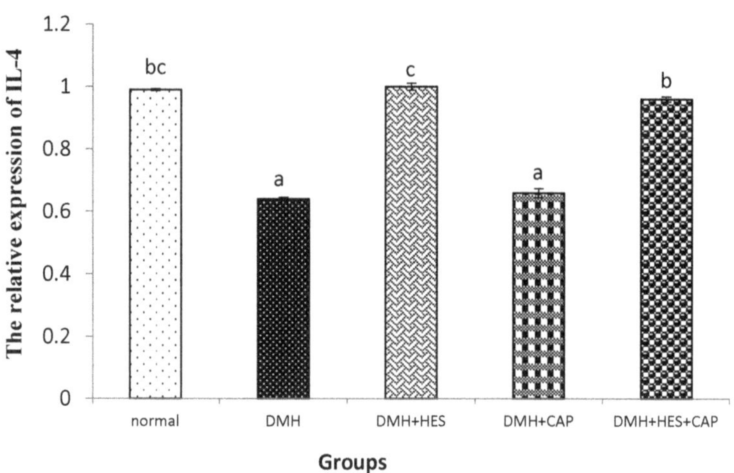

Figure 3. Effects of HES and CAP on IL-4 (B)-mRNA expression in colon tissues of rats given DMH. Data are presented as mean values ± SE with results from 3 independent biological repeats. Means with different symbols (a–c) are significantly different at $p < 0.05$.

The colon-p53-mRNA expression was significantly downregulated in the DMH-administered rats. The treatment of the DMH-administered rats with HES alone and in combination with CAP significantly ($p < 0.05$) suppressed the p53 mRNA expression; the effect of HES seemed to be the most potent (Figure 4).

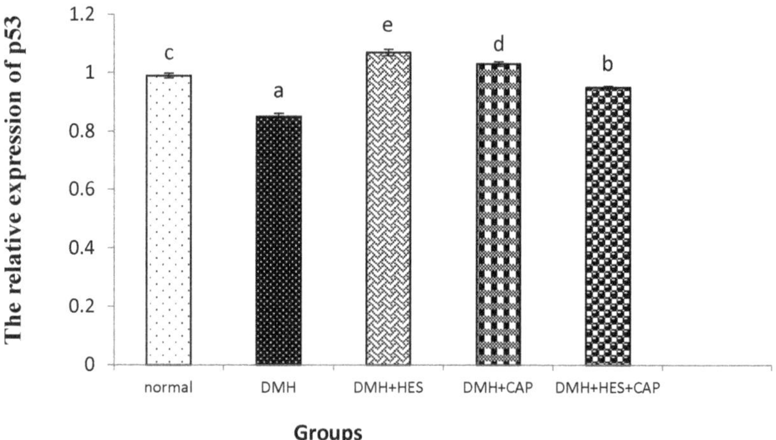

Figure 4. Effects of HES and CAP on p53-mRNA expression in colon tissues of rats given DMH. Data are presented as mean values ± SE with results from 3 independent biological repeats. Means with different symbols (a–e) are significantly different at $p < 0.05$.

3.4. Histopathological Changes

The histological architectures of the colon from the normal rats, DMH-administered rats and DMH-administered rats treated with HES and CAP alone or in combination are shown in Figure 5. The colon sections of the normal control rats showed normal histological architectures with typical histological structures of the digestive tube, including the mucosa, submucosa, muscularis and serosa/adventitia (Figure 5A). The colons of the DMH-administered rats (Figure 5B) exhibited changes, such as hyperactivation of the mucosal glands and hyperplastic polyps, hyperplastic activity of the mucosal glands and the formation of new glandular units, hyperplasia of the epithelial cells and cancerous epithelial cells. The submucosa showed oedema. These alterations were amended in the DMH-administered group treated with HES (Figure 5C), CAP (Figure 5D) and their combination (Figure 5E). The colons of these groups exhibited focal mucosal inflammatory-cell infiltration and submucosal oedema. Submucosal inflammatory-cell infiltration was also observed, as shown in Figure 5E.

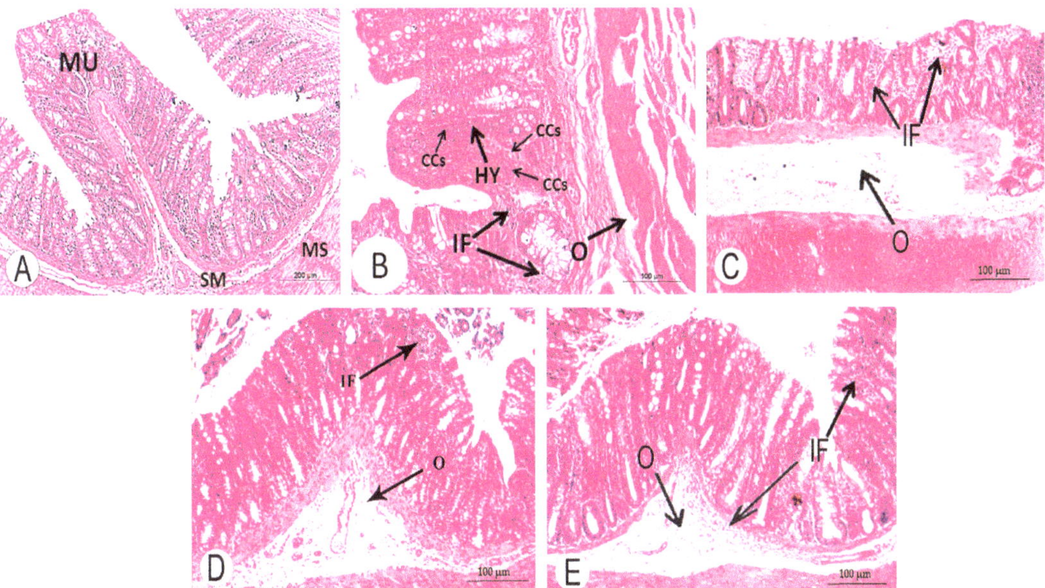

Figure 5. Colon-section photomicrographs of DMH-administered rats treated with HES and CAP, displaying marked improvement in colon architecture. (**A**) Normal control group (CMC), in which the colon has digestive tube with typical histological structures: mucosa (MU), submucosa (SM), muscularis (MS) and serosa/adventitia. (**B**) DMH-administered group, in which the mucosae showed proliferation into the surface epithelial cells (hyperplasia, HY) and cancerous epithelial cells (CCs). Submucosa showed oedema (O). Inflammatory cells (IF) were also observed (H & E × 100). (**C–E**): DMH-administered groups treated with HES (**C**), CAP (**D**) and their mixture (**E**) showed focal mucosal inflammatory cell (IF) infiltration and submucosal O (H & E × 100).

3.5. Effects of HES and CAP on Immunohistochemically Detected TGF-β1

Immunohistochemical staining was used to detect the expressions of the TGF-β1 in the colon tissues of the DMH-administered rats and to evaluate the effects of HES and CAP alone or in combination on DMH-induced colon carcinogenesis. As shown in Figure 6A,B, the colon tissues of the rats in the DMH control group revealed a marked decrease in the number of TGF-β1-positive cells (Figure 6B) compared to the normal controls (Figure 6A). The DMH-administered rats treated with HES and CAP, both individually and in combination (Figure 6C–E), exhibited an increased expression of TGF-β1 when compared to the DMH-administered control.

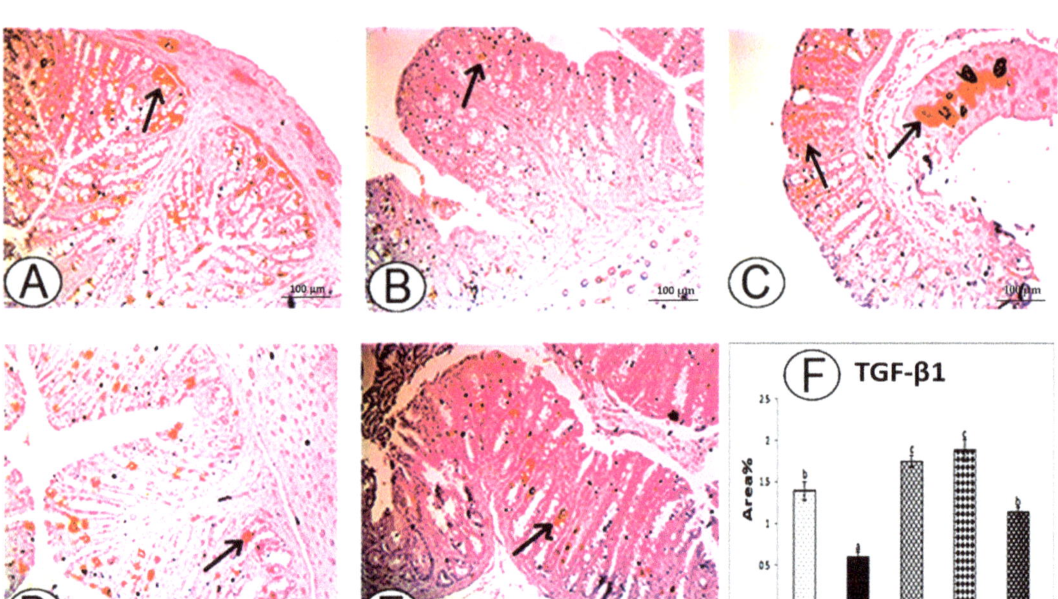

Figure 6. Photomicrographs of rat-colon sections showing the immunohistochemical staining of TGF-β1 in different groups. (**A**) Normal group, showing moderate immunohistochemical staining of TGF-β1 (↑). (**B**) DMH-administered group, showing weak immunohistochemical staining of TGF-β1 (↑). (**C**) DMH-administered group treated with HES, showing strong immunohistochemical staining of TGF-β1 (↑). (**D**) DMH-administered group treated with CAP, showing strong immunohistochemical staining of TGF-β1 (↑). (**E**) DMH-administered group treated with CAP and HES, showing moderate immunohistochemical staining of TGF-β1 (↑). (**F**) Results of image analysis of immunohistochemical staining area percent of TGF-β1 of normal, DMH-administered control and DMH-administered groups treated with HES and/or CAP. Means with different symbols (a–c) are significantly different at $p < 0.05$.

As depicted in Figure 6F, the DMH control group showed a significant decrease ($p < 0.05$) compared with the normal control group. By contrast, the DMH groups treated with HES and CAP, both individually and in combination, showed strong significant immunohistochemical reactions ($p < 0.05$) of TGF-β1 compared to the DMH control group. The treatments with HES and CAP individually were more potent than the treatment with their combination.

4. Discussion

Colorectal cancer is the third most frequent cancer in the world and a leading cause of cancer-related death [67]. Despite the availability of new and innovative medicines, systemic therapy remains the treatment of choice for >25% of patients with metastatic disease [68]. However, the treatment of CRC with chemotherapy results in cytotoxicity and agent resistance [69]. It is thus critical to identify and develop novel compounds with anticancer properties and lower toxicities.

Long-term exposure to DMH has been linked to the development of colon cancer [70]. Azoxymethane (AOM), a metabolite of DMH, is procarcinogen that must undergo metabolic activation in order to produce DNA-reactive byproducts. A reactive metabolite of DMH and AOM called methylazoxymethanol (MAM) rapidly produces the methyldiazonium ion, which can alkylate macromolecules in the liver and colon [71–73].

Hesperetin has a long list of pharmacological and biological activities, including antioxidant, anti-cancer, anti-inflammatory and cardiovascular protection [74,75]. Moreover, HES is also known for its significant therapeutic effects and low levels of toxicity for mammals [76,77]. Hence, the present study was conducted to test the effects of HES, both alone and in combination with CAP, for the treatment of colon cancer induced by DMH in rats.

Tumor markers can be utilized as prospective screening techniques and are commonly employed for the early detection of cancer [78]. For example, CEA is a tumor-antigen glycoprotein that is used as a specific index to diagnose people with colon cancer, as patients with advanced cancer conditions have high levels of CEA [79]. The current study found that giving DMH to rats resulted in a significant increase in the serum levels of CEA when compared to control rats. As a strong carcinogen, the DMH caused damage to the colons, followed by instability in colon-cell metabolism, resulting in a number of variations in the levels of CEA, which is a marker of colon function [80]; these results were in agreement with those obtained by Abdel-Hamid et al. [81]. On the other hand, the administration of HES alone or in combination with CAP significantly reduced the serum levels of CEA.

Oxidative stress is caused by an increase in ROS production and a decrease in antioxidant status [82]. It is one of the primary causes of carcinogenesis due to cell harm [83]. Both in vivo and in vitro, the most important process of free-radical production is lipid peroxidation, which has harmful effects on the membrane system and can destroy cells [84]. Lipid peroxidation can cause structural and functional membrane changes, as well as protein oxidation and the production of oxidation products, such acrolein, crotonaldehyde, MDA and 4-hydroxy-2-nonenal (HNE), which are all powerful carcinogens [85,86].

The flavoprotein oxidoreductase, GR, is responsible for the conversion of oxidized glutathione (GSSG) to its reduced form (GSH), a key component in the ascorbate-glutathione cycle that scavenges H_2O_2 [87,88]. Furthermore, GSH is a low-molecular-weight intracellular antioxidant, which serves as a first line of defense. Along with GSH-dependent enzymes such as GST and GR, it detoxifies free radicals produced endogenously, thus performing a crucial protective role [89]. Superoxide dismutase antioxidants are characterized as first-line-defense antioxidants as they act quickly to reduce superoxide radicals [90]. The past findings are in accordance with the results of this investigation, which found that DMH administration resulted in a high serum level of LPO and a low level of serum GSH, in addition to pronounced antioxidant depletion, evidenced by significant decreases in the activities of SOD, GR and GST in colon tissues; this was in contrast to HES administration, either alone or in combination with CAP, due to its antioxidant nature [91–93]. The HES also reduced colon oxidative stress, as evidenced by the lower colon MDA levels and higher colon GSH levels. According to Parhiz et al. [94], HES has been proven to have antioxidant properties. It works as an antioxidant in two different ways. The first is direct radical scavenging, which involves neutralizing ROS, such as superoxide anions, hydroxyl radicals and peroxynitrite radicals [95]. The second is an increase in antioxidant-defense biomarkers, such as catalase (CAT), SOD, glutathione peroxidase (GPx), GST and GSH [96,97]. In our study, CAP potentiated the effects of HES on GR and GST-antioxidant-enzyme activities in DMH-administered rats. The improvement in the antioxidant-defense systems in the DMH-administered rats due to the treatment with HES and CAP was associated with the return of the colon histological features to near normal levels with the absence of cancer cells; this led us to suggest that the suppression of oxidative stress and the enhancement of the antioxidant defense system may have an important role in producing the anticarcinogenic effects of HES and CAP in DMH-induced colon carcinogenesis in Wistar rats (Figure 7).

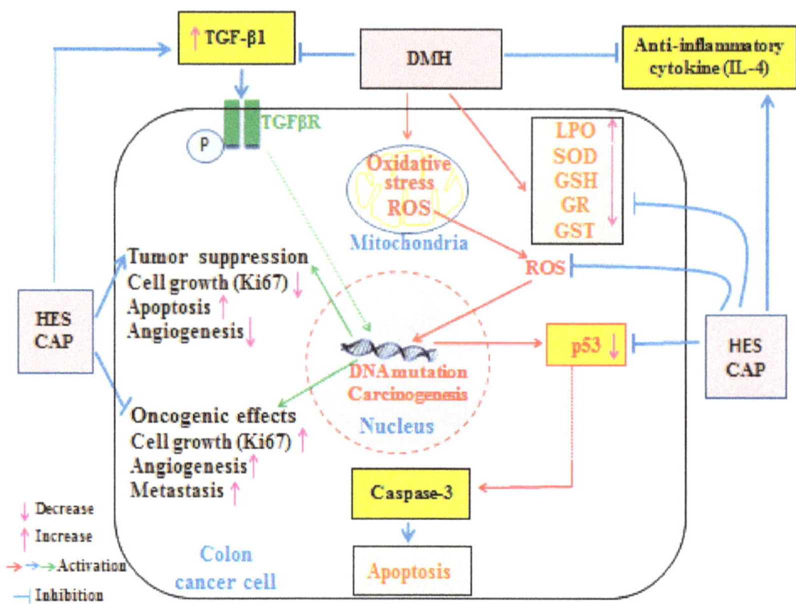

Figure 7. Schematic diagram showing the anticarcinogenic effects of HES and CAP against DMH-induced colon carcinogenenesis via suppression of oxidative stress, inflammation and cell proliferation, as well as induction of cell apoptosis and enhancement of the antioxidant defense system. HES: hesperetin; CAP: capecitabine; DMH: 1,2-dimethylhydrazine; ROS: reactive oxygen species; IL-4: interleukin-4; p53: tumor suppressor protein 53; LPO: lipid peroxides; GSH: glutathione; SOD: superoxide dismutase; GR: glutathione reductase; GST: glutathione-S-transferase; Ki67: proliferator marker.

The anti-inflammatory cytokine, IL-4, is emitted by T cells, mast cells, basophils and a subset of natural killer cells [98]. Many functions of activated macrophages are inhibited by IL-4, including the release of reactive oxygen intermediates [99]. It inhibits the synthesis of TNF-α and IL-1 by macrophages [100] and increases the expression of the IL-1 receptor antagonist. It also increases the activity of macrophage 15-lipoxygenase, which may limit the production of the proinflammatory leukotriene B4 [101]. The present investigation showed a significant decrease in the level of IL-4 mRNA due to the DMH administration, while the expression of this interleukin was increased in the rats administered DMH and treated with HES, both alone and in combination with CAP. Thus, both HES and CAP in DMH-administered rats have potent anti-inflammatory actions, in addition to their efficient antioxidant activities (Figure 7).

The above findings were reinforced by the immunohistochemistry analysis of TGF-β1. A multifunctional cytokine, TGF-β1 influences signaling cascades in tumor cells by regulating the entry of inflammatory/immune cells and cancer-associated fibroblasts into the tumor microenvironment. It can inhibit NF-κB activation by interacting with Smad7 [102], inhibiting proinflammatory TNF-α signals as a major modulator of TGF-β1 signaling [103]. The immunohistochemical analysis of TGF-β1 showed this to be evident in this investigation and indicated that HES alone increased the expression of TGF-β1 and, when combined with CAP, restored the expression of TGF-β1 to normal. In gastrointestinal-tumor development and progression, TGF-β signaling has a dual role, acting as both a tumor suppressor and a tumor promoter (Figure 7) in a stage- and context-dependent manner [104,105]. Furthermore, TGF-β signaling functions as a tumor suppressor by encouraging cell-cycle arrest and death during the early stages of tumor development. On the other hand, TGF-β has been demonstrated to enhance tumor-cell proliferation, epithelial–mesenchymal transition and stem-like activity during tumor progression, as well as inflammation and angiogenesis. The transition of TGF-β's activity from tumor-suppressive to tumor-promoting may be a result of the accumulation of mutations in TGF-β-signaling=pathway components during tumor growth [105]. In the present study, the significant decrease in colon-TGF-β1 expression was associated with a significant increase in the cell-proliferator marker, Ki67 and a decrease in the proapoptotic mediator, p53 in DMH-induced colon cancer. The treatment of the DMH-administered rats with HES and CAP significantly increased the expression of colonic TGF-β1, along with a concomitant decrease in colonic Ki67 and increase in p53. Therefore, TGF-β1 may act as a tumor suppressor under these conditions (Figure 7).

The loss of apoptosis in cancer cells is a critical event in the progression of cancer. Apoptosis is controlled by pro- and anti-apoptotic factor families. Pro-apoptotic (p53 and Bax) and anti-apoptotic genes are involved in cellular growth and apoptosis [106,107]. Cell growth, DNA damage repair and apoptosis are all regulated by the p53 protein [108]. The enhanced malignancy of several major human cancers, including CRC, is associated with an increase in p53 accumulation in the cytoplasm, where the p53 protein is not functional [109]. When compared to normal control rats, those given DMH, in the current study, had colonic cancerous lesions and a significant decrease in the level of colonic p53, which was in agreement with the findings of Gadelmawla et al. [110]. In the present study, the expression of p53 in the colons of the rats given DMH and treated with HES and CAP was high, particularly in the group administered HES. Thus, the induction of apoptosis, as evidenced by the elevated proapoptotic protein, p53, may be involved in the mechanisms of the anticancer actions of HES and CAP (Figure 7).

The proliferation of the cells has been linked to an increased risk of cancer [111]. Furthermore, Ki67 is widely used in pathological investigations to assess cell proliferation in a variety of cancers [112–114]. Although Ki67 is expressed at low levels in benign tumors, it is detected at high levels in a variety of malignant lesions and is closely linked to distant metastasis, resulting in a poor patient prognosis. The current investigation found that the Ki67 expression was much higher in the rats given DMH only than in the healthy control rats, which was consistent with the findings of Tong et al. [115]. The treatments used in this study evoked a significant successful lowering of Ki67 expression, preventing additional harm. Thus, the anticancer effects of HES and CAP in the DMH-administered rats may be attributed to their antiproliferative action secondary to the increase in TGF-β1.

5. Conclusions

Hesperetin, alone or in combination with CAP, exhibited powerful anti-inflammatory, antioxidant and anti-proliferative effects, as well as the amplification of apoptotic actions, thus preventing DMH-induced colon carcinogenesis. The combinatory effect was the most potent in improving the altered serum CEA levels and colon GR and GST activities in the DMH-administered rats. Nevertheless, with the exception of these effects, HES does not add further potential to the anticarcinogenic effects of CAP. Further studies are required to assess the effects of HES alone or in combination with CAP on human CC xenografts and clinical studies are also required to assess the safety and efficacy of these agents in human beings. An important limitation of this study was its focus on the effect on apoptotic protein p53 only and the lack of measurements of other apoptotic mediators, such as caspase-9 in the intrinsic pathway, caspase-8 in the extrinsic pathway and caspase-3, which is a common mediator in both pathways. Thus, further studies are required to assess the effects on mediators other than p53 to elucidate the full effects on the intrinsic and extrinsic pathways of apoptosis.

Author Contributions: Conceptualization: O.M.A., S.M.A.E.-T., A.M.E.-K., M.A.A. and A.K.H.; methodology, O.M.A., A.M.E.-K. and A.K.H.; software, O.M.A., A.M.E.-K. and A.K.H.; validation, O.M.A., S.M.A.E.-T., A.M.E.-K., M.A.A. and A.K.H.; formal analysis, O.M.A., S.M.A.E.-T., A.M.E.-K. and A.K.H.; investigation, O.M.A., A.M.E.-K. and A.K.H.; resources, O.M.A., A.M.E.-K., M.A.A. and A.K.H.; data curation; O.M.A., A.M.E.-K. and A.K.H.; writing—original draft preparation, O.M.A., A.M.E.-K. and A.K.H.; writing—review and editing O.M.A., A.M.E.-K., S.M.A.E.-T., M.A.A. and A.K.H.; visualization, O.M.A., S.M.A.E.-T., A.M.E.-K., M.A.A. and A.K.H.; supervision, O.M.A., S.M.A.E.-T. and A.M.E.-K.; project administration, O.M.A., A.M.E.-K. and A.K.H.; funding acquisition, O.M.A., M.A.A. and A.K.H. All authors read and approved the final manuscript. All authors have read and agreed to the published version of the manuscript.

Funding: This work was funded by Deanship of Scientific Research, Taif University, Saudi Arabia.

Institutional Review Board Statement: The experimental work was approved by the Faculty of Science Experimental Animal Ethics Committee (ethical approval number: BSU/FS/2018/17), Beni-Suef University, Egypt.

Informed Consent Statement: Not applicable.

Data Availability Statement: Data are contained within the article.

Acknowledgments: Researchers would like to acknowledge Deanship of Scientific Research, Taif University, Saudi Arabia for funding and supporting.

Conflicts of Interest: The authors declare no conflict of interest.

Abbreviations

CRC: colorectal cancer; CC: colon cancer; DMH: 1,2-dimethylhydrazine; ROS: reactive oxygen species; DNA: deoxyribonucleic acid; TGF-β: transforming growth factor-beta; CAP: capecitabine; HES: hesperetin; CEA: carcinoembryonic antigen; ELISA: enzyme-linked immunosorbent assay; b.w.: body weight; CMC: carboxymethylcellulose; LPO: lipid peroxides; MDA: malondialdehyde; GSH: reduced glutathione; GR: glutathione reductase; GST: glutathione-S-transferase; SOD: superoxide dismutase; Ki67: proliferator marker; IL-4: interleukin 4, p53: proapoptotic protein 53; RNA: ribonucleic acid; RT-PCR: reverse transcriptase–polymerase chain reaction; MAM: methylazoxymethanol.

References

1. Arnold, M.; Sierra, M.S.; Laversanne, M.; Soerjomataram, I.; Jemal, A.; Bray, F. Global patterns and trends in colorectal cancer incidence and mortality. *Gut* **2017**, *66*, 683–691. [CrossRef] [PubMed]
2. Argilés, G.; Tabernero, J.; Labianca, R.; Hochhauser, D.; Salazar, R.; Iveson, T.; Laurent-Puig, P.; Quirke, P.; Yoshino, T.; Taieb, J.; et al. Localised colon cancer: ESMO clinical practice guidelines for diagnosis, treatment and follow-up. *Ann. Oncol.* **2020**, *31*, 1291–1305. [CrossRef] [PubMed]
3. Balchen, V.; Simon, K. Colorectal cancer development and advances in screening. *Clin. Interv. Aging* **2016**, *11*, 967–976. [CrossRef] [PubMed]
4. Lim, K.G. A review of colorectal cancer research in Malaysia. *Med. J. Malays.* **2014**, *69*, 23–32.
5. Nabil, H.M.; Hassan, B.N.; Tohamy, A.A.; Waaer, H.F.; Abdel Moneim, A.E. Radioprotection of 1, 2-dimethylhydrazine-initiated colon cancer in rats using low-dose γ rays by modulating multidrug resistance-1, cytokeratin 20, and β-catenin expression. *Hum. Exp. Toxicol.* **2016**, *35*, 282–292. [CrossRef]
6. Perše, M.; Cerar, A. Morphological and molecular alterations in 1,2 dimethylhydrazine and azoxymethane induced colon carcinogenesis in rats. *J. Biomed. Biotechnol.* **2010**, *2011*, 473964. [CrossRef]
7. Saini, M.K.; Sharma, P.; Kaur, J.; Sanyal, S.N. The cyclooxygenase-2 inhibitor etoricoxib is a potent chemopreventive agent of colon carcinogenesis in the rat model. *J. Environ. Pathol. Toxicol. Oncol.* **2009**, *28*, 39–46. [CrossRef]
8. Hamiza, O.O.; Rehman, M.U.; Tahir, M.; Khan, R.; Khan, A.Q.; Lateef, A.; Ali, F.; Sultana, S. Amelioration of 1,2 dimethylhydrazine (DMH) induced colon oxidative stress, inflammation and tumor promotion response by tannic acid in Wistar rats. *Asian Pac. J. Cancer Prev.* **2012**, *13*, 4393–4402. [CrossRef]
9. Dalle-Donne, I.; Rossi, R.; Colombo, R.; Giustarini, D.; Milzani, A.D.G. Biomarkers of oxidative damage in human disease. *Clin. Chem.* **2006**, *52*, 601–623. [CrossRef]
10. Agarwal, A.; Aponte-Mellado, A.; Premkumar, B.J.; Shaman, A.; Gupta, S. The effects of oxidative stress on female reproduction: A review. *Reprod. Biol. Endocrinol.* **2012**, *10*, 49. [CrossRef]

11. Tangvarasittichai, S. Oxidative stress, insulin resistance, dyslipidemia and type 2 diabetes mellitus. *World J. Diabetes* **2015**, *6*, 456–480. [CrossRef] [PubMed]
12. Hwang, E.S.; Bowen, P.E. DNA damage, a biomarker of carcinogenesis: Its measurement and modulation by diet and environment. *Crit. Rev. Food Sci. Nutr.* **2007**, *47*, 27–50. [CrossRef] [PubMed]
13. Guyton, K.Z.; Kensler, T.W. Oxidative mechanisms in carcinogenesis. *Br. Med. Bull.* **1993**, *49*, 523–544. [CrossRef] [PubMed]
14. Kumar, B.; Koul, S.; Khandrika, L.; Meacham, R.B.; Koul, H.K. Oxidative stress is inherent in prostate cancer cells and is required for aggressive phenotype. *Cancer Res.* **2008**, *68*, 1777–1785. [CrossRef]
15. Ishikawa, K.; Takenaga, K.; Akimoto, M.; Koshikawa, N.; Yamaguchi, A.; Imanishi, H.; Nakada, K.; Honma, Y.; Hayashi, J.-I. ROS-generating mitochondrial DNA mutations can regulate tumor cell metastasis. *Science* **2008**, *320*, 661–664. [CrossRef]
16. Alpay, M.; Backman, L.R.F.; Cheng, X.; Dukel, M.; Kim, W.-J.; Ai, L.; Brown, K.D. Oxidative stress shapes breast cancer phenotype through chronic activation of ATM-dependent signaling. *Breast Cancer Res. Treat.* **2015**, *151*, 75–87. [CrossRef]
17. Golstein, P. Cell death in us and others. *Science* **1998**, *281*, 1283. [CrossRef]
18. Hotchkiss, R.S.; Strasser, A.; McDunn, J.E.; Swanson, P.E. Cell death. *N. Engl. J. Med.* **2009**, *361*, 1570–1583. [CrossRef]
19. Saikumar, P.; Dong, Z.; Mikhailov, V.; Denton, M.; Weinberg, J.M.; Venkatachalam, M.A. Apoptosis: Definition, mechanisms, and relevance to disease. *Am. J. Med.* **1999**, *107*, 489–506. [CrossRef]
20. Adams, J.M. Ways of dying: Multiple pathways to apoptosis. *Genes Dev.* **2003**, *17*, 2481–2495. [CrossRef]
21. Shi, Y. Mechanical aspects of apoptosome assembly. *Curr. Opin. Cell Biol.* **2006**, *18*, 677–684. [CrossRef] [PubMed]
22. Tang, P.M.-K.; Zhang, Y.-Y.; Mak, T.S.-K.; Tang, P.C.-T.; Huang, X.-R.; Lan, H.-Y. Transforming growth factor-β signalling in renal fibrosis: From Smads to non-coding RNAs. *J. Physiol.* **2018**, *596*, 3493–3503. [CrossRef] [PubMed]
23. Voisin, A.; Damon-Soubeyrand, C.; Bravard, S.; Saez, F.; Drevet, J.R.; Guiton, R. Differential expression and localisation of TGF-β isoforms and receptors in the murine epididymis. *Sci. Rep.* **2020**, *10*, 1–3. [CrossRef]
24. Mourskaia, A.A.; Dong, Z.; Ng, S.; Banville, M.; Zwaagstra, J.C.; O'Connor-McCourt, M.D.; Siegel, P.M. Transforming growth factor-β1 is the predominant isoform required for breast cancer cell outgrowth in bone. *Oncogene* **2008**, *28*, 1005–1015. [CrossRef] [PubMed]
25. Meng, X.M.; Tang, P.M.; Li, J.; Lan, H.Y. TGF-β/Smad signaling in renal fibrosis. *Front. Physiol.* **2015**, *6*, 82. [CrossRef]
26. Prud'Homme, G.J. Pathobiology of transforming growth factor β in cancer, fibrosis and immunologic disease, and therapeutic considerations. *Lab. Investig.* **2007**, *87*, 1077–1091. [CrossRef] [PubMed]
27. Vander Ark, A.; Cao, J.; Li, X. TGF-β receptors: In and beyond TGF-β signaling. *Cell. Signal.* **2018**, *52*, 112–120. [CrossRef]
28. Zhao, H.; Wei, J.; Sun, J. Roles of TGF-β signaling pathway in tumor microenvironment and cancer therapy. *Int. Immunopharmacol.* **2020**, *89*, 107101. [CrossRef]
29. David, C.J.; Massagué, J. Contextual determinants of TGFβ action in development, immunity and cancer. *Nat. Rev. Mol. Cell Biol.* **2018**, *19*, 419–435. [CrossRef]
30. Huang, X.-M.; Yang, Z.-J.; Xie, Q.; Zhang, Z.-K.; Zhang, H.; Ma, J.-Y. Natural products for treating colorectal cancer: A mechanistic review. *Biomed. Pharmacother.* **2019**, *117*, 109142. [CrossRef] [PubMed]
31. Knikman, J.E.; Rosing, H.; Guchelaar, H.; Cats, A.; Beijnen, J.H. A review of the bioanalytical methods for the quantitative determination of capecitabine and its metabolites in biological matrices. *Biomed. Chromatogr.* **2019**, *34*, e4732. [CrossRef]
32. Isidori, M.; Piscitelli, C.; Russo, C.; Smutná, M.; Bláha, L. Teratogenic effects of five anticancer drugs on *Xenopus laevis* embryos. *Ecotoxicol. Environ. Saf.* **2016**, *133*, 90–96. [CrossRef]
33. Huo, Z.; Wang, S.; Shao, H.; Wang, H.; Xu, G. Radiolytic degradation of anticancer drug capecitabine in aqueous solution: Kinetics, reaction mechanism, and toxicity evaluation. *Environ. Sci. Pollut. Res.* **2020**, *27*, 20807–20816. [CrossRef]
34. Deng, P.; Ji, C.; Dai, X.; Zhong, D.; Ding, L.; Chen, X. Simultaneous determination of capecitabine and its three nucleoside metabolites in human plasma by high performance liquid chromatography-tandem mass spectrometry. *J. Chromatogr. B* **2015**, *989*, 71–79. [CrossRef]
35. Reigner, B.; Blesch, K.; Weidekamm, E. Clinical pharmacokinetics of capecitabine. *Clin. Pharmacokinet.* **2001**, *40*, 85–104. [CrossRef]
36. Fernández, J.; Silván, B.; Entrialgo-Cadierno, R.; Villar, C.J.; Capasso, R.; Uranga, J.A.; Lombó, F.; Abalo, R. Antiproliferative and palliative activity of flavonoids in colorectal cancer. *Biomed. Pharmacother.* **2021**, *143*, 112241. [CrossRef] [PubMed]
37. Chen, X.; Ding, H.-W.; Li, H.-D.; Huang, H.-M.; Li, X.-F.; Yang, Y.; Zhang, Y.-L.; Pan, X.-Y.; Huang, C.; Meng, X.-M.; et al. Hesperetin derivative-14 alleviates inflammation by activating PPAR-γ in mice with CCl4-induced acute liver injury and LPS-treated RAW264.7 cells. *Toxicol. Lett.* **2017**, *274*, 51–63. [CrossRef] [PubMed]
38. Chai, T.-T.; Mohan, M.; Ong, H.-C.; Wong, F.-C. Antioxidant, iron-chelating and anti-glucosidase activities of Typha domingensis pers (Typhaceae). *Trop. J. Pharm. Res.* **2014**, *13*, 67. [CrossRef]
39. Lee, J.; Kim, D.H.; Kim, J.H. Combined administration of naringenin and hesperetin with optimal ratio maximizes the anti-cancer effect in human pancreatic cancer via down regulation of FAK and p38 signaling pathway. *Phytomedicine* **2019**, *58*, 152762. [CrossRef]
40. Shirzad, M.; Beshkar, P.; Heidarian, E. The effects of hesperetin on apoptosis induction and inhibition of cell proliferation in the prostate cancer PC3 cells. *J. Herbmed Pharmacol.* **2015**, *4*, 121–124.
41. Li, Q.; Miao, Z.; Wang, R.; Yang, J.; Zhang, D. Hesperetin induces apoptosis in human glioblastoma cells via p38 MAPK ac-tivation. *Nutr. Cancer* **2020**, *72*, 538–545. [CrossRef] [PubMed]

42. Sohel, M.; Sultana, H.; Sultana, T.; Al Amin, M.; Aktar, S.; Ali, M.C.; Bin Rahim, Z.; Hossain, M.R.; Al Mamun, A.; Amin, M.N.; et al. Chemotherapeutic potential of hesperetin for cancer treatment, with mechanistic insights: A comprehensive review. *Heliyon* **2022**, *8*, e08815. [CrossRef]

43. Chen, X.; Wei, W.; Li, Y.; Huang, J.; Ci, X. Hesperetin relieves cisplatin-induced acute kidney injury by mitigating oxidative stress, inflammation and apoptosis. *Chem. Interactions* **2019**, *308*, 269–278. [CrossRef]

44. Nalini, N.; Aranganathan, S.; Kabalimurthy, J. Chemopreventive efficacy of hesperetin (citrus flavonone) against 1,2-dimethylhydrazine-induced rat colon carcinogenesis. *Toxicol. Mech. Methods* **2012**, *22*, 397–408. [CrossRef]

45. Elango, R.; Athinarayanan, J.; Subbarayan, V.P.; Lei, D.K.; Alshatwi, A.A. Hesperetin induces an apoptosis-triggered extrinsic pathway and a p53-independent pathway in human lung cancer H522 cells. *J. Asian Nat. Prod. Res.* **2017**, *20*, 559–569. [CrossRef] [PubMed]

46. Gurushankar, K.; Gohulkumar, M.; Prasad, N.R.; Krishnakumar, N. Synthesis, characterization and in vitro anti-cancer evaluation of hesperetin-loaded nanoparticles in human oral carcinoma (KB) cells. *Adv. Nat. Sci. Nanosci. Nanotechnol.* **2013**, *5*, 15006. [CrossRef]

47. Wu, D.; Zhang, J.; Wang, J.; Li, J.; Liao, F.; Dong, W. Hesperetin induces apoptosis of esophageal cancer cells via mitochondrial pathway mediated by the increased intracellular reactive oxygen species. *Tumor Biol.* **2015**, *37*, 3451–3459. [CrossRef]

48. Coutinho, L.; Oliveira, H.; Pacheco, A.R.; Almeida, L.; Pimentel, F.; Santos, C.; de Oliveira, J.M. Hesperetin-etoposide combinations induce cytotoxicity in U2OS cells: Implications on therapeutic developments for osteosarcoma. *DNA Repair* **2017**, *50*, 36–42. [CrossRef]

49. Mistry, B.; Patel, R.V.; Keum, Y.-S. Access to the substituted benzyl-1,2,3-triazolyl hesperetin derivatives expressing antioxidant and anticancer effects. *Arab. J. Chem.* **2017**, *10*, 157–166. [CrossRef]

50. Patel, P.N.; Yu, X.M.; Jaskula-Sztul, R.; Chen, H. Hesperetin activates the Notch1 signaling cascade, causes apoptosis, and in-duces cellular differentiation in anaplastic thyroid cancer. *Ann. Surg. Oncol.* **2014**, *21*, 497–504. [CrossRef]

51. Adan, A.; Baran, Y. The pleiotropic effects of fisetin and hesperetin on human acute promyelocytic leukemia cells are mediated through apoptosis, cell cycle arrest, and alterations in signaling networks. *Tumor Biol.* **2015**, *36*, 8973–8984. [CrossRef]

52. Thorup, I.; Meyer, O.; Kristiansen, E. Effect of a dietary fiber (beet fiber) on dimethylhydrazine-induced colon cancer in wistar rats. *Nutr. Cancer* **1992**, *17*, 251–261. [CrossRef]

53. Trivedi, P.P.; Kushwaha, S.; Tripathi, D.N.; Jena, G.B. Cardioprotective effects of hesperetin against doxorubicin-induced oxidative stress and DNA damage in rat. *Cardiovasc. Toxicol.* **2011**, *11*, 215–225. [CrossRef]

54. Sanganna, B.; Kulkarni, A.R. Effect of Citrus reticulata essential oil on aberrant crypt foci (acf) development in1, 2-dimetylhydrazine induced colon carcinogenesis rats. *Int. J. Pharmaceut. Appl.* **2013**, *4*, 29–37.

55. Preuss, H.G.; Jarrell, S.T.; Scheckenbach, R.; Lieberman, S.; Anderson, R.A. Comparative effects of chromium, vanadium and gymnema sylvestre on sugar-induced blood pressure elevations in SHR. *J. Am. Coll. Nutr.* **1998**, *17*, 116–123. [CrossRef]

56. Beutler, E.; Duron, O.; Kelly, B.M. Improved method for the determination of blood glutathione. *J. Lab. Clin. Med.* **1963**, *61*, 882–888. [PubMed]

57. Goldberg, D.M. Glutathione reductase. *Methods Enzym. Anal.* **1984**, *3*, 258–265.

58. Mannervik, B.; Guthenberg, C. Glutathione transferase (human placenta). In *Methods in Enzymology*; Academic Press: Cambridge, MA, USA, 1981; Volume 77, pp. 231–235. [CrossRef]

59. Marklund, S.; Marklund, G. Involvement of the superoxide anion radical in the autoxidation of pyrogallol and a convenient assay for superoxide dismutase. *Eur. J. Biochem.* **1974**, *47*, 469–474. [CrossRef]

60. Chomczynski, P.; Sacchi, N. Single-step method of RNA isolation by acid guanidinium thiocyanate-phenol-chloroform ex-traction. *Anal. Biochem.* **1987**, *162*, 156–159. [CrossRef]

61. Yanai, G.; Hayashi, T.; Zhi, Q.; Yang, K.; Shirouzu, Y.; Shimabukuro, T.; Hiura, A.; Inoue, K.; Sumi, S. Electrofusion of mesenchymal stem cells and islet cells for diabetes therapy: A rat model. *PloS ONE* **2013**, *8*, e64499. [CrossRef] [PubMed]

62. Zhou, B.; Luo, G.; Wang, C.; Niu, R.; Wang, J.; Shanxi, P.R. Effects of fluoride on expression of cytokines in the hippocampus of adult rats. *Fluoride* **2014**, *4*, 191–198.

63. Ahmed, O.M.; Abdul-Hamid, M.M.; El-Bakry, A.M.; Mohamed, H.M.; Abdel Rahman, F.E.A. Camellia sinensis and epicatechin abate doxorubicin-induced hepatotoxicity in male Wistar rats via their modulatory effects on oxidative stress, inflammation, and apoptosis. *J. Appl. Pharm. Sci.* **2019**, *9*, 30–44.

64. Banchroft, J.D.; Stevens, A.; Turner, D.R. *Theory and Practice of Histological Techniques*, 4th ed.; Churchill Living Stone: New York, CA, USA; London, UK; San Francisco, CA, USA; Tokyo, Japan, 1996; p. 766.

65. Ahmed, O.M.; Ahmed, R.R. Anti-proliferative and apoptotic efficacies of ulvan polysaccharides against different types of carcinoma cells vitro and in vivo. *J. Cancer Sci. Ther.* **2014**, *6*, 202–208. [CrossRef]

66. IBM Corp. *IBM SPSS Statistics for Windows*; Version 22.0; IBM Corp: Armonk, NY, USA, 2013.

67. Roncucci, L.; Mariani, F. Prevention of colorectal cancer: How many tools do we have in our basket? *Eur. J. Intern. Med.* **2015**, *26*, 752–756. [CrossRef]

68. Mehta, A.; Patel, B.M. Therapeutic opportunities in colon cancer: Focus on phosphodiesterase inhibitors. *Life Sci.* **2019**, *230*, 150–161. [CrossRef] [PubMed]

69. Van Cutsem, E.; Peeters, M.; Siena, S.; Humblet, Y.; Hendlisz, A.; Neyns, B.; Canon, J.-L.; Van Laethem, J.-L.; Maurel, J.; Richardson, G.; et al. Open-label phase III trial of panitumumab plus best supportive care compared with best supportive care alone in patients with chemotherapy-refractory metastatic colorectal cancer. *J. Clin. Oncol.* **2007**, *25*, 1658–1664. [CrossRef] [PubMed]
70. Venkatachalam, K.; Gunasekaran, S.; Jesudoss, V.A.; Namasivayam, N. The effect of rosmarinic acid on 1, 2-dimethylhydrazine induced colon carcinogenesis. *Exp. Toxicol. Pathol.* **2013**, *65*, 409–418. [CrossRef] [PubMed]
71. Notman, J.; Tan, Q.H.; Zedeck, M.S. Inhibition of methylazoxymethanol-induced intestinal tumors in the rat by pyrazole with paradoxical effects on skin and kidney. *Cancer Res.* **1982**, *42*, 1774–1780.
72. Fiala, E.S.; Caswell, N.; Sohn, O.S.; Felder, M.R.; McCoy, G.D.; Weisburger, J.H. Non-alcohol dehydrogenase-mediated metabolism of methylazoxymethanol in the deer mouse, Peromyscus maniculatus. *Cancer Res.* **1984**, *44*, 2885–2891.
73. Fiala, E.S.; Sohn, O.S.; Puz, C.; Czerniak, R. Differential effects of 4-iodopyrazole and 3-methylpyrazole on the metabolic activation of methylazoxymethanol to a DNA methylating species by rat liver and rat colon mucosa in vivo. *J. Cancer Res. Clin. Oncol.* **1987**, *113*, 145–150. [CrossRef]
74. Roohbakhsh, A.; Parhiz, H.; Soltani, F.; Rezaee, R.; Iranshahi, M. Molecular mechanisms behind the biological effects of hesperidin and hesperetin for the prevention of cancer and cardiovascular diseases. *Life Sci.* **2015**, *124*, 64–74. [CrossRef] [PubMed]
75. Ding, H.-W.; Huang, A.-L.; Zhang, Y.-L.; Li, B.; Huang, C.; Ma, T.-T.; Meng, X.-M.; Li, J. Design, synthesis and biological evaluation of hesperetin derivatives as potent anti-inflammatory agent. *Fitoterapia* **2017**, *121*, 212–222. [CrossRef] [PubMed]
76. Roohbakhsh, A.; Parhiz, H.; Soltani, F.; Rezaee, R.; Iranshahi, M. Neuropharmacological properties and pharmacokinetics of the citrus flavonoids hesperidin and hesperetin—A mini-review. *Life Sci.* **2014**, *113*, 1–6. [CrossRef]
77. Shete, G.; Pawar, Y.B.; Thanki, K.; Jain, S.; Bansal, A.K. Oral bioavailability and pharmacodynamic activity of hesperetin nanocrystals generated using a novel bottom-up technology. *Mol. Pharmaceut.* **2015**, *12*, 1158–1170. [CrossRef]
78. Befeler, A.S.; di Bisceglie, A.M. Hepatocellular carcinoma: Diagnosis and treatment. *Gastroenterology* **2002**, *122*, 1609–1619. [CrossRef]
79. Mattar, R.; de Andrade, C.R.A.; DiFavero, G.M.; Gama-Rodrigues, J.J.; Laudanna, A.A. Preoperative serum levels of ca 72-4, cea, ca 19-9, and Alpha-fetoprotein in patients with gastric cancer. *Rev. Hosp. Clin.* **2002**, *57*, 89–92. [CrossRef]
80. Ge, T.; Shen, Q.; Wang, N.; Zhang, Y.; Ge, Z.; Chu, W.; Lv, X.; Zhao, F.; Zhao, W.; Fan, J.; et al. Diagnostic values of alpha-fetoprotein, dickkopf-1, and osteopontin for hepatocellular carcinoma. *Med. Oncol.* **2015**, *32*, 59. [CrossRef]
81. Abdel-Hamid, O.M.; Nafee, A.A.; MA, E.; MA, E. The ameliorative effect of Vitamin C in experimentally induced colon cancer in rats. *Benha Vet. Med. J.* **2018**, *34*, 329–343. [CrossRef]
82. Marrocco, I.; Altieri, F.; Peluso, I. Measurement and clinical significance of biomarkers of oxidative stress in humans. *Oxidative Med. Cell Longev.* **2017**, *2017*, 6501046. [CrossRef] [PubMed]
83. Umesalma, S.; Sudhandiran, G. Chemomodulation of the antioxidative enzymes and peroxidative damage in the colon of 1,2-dimethyl hydrazine-induced rats by ellagic acid. *Phytother. Res.* **2010**, *24* (Suppl. 1), S114–S119. [CrossRef]
84. Cotgreave, I.A.; Moldeus, P.; Orrenius, S. Host biochemical defense mechanisms against prooxidants. *Annu. Rev. Pharmacol. Toxicol.* **1988**, *28*, 189–212. [CrossRef]
85. Bartsch, H.; Nair, J. Chronic inflammation and oxidative stress in the genesis and perpetuation of cancer: Role of lipid peroxidation, DNA damage, and repair. *Langenbeck's Arch. Surg.* **2006**, *391*, 499–510. [CrossRef] [PubMed]
86. Valko, M.; Leibfritz, D.; Moncol, J.; Cronin, M.T.D.; Mazur, M.; Telser, J. Free radicals and antioxidants in normal physiological functions and human disease. *Int. J. Biochem. Cell Biol.* **2007**, *39*, 44–84. [CrossRef]
87. Ding, S.; Wang, L.; Yang, Z.; Lu, Q.; Wen, X.; Lu, C. Decreased glutathione reductase2 leads to early leaf senescence in Ara-bidopsis. *J. Integr. Plant Biol.* **2016**, *58*, 29–47. [CrossRef] [PubMed]
88. Hasanuzzaman, M.; Nahar, K.; Anee, T.I.; Fujita, M. Glutathione in plants: Biosynthesis and physiological role in environ-mental stress tolerance. *Physiol. Mol. Biol. Plants* **2017**, *23*, 249–268. [CrossRef] [PubMed]
89. Sultana, S.; Verma, K.; Khan, R. Nephroprotective efficacy of chrysin against cisplatin-induced toxicity via attenuation of oxidative stress. *J. Pharm. Pharmacol.* **2012**, *64*, 872–881. [CrossRef]
90. Ighodaro, O.M.; Akinloye, O.A. First line defence antioxidants-superoxide dismutase (SOD), catalase (CAT) and glutathione peroxidase (GPX): Their fundamental role in the entire antioxidant defence grid. *Alex. J. Med.* **2018**, *54*, 287–293. [CrossRef]
91. Siddiqi, A.; Hasan, S.K.; Nafees, S.; Rashid, S.; Saidullah, B.; Sultana, S. Chemopreventive efficacy of hesperidin against chemically induced nephrotoxicity and renal carcinogenesis via amelioration of oxidative stress and modulation of multiple molecular pathways. *Exp. Mol. Pathol.* **2015**, *99*, 641–653. [CrossRef]
92. Hanedan, B.; Ozkaraca, M.; Kirbas, A.; Kandemir, F.M.; Aktas, M.S.; Kilic, K.; Comakli, S.; Kucukler, S.; Bilgili, A. Investigation of the effects of hesperidin and chrysin on renal injury induced by colistin in rats. *Biomed. Pharmacother.* **2018**, *108*, 1607–1616. [CrossRef]
93. Afzal, S.M.; Vafa, A.; Rashid, S.; Barnwal, P.; Shahid, A.; Shree, A.; Islam, J.; Ali, N.; Sultana, S. Protective effect of hesperidin against N, N′-dimethylhydrazine induced oxidative stress, inflammation, and apoptotic response in the colon of Wistar rats. *Environ. Toxicol.* **2020**, *36*, 642–653. [CrossRef]
94. Parhiz, H.; Roohbakhsh, A.; Soltani, F.; Rezaee, R.; Iranshahi, M. Antioxidant and anti-inflammatory properties of the citrus flavonoids hesperidin and hesperetin: An updated review of their molecular mechanisms and experimental models. *Phytother. Res.* **2014**, *29*, 323–331. [CrossRef] [PubMed]

95. Kim, J.Y.; Jung, K.J.; Choi, J.S.; Chung, H.Y. Hesperetin: A potent antioxidant against peroxynitrite. *Free Radic. Res.* **2004**, *38*, 761–769. [CrossRef] [PubMed]
96. El-Sayed, E.S.; Abo-Salem, O.M.; Abd-Ellah, M.F.; Abd-Alla, G.M. Hesperidin, an antioxidant flavonoid, prevents acrylonitrile-induced oxidative stress in rat brain. *J. Biochem. Mol. Toxicol.* **2008**, *22*, 268–273. [CrossRef]
97. Aboismaiel, M.G.; El-Mesery, M.; El-Karef, A.; El-Shishtawy, M.M. Hesperetin upregulates Fas/FasL expression and potentiates the antitumor effect of 5-fluorouracil in rat model of hepatocellular carcinoma. *Egypt. J. Basic Appl. Sci.* **2020**, *7*, 20–34. [CrossRef]
98. Paul, W.E.; Seder, R.A. Lymphocyte responses and cytokines. *Cell* **1994**, *76*, 241–251. [CrossRef]
99. Abramson, S.L.; Gallin, J.I. IL-4 inhibits superoxide production by Adams JM, Cory S. The Bcl-2 protein family: Arbiters of cell survival. *Science* **1998**, *281*, 1322–1326.
100. Hart, P.H.; Vitti, G.F.; Burgess, D.R.; Whitty, G.A.; Piccoli, D.S.; Hamilton, J.A. Potential antiinflammatory effects of interleukin 4: Suppression of human monocyte tumor necrosis factor alpha, interleukin 1, and prostaglandin E2. *Proc. Natl. Acad. Sci. USA* **1989**, *86*, 3803–3807. [CrossRef] [PubMed]
101. Katoh, T.; Lakkis, F.G.; Makita, N.; Badr, K.F. Co-regulated expression of glomerular 12/15-lipoxygenase and interleukin-4 mRNAs in rat nephrotoxic nephritis. *Kidney Int.* **1994**, *46*, 341–349. [CrossRef]
102. Monteleone, G.; Mann, J.; Monteleone, I.; Vavassori, P.; Bremner, R.; Fantini, M.; Blanco, G.D.; Tersigni, R.; Alessandroni, L.; Mann, D.; et al. A failure of transforming growth factor-β1 negative regulation maintains sustained NF-κB activation in gut inflammation. *J. Biol. Chem.* **2004**, *279*, 3925–3932. [CrossRef]
103. Hong, S.; Lee, C.; Kim, S.J. Smad7 sensitizes tumor necrosis factor-induced apoptosis through the inhibition of antiapoptotic gene expression by suppressing activation of the nuclear factor-κB pathway. *Cancer Res.* **2007**, *67*, 9577–9783. [CrossRef]
104. Colak, S.; Ten Dijke, P. Targeting TGF-beta signaling in cancer. *Trends Cancer* **2017**, *3*, 56–71. [CrossRef]
105. Luo, J.; Chen, X.-Q.; Li, P. The role of TGF-β and its receptors in gastrointestinal cancers. *Transl. Oncol.* **2019**, *12*, 475–484. [CrossRef] [PubMed]
106. Alabsi, A.M.; Ali, R.; Ali, A.M.; Harun, H.; Al-Dubai, S.A.R.; Ganasegeran, K.; Alshagga, M.A.; Salem, S.D.; Abu Kasim, N.H.B. Induction of caspase-9, biochemical assessment and morphological changes caused by apoptosis in cancer cells treated with Goniothalamin extracted from *Goniothalamus macrophyllus*. *Asian Pac. J. Cancer Prev.* **2013**, *14*, 6273–6280. [CrossRef] [PubMed]
107. Song, M.J.; Bae, S.H. Newer treatments for advanced hepatocellular carcinoma. *Korean J. Intern. Med.* **2014**, *29*, 149–155. [CrossRef] [PubMed]
108. Nguyen, K.T.; Holloway, M.P.; Altura, R.A. The CRM1 nuclear export protein in normal development and disease. *Int. J. Biochem. Mol. Biol.* **2012**, *3*, 137–151.
109. Zhang, J.; Lei, H.; Hu, X.; Dong, W. Hesperetin ameliorates DSS-induced colitis by maintaining the epithelial barrier via blocking RIPK3/MLKL necroptosis signaling. *Eur. J. Pharmacol.* **2020**, *873*, 172992. [CrossRef]
110. Gadelmawla, M.H.A.; Alazzouni, A.S.; Farag, A.H.; Gabri, M.S.; Hassan, B.N. Enhanced effects of ferulic acid against the harmful side effects of chemotherapy in colon cancer: Docking and in vivo study. *J. Basic Appl. Zool.* **2022**, *83*, 28. [CrossRef]
111. Hanahan, D.; Weinberg, R.A. Hallmarks of cancer: The next generation. *Cell* **2011**, *144*, 646–674. [CrossRef]
112. Tian, Y.; Ma, Z.; Chen, Z.; Li, M.; Wu, Z.; Hong, M.; Wang, H.; Svatek, R.; Rodriguez, R.; Wang, Z. Clinicopathological and prognostic value of Ki-67 expression in bladder cancer: A systematic review and meta-analysis. *PLoS ONE* **2016**, *11*, e0158891. [CrossRef]
113. Clay, V.; Papaxoinis, G.; Sanderson, B.; Valle, J.W.; Howell, M.; Lamarca, A.; Krysiak, P.; Bishop, P.; Nonaka, D.; Mansoor, W. Evaluation of diagnostic and prognostic significance of Ki-67 index in pulmonary carcinoid tumours. *Clin. Transl. Oncol.* **2016**, *19*, 579–586. [CrossRef]
114. Berlin, A.; Castro-Mesta, J.F.; Rodriguez-Romo, L.; Hernandez-Barajas, D.; González-Guerrero, J.F.; Rodríguez-Fernández, I.A.; González-Conchas, G.; Verdines-Perez, A.; Vera-Badillo, F.E. Prognostic role of Ki-67 score in localized prostate cancer: A systematic review and meta-analysis. *Urol. Oncol. Semin. Orig. Investig.* **2017**, *35*, 499–506. [CrossRef] [PubMed]
115. Tong, G.; Zhang, G.; Liu, J.; Zheng, Z.; Chen, Y.; Niu, P.; Xu, X. Cutoff of 25% for Ki67 expression is a good classification tool for prognosis in colorectal cancer in the AJCC-8 stratification. *Oncol. Rep.* **2020**, *43*, 1187–1198. [CrossRef] [PubMed]

 *life*

Article

Extracts of *Brocchia cinerea* (Delile) Vis Exhibit In Vivo Wound Healing, Anti-Inflammatory and Analgesic Activities, and Other In Vitro Therapeutic Effects

Abdelkrim Agour [1,*], Ibrahim Mssillou [1], Azeddin El Barnossi [2], Mohamed Chebaibi [3], Amina Bari [2], Manal Abudawood [4], Yazeed A. Al-Sheikh [4], Mohammed Bourhia [5], John P. Giesy [6,7,8,9], Mourad A. M. Aboul-Soud [4,*], Badiaa Lyoussi [1] and Elhoussine Derwich [1]

[1] Laboratory of Natural Substances, Pharmacology, Environment, Modeling, Health, and Quality of Life, Faculty of Sciences Dhar El Mahraz, University Sidi Mohamed Ben Abdellah, Fez 30050, Morocco
[2] Laboratory of Biotechnology, Environment, Agrifood, and Health, Faculty of Sciences Dhar El Mahraz, University of Sidi Mohamed Ben Abdellah, Fez 30050, Morocco
[3] Biomedical and Translational Research Laboratory, Faculty of Medicine and Pharmacy of the Fez, University of Sidi Mohamed Ben Abdellah, BP 1893, Km 22, Road Sidi Harazem, Fez 30070, Morocco
[4] Department of Clinical Laboratory Sciences, College of Applied Medical Sciences, King Saud University, P.O. Box 10219, Riyadh 11433, Saudi Arabia
[5] Laboratory of Chemistry and Biochemistry, Faculty of Medicine and Pharmacy, Ibn Zohr University, Laayoune 70000, Morocco
[6] Toxicology Centre, University of Saskatchewan, Saskatoon, SK S7N 5B3, Canada
[7] Department of Veterinary Biomedical Sciences, University of Saskatchewan, Saskatoon, SK S7N 5B4, Canada
[8] Department of Integrative Biology, Michigan State University, East Lansing, MI 48824, USA
[9] Department of Environmental Sciences, Baylor University, Waco, TX 76706, USA
* Correspondence: abdelkrimagour1@gmail.com (A.A.); maboulsoud@ksu.edu.sa (M.A.M.A.-S.)

Abstract: The plant *Brocchia cinerea* (Delile) (*B. cinerea*) has many uses in traditional pharmacology. Aqueous (BCAE) and ethanolic extracts (BCEE) obtained from the aerial parts can be used as an alternative to some synthetic drugs. In vitro, DPPH, FRAP and TAC are three tests used to measure antioxidant efficacy. Antibacterial activities were determined against one Gram positive and two Gram negative strains of bacteria. The analgesic power was evaluated in vivo using the abdominal contortion model in mice, while carrageenan-induced edema in rats was the model chosen for the anti-inflammatory test; wound healing was evaluated in an experimental second degree burn model. The results of the phytochemical analysis showed that BCEE had the greatest content of polyphenols (21.06 mg AGE/g extract), flavonoids (10.43 mg QE/g extract) and tannins (24.05 mg TAE/g extract). HPLC-DAD reveals the high content of gallic acid, quercetin and caffeic acid in extracts. BCEE has a strong antiradical potency against DPPH (IC_{50} = 0.14 mg/mL) and a medium iron reducing activity (EC_{50} = 0.24 mg/mL), while BCAE inhibited the growth of the antibiotic resistant bacterium, *P. aeruginosa* (MIC = 10 mg/mL). BCAE also exhibited significant pharmacological effects and analgesic efficacy (55.81% inhibition 55.64% for the standard used) and the re-epithelialization of wounds, with 96.91% against 98.60% for the standard. These results confirm the validity of the traditional applications of this plant and its potential as a model to develop analogous drugs.

Keywords: analgesic activity; antibacterial; antioxidant; traditional medicine; wound healing

Citation: Agour, A.; Mssillou, I.; El Barnossi, A.; Chebaibi, M.; Bari, A.; Abudawood, M.; Al-Sheikh, Y.A.; Bourhia, M.; Giesy, J.P.; Aboul-Soud, M.A.M.; et al. Extracts of *Brocchia cinerea* (Delile) Vis Exhibit In Vivo Wound Healing, Anti-Inflammatory and Analgesic Activities, and Other In Vitro Therapeutic Effects. *Life* **2023**, *13*, 776. https://doi.org/10.3390/life13030776

Academic Editors: Efstathia Papada and Stefania Lamponi

Received: 11 January 2023
Revised: 17 February 2023
Accepted: 23 February 2023
Published: 13 March 2023

1. Introduction

Plants have several pharmacological properties which are linked to the presence of various bioactive compounds, including among others, terpenoids, phenolic acids and flavonoids [1]. Several species of plants have been proposed as phytomedical to treat several diseases [2,3]. The family Asteraceae, or Composite, with the common names asters, daisies or sunflowers, contains a number of genera frequently used for therapeutic purposes. Species in this family contain a wide range of phenolic phytochemicals, essential oils, steroids and terpenoids [4].

Since ancient times, plants have been a source of substances used in the treatment of skin wounds. These can be burns caused by heat, electricity, cold, radiation and chemicals [5]. Wound healing requires a complex, orderly and dynamic process that proceeds through the following stages: coagulation, inflammation, proliferation and remodeling [6]. For a patient with burns, microorganisms can penetrate into the tissues under the skin, developing chronic and systemic complications [7,8]. Excessive reactive oxygen species also have a negative role on the healing process [9]. Phytochemicals (alkaloids, terpenes, polyphenols, saponins and essential oils) can optimize the healing process through their antimicrobial, anti-inflammatory and antioxidant activities [6]. Phytochemicals are increasingly sought after as alternatives to synthetic drugs. In addition to their targeted effects, synthetic drugs cause adverse side effects to the human body. Well-known classes of drugs may result in certain levels of nephrotoxicity and hepatotoxicity [10]. Nonsteroidal anti-inflammatory drugs are among the medications used to treat pain, fever and inflammation; however, these drugs can cause significant toxicity [11]. According to previous works, several species belonging to the *Asteraceae* family have been studied for their healing effect [12–14], but the species *B. cinerea* has not yet been evaluated.

In the Saharo-Sindian region, herbaceous plants constitute a large proportion of plants used in traditional medicine. *B. cinerea* is traditionally used for treatment of colic, diarrhea, cough, rheumatism, digestive disorders, urinary tract infections, lung infections, fever and headaches [15]. Extracts of this species are also used for their antimicrobial and insecticidal activities. Essential oil of *B. cinerea* can be used as a biopesticide during integrated pest management against the cowpea weevil, *Callosobruchus maculatus* [16]. Exposure to either of the two extracts of *B. cinerea*, ethyl ether or ethyl acetate, for 24 h was active against the mosquito vector of malaria, *Anopheles labranchiae*, with the LC_{50s} ranging from 28 to 325 ppm [17]. The dichloromethane extract obtained from the aerial parts of *B. cinerea* shows antibiotic activity against the enteric bacterium, *Enterococcus faecalis* [18]. Extracts of *B. cinera* in n-butanol, ethanol, ethyl acetate and petroleum ether at 70% show antimicrobial activity against the bacteria *Escherichia coli* (*E. coli*), *Pseudomonas aeruginosa* (*P. aeruginosa*), *Staphylococcus aureus* (*S. aureus*) and *Klebsiella pneumoniae* [19]. The aqueous extract of *B. cinerea* inhibited two species of fungi of the genus *Fusarium* that affect yields of wheat [20].

B. cinerea has antipyretic effects and reduces fever [21]. The administration of an extract obtained from the dry plant material of *B. cinerea* reduced fever in rats in groups treated with 0.2 and 0.4 g/kg of their body mass (bm). Extracts of *B. cinerea* can be used for treatment of pain, which is defined as a rather uncomfortable physical sensation caused by disease or injury [22].

The present study characterized phytochemically the extracts from the aerial parts of *B. cinerea*, collected from the city of Akka, by the determination of their total polyphenol content (TPC), total flavonoid content (TF) and condensed tannins (CT) content by the use of chromatography and by high performance liquid chromatography, coupled with a diode-array detector (HPLC-DAD). To evaluate their pharmacological potential, in vivo analgesic activity, anti-inflammatory activities and, for the first time, their wound healing potential were characterized. Furthermore, in vitro antioxidant and antibacterial capacities were tested by various methods.

2. Materiel and Methods

2.1. Preparation of BCAE and BCEE

The plant studied (Figure 1) was collected during the second week of February 2021 in the city of Akka, Tata province in southeastern Morocco (latitude: 29.40538437° N (29 N 3253121,303 m N), longitude: 8.27052042° W (29 N 570773,258 m E), elevation: 569,000 m)). The sample was identified by the botanist Professor Bari Amina (Faculty of Sciences, Dhar El Mahraz, University of Sidi Mohamed Ben Abdellah, Fez), the voucher specimen was BC0019220211. In the preparation of the extracts (BCEE and BCAE), 100 g of the powder of the aerial parts (ground by a Waring® blender, New Hartford, CT, USA) was macerated in 1 L of ethanol (Sigma-Aldrich, St. Louis, MO, USA) at 70% for 48 h, while the aqueous extract was prepared by infusion (100 g of the powder in 1 L of boiling water). The filtrates were dried at 37 °C until the total removal of the solvent and then stored at 4 °C [23].

Figure 1. Aerial parts of *B. cinerea* in the flowering stage.

2.2. Characterization of Phytochemicals

2.2.1. Quantification of Total Polyphenol Content (TPC)

A volume of 500 µL of Folin reagent diluted 1/10 was added to 100 µL of the diluted sample. After 4 min, 400 µL of 75 mg sodium carbonate/mL was added to the reaction mixture. After 2 h incubation at laboratory temperature, the optical density was measured at 760 nm in a spectrophotometer. The concentrations of the total polyphenols were calculated by comparison to the calibration range equation established with Gallic acid ($y = 5.3922x + 0.0122$; $R^2 = 0.9933$), expressed as milligram of Gallic acid equivalents per gram of extract (mg GAE/g extract) [24].

2.2.2. Quantification of Total Flavonoids (Fl)

Flavonoids in extracts of *B. cinerea* were quantified by use of previously described methods [25] with slight modifications. A volume of 1.5 mL of the extract was added to 1.5 mL of AlCl3 (2%). The absorbance was measured after 30 min of incubation in the dark spectrophotometrically at 430 nm and compared to a blank and a standard curve ($Y = 13.998X + 0.1133$; $R^2 = 0.99$). The total flavonoids were expressed as mg quercetin equivalents per g extract (mg QE/g extract).

2.2.3. Quantification of Condensed Tannins (CT)

Concentrations of condensed tannins (CT) in extracts of *B. cinerea* were determined by spectrophotometry according to the previously described methods [26]. One hundred (100) µL of the extract was mixed with 500 µL of Folin–Ciocalteu and 1 mL of sodium carbonate (7.5%). After 30 min of incubation in the dark, the absorbance was measured at 760 nm. The concentrations of condensed tannins of the studied extracts were expressed as milligrams of tannic acid equivalence (TAE) per g of extract (mg TAE/g extract) according to an external standard curve ($y = 5.615x - 0.0005$; $R^2 = 0.9871$).

2.2.4. HPLC-DAD Analysis

Chemicals and Standard Preparations

The molecules used as the standards in this analysis were: caffeic acid, gallic acid, rutin, vanillin, quercetin, catechin, arbutin, rosmarinic acid, luteolin and furilic acid (Sigma

Aldrich, Roedermark, Germany). Methanol was the solvent used to prepare the standards at 100 ppm.

HPLC-DAD Conditions

In the reverse phase chromatographic analysis, a Shimadzu HPLC system with an analytical column (C18) SGE 250 × 4.6mm SS Exsil ODS 5 μm was used. A gradient mode separation was completed by using two solvents (X (water/acetic acid) (97.5: 2.5, v/v); Y (methanol/acetonitrile) (50: 50, v/v)), the injection volume was 20 μL and the flow rate was 1 mL/min. Wavelengths used for detection ranged from 200 to 800 nm. The elution gradient was: at 0 min (5% X, 95% Y); at 6.25 min (30% X, 70% Y); at 12.5 min (35% X, 65% Y); at 16.25 min (70% X, 30% Y); at 17.5 min (100% X, 0% Y); at 18.75 min (5% X, 95% Y) [27].

2.3. In Vitro Antioxidant Activity

2.3.1. Trapping of Free Radicals by 2,2′-Diphenyl 1-Picrylhydrazyl Radical (DPPH)

The antioxidant potency of the extracts was measured as a hydrogen donor or free radicals, using stable 2,2′-diphenyl 1-picrylhydrazyl radical 4 mg DPPH/100 mL in 10 test tubes; 100 μL of the various concentrations of the aqueous extract, ethanolic extract and quercetin was added to 750 mL of the DPPH solute on [28]. After 20 min in the dark at room temperature, the absorbance was measured at 517 nm. Quercetin was used as the standard solution. The percentage of DPPH radical inhibition was calculated (Equation (1)).

$$\text{Inhibition\%} = (Ab - Aa/Ab) \times 100 \tag{1}$$

where Ab is the absorption of control (−) and Aa is the absorption of the extract.

2.3.2. Ferric Reducing Antioxidant Power Assay (FRAP)

The ferric reducing power of the extracts was determined by the previously described methods [29]. Briefly, 200 μL of defined concentrations of the extracts was mixed with 500 μL of the phosphate buffer (0.2 M, pH 6.6) and 500 μL of potassium ferricyanide [$K_3Fe(CN)_6$] 1%. The resulting solution was incubated at 50 °C for 20 min, the mixture was acidified with 500 μL of 10% trichloroacetic acid (TCA), 0.5 mL of the supernatant was mixed with 500 μL of distilled water and 100 μL of Fe Cl3 (0.1%) and the absorbance was measured spectroscopically at 700 nm.

2.3.3. Quantification of Total Antioxidant Capacity (TAC)

A total volume of 0.25 mL of a known concentration was added to 2.5 mL of the reagent solution (sulfuric acid (0.6 mol/L); sodium phosphate (28 mmol/L) and ammonium molybdate (4 mmol/L)). The mixtures were incubated in a water bath at 95 °C for one h and 30 min. After incubation, mixtures were cooled to laboratory temperature. An optical density measurement was made without a wavelength at 695 nm. The expression of the total antioxidant capacity of BCEE and BCAE was made by equivalents in mg of ascorbic acid per gram of extract (mg AAE/g extract) [30].

2.4. Antibacterial Activity

2.4.1. Tested Strains

The antibacterial activities of the aqueous and ethanolic extracts of *B. cinerea* were evaluated against three bacterial strains, including the Gram-negative (*E. coli* (ATB:97) BGM and *pseudomonas aeruginosa*, as well as the Gram-positive *Bacillus subtills*, which were provided by the Bacteriology Laboratory of the Centre Hospitalier Universitaire Hassan II of Fez, Morocco.

2.4.2. Disk Diffusion Method

The antibacterial efficacies and potencies of BCAE and BCEE were performed by adaptations of previously published methods [31]. Petri dishes containing a nutrient broth were inoculated with the three bacterial strains. Whatman paper discs (6 mm diameter) were placed on the surface of the inoculated media and impregnated with 0.02 mL of the extract diluted in DMSO at 100 mg extract/mL [32]. The inoculated Petri dishes were incubated at 37 °C, and the inhibition diameter measurement was performed after 24 h.

2.4.3. Determination of the Minimum Inhibitory Concentration

The minimum inhibitory concentrations of the *B. cinerea* extracts against the three bacterial strains were revealed using the microdilution method according to the previously described methods [33]. After 24 h of incubation at 37 °C, the MIC endpoint was determined by the direct observation of the growth in the wells and by using the colorimetric method (TTC 0.2% (w/v)) [34,35].

2.5. Pharmacological Activities
2.5.1. Animal Handling and Housing

In order to characterize anti-inflammatory potency, analgesic and wound healing activity, male mice (0.032 to 0.045 kg) and male rats (0.12 to 0.14 kg) were obtained from the Department of Biology, Faculty of Sciences Dhar El-Mahraz, Sidi Mohamed Ben Abdellah University, Fez, Morocco. The animal housing conditions were: day/night photoperiod (~12/12 h); temperature (28–32 °C); humidity (50–55%). The animals were given free access to food and water, adapted and handled with ethics committee approval (July 2020/LBEAS-08 and 25 February 2021).

2.5.2. Wound Healing Test
Formulation of Ointment

Ointments were prepared from aqueous and ethanolic dry extracts of the aerial parts of *B. cinerea* at 10% in Vaseline® by gentle trituration until the mixture was homogenized. The resulting ointments (Figure 2) were packed in hermetically sealed jars and stored until use.

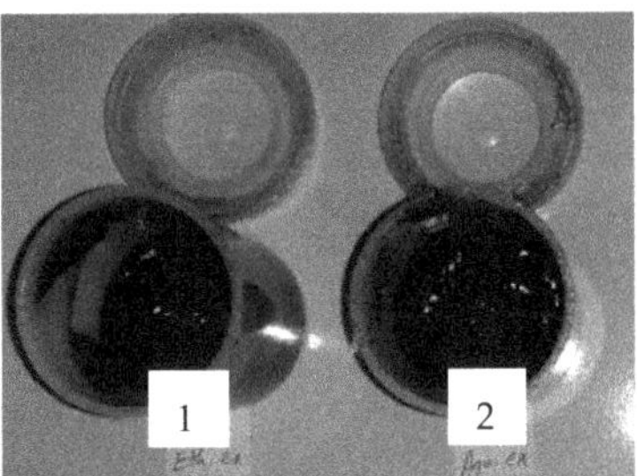

Figure 2. Ointment from BCEE (1) and BCAE (2).

Burn Wound Induction

Wound induction was performed by burning the dorsal area, using the methods described in previous work [36] with slight modifications. Twenty Wistar rats divided into four groups were used: a negative control group (Vaseline® treatment), a second group treated with a healing ointment (Madecassol® at 1%), a third group and a fourth group treated with the BCAE and BCEE ointments, respectively. An electric tensioner was used to shave the dorsal area of the rats, then an intraperitoneal anesthesia (pentobarbital at 50 mg/kg) was conducted before the induction of the burn (placement of a 1.7 cm diameter aluminum rod heated in boiling water on the burn induction area for 10 s). The treatment over the three weeks was performed by the daily covering of the burned dorsal surfaces with the prepared ointments. Photographs were taken of the wound surfaces using a digital camera with the presence of a graduated ruler. The images obtained were analyzed using ImageJ software in order to measure contraction rate of the wound surfaces (Equation (2)).

$$WC\ (\%) = [(WS0 - WSSD)/WS0] \times 100 \tag{2}$$

where WC is the wound contraction, WS0 is the size of the wound on the first day and WSSD is the wound size on each specific day.

2.5.3. Carrageenan-Induced Inflammation of the Right Rat Paw

Evaluation of anti-inflammatory power of BCEE and BCAE (extracts obtained from the aerial parts of *B. cinerea*) was carried out according to the protocol described [37]. Twenty Wistar rats divided into four groups were used in this test: one group received BCEE 500 mg/kg orally, a second group received BCAE 500 mg/kg, a negative control group received 0.9% NaCl and a positive control group received Indomethacin® (10 mg/kg). A measurement of the circumference of the right paw, of each rat was carried out before the injection of the suspension of carrageenan (1%), then measurements of the circumferences of the paws were carried out after the third, fourth, fifth and sixth hour of the injection. An equation was used to calculate the percentage of inflammation inhibition (Equation (3)).

$$PI\% = [(Ct - C0)\ Control - (Ct - C0)\ treated/(Ct - C0)\ Control] \times 100 \tag{3}$$

where C0 is the mean circumference of the paw before injection and Ct is the mean circumference of the paw after a carrageenan injection at a given time.

2.5.4. Analgesic Activity

Four batches, each containing five mice of different weights (32 to 45 g) were used in this study. BCAE, BCEE and the reference drug (Tramadol®) were administered orally to the mice at a dose of 500 mg/kg, bm. The control mice received only physiological water (NaCl 0.9%). After 1 h 30 min, a 0.5% acetic acid solution was injected intraperitoneally (10 mL/kg). Subsequently, 5 min after the acetic acid injection, nociception was assessed for 30 min by counting the number of abdominal contortions [38] (Equation (4)).

$$PI\% = ((Mn - Mt)/Mn) \times 100 \tag{4}$$

where Mn is the mean number of abdominal contractions (negative control group) and Mt is the average number of abdominal contractions (group treated with the extracts or standard compound).

2.5.5. Statistical Analyses

The results were analyzed with the one-way ANOVA test (GraphPad Prism software version 8), and $p < 0.05$ showed statistical significance. The results are expressed as mean $\pm$ SD.

3. Results and Discussion

3.1. TPC, Fl, CT and HPLC-DAD Analysis

The yield obtained by maceration using ethanol (27.34%) was greater than that obtained by diffusion in water (17.50%). The concentrations of TPC, Fl and CT in the extracts of *B. cinerea* extracts were determined (Table 1). The concentrations of phenolic compounds were greater in the ethanolic extract than in the aqueous extract (21.06 and 13.09 mg of GAE/g extract, respectively, for BCEE and BCAE). These extracts are characterized by the presence of large amounts of condensed tannins expressed by mg tannic acid equivalents/g extract; this amount was 24.05 mg in the ethanolic extract. The analysis of BCAE and BCEE by HPLC-DAD allowed the detection of four phenolic compounds (gallic acid and caffeic acid in BCAE; gallic acid, rosmarinic acid and quercetin in BCEE) (Table 2 and Figure 3). Even if there are molecules that are not revealed, the above mentioned compounds were among the main molecules of the *B. cinerea* extracts.

Table 1. Concentrations of total phenolic content (TPC), total flavonoids (Fl) and condensed tannins (CT) in extracts of *B. cinerea*.

	TPC: mg GAE/g of Extract	Fl: mg QE/g of Extract	CT: mg TAE/g of Extract
BCAE	13.09 ± 0.13	9.86 ± 0.41	17.07 ± 0.36
BCEE	21.06 ± 0.04	10.43 ± 0.13	24.05 ± 0.24

Table 2. Compounds identified in *B. cinerea* extracts and chromatographic retention times (RT).

Phenolic Compounds	RT (min)	DO (nm)
Gallic acid	4.69	280
Caffeic acid	7.777	300
Rosmarinic acid	8.300	320
Quercetin	9.403	370

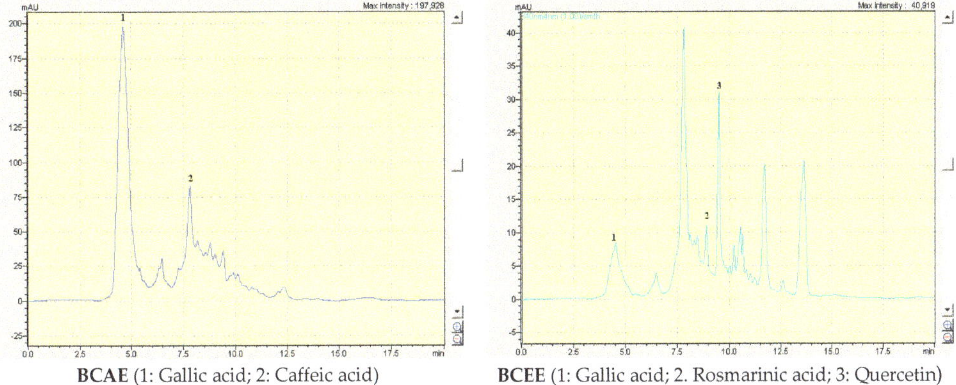

BCAE (1: Gallic acid; 2: Caffeic acid)   **BCEE** (1: Gallic acid; 2. Rosmarinic acid; 3: Quercetin)

Figure 3. Chromatograms of BCAE and BCEE obtained by HPLC-DAD analysis.

The compositions of TPC, Fl and CT in the extracts of *B. cinerea* were variable. According to Chlif et al. (2022) [39], the aqueous extract obtained from the aerial parts of *B. cinerea* contains 23.03 mg of GAE/g extract, 11.69 mg of QE/g extract and 6.41 mg of TAE/g extract. Previous studies have found methanolic extracts in TPC, expressed as mg GAE/g of dry matter (dm), Fl, expressed as mg QE/g of dm and condensed tannins, expressed as mg cyanidin equivalent/g of dm, which were 2.95, 44.58 and 1.44, respectively [40]. In another study, the contents were 22.22 mg GAE/g of dm, TPC 3.93/g of dm and 8.61 catechin equivalents/g of dm for Fl and proanthocyanides, respectively [41]. Among the main phenolic compounds revealed by HPLC-ESI MS in a methanolic extract of the aerial parts of *B. cinerea* are luteolin-4'-*O*-glucoside, 3,5-dicaffeoylquinic acid, 4,5-dicaffeoylquinic acid, 3,4-dicaffeoylquinic acid, cryptochlorogenic acid and chlorogenic acid [42]. The hydroalcoholic extract of the aerial parts of *B. cinerea* contains other phenolic compounds (flavonoid compounds), including chrysospenol-D, chrysosplenetin, oxyayanin-B, axillarin, 3-methylquercetin, pedaletin, isokaempferid, apigenin, luteolin, 6-hydroxyluteolin and others [43]. The remarkable differences in polyphenols between the studied extracts and other extracts reported in the bibliography are strongly explained by the intervention of several factors, among which are the geographical situation and the climate of the harvesting station. The content of phytochemicals, including polyphenols, varies under the action of water deficit (drought stress); this factor leads to differences in the accumulation of polyphenols in the different organs of several plant species [44–46].

3.2. Antioxidant Activity (TAC, DPPH and FRAP Tests)

Extracts of *B. cinerea* (BCEE and BCAE) showed a remarkable antioxidant capacity, (Table 3). The BCEE extract showed a higher total antioxidant capacity than BCAE, 5.312 ± 0.217 mg of AAE/g extract and 2.832 ± 0.148 mg of AAE/g extract, respectively. BCAE was very active against the DPPH radical ($IC_{50} = 4.9 \times 10^{-2}$ mg/mL), but this antiradical power was less than that obtained using Quercitin (7.9×10^{-4} mg/mL). BCEE was the least active against DPPH. Moreover, results of the FRAP assay revealed that BCEE has a moderate ability to reduce iron, while BCAE appeared to be inactive. The richness of BCEE in polyphenolic compounds with high antioxidant power could explain the obtained results.

Table 3. Antioxidant activity of extracts of *B. cinerea*.

TAC (mg AAE/g Extract)		FRAP (EC_{50} mg/mL)			DPPH (IC_{50} mg/mL)		
BCEE	BCAE	BCEE	BCAE	Quercitin	BCEE	BCAE	Quercitin
5.312 ± 0.217	2.832 ± 0.148	0.248668	-	0.007295	0.144	0.049	0.000799

Comparing the results obtained in other studies, the methanolic extract of *B. cinerea* has a total antioxidant capacity of 17.190 ± 1.273 mg ascorbic acid equivalents/g of dm, while this extract has a DPPH free radical scavenging effect ($EC_{50} = 462.19$ mg antioxidant/g DPPH), hydroxyl radical scavenging ($EC_{50} = 0.66 \pm 0.12$ mg/mL) and iron reducing capacity ($IC_{50} = 1.174 \pm 0.05$ mg/mL) [41]. Pure polyphenolic compounds isolated from the aerial parts of *B. cinerea* exert antioxidant activity assessed by the use of the DPPH assay; IC_{50}s are expressed as (μM): luteolin-4′-O-glucoside (30.25), 4,5-dicaffeoylquinic acid (31.49), 3,5-dicaffeoylquinic acid (23.84), 3,4-dicaffeoylquinic acid (30.25), neochlorogenic acid (11.0) and chlorogenic acid (10.5) [47]. In addition, the essential oil of *Cotula cinerea* (a synonym of *B. cinerea*) showed antioxidant activity measured by different assays DPPH, PRAP, B-carotene and ABTS with an IC_{50} (mg/mL) of 1.1 ± 0.15, 2.7 ± 0.09, 0.63 ± 0.16 and 0.91 ± 0.52, respectively [48].

3.3. Antibacterial Activity of BCAE and BCEE

Extracts obtained from the aerial parts of *B. cinerea* exhibited antibacterial activity against both gram-negative and gram-positive bacteria (Table 4 and Figure 4). BCEE was more active against *B. subtills* with a zone of inhibition of 23.16 mm, while BCAE was active against the gram-negative bacteria *P. aeruginosa* (the diameter of the zone of inhibition was 18.66 mm); the MIC values for both the prevalent bacteria were 10 mg/mL. However, *E. coli* was resistant to the extracts used in this study. The MIC values obtained by using streptomycin against the strains used were 25 ± 1.63 and 2.81 ± 0.095 mg/mL, respectively, for *E. coli* and *B. subtilis*, while *P. aeruginosa* was resistant to this antibiotic [16].

Table 4. Antibacterial activity of *B. cinerea* extracts.

	Zone of Inhibition (mm)			CMI mg/mL		
	P. aeruginosa	*E. coli*	*B.subtils*	*P. aeruginosa*	*E. coli*	*B.subtils*
BCEE	-	-	23.16 ± 0.76	-	-	10 ± 0.0
BCAE	18.66 ± 1.6	-	-	10 ± 0.0	-	-

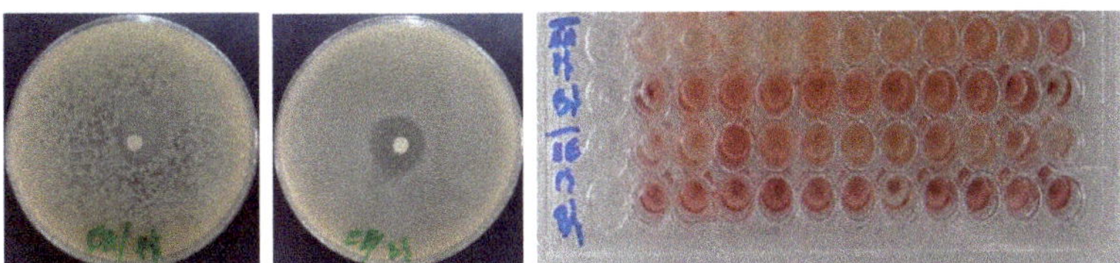

Figure 4. Photos showing the antibacterial effect of extracts of *B. cinerea*.

Compared with the results of previous studies, aqueous extracts obtained from aerial parts, leaves and flowers of *B. cinerea* showed potent activity against gram-positive bacteria (*S. aureus* and *S. faecalis*) with MICs ranging from 0.78 to 12.5 mg/mL [39]. Alternatively, the gram-negative bacteria, *E. coli* and *K. pneumoniae*, were moderately sensitive to the extracts tested, with MICs ranging from 3.13 to 50 mg/mL, with the exception of *P. aeruginosa*, which showed resistance to extracts of the various parts of *B. cinerea*. *Enterococcus faecalis*, a gram-positive bacterium, was inhibited by a dichloromethane extract of the aerial parts of *B. cinerea* [18]. In another study, the inhibition zones obtained against *S. aureus* by using *B. cinerea* extracts were 12 ± 5.20 mm and 11.67 ± 3.79 mm for the n-butanol and ethyl acetate extracts, respectively, whereas the hydroalcoholic extract showed weak antimicrobial activity against the tested strains [19].

Part of the antimicrobial activity exhibited by the extracts obtained from the aerial parts of *B. cinerea* is the result of the antibacterial effect of some phenolic compounds revealed in the studied extracts. Gallic acid may be involved in disrupting the integrity of the membranes of bacterial strains, modifying hydrophobicity, permeability and surface charge [49]. Rosmarinic acid exhibits antimicrobial activity against gram-positive strains. Thus, the lowest blocking concentration against *S. aureus* was 0.8 mg/mL, and the methicillin-resistant *S. aureus* strain was blocked by a concentration of 10 mg/mL [50]. Caffeic acid and other phenylpropanoids have inhibited the growth of certain bacterial strains, including *S. aureus*, *E. coli*, *Listeria monocytogenes* and *Bacillus cereus* [51]. Quercitin can inhibit bacterial growth; its ability to inhibit D-Ala-D-Ala ligation in bacterial cells gave it a bacteriostatic activity [52]. According to Ferreira et al. (2013) [53], in the majority of cases, hydroxybenzoic acids exerted weak antibacterial effects compared to hydroxycinnamic acids.

3.4. Pharmacological Activities

3.4.1. Wound Healing Activity of *B. cinerea* Extracts

Wound healing ability of *B. cinerea* extracts was examined in the Wistar rat burn model. The rate of wound healing in the different groups was different (Figure 5), and the images showing the healing process are shown in Figure 6. The BCAE-based pomade treatment showed a higher rate of healing compared to that obtained by BCEE or by the negative control (Vaseline treatment). At the end of the third week of treatment, the BCAE-based pomade resulted in a very remarkable healing rate (96.91%) close to that of the positive control (98.60%) (*p* value < 0.05), whereas this healing rate was only 90.86% in the negative control group.

To our knowledge, this is the first study to report on the healing activity of extracts of *B. cinerea*, which have not been reported among the 12 species belonging to the Asteraceae family—species that are part of medicinal plants widely used in the treatment of wounds in the Mediterranean region [54]. An evaluation of the healing activity of a plant species belonging to the asteraceae family and using the same burn model with an extract of *Ditrichia viscosa* shows a remarkable wound contraction (99.28%) after 21 days of treatment [27]. The healing efficacies of some species of asteraceae can be attributed to their antibacterial effects, which provide protection of the microenvironment and damaged wound tissue against pathogenic bacterial strains [55].

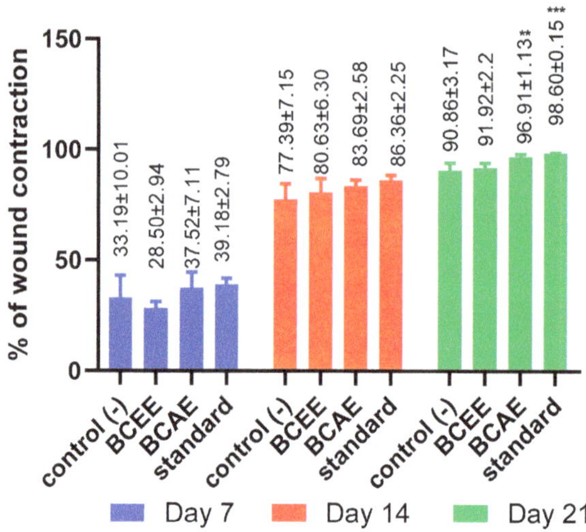

Figure 5. % of wound contraction using *B. cinerea* extracts and standard treatment (Indomethacin). (Dunnett's multiple comparisons test relative to the control: ** $p < 0.01$; *** $p < 0.001$).

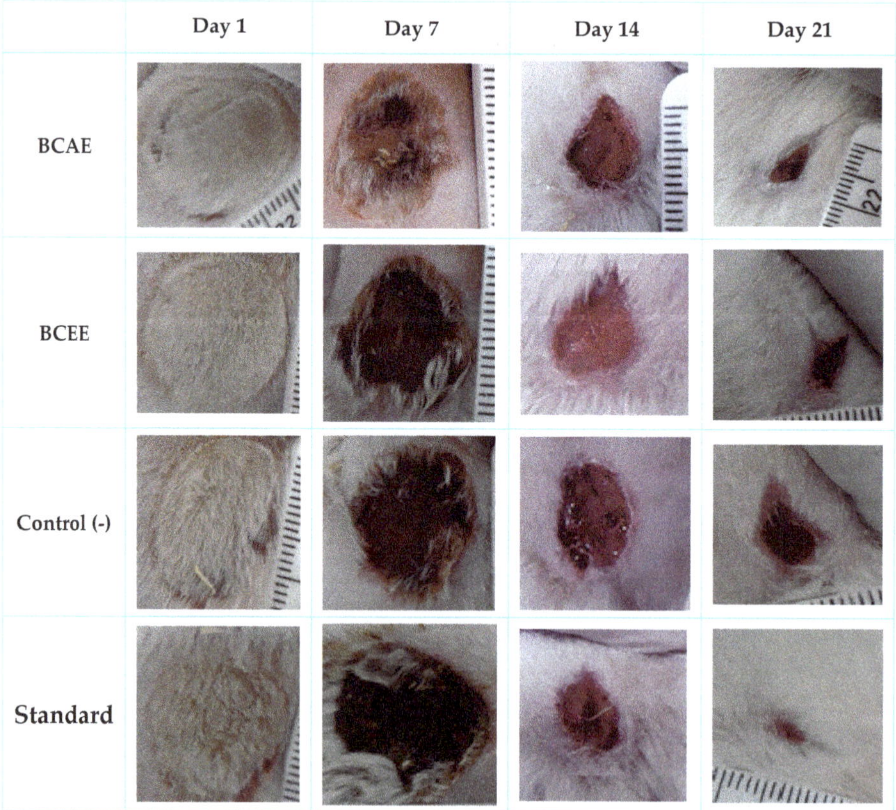

Figure 6. Images of the healing process in treated rats.

3.4.2. Anti-Inflammatory Activity

The anti-inflammatory efficacy and potency of the studied extracts was less than that of the positive control of Indomethacin (Figure 7). Despite its richness in different phenolic types, BCAE has minimal anti-inflammatory efficacy with the greatest % inhibition of 8.1% observed during the sixth hour. BCEE exhibited greater anti-inflammatory efficacy than BCAE. The percentage of inhibition obtained using BCEE varies between 12.82% and 21.62% during the 3 h of measurement (p value < 0.01).

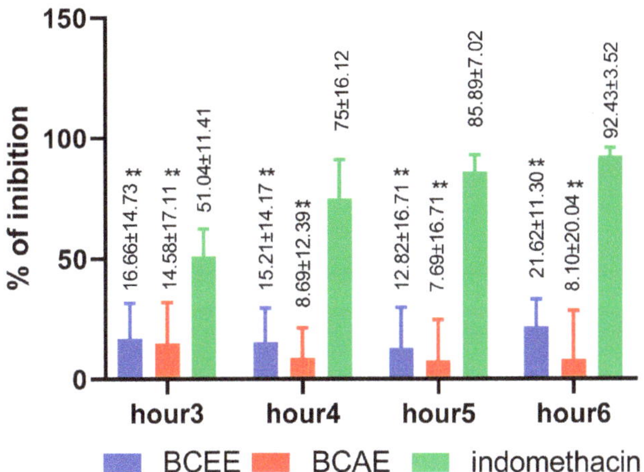

Figure 7. Anti-inflammatory activity of *B. cinerea* extracts and the standard compound (Dunnett's multiple comparisons test relative to the standard: ** p < 0.01).

There are few studies on the anti-inflammatory efficacy or potency of extracts of *B. cinerea*. At a dose of 400 mg/kg bm, aqueous extracts of fresh and dry aerial parts of *B. cinerea* inhibited inflammation by 47.73% and 50.01%, respectively, 5 h after a carrageenan injection, compared to the standard drug, Indomethacin (55.26%) [21]. In another study, the measurement of the anti-inflammatory effect using murine macrophage-like RAW 264.7 cells and quantified by nitric oxide (NO) production shows that the infusion and hydroethanolic extracts revealed a weaker inhibition of NO production, with EC_{50} values of 122 ± 6 and 105 ± 9 µg/mL, respectively [56]. Iluteolin, apigenin, kaempferol and caffeoylquinic acid derivatives are known to have a potent anti-inflammatory effect [57,58].

3.4.3. Peripheral Analgesic Activity

BCAE exerts remarkable analgesic activity with a percentage of inhibition of $51.60 \pm 15.18\%$, while BCEE exerts weak activity with a percent of inhibition of $34.33 \pm 6.60\%$. BCEE also shows low analgesic activity compared to Tramadol ($55.64 \pm 13.83\%$) with a significant difference (Figure 8). BCAE exhibited analgesic activity, similar to that of Tramadol with no significant difference.

The results of a recent study found that a dose of 400 mg/kg, bm of aqueous extracts of fresh and dry aerial parts of *B. cinerea* were effective at inhibiting spasms caused by the injection of acetic acid (0.6%) by 50.71% and 45.51%, respectively [21]. Alternatively, it has been reported that the use of n-butanol, ethyl acetate and ethyl ether extracts induced a percentage of inhibition of constriction of 40.21%, 50% and 62.49%, respectively [59]. The analgesic efficacies of extracts of fresh and dry aerial parts of *B. cinerea* could be related to an alteration of the biosynthesis of the prostaglandins by the inhibition of the cyclooxygenase [21].

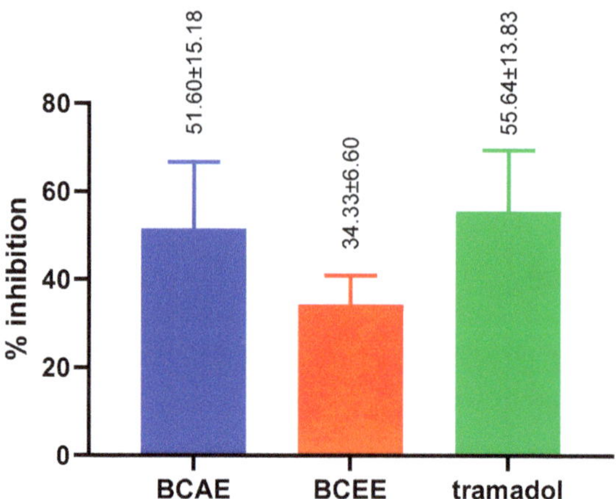

Figure 8. Analgesic efficacy of extracts of *B. cinerea* and the standard compound, Tramadol (Bartlett's statistic (corrected): *p* value = 3031; are SDs significantly different: no).

4. Conclusions

The chemical composition of extracts of *B. cinerea* is consistent with its use as a traditional pharmacological treatment. The application of BCAE and BCEE on rats and mice demonstrated the ability of the aqueous extract to minimize abdominal contortions and to increase the rate of re-epithelialization of burn-induced wounds. The healing power revealed by this study, through the use of the aqueous extract of *B. cinerea*, promotes the application of this extract in the development of treatments that combine the use of modern products and practices with herbal healing agents. Moreover, the results of the antioxidant activities showed that the ethanolic extract is recommended as an antioxidant agent due to its high content of polyphenols, flavonoids and tannins. Thus, this work provides useful results for further pharmacological studies and the design of new drugs based on the species of *B. cinerea*.

Author Contributions: Conceptualization, A.A. and E.D.; methodology, I.M., M.C. and A.A.; formal analysis, A.A., M.A.M.A.-S., M.A. and A.E.B.; investigation, A.A. and M.B.; visualization, M.A.M.A.-S.; data curation, A.A. and I.M.; writing—original draft preparation, A.A. and I.M.; writing—review and editing J.P.G., Y.A.A.-S., M.A.M.A.-S. and A.B.; supervision, B.L. and E.D.; funding acquisition, Y.A.A.-S. All authors have read and agreed to the published version of the manuscript.

Funding: This research received no external funding.

Institutional Review Board Statement: The institutional ethical committee of care and use of the animals at the laboratory of Biotechnology, Health, Agrofood, and Environment of Sidi MohamedBen Abdallah Fez University, Morocco, reviewed and approved the present study (July 2020/LBEAS-08 and 25 February 2021).

Informed Consent Statement: Not applicable.

Data Availability Statement: All data pertinent to the work are included in the paper.

Acknowledgments: The authors extend their appreciation to Researchers Supporting Project number (RSP-2022R520), King Saud University, Riyadh, Saudi Arabia.

Conflicts of Interest: The authors declare no conflict of interest.

References

1. Bouyahya, A.; Guaouguaou, F.E.; El Omari, N.; El Menyiy, N.; Balahbib, A.; El-Shazly, M.; Bakri, Y. Anti-Inflammatory and Analgesic Properties of Moroccan Medicinal Plants: Phytochemistry, in vitro and in vivo Investigations, Mechanism Insights, Clinical Evidences and Perspectives. *J. Pharm. Anal.* **2022**, *12*, 35–57. [CrossRef]
2. Agour, A.; Mssillou, I.; Es-Safi, I.; Conte, R.; Mechchate, H.; Slighoua, M.; Amrati, F.E.; Parvez, M.K.; Numan, O.; Bari, A.; et al. The Antioxidant, Analgesic, Anti-Inflammatory, and Wound Healing Activities *of Haplophyllum tuberculatum* (Forsskal) A. Juss Aqueous and Ethanolic Extract. *Life* **2022**, *12*, 1553. [CrossRef] [PubMed]
3. Allali, A.; Rezoukin, S.; Fadli, A.; El Moussaoui, A.; Bourhia, M.; Salamatullah, M.; Alzahrani, A.; Abdulhakeem, A.; Khalil Albadr, H.; Albadr, A.N.; et al. Chemical Characterization and Antioxidant, Antimicrobial, and Insecticidal Properties of Essential Oil from *Mentha pulegium* L. *Evid.-Based Complement. Altern. Med.* **2021**, *2012*, 1108133. [CrossRef]
4. Rolnik, A.; Olas, B. The Plants of the Asteraceae Family as Agents in the Protection of Human Health. *Int. J. Mol. Sci.* **2021**, *22*, 3009. [CrossRef] [PubMed]
5. Abazari, M.; Ghaffari, A.; Rashidzadeh, H.; Badeleh, S.M.; Maleki, Y. A Systematic Review on Classification, Identification, and Healing Process of Burn Wound Healing. *Int. J. Low. Extrem. Wounds* **2022**, *21*, 18–30. [CrossRef]
6. Criollo-Mendoza, M.S.; Contreras-Angulo, L.A.; Leyva-López, N.; Gutiérrez-Grijalva, E.P.; Jiménez-Ortega, L.A.; Heredia, J.B. Wound Healing Properties of Natural Products: Mechanisms of Action. *Molecules* **2023**, *28*, 598. [CrossRef]
7. Daou, A.; Baydoun, E.; Nasser, M.; Albahri, G.; Badran, A.; Hijazi, A. The Therapeutic Wound Healing Bioactivities of Various Medicinal Plants. *Life* **2023**, *13*, 317.
8. Markiewicz-Gospodarek, A.; Kozioł, M.; Tobiasz, M.; Baj, J.; Radzikowska-Büchner, E.; Przekora, A. Burn Wound Healing: Clinical Complications, Medical Care, Treatment, and Dressing Types: The Current State of Knowledge for Clinical Practice. *Int. J. Environ. Res. Public Health* **2022**, *19*, 1338. [CrossRef]
9. Wang, G.; Yang, F.; Zhou, W.; Xiao, N.; Luo, M.; Tang, Z. The Initiation of Oxidative Stress and Therapeutic Strategies in Wound Healing. *Biomed. Pharmacother.* **2023**, *157*, 114004. [CrossRef]
10. Wu, Z.; Chen, L. Similarity-Based Method with Multiple-Feature Sampling for Predicting Drug Side Effects. *Comput. Math. Methods Med.* **2022**, *2022*, 9547317. [CrossRef]
11. Miranda, G.M.; Ramos, E.; Santos, V.O.; Bessa, J.R.; Teles, Y.C.F.; Yahouédéhou, S.C.M.A.; Goncalves, M.S.; Ribeiro-Filho, J. Inclusion Complexes of Non-Steroidal Anti-Inflammatory Drugs with Cyclodextrins: A Systematic Review. *Biomolecules* **2021**, *11*, 361. [CrossRef] [PubMed]
12. Komakech, R.; Matsabisa, M.G.; Kang, Y. The Wound Healing Potential of Aspilia Africana (Pers.) CD Adams (Asteraceae). *Evid.-Based Complement. Altern. Med.* **2019**, *2019*, 7957860. [CrossRef] [PubMed]
13. Carvalho, A.R.; Diniz, R.M.; Suarez, M.A.M.; Figueiredo, C.S.S.S.; Zagmignan, A.; Grisotto, M.A.G.; Fernandes, E.S.; da Silva, L.C.N. Use of Some Asteraceae Plants for the Treatment of Wounds: From Ethnopharmacological Studies to Scientific Evidences. *Front. Pharmacol.* **2018**, *9*, 784. [CrossRef] [PubMed]
14. Lambebo, M.K.; Kifle, Z.D.; Gurji, T.B.; Yesuf, J.S. Evaluation of Wound Healing Activity of Methanolic Crude Extract and Solvent Fractions of the Leaves of *Vernonia auriculifera* Hiern (Asteraceae) in Mice. *J. Exp. Pharmacol.* **2021**, *13*, 677–692. [CrossRef]
15. Lakhdar, M. Traditional Uses, Phytochemistry and Biological Activities of *Cotula cinerea* Del: A Review. *Trop. J. Pharm. Res.* **2018**, *17*, 365–373. [CrossRef]
16. Agour, A.; Mssillou, I.; Mechchate, H.; Es-Safi, I.; Allali, A.; El Barnossi, A.; Al Kamaly, O.; Alshawwa, S.Z.; El Moussaoui, A.; Bari, A.; et al. *Brocchia cinerea* (Delile) Vis. Essential Oil Antimicrobial Activity and Crop Protection against Cowpea Weevil *Callosobruchus maculatus* (Fab.). *Plants* **2022**, *11*, 583. [CrossRef]
17. Markouk, M.; Bekkouche, K.; Larhsini, M.; Bousaid, M.; Lazrek, H.B.; Jana, M. Evaluation of Some Moroccan Medicinal Plant Extracts for Larvicidal Activity. *J. Ethnopharmacol.* **2000**, *73*, 293–297. [CrossRef]
18. Cimmino, A.; Roscetto, E.; Masi, M.; Tuzi, A.; Radjai, I.; Gahdab, C.; Paolillo, R.; Guarino, A.; Catania, M.R.; Evidente, A. Sesquiterpene Lactones from *Cotula cinerea* with Antibiotic Activity against Clinical Isolates of Enterococcus Faecalis. *Antibiotics* **2021**, *10*, 819. [CrossRef]
19. Bensizerara, D.; Menasria, T.; Melouka, M.; Cheriet, L.; Chenchouni, H. Antimicrobial Activity of Xerophytic Plant (*Cotula cinerea* Delile) Extracts Against Some Pathogenic Bacteria and Fungi. *Jordan J. Biol. Sci.* **2013**, *6*, 266–271. [CrossRef]
20. Salhi, N.; Mohammed Saghir, S.A.; Terzi, V.; Brahmi, I.; Ghedairi, N.; Bissati, S. Antifungal Activity of Aqueous Extracts of Some Dominant Algerian Medicinal Plants. *BioMed Res. Int.* **2017**, *2017*, 7526291. [CrossRef]
21. Chlif, N.; Bouymajane, A.; Oulad El Majdoub, Y.; Diouri, M.; Rhazi Filali, F.; Bentayeb, A.; Altemimi, A.B.; Mondello, L.; Cacciola, F. Phenolic Compounds, in vivo Anti-Inflammatory, Analgesic and Antipyretic Activities of the Aqueous Extracts from Fresh and Dry Aerial Parts of *Brocchia cinerea* (Vis.). *J. Pharm. Biomed. Anal.* **2022**, *213*, 114695. [CrossRef] [PubMed]
22. Aydin, M.E.; Tekin, E.; Ahiskalioglu, E.O.; Ates, I.; Karagoz, S.; Aydin, O.F.; Ozkaya, F.; Ahiskalioglu, A. Erector Spinae Plane Block vs Non-Steroidal Anti-Inflammatory Drugs for Severe Renal Colic Pain: A Pilot Clinical Feasibility Study. *Int. J. Clin. Pract.* **2021**, *75*, e13789. [CrossRef] [PubMed]
23. Mssillou, I.; Agour, A.; Slighoua, M.; Tourabi, M.; Nouioura, G.; Lyoussi, B.; Derwich, E. Phytochemical Characterization, Antioxidant Activity, and in vitro Investigation of Antimicrobial Potential of *Dittrichia viscosa* L. Leaf Extracts against Nosocomial Infections. *Acta Ecol. Sin.* **2021**, *42*, 661–669. [CrossRef]

24. Boizot, N.; Charpentier, J.; Boizot, N.; Méthode, J.C. *Méthode Rapide d' Évaluation du Contenu en Composés Phénoliques Des Organes d' Un Arbre Forestier to Cite This Version: HAL Id: Hal-02669118 Méthode Rapide d' Évaluation du Contenu en Composés Phénoliques Des Organes d' Un Arbre Forestier;* INRA: Paris, France, 2020.
25. Bahorun, T.; Gressier, B.; Trotin, F.; Brunet, C.; Dine, T.; Luyckx, M.; Vasseur, J.; Cazin, M.; Cazin, J.C.; Pinkas, M. Oxygen Species Scavenging Activity of Phenolic Extracts from Hawthorn Fresh Plant Organs and Pharmaceutical Preparations. *Arzneimittelforschung* **1996**, *46*, 1086–1089.
26. Joslyn, M.A.; Glick, Z. Comparative Effects of Gallotannic Acid and Related Phenolics on the Growth of Rats. *J. Nutr.* **1969**, *98*, 119–126. [CrossRef]
27. Mssillou, I.; Agour, A.; Slighoua, M.; Chebaibi, M.; Amrati, F.E.Z.; Alshawwa, S.Z.; Al Kamaly, O.; El Moussaoui, A.; Lyoussi, B.; Derwich, E. Ointment-Based Combination of *Dittrichia viscosa* L. and *Marrubium vulgare* L. Accelerate Burn Wound Healing. *Pharmaceuticals* **2022**, *15*, 289. [CrossRef]
28. Sahin, F.; Gulluce, M.; Daferera, D.; Sokmen, A.; Sokmen, M.; Polissiou, M.; Agar, G.; Ozer, H. Biological Activities of the Essential Oils and Methanol Extract of *Origanum culgare* ssp. Vulgare in the Eastern Anatolia Region of Turkey. *Food Control* **2004**, *15*, 549–557. [CrossRef]
29. Oyaizy, M. Studies on Product of Browning Reaction Prepared from Glucose Amine. *Jpn. J. Nutr.* **1986**, *44*, 307–315. [CrossRef]
30. Prieto, P.; Pineda, M.; Aguilar, M. Spectrophotometric Quantitation of Antioxidant Capacity through the Formation of a Phosphomolybdenum Complex: Specific Application to the Determination of Vitamin E. *Anal. Biochem.* **1999**, *269*, 337–341. [CrossRef]
31. Balouiri, M.; Bouhdid, S.; Harki, E.H.; Sadiki, M.; Ouedrhiri, W.; Ibnsouda, S.K. Antifungal Activity of *Bacillus* spp. Isolated from *Calotropis procera* Ait. Rhizosphere against *Candida Albicans*. *Asian J. Pharm. Clin. Res.* **2015**, *8*, 213–217.
32. El Barnossi, A.; Moussaid, F.; Iraqi Housseini, A. Antifungal Activity of Bacillussp. Gn-A11-18isolated from Decomposing Solid Green Household Waste in Water and Soil against Candida Albicans and Aspergillus Niger. *E3S Web Conf.* **2020**, *150*, 02003. [CrossRef]
33. Balouiri, M.; Sadiki, M.; Ibnsouda, S.K. Methods for in vitro Evaluating Antimicrobial Activity: A Review. *J. Pharm. Anal.* **2016**, *6*, 71–79. [CrossRef] [PubMed]
34. Eloff, J.N. A Sensitive and Quick Microplate Method to Determine the Minimal Inhibitory Concentration of Plant Extracts for Bacteria. *Planta Med.* **1998**, *64*, 711–713. [CrossRef]
35. Chebaibi, A.; Marouf, Z.; Rhazi-Filali, F.; Fahim, M.; Ed-Dra, A. Evaluation of Antimicrobial Activity of Essential Oils from Seven Moroccan Medicinal Plants. *Phytotherapie* **2016**, *14*, 355–362. [CrossRef]
36. Heidari, M.; Bahramsoltani, R.; Abdolghaffari, A.H.; Rahimi, R.; Esfandyari, M.; Baeeri, M.; Hassanzadeh, G.; Abdollahi, M.; Farzaei, M.H. Efficacy of Topical Application of Standardized Extract of Tragopogon Graminifolius in the Healing Process of Experimental Burn Wounds. *J. Tradit. Complement. Med.* **2019**, *9*, 54–59. [CrossRef]
37. Winter, C.A.; Risley, E.A.; Nuss, G.W. Carrageenin-Induced Edema in Hind Paw of the Rat as an Assay for Antiinflammatory Drugs. *Proc. Soc. Exp. Biol. Med.* **1962**, *111*, 544–547. [CrossRef] [PubMed]
38. Bentley, G.A.; Newton, S.H.; Starr, J. Evidence for an action of morphine and the enkephalins on sensory nerve endings in the mouse peritoneum. *Pharmacology* **1981**, *73*, 325–332. [CrossRef] [PubMed]
39. Chlif, N.; Diouri, M.; El Messaoudi, N.; Ed-Dra, A.; Filali, F.R.; Bentayeb, A. Phytochemical Composition, Antioxidant and Antibacterial Activities of Extracts from Different Parts of *Brocchia cinerea* (Vis.). *Biointerface Res. Appl. Chem.* **2022**, *12*, 4432–4447. [CrossRef]
40. Ben moussa, M.T.; Cherif, R.A.; Lekhal, S.; Bounab, A.; Youcef, H. Dosage Des Composés Phénoliques et Détermination de l'activité Antioxydante Des Extraits Méthanoliques de *Brocchia cinerea* VIS del' Algérie (Sud-Est). *Alger. J. Pharm.* **2022**, *4*, 49–59.
41. Belyagoubi-Benhammou, N.; Belyagoubi, L.; Bekkara, F.A. Phenolic Contents and Antioxidant Activities in vitro of Some Selected Algerian Plants. *J. Med. Plants Res.* **2014**, *8*, 1198–1207. [CrossRef]
42. Bouziane, M.; Badjah-Hadj-Ahmed, Y.; Hadj-Mahammed, M. Chemical Composition of the Essential Oil of *Brocchia cinerea* Grown in South Eastern of Algeria. *Asian J. Chem.* **2013**, *25*, 3917–3921. [CrossRef]
43. Dendougui, H.; Seghir, S.; Jay, M.; Benayache, F.; Benayache, S. Flavonoids from *Cotula cinerea* Del. *Int. J. Med. Aromat. Plant* **2012**, *2*, 2249–4340.
44. Arabsalehi, F.; Rahimmalek, M.; Sabzalian, M.R.; Ghanadian, M.; Matkowski, A.; Szumny, A. Changes in Polyphenolic Composition, Physiological Characteristics, and Yield-Related Traits of Moshgak (*Ducrosia anethifolia* Boiss.) Populations in Response to Drought Stress. *Protoplasma* **2022**, 1–19. [CrossRef] [PubMed]
45. Mirniyam, G.; Rahimmalek, M.; Arzani, A.; Matkowski, A.; Gharibi, S.; Szumny, A. Changes in Essential Oil Composition, Polyphenolic Compounds and Antioxidant Capacity of Ajowan (*Trachyspermum ammi* L.) Populations in Response to Water Deficit. *Foods* **2022**, *11*, 3084. [CrossRef]
46. Gharibi, S.; Sayed Tabatabaei, B.E.; Saeidi, G.; Talebi, M.; Matkowski, A. The Effect of Drought Stress on Polyphenolic Compounds and Expression of Flavonoid Biosynthesis Related Genes in *Achillea pachycephala* Rech.F. *Phytochemistry* **2019**, *162*, 90–98. [CrossRef]
47. Khallouki, F.; Sellam, K.; Koyun, R.; Ricarte, I.; Alem, C.; Elrhaffari, L.; Owen, R.W. Phytoconstituents and in vitro Evaluation of Antioxidant Capacities of *Cotula cinerea* (Morocco) Methanol Extracts. *Rec. Nat. Prod.* **2015**, *9*, 572–575.
48. Kasrati, A.; Alaoui Jamali, C.; Bekkouche, K.; Wohlmuth, H.; Leach, D.; Abbad, A. Comparative Evaluation of Antioxidant and Insecticidal Properties of Essential Oils from Five Moroccan Aromatic Herbs. *J. Food Sci. Technol.* **2015**, *52*, 2312–2319. [CrossRef]

49. Teodoro, G.; Ellepola, K.; Seneviratne, C.; Koga-ito, C. Potential Use of Phenolic Acids as Anti-*Candida* Agents: A Review. *Front. Microbiol.* **2015**, *6*, 1420. [CrossRef]
50. Ekambaram, S.P.; Perumal, S.S.; Balakrishnan, A.; Marappan, N.; Gajendran, S.S.; Viswanathan, V. Antibacterial Synergy between Rosmarinic Acid and Antibiotics against Methicillin-Resistant *Staphylococcus aureus*. *J. Intercult. Ethnopharmacol.* **2016**, *5*, 358. [CrossRef]
51. Campos, F.M.; Couto, J.A.; Hogg, T.A. Influence of Phenolic Acids on Growth and Inactivation of *Oenococcus oeni* and *Lactobacillus hilgardii*. *J. Appl. Microbiol.* **2000**, *94*, 167–174. [CrossRef]
52. Wu, D.; Kong, Y.; Han, C.; Chen, J.; Hu, L.; Jiang, H.; Shen, X. D-Alanine: D-Alanine Ligase as a New Target for the Flavonoids Quercetin and Apigenin. *Int. J. Antimicrob. Agents* **2008**, *32*, 421–426. [CrossRef] [PubMed]
53. Ferreira, C.; Saavedra, M.J.; Simo, M. Antibacterial Activity and Mode of Action of Ferulic. *Microb. Drug Resist.* **2013**, *19*, 256–265. [CrossRef]
54. Mssillou, I.; Bakour, M.; Slighoua, M.; Laaroussi, H.; Saghrouchni, H.; Amrati, F.E.-Z.; Lyoussi, B.; Derwich, E. Investigation on Wound Healing Effect of Mediterranean Medicinal Plants and Some Related Phenolic Compounds: A Review. *J. Ethnopharmacol.* **2022**, *298*, 115663. [CrossRef]
55. Jawhari, F.Z.; El Moussaoui, A.; Bourhia, M.; Imtara, H.; Mechchate, H.; Es-Safi, I.; Ullah, R.; Ezzeldin, E.; Mostafa, G.A.; Grafov, A.; et al. *Anacyclus pyrethrum* (L): Chemical Composition, Analgesic, Anti-Inflammatory, and Wound Healing Properties. *Molecules* **2020**, *25*, 5469. [CrossRef]
56. Ghouti, D.; Rached, W.; Abdallah, M.; Pires, T.C.S.P.; Calhelha, R.C.; Alves, M.J.; Abderrahmane, L.H.; Barros, L.; Ferreira, I.C.F.R. Phenolic Profile and in vitro Bioactive Potential of Saharan *Juniperus phoenicea* L. and *Cotula cinerea* (Del) Growing in Algeria. *Food Funct.* **2018**, *9*, 4664–4672. [CrossRef]
57. Tian, C.; Liu, X.; Chang, Y.; Wang, R.; Lv, T.; Cui, C.; Liu, M. Investigation of the Anti-Inflammatory and Antioxidant Activities of Luteolin, Kaempferol, Apigenin and Quercetin. *S. Afr. J. Bot.* **2021**, *137*, 257–264. [CrossRef]
58. Peng, W.; Han, P.; Yu, L.; Chen, Y.; Ye, B.; Qin, L.; Xin, H.; Han, T. Anti-Allergic Rhinitis Effects of Caffeoylquinic Acids from the Fruits of *Xanthium strumarium* in Rodent Animals via Alleviating Allergic and Inflammatory Reactions. *Rev. Bras. Farmacogn.* **2019**, *29*, 46–53. [CrossRef]
59. Kang, J.Y.; Khan, M.N.A.; Park, N.H.; Cho, J.Y.; Lee, M.C.; Fujii, H.; Hong, Y.K. Antipyretic, Analgesic, and Anti-Inflammatory Activities of the Seaweed *Sargassum fulvellum* and *Sargassum thunbergii* in Mice. *J. Ethnopharmacol.* **2008**, *116*, 187–190. [CrossRef]

Article

Antioxidant, Anti-Inflammatory, and Analgesic Properties of Chemically Characterized Polyphenol-Rich Extract from *Withania adpressa* Coss. ex Batt

Ahmad Mohammad Salamatullah

Department of Food Science & Nutrition, College of Food and Agricultural Sciences, King Saud University, P.O. Box 2460, Riyadh 11451, Saudi Arabia; asalamh@ksu.edu.sa

Abstract: The current work was undertaken to investigate the chemical composition, antioxidant, anti-inflammatory, and analgesic properties of a polyphenol-rich fraction from *Withania adpressa* Coss. ex Batt. After being extracted, the polyphenol-rich fraction was chemically characterized through use of high-performance liquid chromatography (HPLC). Antioxidant potency was assessed through the use of 2,2-diphenyl-1-picrylhydrazyl (DPPH) and total antioxidant capacity (TAC). Inflammatory and analgesic properties were assessed in vivo through the use of carrageenan and heat stimulus assays, respectively. Chromatographic analysis of polyphenol-rich fraction revealed the presence of potentially bioactive phenols including epicatechin, apigenin, luteolin, quercetin, caffeic acid, p-coumaric acid, and rosmarinic acid. The polyphenol-rich fraction showed interesting anti-free-radical potency with a calculated IC_{50} value of 27.84 ± 1.48 µg/mL. At the highest dose used (1000 µg/mL), the polyphenol-rich fraction scored good total antioxidant capacity with a calculated value of 924.0 ± 28.29 µg EAA/mg. The polyphenol-rich fraction strongly alleviated the inflammatory effect of carrageenan injected into the plantar fascia of rats resulting in inhibition up to $89.0 \pm 2.08\%$ at the highest tested dose (500 mg/kg). The polyphenol-rich fraction showed a good analgesic effect wherein the delay in reaction time to a thermal stimulus caused by 500 mg/kg had a highly similar effect to that induced by Tramadol used as a positive control. The findings of the current work highlight the importance of polyphenol-rich fractions from *W. adpressa* Coss. ex Batt. as an alternative source of natural antioxidant, inflammatory, and analgesic drugs to control relative diseases.

Keywords: plants; natural products; free radicals; inflammation; medicinal; caffeic acid

Citation: Salamatullah, A.M. Antioxidant, Anti-Inflammatory, and Analgesic Properties of Chemically Characterized Polyphenol-Rich Extract from *Withania adpressa* Coss. ex Batt. *Life* **2023**, *13*, 109. https:// doi.org/10.3390/life13010109

Academic Editors: Efstathia Papada, Charalampia Amerikanou, Marisa Colone and Giorgio Lenaz

Received: 4 November 2022
Revised: 13 December 2022
Accepted: 27 December 2022
Published: 30 December 2022

1. Introduction

It is well known that plants have been utilized for medical reasons, cosmetic purposes, and as a dietary component all over the globe for hundreds of years [1,2]. The presence of phytochemicals in herbs, especially secondary metabolites, is the most important factor contributing to their beneficial characteristics [3]. Higher plants synthesize these organic molecules, which, in most instances, are not required for growth and development but are instead formed in reaction to biotic and abiotic environmental conditions [4]. In plants, secondary metabolites consist of terpenoids, alkaloids, and flavonoids, which have been scientifically shown to be promising bioactive agents with antioxidant, antibiotic, anti-inflammatory, anti-aging, and antitumor properties [5–7]. Depending on the use, plants may be used fresh, dried, or processed into essential oils or crude extracts [7]. The compounds responsible for particular biological potential in a plant have been examined in many studies [8].

Diets high in antioxidants have been shown to protect humans against degenerative illnesses including cancer and cardiovascular disease [9]. To avoid the oxidative degradation of foodstuffs caused by free radicals, natural antioxidants tend to be favored by users in the food business over synthetic antioxidants, according to recent research [10]. Free radicals have the potential to generate cytotoxic effects and tissue lesions, as well as DNA damage. The human body requires antioxidant agents in order to fight itself against free radicals. These antioxidant agents are found in fruits and vegetables and almost all plants. Because of their secondary metabolites, plants are able to supply potent antioxidant agents that help to manage and alleviate the effects of free radicals [11].

Inflammatory disorders are growing more widespread throughout the globe [12]. Inflammation may be triggered by several factors, including physical injury, ultraviolet irradiation, microbial invasion, and immunological responses. Sclerosis, inflammatory bowel disease, chronic asthma, and psoriasis are among disorders that may be caused by inflammation cascades. Inflammation is also involved in the development of chronic fatigue syndrome [13,14]. The disadvantages of clinically utilized anti-inflammatory medications include the presence of side effects as well as the high expense of treatment [12]. Traditional remedies and natural products may be used as an alternative to these treatments, and they hold significant promise in the discovery of bioactive lead compounds into therapeutics for the treatment of inflammatory illnesses. Traditional remedies and phytopharmaceuticals have been utilized for the treatment of inflammatory and other illnesses for many years [12].

W. adpressa Coss. ex Batt (*Solanaceae*), which is a herb commonly known by its name Winter Cherry, grows in North Africa and the Mediterranean basin; it has been shown to have pharmacological properties, including anti-tumor, immunomodulatory, anti-convulsant, and anti-stress properties. Importantly, diseases such as conjunctivitis, inflammation, anxiety, nervous system diseases, bronchitis, ulcers, liver disease, and Parkinson's disease have been treated with plants in the genus of Withania for a long time [15]. Previous reports showed that the gnus *withania* possesses many phenols, notably glycowithanolides and withanolides, as well as volatile chemicals [16].

The current study was conducted to investigate the chemical composition, antioxidant, anti-inflammatory, and analgesic properties of a polyphenol-rich fraction from leaves *W. adpressa* Coss. ex Batt.

2. Materials and Methods

2.1. Plant Material

From the Sahara area (29.7519° N, 7.9756° W) in March 2021, *W. adpressa* Coss. ex Batt was collected. Following the confirmation of the plant's identity by a botanist, it was placed in the University Herbarium under the reference A2/WDBF21. Consequently, the leaves were washed and dried for seven days in the dark and a well-ventilated environment before being extracted.

2.2. Extraction of Phenols

The extraction of phenols was successfully conducted by using maceration as previously described [17]. To summarize, a total of 100 g of *W. adpressa* leaves was macerated with 300 mL of methanol. Following that, the solvent was removed through the use of a rotary evaporator at decreased pressure and low temperature to obtain the concentrated extract. The obtained extract was solubilized in 500 mL of distilled water before extracting it three times further through the use of liquid–liquid extraction using 200 mL of each of the following solvents: hexane, chloroform, and ethyl acetate. Following that, the ethyl acetate layer was evaporated at decreased pressure using a rotary evaporator, which was used to remove the solvent. The residue was dissolved in 300 mL water once more and freeze-dried in order to obtain a polyphenol-rich fraction [12].

2.3. HPLC Analysis

The polyphenol-rich fraction was phytochemically characterized using the HPLC as described by Amrati [17], with minor modifications. The HPLC system was used for separation and identification of compounds. Polyphenol and standard samples were filtered using a 0.2 m membrane filter to eliminate particle residues before being injected. After that, a volume of 5 mL of polyphenol extract was injected over a C18 ZORBAX

Eclipse column at an injection rate of 0.7 mL/min and a column temperature of 30 °C. Acidified water (acetic acid 0.1%) (A) and acetonitrile (B) were utilized as the mobile phase in this experiment, and the reaction was allowed to proceed for 65 min. Compounds were recognized by comparing their spectra to those of reference compounds under the same conditions.

2.4. In Vitro Antioxidant Activity of Polyphenol-Rich Fraction

In the current work, assessment of the antioxidant activity of polyphenol-rich extract was carried out through the use of DPPH and molybdate in triplicate assays.

2.4.1. Antioxidant Power of Polyphenol-Rich Fraction Using DPPH Assay

This test was performed according to the methodology described by Kuramasamy et al. [18]. Briefly, 1000 μL of a methanolic DPPH (0.2 mM) solution was combined with the polyphenol-rich fraction (0–1 mg/mL). The obtained mixture was then held at room temperature for 30 min in the darkness, and the absorbance was measured at 517 nm. A blank solution consisting of 1000 μL of DPPH solution and 1000 μL of methanol was used to serve as a negative control. The blank solution, as well as samples and positive controls (quercetin and BHT), was produced under identical working conditions. Next, a spectrophotometer was used to quantify the absorbance decline, and the inhibition percentage was determined through the use of the following formula:

$$\text{Inhibition (\%)} = [1 - (\text{sample}/\text{control})] \times 100$$

2.4.2. Total Antioxidant Capacity of Polyphenol-Rich Fraction

A reagent was prepared by mixing H2SO4, (0.6 M), Na2PO4 (28 mM), and ammonium molybdate (4 mM) to measure the total antioxidant capacity of the polyphenol-rich fraction. Briefly, one milliliter of reagent was added to 0.1 mL of the polyphenol-rich fraction at various concentrations (0.2, 0.5, and 1 mg/mL). Thereafter, a blank solution composed of 1 mL reagent and 0.1 mL methanol was incubated in a water bath set to 95 °C for 90 min. Next, the absorbance was measured at 695 nm. The results were expressed in mg ascorbic acid equivalent per gram of dry extract (μg EAA/mg) [19].

2.5. Animal Material

Male rats weighing between 100 and 150 g were during the two-week acclimatization period; the animals were housed in cages with five rats each and maintained at 22 °C with a 12 h light–dark cycle. The method used in the present study complied with the globally recognized Guide for the Care and Use of Laboratory Animals. The animals were given unrestricted access to food and water at all times [20].

2.6. Anti-Inflammatory Activity

The anti-inflammatory activity of the polyphenol-rich fraction was assessed as described in earlier work. Animals were divided into groups of five rats each, of which two groups served as negative and positive controls, which received 0.9% saline and Diclofenac (1%), respectively, while other groups served as treatments, which received the polyphenol-rich fraction. Ninety minutes after local applications or one hour after oral administration of the polyphenol-rich fraction at different doses (200, 400, and 500 mg/kg), the plantar fascia of the right hind leg of rat was injected with 0.1 mL of carrageenan intradermally. The circumference of the applied sample was measured before the injection of carrageenan and then after every hour from the third hour until the sixth hour after the administration of carrageenan. The following formula was used to compute the % inhibition of inflammation:

$$\% \text{ inhibition} = [((S_t - S_0) \text{ control} - (S_t - S_0) \text{ sample})/((S_t - S_0) \text{ control})] \times 100$$

2.7. Analgesic Activity

The animals were divided into groups of five rats each, of which two groups served as negative and positive controls, which received 0.9% saline and Tramadol, respectively. Each animal was individually placed in an enclosed space on a glass surface (L × W × H = 10 × 20 × 14). After 10 min of adaptation, rats received oral administration of polyphenol-rich fraction before being subjected to the heat stimulus (50 °C) onto the plantar surface of each hindpaw. The increase in temperature under the plantar fascia of the right hind leg resulted in rat movement. The delay in reaction time to the thermal stimulus was recorded.

2.8. Statistical Analysis

Data were expressed in means with standard deviations of triplicate tests using Graph-Pad Prism software (version.7). Normality of distributions was tested by the use of Shapiro–Wilk's test, whilst the homogeneity of variances was checked by the use of Levene's test. Analysis of variance (ANOVA) was performed, with Tukey's HSD test as a post hoc test for multiple comparisons. A significant difference was considered at $p < 0.05$.

3. Results

3.1. Chemical Characterization

The chemical characterization of the polyphenol-rich fraction from leaves of *W. adpressa* allowed the identification of seven major compounds including flavonoid compounds; epicatechin, apigenin, luteolin, and quercetin; phenolic acids; caffeic acid and p-coumaric acid; and polyphenols derived from hydroxycinnamic-acid-like rosmarinic acid (Figure 1, Table 1). The chemical composition of different extracts from the genus Withania has been widely investigated. Notably, Jain et al. (2012) revealed that extracts from *Withania somnifera* and *Withania coagulans* possessed alkaloids; isopelletierine, anaferine, and saponins with an additional acyl group; sitoindoside VII and VIII; withanolides with glucose at carbon 27; withanolides; and withaferines [21]. Matsuda and co-authors (2001) showed that *W. somnifera* possessed withanolide glycosides and withanosides [22]. The genus Withania possessed various fatty acids; octacosan; oleic and stearic fatty acids; steroids; and oleanolic acid as reported in earlier work [23].

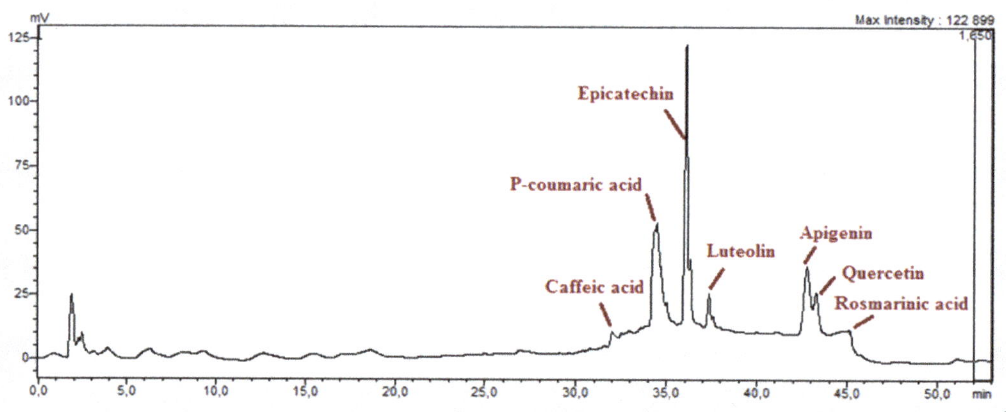

Figure 1. Chromatographic analysis of polyphenol-rich fraction using HPLC.

Table 1. Chemical structure of major compounds identified in polyphenol-rich fraction using HPLC.

RT	Identification Compound	Standard Use	Concentration in µg/mg	Molecular Structure
31.91	Caffeic acid	Cafeic acid	71.28	Caffeic acid
34.76	Coumaric acid	p-coumaric acid	30.46	Coumaric acid
36.09	Epicatechin	Epicatechin	40.53	Epicatechin
37.41	Chrysoeriol-7-diglucuronide	Luteolin	33.62	chrysoeriol-7-diglucuronide
42.78	Acacetin-7-diglucuronide	Apigenin	34.20	acacetin-7-diglucuronide
43.35	Quercetin-3-*O*-glucuronide	Quercetin	64.25	Quercetin-3-O-glucuronide
45.19	Rosmarinic acid	Rosmarinic acid	39.81	Rosmarinic acid

3.2. Antioxidant Activity

The antioxidant activity of the polyphenol-rich fraction using the DPPH method, as represented in Figure 1, showed that the polyphenol-rich fraction exhibited good antioxidant activity in a dose-dependent manner. From this figure, it can be seen that increasing the concentration of the polyphenol-rich fraction increased the inhibition percentage of DPPH free radicals, i.e., 10 and 100 µg/mL of the polyphenol-rich fraction inhibited 42% and 72%, respectively (Table 1). The polyphenol-rich fraction recorded an IC_{50} value of 27.84 ± 1.48 µg/mL, which can be considered important when compared to that obtained with BHT (13.42 ± 0.87 µg/mL) and quercetin (14.27 ± 0.59 µg/mL) (Table 2). Antioxidant capacity evaluated by the use of ammonium molybdate showed that the polyphenol-rich fraction possessed important antioxidant capacity as shown in (Figure 2b). From this figure, it can see that 1000 µg/mL of the polyphenol-rich fraction recorded antioxidant capacity in the order of 924.0 ± 28.29 µg EAA/mg and 500 µg/ml recorded antioxidant capacity in the order of 387.1 ± 25.45 µg EAA/mg (Figure 2b).

Table 2. Antioxidant power of polyphenol-rich fraction tested by the use of DPPH bioassay.

Samples	Anti-Radical Activity by the DPPH Method			IC-50 in µg/mL
	10 µg/mL	100 µg/mL	1000 µg/mL	
Polyphenol-rich fraction	42%	72%	91%	27.84 ± 1.48
Quercetin	46%	87%	94%	14.27 ± 0.59
BHT	48%	89%	96%	13.42 ± 0.87

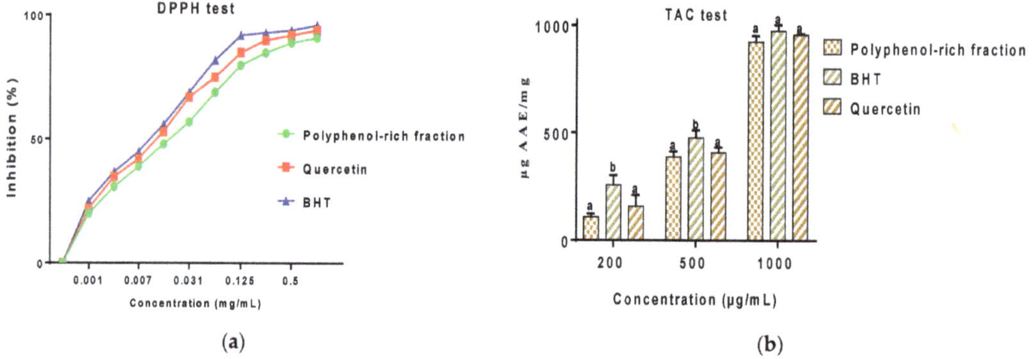

Figure 2. Antioxidant activity of polyphenol-rich fraction by the use of DPPH (**a**) and ammonium molybdate (**b**). The graphs have the same letter do not present a significant difference.

The results of antioxidant activity showed that the polyphenol-rich fraction possessed excellent antioxidant power, which can be explained by its richness in phenols with antioxidant power such as epicatechin, apigenin, luteolin, quercetin, caffeic acid, p-coumaric acid, and rosmarinic acid (Figure 1). Previous work on caffeic acid (phenolic acid) revealed an inhibition percentage of DPPH radicals in the order of 93.9% (DPPH), even at a low concentration (20 µg/mL) [24]. This antioxidant power may be due to the richness of polyphenol-rich fractions in polyphenols, which could react with free radicals, whether separately or in synergy, resulting in an antioxidant effect [25]. Catechin and epicatechin contained in the polyphenol-rich fraction are the predominant compounds with antioxidant power as reported in earlier work [26]. Our findings are in agreement with those reported by El Moussawi and co-authors who showed that the genus Withania possessed antioxidant potency, particularly *Withania frutescens*, which revealed an anti-free-radical activity of the order of 177.65 µg EAA/mg [27].

3.3. Anti-Inflammatory Activity of Polyphenol-Rich Fraction

The anti-inflammatory activity of the polyphenol-rich fraction was studied by measuring the amount of edema induced by carrageenan in rats. The results are presented in Figure 3, which shows the percentage of anti-inflammatory evolution of edema as a function of time. From this table, it can be seen that the polyphenol-rich fraction given to mice via oral administration alleviated the inflammatory effect of carrageenan injected into the plantar fascia of the right posterior leg of rats within three hours following injection, resulting in inhibition of 13.02 ± 1.27 % at 400 mg/kg and 18 ± 1.20% at 500 mg/kg. The anti-inflammatory effect increased progressively and reached a maximum inhibition after six hours of post-treatment, which reached 57.32 ± 2.05% at 200 mg/kg, 69.46 ± 2.13% at 400 mg/kg, and $89. 0 \pm 2.08$% at 500 mg/kg, while the positive control (Diclofenac 1%) inhibited edema by 91.51 ± 2.41% (Figure 3). The pretreatment of rats with the different doses of polyphenol-rich fraction induced a strong inhibition of inflammation at the sixth hour when compared to Diclofenac used as a drug reference.

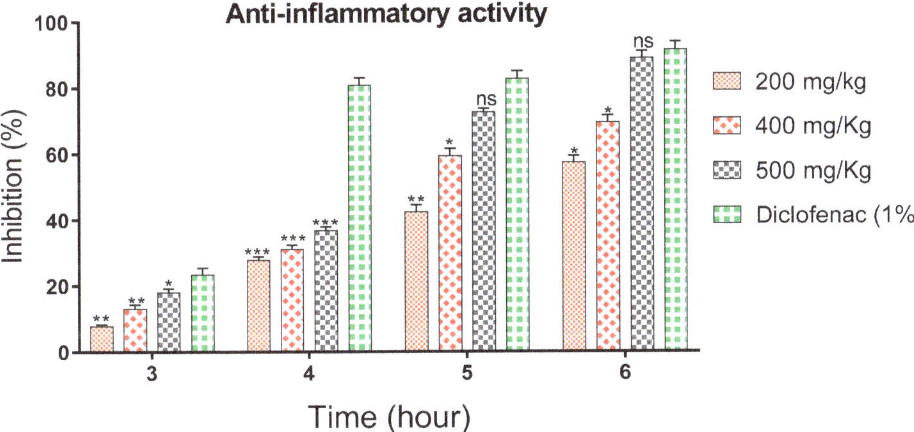

Figure 3. The anti-inflammatory activity of polyphenol-rich fraction at different doses (200, 400, and 500 mg/Kg) and Diclofenac (1%). $p \leq 0.05$ (*); $p \leq 0.005$ (**); $p \leq 0.001$ (***); ns: no significant difference.

Carrageenan injection triggers an increase in the levels of cyclooxygenase 2 (COX-2) mRNA synthesis, resulting in an increased concentration of this enzyme, which peaks at 1 h [28]. This is accompanied by an increase in the synthesis of prostaglandins (PGs), mainly prostaglandin E2 (PGE2) (maximum at 2 h) which is mainly involved in certain pain and inflammation processes [29]. This feature helps explain why non-steroidal anti-inflammatory drugs (NSAIDs), such as aspirin, have no effect at 1 h. This delay is due to their mechanisms of operation, namely the inhibition of PGs by simultaneously stimulating the two enzymes COX-1 and COX-2, for which the inhibition curve consolidates after 3 h, reflecting the stabilization of the mediators [30]. As for steroidal anti-inflammatory drugs (AIS) such as dexamethasone, their action is apparent from the first hour, thanks to their direct interaction with DNA, whose effect is associated with the action of several molecular pathways including pro inflammatory cytokines, phospholipase, and COX, mainly through the nuclear transcription factor NF-kB [30,31].

In the present work, the anti-inflammatory effect of the polyphenol-rich fraction by the carrageenan-induced edema test showed interesting results, whereby the inflammation was reduced by 89% at 500 mg/kg. These results are in agreement with those reported by Elmoussaoui et al. [29], who revealed $82.20\% \pm 8.69$ as a percentage of inhibition induced by 450 mg/kg *Withania frutescens* extract [29]. These results could be explained by various polar phenolic compounds identified by HPLC in polyphenol-rich fractions such as epicatechin, apigenin, luteolin, quercetin, caffeic acid, p-coumaric acid, and rosmarinic acid. Other phytoconstituents in the genus Withania, particularly tannins, mucilages, alkaloids, coumarins, and free quinone, could be also responsible for the anti-inflammatory activity investigated in the present work. These compounds can act by preventing the synthesis of prostaglandins through cyclooxygenase [14,17,32,33].

Withania somnifera exerted an anti-inflammatory role by repressing the expression of certain cytokines including tumor necrosis factor-(TNF-) α, interleukin-(IL-) 8 and 1, nitric oxide, and reactive oxygen species. Leukocyte adhesion and migration, as well as cell adhesion molecules, the production of IL-6 and TNF-a, and the activation of NF-k (nuclear factor kappa-luminous chain-enhancer of activated B-cells), were successfully blocked by withaferin A, one of the active components in *W. somnifera*. [18,20]. In addition, this compound blocked the activation of PMA-induced phosphorylation of the transcription factor p38, extracellular-regulated kinases (ERK 12), and the transcription factor c-Jun N terminal kinase (c-Jun N-terminal kinase) (JNK) [18,21].

3.4. Analgesic Activity of Polyphenol-Rich Fraction

Regarding analgesic activity, Figure 4 provides the result of the paw withdrawal test performed using the plantar thermal hyperalgesia model. The results showed that the paw withdrawal latency of the animal treated with the polyphenol-rich fraction was significantly higher than that of the negative control group which received only 0.9% NaCl physiological solution. The delay in reaction time to the thermal stimulus was directly related to the increase in doses of the sample. Notably, the delay in reaction time for 400 mg/kg and 500 mg/kg was 35.51 ± 1.04 s and 39.64 ± 1.17 s, respectively, while the delay in reaction time for the negative control group was 12.38 ± 1.28 s. Rats treated with Tramadol (1%), the reference analgesic, had a longer delay in reaction time to the thermal stimulus than the groups treated with the different doses of the polyphenol-rich fraction, which was 42.68 ± 0.78 s.

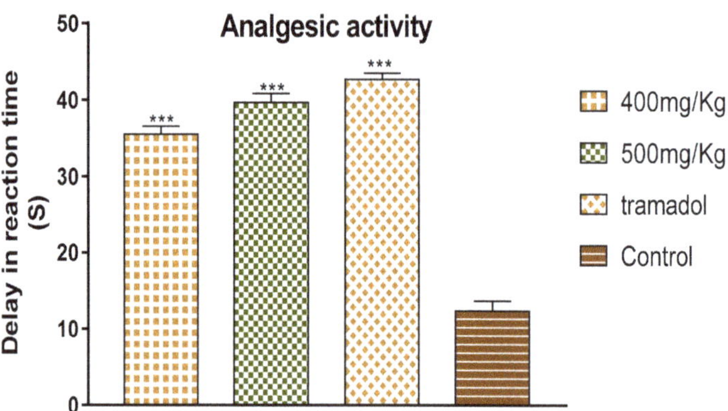

Figure 4. Analgesic activity of polyphenol-rich fraction. p *** ≤ 0.001.

The results show that the delay in reaction time to the thermal stimulus caused by 500 mg/kg of the polyphenol-rich fraction had a similar effect to that induced by the drug used as a positive control. This result can be considered a strong indication of the analgesic effect of the plant extract. The analgesic effect of the polyphenol-rich fraction can be explained by the property of phytochemicals identified using HPLC without excluding other phytochemicals in the genus Withania such as withanolides, withaferin A, withanolide F, coagulin L, and nicotiflorin, which are known to possess a notable anti-nociceptive property [34–38]. All these results are in agreement with those reported by Moussaoui et al. (2020) who revealed an important analgesic effect of the methanolic extract of *Withania frutescens* leaves at the dose of 450 mg/kg using the acetic acid method [29]. In addition, the current study was in accordance with Kumar et al. (2015), who demonstrated that *Withania somnifera* extract significantly increased the potency and time of the pain threshold as well as the potency and time of pain tolerance compared to placebo, reflecting a significant analgesic effect [39,40]. The analgesic effect presented here can be explained by the fact that phenols exhibited a central anti-nociceptive effect via the activation of opioid receptors whose activation results in a decrease in the release of pain mediators such as substance P [5,22].

4. Conclusions

The current work highlighted the antioxidant, anti-inflammatory, and analgesic properties of chemically characterized polyphenols from *Withania adpressa* Coss. This study concluded that polyphenols from *Withania adpressa* Coss. possess a potential for being used as an alternative reservoir of natural antioxidant, inflammatory, and analgesic drugs to control relative diseases. Even though the outcome of the present work is highly interesting, further pre-clinical and clinical investigations on nonhuman primates and humans will be required before any possible use of polyphenol-rich fractions as a natural drug.

Funding: This work was funded by the Researchers Supporting Project number (RSP-2022R437), King Saud University, Riyadh, Saudi Arabia, and the authors would like to extend their appreciation for their support.

Institutional Review Board Statement: Not applicable.

Informed Consent Statement: Not applicable.

Data Availability Statement: Not applicable.

Conflicts of Interest: The authors declare no conflict of interest.

References

1. Bourhia, M.; Elmahdaoui, H.; Moussa, S.I.; Ullah, R.; Bari, A. Potential Natural Dyes Food from the Powder of Prickly Pear Fruit Peels (*Opuntia* spp.) Growing in the Mediterranean Basin under Climate Stress. *BioMed Res. Int.* **2020**, *2020*, 7579430. [CrossRef]
2. Amrati, F.E.-Z.; Bourhia, M.; Saghrouchni, H.; Slighoua, M.; Grafov, A.; Ullah, R.; Ezzeldin, E.; Mostafa, G.A.; Bari, A.; Ibenmoussa, S. *Caralluma europaea* (Guss.) NE Br.: Anti-Inflammatory, Antifungal, and Antibacterial Activities against Nosocomial Antibiotic-Resistant Microbes of Chemically Characterized Fractions. *Molecules* **2021**, *26*, 636. [CrossRef] [PubMed]
3. Dillard, C.J.; German, J.B. Phytochemicals: Nutraceuticals and Human Health. *J. Sci. Food Agric.* **2000**, *80*, 1744–1756. [CrossRef]
4. Jain, C.; Khatana, S.; Vijayvergia, R. Bioactivity of Secondary Metabolites of Various Plants: A Review. *Int. J. Pharm. Sci. Res.* **2019**, *10*, 494–504.
5. Leja, K.B.; Czaczyk, K. The Industrial Potential of Herbs and Spices? A Mini Review. *Acta Sci. Pol. Technol. Aliment.* **2016**, *15*, 353–365. [CrossRef] [PubMed]
6. Wu, L.; Hsu, H.-W.; Chen, Y.-C.; Chiu, C.-C.; Lin, Y.-I.; Ho, J.A. Antioxidant and Antiproliferative Activities of Red Pitaya. *Food Chem.* **2006**, *95*, 319–327. [CrossRef]
7. Bourhia, M.; Bari, A.; Ali, S.S.; Benbacer, L. Phytochemistry and Toxicological Assessment of *Bryonia dioica* Roots Used in North-African Alternative Medicine. *Open Chem.* **2019**, *17*, 1403–1411. [CrossRef]
8. El Fakir, L.; Bouothmany, K.; Alotaibi, A.; Bourhia, M.; Ullah, R.; Zahoor, S.; El Mzibri, M.; Gmouh, S.; Alaoui, T.; Zaid, A. Antioxidant and Understanding the Anticancer Properties in Human Prostate and Breast Cancer Cell Lines of Chemically Characterized Methanol Extract from *Berberis hispanica* Boiss. & Reut. *Appl. Sci.* **2021**, *11*, 3510.
9. Rossi, R.; Corino, C.; Pastorelli, G.; Durand, P.; Prost, M. Assessment of Antioxidant Activity of Natural Extracts. *Ital. J. Anim. Sci.* **2009**, *8*, 655–657. [CrossRef]
10. Vercellotti, J.R.; St. Angelo, A.J.; Spanier, A.M. Lipid Oxidation in Foods: An Overview. In *Lipid Oxidation in Food*; American Chemical Society: Washington, DC, USA, 1992.
11. Bourhia, M.; Elmahdaoui, H.; Ullah, R.; Bari, A.; Benbacer, L. Promising Physical, Physicochemical, and Biochemical Background Contained in Peels of Prickly Pear Fruit Growing under Hard Ecological Conditions in the Mediterranean Countries. *BioMed Res. Int.* **2019**, *2019*, 9873146. [CrossRef]
12. Gautam, R.; Jachak, S.M. Recent Developments in Anti-Inflammatory Natural Products. *Med. Res. Rev.* **2009**, *29*, 767–820. [CrossRef] [PubMed]
13. Woolf, A.D.; Pfleger, B. Burden of Major Musculoskeletal Conditions. *Bull. World Health Organ.* **2003**, *81*, 646–656. [PubMed]
14. Smith, R.J. Therapies for Rheumatoid Arthritis: Hope Springs Eternal. *Drug Discov. Today* **2005**, *23*, 1598–1606. [CrossRef] [PubMed]

15. Salamatullah, A.M. Antioxidant and Antimicrobial Properties of Polyphenolics from *Withania adpressa* (Coss.) Batt. against Selected Drug-Resistant Bacterial Strains. *Open Chem.* **2022**, *20*, 474–483. [CrossRef]

16. Salamatullah, A.M. Promising Antioxidant and Insecticidal Properties of Chemically Characterized Hydroethanol Extract from *Withania adpressa* Coss. Ex Batt. *Horticulturae* **2022**, *8*, 698. [CrossRef]

17. Amrati, F.E.-Z.; Bourhia, M.; Slighoua, M.; Boukhira, S.; Ullah, R.; Ezzeldin, E.; Mostafa, G.A.; Grafov, A.; Bousta, D. Protective Effect of Chemically Characterized Polyphenol-Rich Fraction from *Apteranthes europaea* (Guss.) Murb. subsp. maroccana (Hook. f.) Plowes on Carbon Tetrachloride-Induced Liver Injury in Mice. *Appl. Sci.* **2021**, *11*, 554. [CrossRef]

18. Kumarasamy, Y.; Byres, M.; Cox, P.J.; Jaspars, M.; Nahar, L.; Sarker, S.D. Screening Seeds of Some Scottish Plants for Free Radical Scavenging Activity. *Phytother. Res.* **2007**, *21*, 615–621. [CrossRef]

19. Prieto, P.; Pineda, M.; Aguilar, M. Spectrophotometric Quantitation of Antioxidant Capacity through the Formation of a Phosphomolybdenum Complex: Specific Application to the Determination of Vitamin E. *Anal. Biochem.* **1999**, *269*, 337–341. [CrossRef]

20. Bourhia, M.; Haj Said, A.A.; Chaanoun, A.; El Gueddari, F.; Naamane, A.; Benbacer, L.; Khlil, N. Phytochemical Screening and Toxicological Study of *Aristolochia baetica* Linn Roots: Histopathological and Biochemical Evidence. *J. Toxicol.* **2019**, *2019*, 8203832. [CrossRef]

21. Jain, R.; Kachhwaha, S.; Kothari, S. Phytochemistry, Pharmacology, and Biotechnology of *Withania somnifera* and *Withania coagulans*: A Review. *J. Med. Plants Res.* **2012**, *6*, 5388–5399.

22. Matsuda, H.; Murakami, T.; Kishi, A.; Yoshikawa, M. Structures of Withanosides I, II, III, IV, V, VI, and VII, New Withanolide Glycosides, from the Roots of Indian *Withania somnifera* DUNAL. and Inhibitory Activity for Tachyphylaxis to Clonidine in Isolated Guinea-Pig Ileum. *Bioorganic Med. Chem.* **2001**, *9*, 1499–1507. [CrossRef] [PubMed]

23. Misra, L.; Mishra, P.; Pandey, A.; Sangwan, R.S.; Sangwan, N.S. 1,4-Dioxane and Ergosterol Derivatives from *Withania somnifera* Roots. *J. Asian Nat. Prod. Res.* **2012**, *14*, 39–45. [CrossRef] [PubMed]

24. Gülçin, I. Antioxidant Activity of Caffeic Acid (3,4-Dihydroxycinnamic Acid). *Toxicology* **2006**, *217*, 213–220. [CrossRef] [PubMed]

25. Jawhari, F.Z.; Moussaoui, A.E.; Bourhia, M.; Imtara, H.; Saghrouchni, H.; Ammor, K.; Ouassou, H.; Elamine, Y.; Ullah, R.; Ezzeldin, E. *Anacyclus pyrethrum* var. *pyrethrum* (L.) and *Anacyclus pyrethrum* var. *depressus* (Ball) Maire: Correlation between Total Phenolic and Flavonoid Contents with Antioxidant and Antimicrobial Activities of Chemically Characterized Extracts. *Plants* **2021**, *10*, 149. [CrossRef]

26. Katz, D.L.; Doughty, K.; Ali, A. Cocoa and Chocolate in Human Health and Disease. *Antioxid. Redox Signal.* **2011**, *15*, 2779. [CrossRef]

27. El Moussaoui, A.; Jawhari, F.Z.; Almehdi, A.M.; Elmsellem, H.; Fikri Benbrahim, K.; Bousta, D.; Bari, A. Antibacterial, Antifungal and Antioxidant Activity of Total Polyphenols of *Withania frutescens* L. *Bioorganic Chem.* **2019**, *93*, 103337. [CrossRef]

28. Posadas, I.; Bucci, M.; Roviezzo, F.; Rossi, A.; Parente, L.; Sautebin, L.; Cirino, G. Carrageenan-Induced Mouse Paw Oedema Is Biphasic, Age-Weight Dependent and Displays Differential Nitric Oxide Cyclooxygenase-2 Expression. *Br. J. Pharmacol.* **2004**, *142*, 331–338. [CrossRef]

29. Moussaoui, A.E.; Jawhari, F.Z.; Bourhia, M.; Maliki, I.; Sounni, F.; Mothana, R.A.; Bousta, D.; Bari, A. *Withania frutescens*: Chemical Characterization, Analgesic, Anti-Inflammatory, and Healing Activities. *Open Chem.* **2020**, *18*, 927–935. [CrossRef]

30. Chatter, R.; Tarhouni, S.; Kharrat, R.R. Criblage de l'effet anti-inflammatoire et analgésique des algues marines de la mer méditerranée. *Arch. Inst. Pasteur Tunis* **2012**, *88*, 19–28.

31. Fiorucci, S.; Meli, R.; Bucci, M.; Cirino, G. Dual Inhibitors of Cyclooxygenase and 5-Lipoxygenase. A New Avenue in Anti-Inflammatory Therapy? *Biochem. Pharmacol.* **2001**, *62*, 1433–1438. [CrossRef]

32. Martel-Pelletier, J.; Lajeunesse, D.; Reboul, P.; Pelletier, J. Therapeutic Role of Dual Inhibitors of 5-LOX and COX, Selective and Non-Selective Non-Steroidal Anti-Inflammatory Drugs. *Ann. Rheum. Dis.* **2003**, *62*, 501–509. [CrossRef] [PubMed]

33. Zomborszki, Z.; Peschel, W.; Boros, K.; Hohmann, J.; Dezső, C. Development of an Optimized Processing Method for *Withania frutescens*. *Acta Aliment.* **2016**, *45*, 452–456. [CrossRef]

34. Naidoo, D.B.; Chuturgoon, A.A.; Phulukdaree, A.; Guruprasad, K.P.; Satyamoorthy, K.; Sewram, V. *Withania somnifera* Modulates Cancer Cachexia Associated Inflammatory Cytokines and Cell Death in Leukaemic THP-1 Cells and Peripheral Blood Mononuclear Cells (PBMC's). *BMC Complement. Altern. Med.* **2018**, *18*, 126. [CrossRef]

35. Saleem, S.; Muhammad, G.; Hussain, M.A.; Altaf, M.; Bukhari, S.N.A. *Withania somnifera* L.: Insights into the Phytochemical Profile, Therapeutic Potential, Clinical Trials, and Future Prospective. *Iran. J. Basic Med. Sci.* **2020**, *23*, 1501–1526. [CrossRef] [PubMed]

36. Sharifi-Rad, J.; Quispe, C.; Ayatollahi, S.A.; Kobarfard, F.; Staniak, M.; Stępień, A.; Czopek, K.; Sen, S.; Acharya, K.; Matthews, K.R.; et al. Chemical Composition, Biological Activity, and Health-Promoting Effects of *Withania somnifera* for Pharma-Food Industry Applications. *J. Food Qual.* **2021**, *2021*, e8985179. [CrossRef]

37. Mukherjee, P.K.; Banerjee, S.; Biswas, S.; Das, B.; Kar, A.; Katiyar, C.K. *Withania somnifera* (L.) Dunal—Modern Perspectives of an Ancient Rasayana from Ayurveda. *J. Ethnopharmacol.* **2021**, *264*, 113157. [CrossRef]

38. Ben Bakrim, W.; El Bouzidi, L.; Nuzillard, J.-M.; Cretton, S.; Saraux, N.; Monteillier, A.; Christen, P.; Cuendet, M.; Bekkouche, K. Bioactive Metabolites from the Leaves of *Withania adpressa*. *Pharm. Biol.* **2018**, *56*, 505–510. [CrossRef]

39. Kumar, C.U.; Pokuri, V.K.; Pingali, U. Evaluation of the Analgesic Activity of Standardized Aqueous Extract of *Terminalia chebula* in Healthy Human Participants Using Hot Air Pain Model. *J. Clin. Diagn. Res.* **2015**, *9*, FC01–FC04. [CrossRef]
40. EvanPrince, S.; Sonal, C.; Rasool, M. Evaluation of Analgesic, Antipyretic and Ulcerogenic Effect of Withaferin A. *Int. J. Integr. Biol.* **2009**, *6*, 52–56.

Review

Discovery and Anticancer Activity of the Plagiochilins from the Liverwort Genus *Plagiochila*

Christian Bailly [1,2,3]

[1] Institut de Chimie Pharmaceutique Albert Lespagnol (ICPAL), Faculté de Pharmacie, University of Lille, 3 rue du Professeur Laguesse, F-59006 Lille, France; christian.bailly@oncowitan.com

[2] CNRS, Inserm, CHU Lille, UMR9020-U1277-CANTHER-Cancer Heterogeneity Plasticity and Resistance to Therapies, University of Lille, F-59000 Lille, France

[3] OncoWitan, Consulting Scientific Office, Wasquehal, F-59290 Lille, France

Abstract: The present analysis retraces the discovery of plagiochilins A-to-W, a series of seco-aromadendrane-type sesquiterpenes isolated from diverse leafy liverworts of the genus *Plagiochila*. Between 1978, with the first isolation of the leader product plagiochilin A from *P. yokogurensis*, and 2005, with the characterization of plagiochilin X from *P. asplenioides*, a set of 24 plagiochilins and several derivatives (plagiochilide, plagiochilal A-B) has been isolated and characterized. Analogue compounds recently described are also evoked, such as the plagiochianins and plagicosins. All these compounds have been little studied from a pharmacological viewpoint. However, plagiochilins A and C have revealed marked antiproliferative activities against cultured cancer cells. Plagiochilin A functions as an inhibitor of the termination phase of cytokinesis: the membrane abscission stage. This unique, innovative mechanism of action, coupled with its marked anticancer action, notably against prostate cancer cells, make plagiochilin A an interesting lead molecule for the development of novel anticancer agents. There are known options to increase its potency, as deduced from structure–activity relationships. The analysis shed light on this family of bryophyte species and the little-known group of bioactive terpenoid plagiochilins. Plagiochilin A and derivatives shall be further exploited for the design of novel anticancer targeting the cytokinesis pathway.

Keywords: anticancer agents; aromadendrane; bryophytes; cytokinesis inhibitor; *Plagiochila* species; plagiochilins

Citation: Bailly, C. Discovery and Anticancer Activity of the Plagiochilins from the Liverwort Genus *Plagiochila*. *Life* **2023**, *13*, 758. https://doi.org/10.3390/life13030758

Academic Editors: Marisa Colone, Charalampia Amerikanou and Efstathia Papada

Received: 16 January 2023
Revised: 4 March 2023
Accepted: 9 March 2023
Published: 10 March 2023

1. Introduction

Bryophytes are non-vascular plants which include thalloid and leafy liverworts, mosses and hornworts. These three lineages form a unique part of the vegetation. They are small-sized, structurally simple diversified plants able to adapt to most ecosystems on Earth [1,2]. Bryophytes and tracheophytes (non-vascular and vascular plants, respectively) derive from an ancestral land plant and diverged during the Cambrian, some 500 million years ago [3]. Bryophytes are collectively divided into three main groups: Bryophyta (mosses), Marchantiophyta (liverworts) and Anthocerotophyta (hornworts). They represent the second-largest group of land plants after angiosperms. Liverworts are particularly abundant, with some 7300 extant species [4]. The first representations of liverworts date from late antiquity [5].

Leafy (or scaly) liverworts are particularly abundant and diversified (order: Jungermanniales). They grow commonly on moist soil or damp rocks (such as thallose liverworts). In 2016, a worldwide checklist for liverworts and hornworts included 7486 species in 398 genera representing 92 families from the two phyla [6]. The genus *Plagiochila* (Plagiochilaceae) represents one of the largest groups of leafy liverworts, with more than 500 species distributed worldwide and a broad geographical amplitude, mostly in the humid tropics [7]. World Flora Online refers to 556 accepted names of *Plagiochila* species and more than 950 species including synonyms and unchecked species [8]. Another recent study refers to 1600 validly published *Plagiochila* names [9].

Despite their number and high adaptative capacities, the medicinal use of bryophytes remains relatively limited, probably because of their small size, lack of conspicuous organs such as colored fruits and flowers, and the difficulty of identification. There are, however, species with medicinal properties, such as *Conocephalum conicum* (L.) Dumort., *Polytrichum commune* and *Marchantia chenopoda* L. [10]. In the genus *Plagiochila*, a few species have also been used ethnomedicinally, such as *P. beddomei* Steph. used in the form of paste by tribe Melghat Region (India) for treating skin diseases [11,12] and *P. disticha* (Lehm. & Lindenb.) Lindenb used traditionally in Peru to treat rheumatism or to regulate menstruation [13].

Diverse bioactive products have been isolated from *Plagiochila* species, including antitumor agents [14], antifungal molecules [15,16], insecticidal compounds [17] and antimicrobial products [18]. Most of the isolated bioactive compounds are terpenoids such as the antifungal products plagicosins A-N, or alkaloids such as plagiochianins A-B from the Chinese liverworts *P. fruticosa* Mitt. and *P. duthiana* Steph., respectively [18,19]. However, the leading product isolated from *Plagiochila* species is without doubt the sesquiterpenoid plagiochilin A, first isolated from several *Plagiochila* species in the 1970s, together with its congeners plagiochilins B and C, and precursors plagiochilide and plagiochilal [20]. Over the past 43 years, different analogues have been isolated leading to a series of 24 derivatives, designated plagiochilins A-to-X, and related compounds (Figure 1). The present review deals the identification of these compounds and their pharmacological properties. Information about their mechanism of action is often very limited, but important observations have been made, leading to the identification of potential targets for some of these compounds, in particular for the leader product plagiochilin A (Figure 2).

Several scientific databases (mostly PubMed, Science Direct and Scopus) and internet search engines (Google, Bing) were used to execute a systematic search of the existing literature, considering all publications published up until January 2023, without any language restriction. Databases were queried using specific keywords such as "bryophytes", "*Plagiochila*", "natural products", "aromadendrane", and "plagiochilin". The articles were searched using a Boolean logic operator (and/or/not) combined with Medical Subject Headings (MeSH) terms and keywords. The relevance of the collected articles was determined (individual expertise), and then the data were extracted and analyzed.

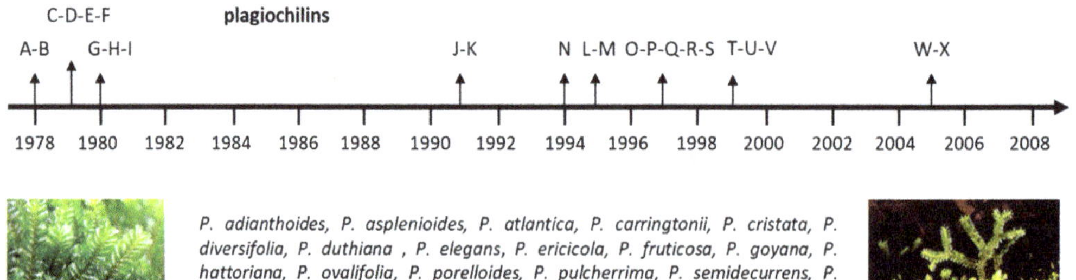

Figure 1. History of plagiochilins discovery. The 24 plagiochilins (A–W) have been identified and structurally characterized aver a period of 30 years. They are produced by several *Plagiochila* species, such as those indicated (a non-exhaustive list).

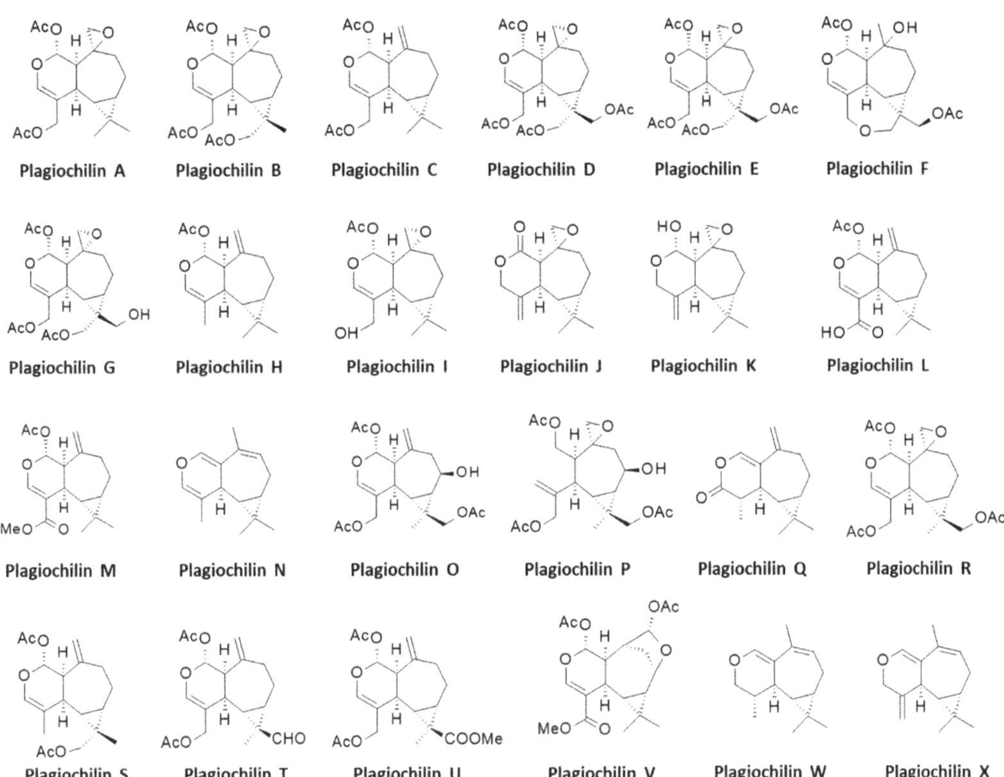

Figure 2. Structures of the 24 plagiochilins (A-to-W).

2. Discoveries of the Plagiochilins

It all started in 1978 (Figure 1) when Asakawa and coworkers reported the isolation of the sesquiterpene plagiochilin A from *P. yokogurensis* Stephani, as the epoxide counterpart of plagiochilide (Figure 3) [21]. The two products with a seco-aromadendrane skeleton have been found also in *P. fruticosa* [22]. The aromadendrene scaffold is not rare in plants. Aromadendrene is an antibacterial product found in the medicinal plant *Lophostemon suaveolens* [23] and in the common hop (*Humulus lupulus* 'Nordbrau') [24], for example. However, epoxide derivatives with an aromadendrene scaffold are quite rare. A compound designated aromadendrene oxide (an epoxide derivative) has been isolated from the plants *Tetradenia riparia* (Hochst.) Codd, (Lamiaceae) and *Kickxia aegyptiaca* (L.) N. (Plantaginaceae) and shown to display antibacterial effects [25,26]. The 2,3-seco-aromadendrane unit is typical of *Plagiochila* species (Figure 3). The terpenoids plagiochilin A from *P. yokogurensis* and plagicosin G from *P. fruticosa* [18], and the alkaloids plagiochianins A and B from *P. duthiana* are all seco-aromadendrane derivatives [19] (Figure 3).

Following the isolation of plagiochilin A from *P. yokogurensis*, Asakawa and coworkers discovered a variety of seco-aromadendrane-type sesquiterpenoids from several *Plagiochila* species, including plagiochilins C, D, E and F in *P. asplenioides*, plagiochilins A and C in *P. semidecurrens* [27] and related products in other *Plagiochila* species, such as *P. fructicosa*, *P. ovalifolia*, and *P. porelloides* [28]. Notably, they identified plagiochilins A and B from *P. hattoriana* [21] and from *P. pulcherrima* [29], followed with plagiochilin C from both *P. ovalifolia* and *P. asplenioides*. They also identified plagiochilins D, E and F from *P. asplenioides*, together with a few related compounds such as furanoplagiochilal, plagiochilal A-B and plaigiochilide [28] (Figure 3). Similarly, plagiochilins A and B have been identified from

P. semidecurrens [30] and *P. diversifolia* [31]. Plagiochilin A was found also in *P. elegans*, together with isoplagiochilide [32].

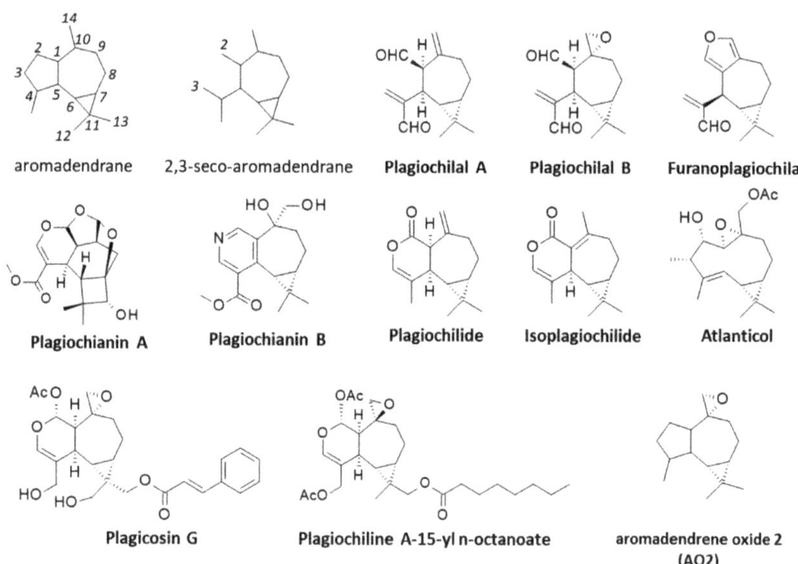

Figure 3. Structures of the aromadendrane and 2,3-seco-aromadendrane scaffolds (with the numbering scheme), and other products isolated from *Plagiochila* species and structurally related to the plagiochilins.

The species *P. porelloides* has afforded 2,3-secoaromadendrane-type sesquiterpene esters derived from plagiochilin D [33]. Seco-aromadendrane-type sesquiterpenoids are considered as chemosystematic markers in the Plagiochilaceae [34]. In the early 1980s, studies were essentially concerned with the isolation of the compounds and their structural characterization. Nevertheless, the insecticidal activity of plagiochilin A was evidenced early on, notably its capacity to inhibit the feeding of the army worm *Spodoptera exempta* (Fabricius, 1775) [35]. Plagiochilin A exhibits a strong pungent taste; it can be converted into plagiochilal B and furanoplagiochilal upon exposure to human saliva [36].

Plagiochilin G was isolated subsequently from *P. ovalifolia*, together with plagiochilins H and I from *P. yokogurensis* [37]. Plagiochilin H is a close analogue of plagiochilin C, whereas plagiochilins G and I are structurally close to plagiochilins A-B (Figure 2). *P. ovalifolia* provides a rich source of seco-aromadendrane sesquiterpenoids, notably ester derivatives of plagiochilin A endowed with cytotoxic properties [38]. From this species, the isolation and structural characterization of plagiochilins C and N were reported, together with the derivative acetoxyisoplagiochilide [39,40]. Later, a total synthesis of plagiochilin N was proposed from the natural precursor santonin, a common anthelmintic sesquiterpene lactone readily available [41] (Figure 3). It is a long and difficult synthesis (16 steps), useful to confirm the stereochemistry of plagiochilin N [42] (Figure 4). Plagiochilin N is the only compound in the series for which a total synthesis has been proposed. All the other plagiochilins are natural products. There is a need for efficient syntheses of compounds related to plagiochilin A. Plagiochilins J and K have been isolated from *P. fruticosa* in 1991, together with the sesquiterpene dialdehyde plagiochilal B which is considered as the precursor for these two plagiochilins [43]. Plagiochilins L and M were described four years later, but they derive from a totally distinct liverwort species, *Heteroscyphus planus*, which is however chemically similar to *Plagiochila* species. It contained also plagiochilin C. *Plagiochila* and *Heteroscyphus* belong to two families of the same sub-order (Lophocoleineae). The presence of these compounds may be more widespread between the species. The two compounds

L and M bear the seco-aromadendrane skeleton typical of the product family [44]. Plagiochilin M has been isolated also from the andine species *Plagiochila tabinensis* Steph. [45].

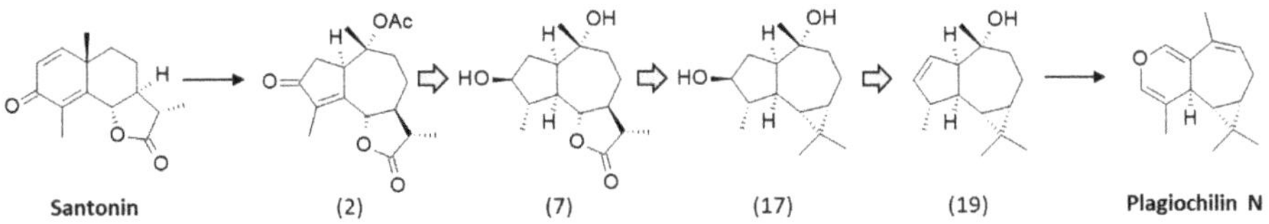

Figure 4. The total synthesis of plagiochilin N has been achieved from sesquiterpene lactone santonin (in fact from O-acetylisophotosantonin (2), obtained by photochemical rearrangement of santonin). The total synthesis included 16 steps. Only selected key intermediates are shown here. See [42] for the detailed synthesis which confirmed the stereochemistry of the compound.

Then, five other plagiochilins (O-P-Q-R-S) were discovered from a diethyl ether extract of the Colombian liverwort *P. cristata* (O-P-Q), from the *P. ericicola* (R) and from a dichloromethane extract prepared from an axenic culture of *P. adianthoides* (S) [46]. Plagiochilin P can also be found in *P. asplenioides* [47]. The species *P. cristata* and *P. ericicola* also contained plagiochilins C and H which are both structurally close to plagiochilin O, whereas plagiochilin R is a close analogue to plagiochilin B (Figure 2). Plagiochilins T and U were disclosed two years later, isolated from a specimen of *P. carringtonii* collected in Scotland [48]. They correspond to C-13 oxidation products derived from plagiochilin C (Figure 1). This particular *Plagiochila* species, a leafy liverwort officially named *Plagiochila carringtonii* (Balf. ex Carrington) Grolle (commonly called Carrington's Featherwort) is well distributed in Scotland and Ireland [49]. It is distinct from the species *P. atlantica* also abundant in the West of Scotland and which was shown to contain plagiochilin C and a structurally close product designated atlanticol (Figure 3) [50] (not to be confused with another product also called atlanticol, an alkaloid from the Rutaceae *Spiranthera atlantica* [51]). Compound plagiochilin V has been rarely mentioned. It is cited in a single publication about the species *P. porelloides* from Changbai mountain in China [52]. This species has been used to isolate 2,3-secoaromadendrane-type esters derived from plagiochilin D [32]. It has been used recently as a model to study desiccation tolerance [53]. To our knowledge, the chemical structure of plagiochilin V, proposed 25 years ago, has not been confirmed. The series ends up with plagiochilins W-X, both isolated from a sample of *P. asplenioides* collected in Germany, together with the acetylated hemiacetal plagiochilin H [54].

Altogether, the plagiochilins represent a series of 24 natural products isolated from diverse *Plagiochila* species between 1978 and 2005 [55–57]. A few other 2,3-seco-aromadendrane derivatives were isolated subsequently but named differently, such as plagiocosin G with a phenylpropanoyloxy side chain (Figure 3), isolated from *P. fruticosa* [18]. The majority of these plagiochilin compounds has been structurally characterized, but pharmacologically neglected. In rare cases, individual properties have been reported, as described below.

3. Pharmacological Properties and Mechanism of Action of the Plagiochilins

3.1. Pharmacological Properties of Plagiochilins A and C

The pharmacological effects of the plagiochilins have been rarely investigated. Nevertheless, a few types of bioactivities have been reported with plagiochilins A and C. The data in Table 1 illustrate the potential of the two compounds, but there is no systematic study comparing activities of all compounds, or activities of a given compound across multiple indications or pathologies. Initially, it was shown that plagiochilin A displays a modest antifeedant action against the armyworm *Spodoptera exempta* Walker (Lepidoptera: Noctuidae) which is an episodic migratory pest of cereal crops in sub-Saharan Africa. The level of activity is quite modest [35]. Another study has investigated the insecticidal activity of natural products isolated from *P. diversifolia*, but in that case the only plagiochilin tested was plagiochilin B and no activity was reported. In this case, a marked insecticidal action was observed with another epoxide-containing compound called fusicogigantone B, a fusicoccane-type diterpenoid [31]. Both fusicogigantones A and B, isolated from *P. bursata* and *P. diversifolia*, respectively, have been shown to inhibit the growth of another species of Lepidoptera (*Spodoptera frugiperda*) [17,31]. Plagiochilin A was shown to display a noticeable antiprotozoal activity, reducing the growth of the amastigote form of *Leishmania amazonensis*, with an IC_{50} value of 7.1 µM. However, the level of activity is quite modest compared to that of the control drug amphotericin B (IC_{50} = 0.13 µM). In the same study, no activity was observed against the fungus *Mycobacterium tuberculosis* [13].

Table 1. Bioactivities reported with plagiochilins.

Compounds	Bioactivities	Tests/Species	End Points	Ref.
Plagiochilin A	Antifeedant	African armyworm *Spodoptera exempta*	Activity observed at 1–10 ng/cm^2	[35]
Plagiochilin A	Antiparasitic	*Leishmania amazonensis* axenic amastigotes	IC_{50} = 7.1 µM	[13]
Plagiochilin A	Antiparasitic	*Trypanosoma cruzi* trypomastigotes	MIC = 14.5 µM	[13]
Plagiochilin A	Anti-proliferative	P-388 murine leukemia cells	IC_{50} = 3.0 µg/mL	[38]
Plagiochilin A	Anti-proliferative	A172 glioblastoma cells	IC_{50} = 19.4 µM.	[56]
Plagiochilin-A-15-yl n-octanoate	Anti-proliferative	P-388 murine leukemia cells	IC_{50} = 0.05 µg/mL	[38]
Plagiochilin C	Antiplatelet	Inhibition of arachidonate-induced rabbit platelet aggregation	95% and 45% inhibition at 100 and 50 µg/mL, respectively.	[32]
Plagiochilin C	Anti-proliferative	A172 glioblastoma cells	IC_{50} = 4.3 µM	[58]

In a more interesting way, plagiochilin A was characterized as an antiproliferative agent, reducing the growth of different cultured cancer cell lines. Both plagiochilins A and C display marked antiproliferative activities and there are known options to further increase their anticancer potency. An option is to introduce a methoxy group at position C-3, as observed with the derivative methoxyplagiochilin A2 which has been shown to be more potent than plagiochilin C against H460 lung cancer cells (IC_{50} = 6.7 and 13.1 µM, respectively) [58]. Another option is to introduce a side chain at the C-14/C-15 position, either an octanoyl side chain or a dodecadienoate side chain, for example. In both cases, the resulting compounds were found to be 60 times more potent against P-388 leukemia cells than the parent compound plagiochilin A [38]. The extraordinary potency of plagiochilin A-15-yl n-octanoate (Figure 3) raises questions (solubility, stability) and opens perspectives. The octanoate moiety may serve only as a "lipophilic carrier" (bioavailability enhancement), and may not be directly implicated in the target interaction. Novel C-12/C-13-substituted derivatives of plagiochilin A should be designed.

Plagiochilin A exhibits antiproliferative activities against different types of cancer cells. The growth inhibition GI_{50} values ranged from 1.4 to 6.8 µM with a range of tumor cell lines, including prostate (DU145), breast (MCF-7), lung (HT-29) and leukemia (K562) cells for example [13]. The level of activity against DU145 prostate cancer cell is interesting (GI_{50} = 1.4 µM) because it is superior to that observed with the reference anticancer drug fludarabine phosphate (GI_{50} = 3.0 µM). The sensitivity of prostate cancer cells toward plagiochilin A warranted further investigation. In 2018, Bates and coworkers analyzed the effect of plagiochilin A on the cell cycle progression of DU145 cells and their capacity to complete cytokinesis, the part of the cell division process during which the cytoplasm of a single eukaryotic cell divides into two daughter cells. Interestingly, it was observed that the compound (at 5 µM for 24–48 h) could block cell division by preventing completion of cytokinesis, and thereby inducing cell death [59]. The treated DU145 cells accumulated at the G2/M phase, notably cells still connected with intercellular bridges, corresponding to a late cytokinesis stage, the so-called membrane abscission stage (stained with an α-anti-tubulin antibody). The compound induced specific mitotic figures and reduced significantly the number and size of DU145 cell colonies. The failure of the cells to complete cytokinesis triggered apoptosis [59]. Altogether, these data indicated that plagiochilin A exerts an effect on the cytoskeleton, with a rearrangement of α-tubulin characteristic of cytokinetic membrane abscission, which is a spatially and temporally regulated process [59] (Figure 5).

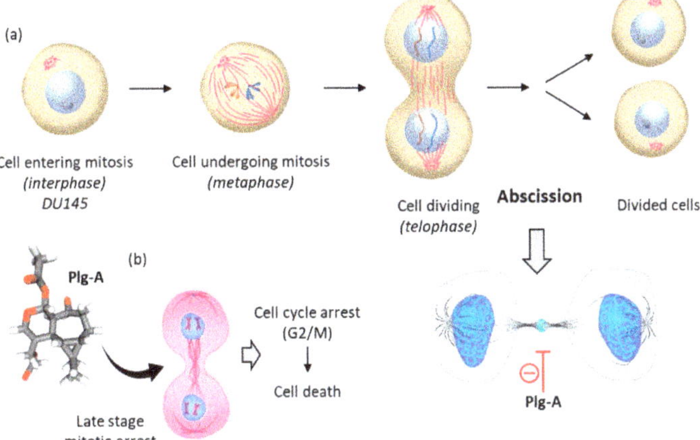

Figure 5. Mechanism of action of plagiochilin A (Plg-A). (**a**) The cell division process, to illustrate DU145 prostate cancer cells undergoing mitosis. Prior to cell division, at the telophase stage of mitosis, the two cells are connected by an intercellular bridge (with a central midbody). Plg-A inhibits cell division by preventing completion of cytokinesis, particularly at the final abscission stage. (**b**) Inhibition of the late stage of cytokinesis leads to cell cycle arrest (G2/M) and subsequently to inhibition of cell colony formation and induction of cell death [59].

3.2. Hypothesized Mechanism of Action of Plagiochilin A

The mechanics implicated in the regulation of abscission is relatively well-known. This process leads to the physical cut of the intercellular bridge which connects two daughter cells and concludes cell division. The process is tightly regulated in cells, with intervention of multiple protein effectors including different kinases (e.g., PLK4, Aurora B) and proteins containing microtubule-interacting and trafficking (MIT) domains [60–62]. The process opens perspectives to comprehend the mechanism of action of plagiochilin A. the compound may target MIT-containing proteins, or more directly it may associate with alpha-tubulin during mitosis, for example. There exist small molecules which induced cytokinesis failure at the point of abscission, such as a series of dynamin GTPase inhibitors called dynoles [63,64]. These products induce apoptosis following cytokinesis failure, as observed with plagiochilin A. Therefore, we can imagine that the natural product acts as an inhibitor of termination of cytokinesis (abscission) by blocking one or several proteins implicated in the process, or by directly altering the microtubule-organizing center which recruits α- and β-tubulins for microtubule nucleation. In this context, one of the potential mechanisms could be a direct binding of the compound to α-tubulin, in particular to the pironetin-binding site which is known to accommodate compounds bearing a dihydro-pyrone moiety [65,66]. This moiety can be found in plagiochilin Q, for example, and recently, we have shown that natural products with a 5,6-dihydro-α-pyrone unit (cryptoconcatones) can function as α-tubulin-binding agents [67]. Based on these considerations, we have initiated binding studies of plagiochilins to α-tubulin and the first information obtained by molecular docking look interesting. For the docking analysis, the high-resolution crystal structure of the reference product pironetin (a dihydropyrone derivative with an α,β-unsaturated lactone acting as a plant growth regulator) bound to α/β-tubulin dimer was used as a template (PDB: 5FNV) and the binding of plagiochilin A to the pironetin site was modeled. Apparently, plagiochilin A could form stable complexes with α-tubulin, via binding to the pironetin site, as represented in Figure 6. These are preliminary, but promising information. We are now comparing various plagiochilins for their capacity to bind to α-tubulin, using molecular modeling. The mechanism whereby plagiochilin A specifically blocks abscission warrant further investigation. The compound is an atypical inhibitor of cytokinesis. The panoply of plagiochilins shall be further exploited to identify the best inhibitors and to delineate the structure–activity relationships in the series. The modeling analysis shall help also to delineate the mechanism of action of other aromadendrane derivatives, such as the related natural products hanegokedial, ovalifolienal, ovalifolienalone (from *P. semidecurrens*) and others [30,68].

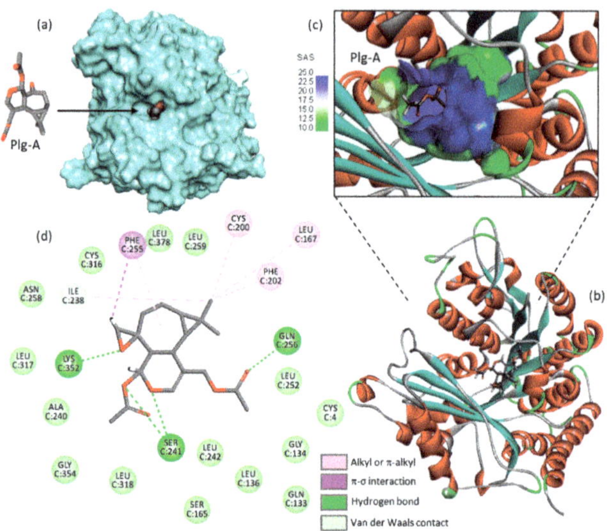

Figure 6. Molecular model of plagiochilin A (Plg-A) bound to the pironetin site of α-tubulin (PDB: 5FNV). (**a**) Plg-A fits into a central, deep cavity of the protein. (**b**) Ribbon model of α-tubulin with bound Plg-A, with α-helices (in red) and β-sheets (in cyan). (**c**) A close-up view of Plg-A inserted into the binding cavity, with the solvent-accessible surface (SAS) surrounding the drug binding zone (color code indicated). (**d**) Binding map contacts for Plg-A bound to α-tubulin (color code indicated). The docking model was kindly provided by Prof. Gérard Vergoten (University of Lille, France). The docking analysis was performed as recently described [65].

4. Discussion

Bryophytes are known to produce bioactive compounds with a broad range of therapeutic potential [69]. For example, extracts prepared from a range of mosses and liverworts have been screened recently for their anti-inflammatory properties. Two particular species, *Dicranum majus* Sm. And *Thuidium delicatulum* (Hedw.) Schimp., were found to inhibit production of nitric oxide in lipopolysaccharide-stimulated Raw 264.7 murine macrophages [70]. In this study, an extract of *Plagiochila asplenioides* (L.) Dumort. Was tested but it did not show a marked activity. Other bryophyte extracts have revealed antioxidant and/or antimicrobial activities [71,72]. The use of extracts from *Plagiochila* species is relatively rare. However, there are noticeable exceptions, for example, the use of methanol extract of the liverwort *Plagiochila beddomei* Steph. which has revealed marked antimicrobial activities against a wide group of bacteria and fungi, including the pathogen *Candida albicans* (MIC = 0.75 mg/mL). The extract contained flavonoids, saponins, tannins and phenols [11,12,73]. In most cases, bioactivities with bryophytic extracts have been attributed to the presence of flavonoids, terpenoids or alkaloids, such as macrocyclic bibenzyls and bis(bibenzyls), and many sesqui- and diterpenoids [74–79]. However, *Plagiochila* species and plagiochilins are rarely evoked.

The different *Plagiochila* species mentioned here produce 2,3-secoaromadendrane-type sesquiterpenes, chiefly the cytotoxic product plagiochilin A [56,80]. This compound is considered as being responsible for the pungent and bitter taste of *Plagiochila* liverworts, at least in part [80]. However, it is also a robust anticancer agent, insufficiently considered despite its very innovative mechanism of action as a late-stage cytokinesis inhibitor. Whether the compound target α-tubulin or other proteins implicated in membrane abscission remains to be determined experimentally. However, whatever the exact target, the mode of action of the compound is particularly attractive and inspiring. There are not many products acting at the abscission checkpoint, and it is probably the only natural product known to regulate cytokinetic abscission. Plagiochilin A could serve as a useful tool to dissect the abscission mechanism which is a multi-step process which influences cell fate and tissue growth [81,82]. There is a need for pharmacological tools to dissect the mechanism of abscission of nascent daughter cells, to stop or halt the process and to study the implication of specific protein complexes and cellular structures, such as midbody proteins and the midbody organelle [83]. There is a rearrangement of α-tubulin and actin, and important cytoskeleton modification during abscission. The process can be modulated with the use of histone deacetylase (HDAC) inhibitors [84], actin polymerization inhibitors [85], and various kinase inhibitors [86]. New tools are needed to dissect the process, notably to help understanding the function of the residual midbody remnant generated at the end of the process and which affects cell fate and tumorigenesis [87]. Plagiochilin A has the capacity to induce an accumulation of cancer cells (at least DU145 prostate cancer cells) connected by intercellular bridges, probably via a selective action on the sequential assembly of the endosomal sorting complexes required for transport (ESCRT) machinery [59]. The compound affords a useful tool to dissect the process, to study ESCRT assembly and regulation. In parallel, the use of high-content phenotypic screens focused on cytokinesis would be useful to further identify the targets of the compound and the fine tuning of its mechanism of action [88].

Plagiochilin A deserves further studies as an anticancer agent owing to its mechanism of action and its level of activity. It could be a useful starting point to elaborate more potent analogues because there are strategies to enhance the antiproliferative potency of plagiochilin A, notably via C-12/C-13 substitution [38]. There are also many naturally occurring derivatives (plagiochilins, plagiochilide, plagiochilal, and others) which could be exploited to determine structure–activity relationships. Moreover, the chemical diversity of the natural products can be enhanced upon biotransformation with microorganisms. For example, plagiochilide can be chemically modified to afford 12-hydroxyplagiochilide and plagiochilide-12-oic acid in the presence of the fungus *Aspergillus niger* [22]. The difficulty is to obtain the compounds. There is a possibility to produce biomass under laboratory conditions through bryo-reactors and molecular farming (thus decreasing pressure for natural populations) [89], but the subsequent purification of natural products would not be an easy ride. Axenic cultures of *Plagiochilla* have been developed in rare cases only, notably for *P. arctica* Bryhn and Kaal. [90].

Apart from plagiochilin N for which a total synthesis has been reported [40], the other compounds have to be isolated from plants or novel syntheses have to be designed for these compounds. An alternative option is to access the related compound called aromadendrene oxide 2 (AO2) which is an analogue of plagiochilin A (Figure 3). This sesquiterpene can be found in essential oils from different plants [25,91,92]. The compound has been shown to induce apoptosis of skin epidermoid A431 cancer cells, via activation of the mitochondrial pathway [93,94]. Its activity further supports the interest for plagiochilin A and more globally the use of the 2,3-secoaromadendrane as a template to design novel anticancer agents. There are other 2,3-secoaromadendrane-type sesquiterpenoids of interest, with an unknown mechanism of action, such as the two products psilosamuiensins A and B isolated from the broth of the psychoactive fungus *Psilocybe samuiensis* [95]. This family of terpenoids with a 2,3-secoaromadendrane scaffold should be investigated further as a source of anticancer agents. More generally, this work also underlines the interest of bryophytes as a source of bioactive compounds, and anticancer agents in particular. In recent years, different bryophytes species have revealed marked antitumor effects, including a capacity to kill chemo-resistant cancer stem cells [71,96]. The use and study of bryophytes, in particular liverworts (Hepaticae or hepatics), shall be encouraged [97,98]. Liverworts lack roots, seeds, fruit and flowers, but they do not lack major interest as a source of bioactive compounds. It is time to get these simple, single cellular plants back into the limelight, and in particular the *Plagiochila* group of leafy liverworts which can offer an array of captivating compounds.

5. Conclusions

The analysis shed light on a little-known family of 24 terpenoids designated plagiochilins A-to-W, isolated from various Plagiochila species. Leafy liverworts of the genus *Plagiochila* produce a variety of seco-aromadendrane-type sesquiterpenes, among which plagiochilin A is a lead product endowed with marked antiproliferative activities against cancer cells. This epoxide-containing natural product functions has been characterized as an inhibitor of the termination phase of cytokinetic abscission, a somewhat unique action at the origin of its capacity to delay cell cycle progression and to induce cell death. This atypical mechanism of action makes plagiochilin A an interesting tool to study the abscission process and a lead compound to design more potent analogues. There are known options to increase the potency of the compound. Preliminary structure–activity relationships have been delineated in the plagiochilin series. The *Plagiochila* genus of briophytes deserves further studies as a source of bioactive compounds.

Funding: This research received no external funding.

Institutional Review Board Statement: Not applicable.

Informed Consent Statement: Not applicable.

Data Availability Statement: Data are contained within the article.

Acknowledgments: The in silico molecular modeling presented in Figure 6 was kindly performed by Gérard Vergoten (University of Lille, Inserm U995, Lille, France). His contribution is greatly appreciated.

Conflicts of Interest: The author declares no conflict of interest.

References

1. Wang, Q.H.; Zhang, J.; Liu, Y.; Jia, Y.; Jiao, Y.N.; Xu, B.; Chen, Z.D. Diversity, phylogeny, and adaptation of bryophytes: Insights from genomic and transcriptomic data. *J. Exp. Bot.* **2022**, *73*, 4306–4322. [CrossRef]
2. Kulshrestha, S.; Jibran, R.; van Klink, J.W.; Zhou, Y.; Brummell, D.A.; Albert, N.W.; Schwinn, K.E.; Chagné, D.; Landi, M.; Bowman, J.L.; et al. Stress, senescence, and specialized metabolites in bryophytes. *J. Exp. Bot.* **2022**, *73*, 4396–4411. [CrossRef] [PubMed]
3. Harris, B.J.; Clark, J.W.; Schrempf, D.; Szöllősi, G.J.; Donoghue, P.C.J.; Hetherington, A.M.; Williams, T.A. Divergent evolutionary trajectories of bryophytes and tracheophytes from a complex common ancestor of land plants. *Nat. Ecol. Evol.* **2022**, *6*, 1634–1643. [CrossRef] [PubMed]
4. Dong, S.; Yu, J.; Zhang, L.; Goffinet, B.; Liu, Y. Phylotranscriptomics of liverworts: Revisiting the backbone phylogeny and ancestral gene duplications. *Ann. Bot.* **2022**, *130*, 951–964. [CrossRef] [PubMed]
5. Bowman, J.L. A Brief History of Marchantia from Greece to Genomics. *Plant Cell Physiol.* **2016**, *57*, 210–229. [CrossRef]
6. Söderström, L.; Hagborg, A.; von Konrat, M.; Bartholomew-Began, S.; Bell, D.; Briscoe, L.; Brown, E.; Cargill, D.C.; Costa, D.P.; Crandall-Stotler, B.J.; et al. World checklist of hornworts and liverworts. *PhytoKeys* **2016**, *59*, 1–828. [CrossRef]
7. Heinrichs, J.; Hentschel, J.; Feldberg, K.; Bombosch, A.; Schneider, H. Phylogenetic biogeography and taxonomy of disjunctly distributed bryophytes. *J. Syst. Evol.* **2009**, *47*, 497–508. [CrossRef]
8. The World Flora Online (WFO). Available online: http://www.worldfloraonline.org (accessed on 13 January 2023).
9. Renner, M.A.M. The typification of Australasian *Plagiochila* species (Plagiochilaceae: Jungermanniidae): A review with Recommendations. *N. Z. J. Bot.* **2021**, *59*, 323–375. [CrossRef]
10. Drobnik, J.; Stebel, A. Four Centuries of Medicinal Mosses and Liverworts in European Ethnopharmacy and Scientific Pharmacy: A Review. *Plants* **2021**, *10*, 1296. [CrossRef]
11. Manoj, G.S.; Murugan, K. Wound healing activity of methanolic and aqueous extracts of *Plagiochila beddomei* Steph. thallus in rat model. *Indian J. Exp. Biol.* **2012**, *50*, 551–558.
12. Manoj, G.S.; Murugan, K. Phenolic profiles, antimicrobial and antioxidant potentiality of methanolic extract of a liverwort, *Plagiochila beddomei* Steph. *Indian J. Nat. Prod. Resour.* **2012**, *3*, 173–183.

13. Aponte, J.C.; Yang, H.; Vaisberg, A.J.; Castillo, D.; Málaga, E.; Verástegui, M.; Casson, L.K.; Stivers, N.; Bates, P.J.; Rojas, R.; et al. Cytotoxic and anti-infective sesquiterpenes present in *Plagiochila disticha* (Plagiochilaceae) and *Ambrosia peruviana* (Asteraceae). *Planta Med.* **2010**, *76*, 705–707. [CrossRef] [PubMed]

14. Morita, H.; Tomizawa, Y.; Tsuchiya, T.; Hirasawa, Y.; Hashimoto, T.; Asakawa, Y. Antimitotic activity of two macrocyclic bis(bibenzyls), isoplagiochins A and B from the Liverwort *Plagiochila fruticosa*. *Bioorg. Med. Chem. Lett.* **2009**, *19*, 493–496. [CrossRef] [PubMed]

15. Lorimer, S.D.; Perry, N.B.; Tangney, R.S. An antifungal bibenzyl from the New Zealand liverwort, *Plagiochila stephensoniana*. Bioactivity-directed isolation, synthesis, and analysis. *J. Nat. Prod.* **1993**, *56*, 1444–1450. [CrossRef] [PubMed]

16. Lorimer, S.D.; Perry, N.B. Antifungal hydroxy-acetophenones from the New Zealand liverwort, *Plagiochila fasciculata*. *Planta Med.* **1994**, *60*, 386–387. [CrossRef] [PubMed]

17. Ramírez, M.; Kamiya, N.; Popich, S.; Asakawa, Y.; Bardón, A. Insecticidal constituents from the argentine liverwort *Plagiochila bursata*. *Chem. Biodivers.* **2010**, *7*, 1855–1861. [CrossRef]

18. Qiao, Y.N.; Jin, X.Y.; Zhou, J.C.; Zhang, J.Z.; Chang, W.Q.; Li, Y.; Chen, W.; Ren, Z.J.; Zhang, C.Y.; Yuan, S.Z.; et al. Terpenoids from the Liverwort *Plagiochila fruticosa* and Their Antivirulence Activity against *Candida albicans*. *J. Nat. Prod.* **2020**, *83*, 1766–1777. [CrossRef]

19. Han, J.J.; Zhang, J.Z.; Zhu, R.X.; Li, Y.; Qiao, Y.N.; Gao, Y.; Jin, X.Y.; Chen, W.; Zhou, J.C.; Lou, H.X. Plagiochianins A and B, Two ent-2,3-seco-Aromadendrane Derivatives from the Liverwort *Plagiochila duthiana*. *Org. Lett.* **2018**, *20*, 6550–6553. [CrossRef]

20. Asakawa, Y.; Toyota, M.; Takemoto, T. Plagiochilide et plagiochilin a, secoaromadendrane-type sesquiterpenes de la mousse, *plagiochila yokogurensis* (plagiochilaceae). *Tetrahedron Lett.* **1978**, *19*, 1553–1556. [CrossRef]

21. Asakawa, Y.; Toyota, M.; Takemoto, T. La plagiochilin a et la plagiochilin b, les sesquiterpenes du type secoaromadendrane de la mousse, *Plagiochila hattoriana*. *Phytochemistry* **1978**, *17*, 1794. [CrossRef]

22. Furusawa, M.; Hashimoto, T.; Noma, Y.; Asakawa, Y. Biotransformation of aristolane- and 2,3-secoaromadendrane-type sesquiterpenoids having a 1,1-dimethylcyclopropane ring by *Chlorella fusca* var. *vacuolata*, *mucor* species, and *Aspergillus niger*. *Chem. Pharm. Bull.* **2006**, *54*, 861–868. [CrossRef] [PubMed]

23. Naz, T.; Packer, J.; Yin, P.; Brophy, J.J.; Wohlmuth, H.; Renshaw, D.E.; Smith, J.; Elders, Y.C.; Vemulpad, S.R.; Jamie, J.F. Bioactivity and chemical characterisation of *Lophostemon suaveolens*—An endemic Australian Aboriginal traditional medicinal plant. *Nat. Prod. Res.* **2016**, *30*, 693–696. [CrossRef] [PubMed]

24. Paguet, A.S.; Siah, A.; Lefèvre, G.; Moureu, S.; Cadalen, T.; Samaillie, J.; Michels, F.; Deracinois, B.; Flahaut, C.; Alves Dos Santos, H.; et al. Multivariate analysis of chemical and genetic diversity of wild *Humulus lupulus* L. (hop) collected in situ in northern France. *Phytochemistry* **2023**, *205*, 113508. [CrossRef] [PubMed]

25. de Melo, N.I.; de Carvalho, C.E.; Fracarolli, L.; Cunha, W.R.; Veneziani, R.C.; Martins, C.H.; Crotti, A.E. Antimicrobial activity of the essential oil of *Tetradenia riparia* (Hochst.) Codd. (Lamiaceae) against cariogenic bacteria. *Braz. J. Microbiol.* **2015**, *46*, 519–525. [CrossRef]

26. Abd-ElGawad, A.M.; El-Amier, Y.A.; Bonanomi, G.; Gendy, A.E.G.E.; Elgorban, A.M.; Alamery, S.F.; Elshamy, A.I. Chemical Composition of *Kickxia aegyptiaca* Essential Oil and Its Potential Antioxidant and Antimicrobial Activities. *Plants* **2022**, *11*, 594. [CrossRef] [PubMed]

27. Asakawa, Y.; Toyota, M.; Takemoto, T.; Suire, C. Plagiochilins C, D, E and F, four novel secoaromadendrane-type sesquiterpene hemiacetals from *Plagiochila asplenioides* and *Plagiochila semidecurrens*. *Phytochemistry* **1979**, *18*, 1355–1357. [CrossRef]

28. Asakawa, Y.; Inoue, H.; Toyota, M.; Takemoto, T. Sesquiterpenoids of fourteen *Plagiochila* species. *Phytochemistry* **1980**, *19*, 623–2626. [CrossRef]

29. Fukuyama, Y.; Toyota, M.; Asakawa, Y. Ent-kaurene diterpene from the liverwort *Plagiochila pulcherrima*. *Phytochemistry* **1988**, *27*, 1425–1427. [CrossRef]

30. Matsuo, A.; Atsumi, K.; Nakayama, M. Structures of ent-2,3-Secoalloaromadendrane Sesquiterpenoids, which have plant growth inhibitory Activity, from *Plagiochila semidecurrens* (Liverwort). *J. Chem. Soc. Perkin Trans.* **1981**, *1*, 2816–2824. [CrossRef]

31. Ramírez, M.; Kamiya, N.; Popich, S.; Asakawa, Y.; Bardón, A. Constituents of the Argentine Liverwort *Plagiochila diversifolia* and Their Insecticidal Activities. *Chem. Biodivers.* **2017**, *14*, e1700229. [CrossRef]

32. Lin, S.J.; Wu, C.L. Isoplagiochilide from the liverwort *Plagiochila elegans*. *Phytochemistry* **1996**, *41*, 1439–1440.

33. Toyota, M.; Nakamura, I.; Huneck, S.; Asakawa, Y. Sesquiterpene esters from the liverwort *Plagiochila porelloides*. *Phytochemistry* **1994**, *37*, 1091–1093. [CrossRef]

34. Asakawa, Y. Chemosystematics of the hepaticae. *Phytochemistry* **2004**, *65*, 623–669. [CrossRef] [PubMed]

35. Asakawa, Y.; Toyota, M.; Takemoto, T.; Kubo, I.; Nakanishi, K. Insect antifeedant secoaromadendrane-type sesquiterpenes from *Plagiochila* species. *Phytochemistry* **1980**, *19*, 2147–2154. [CrossRef]

36. Hashimoto, T.; Tanaka, H.; Asakawa, Y. Stereostructure of plagiochilin A and conversion of plagiochilin A and stearoylvelutinal into hot-tasting compounds by human saliva. *Chem. Pharm. Bull.* **1994**, *42*, 1542–1544. [CrossRef]

37. Asakawa, Y.; Toyota, M.; Takemoto, T. Three ent-secoaromadendrane-type sesquiterpene hemiacetals and a bicyclogermacrene from *Plagiochila ovalifolia* and *Plagiochila yokogurensis*. *Phytochemistry* **1980**, *19*, 2141–2145. [CrossRef]

38. Toyota, M.; Tanimura, K.; Asakawa, Y. Cytotoxic 2,3-secoaromadendrane-type sesquiterpenoids from the liverwort *Plagiochila ovalifolia*. *Planta Med.* **1998**, *64*, 462–464. [CrossRef]

39. Nagashima, F.; Tanaka, H.; Toyota, M.; Hashimoto, T. Sesqui- and diterpenoids from *Plagiochila* species. *Phytochemistry* **1994**, *36*, 1425–1430. [CrossRef]
40. Nagashima, F.; Tanaka, H.; Toyota, M.; Hashimoto, T.; Okamoto, Y.; Tori, M.; Asakawa, Y. 2,3-Secoaromadendrane-, Aromadendrane and Maaliane-Type Sesquiterpenoids from Liverwort. *J. Essent. Oil Res.* **1995**, *7*, 343–345. [CrossRef]
41. Birladeanu, L. The stories of santonin and santonic acid. *Angew. Chem. Int. Ed. Engl.* **2003**, *42*, 1202–1208. [CrossRef]
42. Blay, G.; Cardona, L.; García, B.; Lahoz, L.; Pedro, J.R. Synthesis of plagiochilin N from santonin. *J. Org. Chem.* **2001**, *66*, 7700–7705. [CrossRef] [PubMed]
43. Fukuyama, Y.; Asakawa, Y. Neurotrophic secoaromadendrane-type sesquiterpenes from the liverwort *Plagiochila fruticosa*. *Phytochemistry* **1991**, *30*, 4061–4065. [CrossRef]
44. Hashimoto, T.; Nakamura, I.; Tori, M.; Takaoka, S.; Asakawa, Y. Epi-neoverrucosane- and *ent*-clerodane-type diterpenoids and *ent*-2,3-secoaromadendrane- and calamenene-type sesquiterpenoids from the liverwort *heteroscyphus planus*. *Phytochemistry* **1995**, *38*, 119–127. [CrossRef]
45. Heinrichs, J.; Anton, H.; Holz, I.; Grolle, R. The andine *Plagiochila tabinensis* Steph. and the identity of *Acrobolbus laceratus* R.M. Schust. (Hepaticae). *Nova Hedwig.* **2001**, *73*, 445–452. [CrossRef]
46. Valcic, S.; Zapp, J.; Becker, H. Plagiochilins and other sesquiterpenoids from *Plagiochila* (Hepaticae). *Phytochemistry* **1997**, *44*, 89–99. [CrossRef]
47. Kraut, L.; Mues, R. The First Biflavone Found in Liverworts and Other Phenolics and Terpenoids from *Chandonanthus hirtellus ssp. giganteus* and *Plagiochila asplenioides*. *Z. Naturforsch.* **1999**, *54*, 6–10. [CrossRef]
48. Rycroft, D.S.; Cole, W.J.; Lamont, Y.M. Plagiochilins T and U, 2,3-secoaromadendranes from the liverwort *Plagiochila carringtonii* from Scotland. *Phytochemistry* **1999**, *51*, 663–667. [CrossRef]
49. Rycroft, D.S. *Plagiochila carringtonii*: Carrington's Featherwort. In *Mosses and Liverworts of Britain and Ireland: A Field Guide*; Atherton, I., Bosanquet, S.D.S., Lawley, M., Eds.; British Bryological Society: Middlewich, UK, 2010; p. 197. ISBN 9780956131010.
50. Rycroft, D.S.; Cole, W.J. Atlanticol, an epoxybicyclogermacrenol from the liverwort *Plagiochila atlantica*F. Rose. *Phytochemistry* **1998**, *49*, 1641–1644. [CrossRef]
51. Rodrigues e Rocha, M.; da Cunha, C.P.; Filho, R.B.; Vieira, I.J. A Novel Alkaloid Isolated from *Spiranthera atlantica* (Rutaceae). *Nat. Prod. Commun.* **2016**, *11*, 393–395.
52. Söderström, L.; Rycroft, D.S.; Cole, W.J.; Wei, S. *Plagiochila porelloides* (Plagiochilaceae, Hepaticae) from Changbai Moutain, new to China, with chemical characterization and chromosome measurements. *Bryothera* **1999**, *5*, 195–201.
53. Silva-E-Costa, J.D.C.; Luizi-Ponzo, A.P.; McLetchie, D.N. Sex Differences in Desiccation Tolerance Varies by Colony in the Mesic Liverwort *Plagiochila porelloides*. *Plants* **2022**, *11*, 478. [CrossRef] [PubMed]
54. Adio, A.M.; König, W.A. Sesquiterpene constituents from the essential oil of the liverwort *Plagiochila asplenioides*. *Phytochemistry* **2005**, *66*, 599–609. [CrossRef] [PubMed]
55. Asakawa, Y. Phytochemistry of Bryophytes. In *Phytochemicals in Human Health Protection, Nutrition, and Plant Defense*; Recent Advances in Phytochemistry; Romeo, J.T., Ed.; Springer: Boston, MA, USA, 1999; Volume 33. [CrossRef]
56. Asakawa, Y.; Ludwiczuk, A.; Nagashima, F. Phytochemical and biological studies of bryophytes. *Phytochemistry* **2013**, *91*, 52–80. [CrossRef] [PubMed]
57. Ludwiczuk, A.; Asakawa, Y. Bryophytes as a source of bioactive volatile terpenoids—A review. *Food Chem. Toxicol.* **2019**, *132*, 110649. [CrossRef]
58. Wang, S.; Liu, S.S.; Lin, Z.M.; Li, R.J.; Wang, X.N.; Zhou, J.C.; Lou, H.X. Terpenoids from the Chinese liverwort *Plagiochila pulcherrima* and their cytotoxic effects. *J. Asian Nat. Prod. Res.* **2013**, *15*, 473–481. [CrossRef]
59. Stivers, N.S.; Islam, A.; Reyes-Reyes, E.M.; Casson, L.K.; Aponte, J.C.; Vaisberg, A.J.; Hammond, G.B.; Bates, P.J. Plagiochilin A Inhibits Cytokinetic Abscission and Induces Cell Death. *Molecules* **2018**, *23*, 1418. [CrossRef]
60. Andrade, V.; Echard, A. Mechanics and regulation of cytokinetic abscission. *Front. Cell Dev. Biol.* **2022**, *10*, 1046617. [CrossRef]
61. Sechi, S.; Piergentili, R.; Giansanti, M.G. Minor Kinases with Major Roles in Cytokinesis Regulation. *Cells* **2022**, *11*, 3639. [CrossRef]
62. Wenzel, D.M.; Mackay, D.R.; Skalicky, J.J.; Paine, E.L.; Miller, M.S.; Ullman, K.S.; Sundquist, W.I. Comprehensive analysis of the human ESCRT-III-MIT domain interactome reveals new cofactors for cytokinetic abscission. *Elife* **2022**, *11*, e77779. [CrossRef]
63. Chircop, M.; Perera, S.; Mariana, A.; Lau, H.; Ma, M.P.; Gilbert, J.; Jones, N.C.; Gordon, C.P.; Young, K.A.; Morokoff, A.; et al. Inhibition of dynamin by dynole 34-2 induces cell death following cytokinesis failure in cancer cells. *Mol. Cancer Ther.* **2011**, *10*, 1553–1562. [CrossRef]
64. Tremblay, C.S.; Chiu, S.K.; Saw, J.; McCalmont, H.; Litalien, V.; Boyle, J.; Sonderegger, S.E.; Chau, N.; Evans, K.; Cerruti, L.; et al. Small molecule inhibition of Dynamin-dependent endocytosis targets multiple niche signals and impairs leukemia stem cells. *Nat. Commun.* **2020**, *11*, 6211. [CrossRef] [PubMed]
65. Huang, D.S.; Wong, H.L.; Georg, G.I. Synthesis and Cytotoxicity Evaluation of C4- and C5-Modified Analogues of the α,β-Unsaturated Lactone of Pironetin. *ChemMedChem* **2017**, *12*, 520–528. [CrossRef] [PubMed]
66. Coulup, S.K.; Georg, G.I. Revisiting microtubule targeting agents: α-Tubulin and the pironetin binding site as unexplored targets for cancer therapeutics. *Bioorg. Med. Chem. Lett.* **2019**, *29*, 1865–1873. [CrossRef] [PubMed]
67. Vergoten, G.; Bailly, C. Molecular Docking of Cryptoconcatones to α-Tubulin and Related Pironetin Analogues. *Plants* **2023**, *12*, 296. [CrossRef]

68. Durán-Peña, M.J.; Botubol Ares, J.M.; Hanson, J.R.; Collado, I.G.; Hernández-Galán, R. Biological activity of natural sesquiterpenoids containing a gem-dimethylcyclopropane unit. *Nat. Prod. Rep.* **2015**, *32*, 1236–1248. [CrossRef]
69. Sabovljević, M.S.; Sabovljević, A.D.; Ikram, N.K.K.; Peramuna, A.; Bae, H.; Simonsen, H.T. Bryophytes—An emerging source for herbal remedies and chemical production. (Special Issue 4: Evolving trends in plant-based drug discovery). *Plant Genet. Resour. Charact. Util.* **2016**, *14*, 314–327. [CrossRef]
70. Marques, R.V.; Sestito, S.E.; Bourgaud, F.; Miguel, S.; Cailotto, F.; Reboul, P.; Jouzeau, J.Y.; Rahuel-Clermont, S.; Boschi-Muller, S.; Simonsen, H.T.; et al. Anti-Inflammatory Activity of Bryophytes Extracts in LPS-Stimulated RAW264.7 Murine Macrophages. *Molecules* **2022**, *27*, 1940. [CrossRef]
71. Vollár, M.; Gyovai, A.; Szűcs, P.; Zupkó, I.; Marschall, M.; Csupor-Löffler, B.; Bérdi, P.; Vecsernyés, A.; Csorba, A.; Liktor-Busa, E.; et al. Antiproliferative and Antimicrobial Activities of Selected Bryophytes. *Molecules* **2018**, *23*, 1520. [CrossRef]
72. Wolski, G.J.; Sadowska, B.; Fol, M.; Podsędek, A.; Kajszczak, D.; Kobylińska, A. Cytotoxicity, antimicrobial and antioxidant activities of mosses obtained from open habitats. *PLoS ONE* **2021**, *16*, e0257479. [CrossRef]
73. Manoj, G.S.; Murugan, K. Wound healing potential of aqueous and methnolic extracts of *Plagiochila beddomei* Steph.—A briophyte. *Int. J. Pharm. Pharm. Sci.* **2012**, *4*, 222–227.
74. Nandy, S.; Dey, A. Bibenzyls and bisbybenzyls of bryophytic origin as promising source of novel therapeutics: Pharmacology, synthesis and structure-activity. *Daru* **2020**, *28*, 701–734. [CrossRef] [PubMed]
75. Métoyer, B.; Benatrehina, A.; Rakotondraibe, L.H.; Thouvenot, L.; Asakawa, Y.; Nour, M.; Raharivelomanana, P. Dimeric and esterified sesquiterpenes from the liverwort *Chiastocaulon caledonicum*. *Phytochemistry* **2020**, *179*, 112495. [CrossRef] [PubMed]
76. Wang, X.; Qian, L.; Qiao, Y.; Jin, X.; Zhou, J.; Yuan, S.; Zhang, J.; Zhang, C.; Lou, H. Cembrane-type diterpenoids from the Chinese liverwort *Chandonanthus birmensis*. *Phytochemistry* **2022**, *203*, 113376. [CrossRef] [PubMed]
77. Asakawa, Y.; Nagashima, F.; Ludwiczuk, A. Distribution of Bibenzyls, Prenyl Bibenzyls, Bis-bibenzyls, and Terpenoids in the Liverwort Genus *Radula*. *J. Nat. Prod.* **2020**, *83*, 756–769. [CrossRef] [PubMed]
78. Asakawa, Y.; Ludwiczuk, A.; Novakovic, M.; Bukvicki, D.; Anchang, K.Y. Bis-bibenzyls, Bibenzyls, and Terpenoids in 33 Genera of the Marchantiophyta (Liverworts): Structures, Synthesis, and Bioactivity. *J. Nat. Prod.* **2022**, *85*, 729–762. [CrossRef]
79. Kirisanth, A.; Nafas, M.N.M.; Dissanayake, R.K.; Wijayabandara, J. Antimicrobial and Alpha-Amylase Inhibitory Activities of Organic Extracts of Selected Sri Lankan Bryophytes. *Evid. Based Complement. Altern. Med.* **2020**, *2020*, 3479851. [CrossRef]
80. Asakawa, Y.; Nagashima, F.; Hashimoto, T.; Toyota, M.; Ludwiczuk, A.; Komala, I.; Ito, T.; Yagi, Y. Pungent and bitter, cytotoxic and antiviral terpenoids from some bryophytes and inedible fungi. *Nat. Prod. Commun.* **2014**, *9*, 409–417. [CrossRef]
81. Mierzwa, B.; Gerlich, D.W. Cytokinetic abscission: Molecular mechanisms and temporal control. *Dev. Cell* **2014**, *31*, 525–538. [CrossRef]
82. McNeely, K.C.; Dwyer, N.D. Cytokinetic Abscission Regulation in Neural Stem Cells and Tissue Development. *Curr. Stem Cell Rep.* **2021**, *7*, 161–173. [CrossRef]
83. Halcrow, E.F.J.; Mazza, R.; Diversi, A.; Enright, A.; D'Avino, P.P. Midbody Proteins Display Distinct Dynamics during Cytokinesis. *Cells* **2022**, *11*, 3337. [CrossRef]
84. Chatterjee, N.; Wang, W.L.; Conklin, T.; Chittur, S.; Tenniswood, M. Histone deacetylase inhibitors modulate miRNA and mRNA expression, block metaphase, and induce apoptosis in inflammatory breast cancer cells. *Cancer Biol. Ther.* **2013**, *14*, 658–671. [CrossRef] [PubMed]
85. Gill, M.R.; Jarman, P.J.; Hearnden, V.; Fairbanks, S.D.; Bassetto, M.; Maib, H.; Palmer, J.; Ayscough, K.R.; Thomas, J.A.; Smythe, C. A Ruthenium(II) Polypyridyl Complex Disrupts Actin Cytoskeleton Assembly and Blocks Cytokinesis. *Angew. Chem. Int. Ed. Engl.* **2022**, *61*, e202117449. [CrossRef] [PubMed]
86. Petsalaki, E.; Zachos, G. An ATM-Chk2-INCENP pathway activates the abscission checkpoint. *J. Cell Biol.* **2021**, *220*, e202008029. [CrossRef] [PubMed]
87. Sardina, F.; Monteonofrio, L.; Ferrara, M.; Magi, F.; Soddu, S.; Rinaldo, C. HIPK2 Is Required for Midbody Remnant Removal Through Autophagy-Mediated Degradation. *Front. Cell Dev. Biol.* **2020**, *8*, 572094. [CrossRef] [PubMed]
88. Boullé, M.; Davignon, L.; Nabhane Saïd Halidi, K.; Guez, S.; Giraud, E.; Hollenstein, M.; Agou, F. High-Content RNAi Phenotypic Screening Unveils the Involvement of Human Ubiquitin-Related Enzymes in Late Cytokinesis. *Cells* **2022**, *11*, 3862. [CrossRef] [PubMed]
89. Sabovljević, M.S.; Ćosić, M.V.; Jadranin, B.Z.; Pantović, J.P.; Giba, Z.S.; Vujičić, M.M.; Sabovljević, A.D. The Conservation Physiology of Bryophytes. *Plants* **2022**, *11*, 1282. [CrossRef]
90. Basile, D.V.; Lin, J.J.; Varner, J.E. The metabolism of exogenous hydroxyproline by gametophytes of *Plagiochila arctica* Bryhn et Kaal. (Hepaticae). *Planta* **1988**, *175*, 539–545. [CrossRef]
91. Ugur, A.; Sarac, N.; Duru, M.E. Antimicrobial activity and chemical composition of *Senecio sandrasicus* on antibiotic resistant staphylococci. *Nat. Prod. Commun.* **2009**, *4*, 579–584. [CrossRef]
92. Silva, E.A.J.; Estevam, E.B.B.; Silva, T.S.; Nicolella, H.D.; Furtado, R.A.; Alves, C.C.F.; Souchie, E.L.; Martins, C.H.G.; Tavares, D.C.; Barbosa, L.C.A.; et al. Antibacterial and antiproliferative activities of the fresh leaf essential oil of *Psidium guajava* L. (Myrtaceae). *Braz. J. Biol.* **2019**, *79*, 697–702. [CrossRef]
93. Pavithra, P.S.; Mehta, A.; Verma, R.S. Aromadendrene oxide 2, induces apoptosis in skin epidermoid cancer cells through ROS mediated mitochondrial pathway. *Life Sci.* **2018**, *197*, 19–29. [CrossRef]

94. Pavithra, P.S.; Mehta, A.; Verma, R.S. Synergistic interaction of β-caryophyllene with aromadendrene oxide 2 and phytol induces apoptosis on skin epidermoid cancer cells. *Phytomedicine* **2018**, *47*, 121–134. [CrossRef] [PubMed]
95. Pornpakakul, S.; Suwancharoen, S.; Petsom, A.; Roengsumran, S.; Muangsin, N.; Chaichit, N.; Piapukiew, J.; Sihanonth, P.; Allen, J.W. A new sesquiterpenoid metabolite from *Psilocybe samuiensis*. *J. Asian Nat. Prod. Res.* **2009**, *11*, 12–17. [CrossRef] [PubMed]
96. Özerkan, D.; Erol, A.; Altuner, E.M.; Canlı, K.; Kuruca, D.S. Some Bryophytes Trigger Cytotoxicity of Stem Cell-like Population in 5-Fluorouracil Resistant Colon Cancer Cells. *Nutr. Cancer* **2022**, *74*, 1012–1022. [CrossRef] [PubMed]
97. Chandra, S.; Chandra, D.; Barh, A.; Pandey, R.K.; Sharma, I.P. Bryophytes: Hoard of remedies, an ethno-medicinal review. *J. Tradit. Complement. Med.* **2016**, *7*, 94–98. [CrossRef] [PubMed]
98. Commisso, M.; Guarino, F.; Marchi, L.; Muto, A.; Piro, A.; Degola, F. Bryo-Activities: A Review on How Bryophytes Are Contributing to the Arsenal of Natural Bioactive Compounds against Fungi. *Plants* **2021**, *10*, 203. [CrossRef]

Review

Pharmacological Activities of Mogrol: Potential Phytochemical against Different Diseases

Varun Jaiswal [1] and Hae-Jeung Lee [1,2,3,*]

[1] Department of Food and Nutrition, College of BioNano Technology, Gachon University, 1342 Seongnam-daero, Sujeong-gu, Seongnam-si 13120, Republic of Korea

[2] Institute for Aging and Clinical Nutrition Research, Gachon University, Seongnam-si 13120, Republic of Korea

[3] Department of Health Sciences and Technology, GAIHST, Gachon University, Incheon 21999, Republic of Korea

* Correspondence: skysea@gachon.ac.kr; Tel.: +82-31-750-5968; Fax: +82-31-724-4411

Abstract: Recently, mogrol has emerged as an important therapeutic candidate with multiple potential pharmacological properties, including neuroprotective, anticancer, anti-inflammatory, antiobesity, antidiabetes, and exerting a protective effect on different organs such as the lungs, bone, brain, and colon. Pharmacokinetic studies also highlighted the potential of mogrol as a therapeutic. Studies were also conducted to design and synthesize the analogs of mogrol to achieve better activities against different diseases. The literature also highlighted the possible molecular mechanism behind pharmacological activities, which suggested the role of several important targets, including AMPK, TNF-α, and NF-κB. These important mogrol targets were verified in different studies, indicating the possible role of mogrol in other associated diseases. Still, the compilation of pharmacological properties, possible molecular mechanisms, and important targets of the mogrol is missing in the literature. The current study not only provides the compilation of information regarding pharmacological activities but also highlights the current gaps and suggests the precise direction for the development of mogrol as a therapeutic against different diseases.

Keywords: mogrol; mogroside; phytochemical; pharmacology; anticancer; osteoporosis; antiobesity; anti-inflammation; neurodegenerative disorders

Citation: Jaiswal, V.; Lee, H.-J. Pharmacological Activities of Mogrol: Potential Phytochemical against Different Diseases. *Life* **2023**, *13*, 555. https://doi.org/10.3390/life13020555

Academic Editors: Marisa Colone, Charalampia Amerikanou and Efstathia Papada

Received: 21 January 2023
Revised: 11 February 2023
Accepted: 14 February 2023
Published: 16 February 2023

1. Introduction

In therapeutics, plant-derived drugs have gained significant importance in recent years, with advancements in the technologies such as analytical techniques, genomics, and engineering strategies [1]. Triterpenes are among the important phytochemicals that are known to have different important biological properties that may be useful as therapeutics [2,3]. Triterpene, mogrol found in the fruits of *Siraitia grosvenorii* recently stretched its importance as a potential therapeutic candidate as it was observed to have different pharmacological activities which can be utilized in the development of therapeutics among different diseases or conditions. These activities of mogrol include neuroprotective, anticancer, anti-inflammatory, antiobesity, antidiabetes, anti-osteoporosis, and anti-colitis activities, and protective activity in lung diseases such as lung injury and pulmonary fibrosis [4–7]. Mogrol is the aglycone of mogrosides that are the main constituents of the extract of the fruit of *Siraitia grosvenorii* [8]. *S. grosvenorii* is a traditional Chinese medicinal food known for various pharmacological activities, which may be attributed to its several bioactive components, such as mogrol and mogrosides [9]. Some studies have shown that biological activities, such as reduction in triglyceride accumulation and activation of AMPK, were only attributed to mogrol, not to mogroside V, the most abundant mogroside of *Siraitia grosvenorii* [4,6]. Recently, studies have shown that most mogrosides are converted through intestinal digestive enzymes and microflora to mogrol [10–12]. Therefore, mogrol may also be the main functional component of mogrosides responsible for pharmacological effects.

Recently, considering the high application prospects of mogrol in medicine and the low yield in extraction from plant extract, a de novo biosynthesis pathway was constructed in yeast (*Saccharomyces cerevisiae*) that can be developed as an alternate source for the production of mogrol in place its extraction from the plant [13]. Recently, a multigene stacking strategy is developed to synthesize different mogrosides through metabolic engineering in other plants [14]. Similarly, considering the role of plant-based sweeteners in the environment and economic benefits [15], researchers have created UDP-glycosyltransferase enzymes to synthesize high-intensity sweetener mogrosides from mogrol [16]. Researchers have studied the quantification method of mogrol along with mogroside V in rat plasma after oral and intravenous administration for pharmacokinetic studies [17–19]. Along with the different pharmacological activities of mogrol, researchers have also been elucidating the mechanism behind the pharmacological activities to understand and optimize the pharmacological activities of mogrol. In these studies, several important targets such as AMPK, TNF-α, STAT3, P53, and NF-κB were found to be associated with the activity of mogrol, which suggests the role of mogrol may be studied against other important diseases associated with these important targets [20–22]. Researchers have also begun to design and test the analogues of mogrol to achieve better pharmacological activity against important diseases such as cancer and targets such as AMPK [23]. However, a comprehensive compilation of the pharmacological activities, associated mechanisms, and targets of mogrol, which could serve as a resource to highlight the current gaps and propose future directions for its development as a therapeutic candidate, is not available.

2. Literature Search

A literature search was conducted using online databases, including Scopus, PubMed, Google Scholar, Google, and ResearchGate. In the search, important keywords such as mogrol and its combinations with pharmacology, disease, biological activity, anticancer, antidiabetes, antiobesity, anti-inflammatory, in vivo, and in vitro studies were used. The resulting research articles, review articles, book chapters, and books published in the English language until December 2022 were considered, and the important literature was included in this study.

3. Mogrol Structure and Properties

Mogrol and mogrosides are mainly found in the fruits of *S. grosvenorii* (also known as monk fruit), and the content of these compounds is in the range of 0.55–0.65% in fresh fruit [24]. Mogrol is a cucurbitane-type tetracyclic triterpenoid, and the mogrosides are the group of compounds that contain different glycosylated compounds of mogrol. Mogrol is the precursor of mogrosides in the fruits of *S. grosvenorii*, and mogrol is converted into mogrosides with the growth and ripening of the fruit through the enzyme glycosyltransferases [16]. Several mogrosides were identified from the monk fruit named according to the number of glycosylated sugar moieties which form β-linkages with mogrol. The presence of no sugar moiety in mogrol and more sugar moieties in mogroside V is considered the reason behind the bitterness of mogrol and the high sweetness of mogroside V [16]. The study has shown that the mogrosides on digestion are mostly converted to mogrol. The structure of mogrol and important mogrosides (such as mogrosides IE, IIE, IIIE, IVE, and V) are provided in Figure 1. The molecular weight of mogrol is 476.7 Da, and it contains four hydrogen bond donors and four hydrogen bond acceptors in the structure. Limited toxicology studies have been conducted on mogrol and its synthetic derivatives, but toxicology studies on mogrosides extracted from the fruits are available. In cell proliferation assays, mogrol has no effect on the viability of 3T3-L1, mouse type II alveolar epithelial cell MLE-12, and bone marrow macrophages at the concentration up to 50, 40, and 10 μM, respectively [4,25,26]. Similarly, mogrosides were found to be safe in the study conducted on dogs for 28 and 90 days of oral administration at a dose (3000 mg/kg bw/day). In this study, there was no adverse effect of mogrosides from the monk fruits observed in clinical observations, body weight, food consumption, hematology, blood chemistry, urinalysis, gross necropsy, organ weight, and histopathology [27]. Additionally, in the Ames mutagenicity test, the mogrosides were negative, indicating that they do not have the mutagenicity effect [28]. The pharmacokinetic parameter, such as absolute oral bioavailability and elimination half-life ($t_{1/2}$) of mogrol, were approximately $10.3 \pm 2.15\%$ and 2.41 ± 0.11 h, respectively [18].

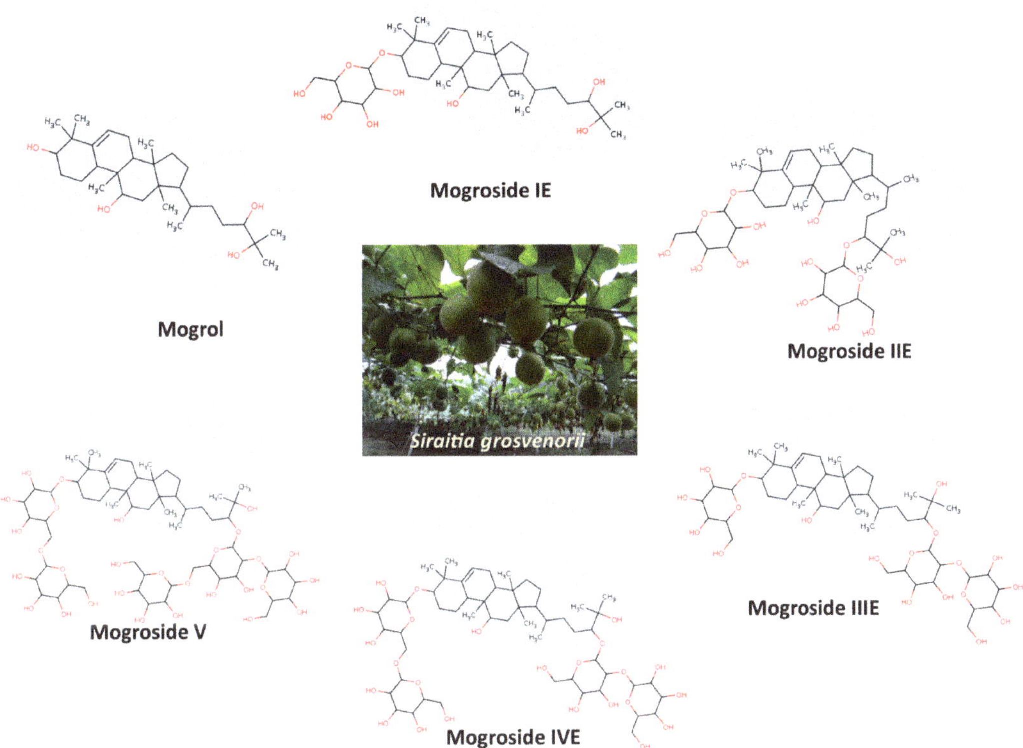

Figure 1. Structure of mogrol and mogrosides found in monk fruit (*S. grosvenorii*). The oval shape highlights the mogrol moiety present in mogrosides.

4. Pharmacological Activities of Mogrol

4.1. In Vitro Antiobesity and Antidiabetes Activity of Mogrol

The antiobesity and antidiabetes activity of mogrol was studied in different in vitro experiments, and the antiobesity and antidiabetic properties of the crude extract of *S. grosvenorii*, rich in several triterpenes (mainly mogrol) and triterpene glycosides (mogroside V), have been observed in different studies [8,29,30]. To precisely identify the compounds in the extract important for antidiabetic effect, the compounds were derived from the crude extract and studied for the activator activity of adenosine monophosphate-activated protein kinase (AMPK) in the HepG2 cell line [6]. AMPK activation is considered a potential therapeutic strategy for the treatment of metabolic disorders, including diabetes and obesity [31]. Mogrol derived from the extract was among the compounds that can cause the highest increase in AMPK phosphorylation better than the positive control compound (berberine) at all studied concentrations (1, 10, and 20 µM) [6], while the main ingredient of the extract, i.e., mogroside V, had no effect on the phosphorylation of AMPK. The results suggested that the mogrol-induced activation of AMPK has been shown to contribute to the anti-hyperglycemic and antilipidemic properties of the *S. grosvenorii* extract [6].

Later, the effect of mogrol and different mogrosides of *S. grosvenorii* on adipogenesis in 3T3-L1 cells was studied. Mogrol, not mogrosides, was able to suppress the differentiation of 3T3L1 preadipocytes to adipocytes. In the study, only mogrol was found to be effective in reducing the accumulation of triglycerides at both 10 and 20 μM concentrations. The effect of mogrol on AMPK phosphorylation was analyzed with the positive control (acadesine). As in the previous study, mogrol increased AMPK phosphorylation [6]. Similarly, the mRNA level of C/EBPβ was also suppressed by the mogrol treatment observed in the study [4].

cAMP response element (CRE)-mediated transcriptional activity was also suppressed by mogrol treatment. Similarly, phosphorylation of cAMP response element-binding protein (CREB) was also suppressed by mogrol. The study concluded that mogrol might be suppressing the lipid accumulation in 3T3L1 adipocytes through activation of AMPK, repression of CREB phosphorylation, and CRE-mediated transcription activity [4].

In the same year, the researcher studied the potential of the mogrol- and cucurbitane-type triterpenoid (mogroside V) for the activation of AMPK. Mogrol was found to be the potent AMPK activator that can activate the AMPK heterotrimer, AMPKα2β1γ1, with an EC_{50} of 4.2 μM, which was better than the mogroside V (EC_{50} of 20.4 μM). Further, mogrol and mogroside V were subjected to pharmacokinetics analysis in rats [17].

Inspired by their earlier studies of the allosteric activation of AMPK by mogrol, the researcher designed the study to increase the allosteric activation of AMPK through mogrol derivatives at the 24 position. The allosteric activation of AMPK was studied for mogrol and its 21 derivatives. Finally, the most potent compound (derivative with the aliphatic-acyclic group at the 24 position in mogrol) was found to be 20 times more effective than mogrol (EC_{50} of 3.0 ± 0.20 μM) [23]. Overall, the different studies suggest that the antiobesity effect of mogrol may be through reducing CREB phosphorylation, suppressing CRE-mediated transcription, and activating AMPK signaling [4,6,17,23].

4.2. In Vitro Anti-Inflammatory Activity of Mogrol

The anti-inflammatory activity of mogrol was observed in both in vitro and in vivo studies. Recently, the anti-inflammatory activity of mogrol was evaluated through an in vitro experiment on lipopolysaccharide (LPS)-induced RAW 264.7 cells (Table 1). The application of mogrol at 10 μM concentration significantly reduced the level of pro-inflammatory cytokines (TNF-α, IL-6) and NO production [7]. The anti-inflammatory activity of mogrol was also observed in various in vivo studies in different organs such as the brain, lungs, and colon (Figure 2); these studies are presented in the respective section of the manuscript.

Table 1. Pharmacological activities of mogrol studied in the in vitro experiments.

Activity (and Probable Mechanism)	Dose	Method	Result	Ref.
Antiobesity/antidiabetes (through reducing CREB activation and promoting AMPK activation)	1, 10, and 20 μM	HepG2 cell line AMPK activator (phosphorylation of AMPK)	p-AMPK/AMPK $\uparrow$ more efficiently than berberine	[6]
	1, 5, 10, and 20 μM	3T3-L1 cells Oil red O staining and classical Folch method	At 20 μM lipid accumulation $\downarrow$ and cellular TG levels $\downarrow$	[4]
	20 μM	WB	p-AMPK/AMPK $\uparrow$ and p-CREB/CREB $\downarrow$	[4]
	20 μM	Real-time PCR	C/EBPβ $\downarrow$	[4]
	20 μM	Reporter assay	CRE-mediated transcription $\downarrow$	[4]
	0.625–40 μM	AMPK activations through HTRF assay	A769662 (EC$_{50}$ of 24.9 nM) and AMP (EC$_{50}$ of 1.4 nM) EC$_{50}$ of 4.2 μM.	[17]
	mogrol	HepG2 cells AMPK allosteric activator screening based on SPA assay	EC$_{50}$ = 3.0 $\pm$ 0.20 for mogrol and 1.4 $\pm$ 0.10 μM for AMP	[23]
Anti-inflammatory (inhibition of TNF-α mediated inflammation)	10 μM of mogrol	RAW 264.7 cells NO production	NO production $\downarrow$ (17%)	[7]
	10 μM of mogrol	ELISA	The levels of TNF-α $\downarrow$ and IL-6 $\downarrow$	[7]
Anticancer (autophagy and autophagic cell death via activating AMPK signaling pathway, ERK, and STAT3 inhibition)	0.1, 1, 10, 100, and 250 μM	Cell proliferation (K562 cells) study through MTT assay	Growth inhibition 88% at 250 μM.	[32]
	0.1, 1, 10, 100, and 250 μM	WB	P21 $\uparrow$, Bcl2 $\downarrow$, p-ERK/ERK $\downarrow$, and p-STAT3/STAT3 $\downarrow$	[32]
	50 and 100 μM	The A549 and CNE1 cell lines MTT assay	IC$_{50}$ (μM) 89.51 $\pm$ 3.95 (A549) 81.48 $\pm$ 4.73 (CNE1)	[33]
	30 μM	CCK8 against lung cancer cell lines (A549 and NCI-H460)	IC$_{50}$ (μM) 27.78 $\pm$ 0.98 (A549) >100 (NCI-H460)	[34]
		A549, NCI-H460, H1299 and H1975	IC$_{50}$ (μM) 28.56 $\pm$ 1.98 (A549) >100 (NCI-H460) >100 (H1299) 87.14 $\pm$ 2.56 (H1975)	[7]
		A549, NCI-H460, and CNE1 CCK8 assay	IC$_{50}$ (μM) 27.78 $\pm$ 0.98 (A549) >100 (NCI-H460) >100 (CNE1)	[35]
	50 μM	A549, H1299, H1975, and SK-MES-1 analyzed through CCK-8 kit and celigo cell counter	IC$_{50}$ (μM) <25 μM	[36]
	50 μM	Inverted confocal microscopy and surface area were measured by ImageJ	Surface area $\downarrow$ and the number of lung cancer $\downarrow$ in A549, H1299, and SK-MES-1 cells	[36]
	50 μM	Scratch-wound migration assay	Migration of all studied cells $\downarrow$	[36]
	50 μM	WB	LC3-II $\uparrow$	[36]
	50 μM	Adenovirus expressing fluorescent-mRFP-GFP-LC3	Autophagosomes $\uparrow$, as well as autolysosome $\uparrow$	[36]
	50 μM	A549 and SK-MES-1 cells WB	Cleaved caspase3 $\uparrow$, LC3-II $\uparrow$, p-AMPK/AMPK $\uparrow$, p-P53/P53 $\uparrow$, PUMA $\uparrow$, p21 $\uparrow$, p27 $\uparrow$, and Bcl-2 $\downarrow$	[36]

Table 1. *Cont.*

Activity (and Probable Mechanism)	Dose	Method	Result	Ref.
Anti-colitis (through promoting AMPK activation and inhibition of TNF-α mediated inflammation)	pre-treated with mogrol (1 or 10 μM)	NCM460 human intestinal epithelial cells stimulated through TNF-α (metformin (the positive control drug), 2 mM used as control in the study)	Occludin ↑, ZO-1 ↑, bcl-2/bax ↑, TNF-α ↓, and p-AMPK/AMPK ↑ at (10 μM of mogrol)	[37]
	10 μM	THP-1 cells were stimulated by PMA to macrophages and stimulated by LPS and ATP (metformin as positive control drug used in the study)	Cleaved-caspase1/pro-caspase1 ↓, and IL-1β ↓	
Antifibrotic activity (through activation of AMPK-mediated signaling pathways and inhibition of NF-κB signaling pathways)	1, 5, and 10 μM of mogrol	TGF-β1-treated mouse type II alveolar epithelial cells (MLE-12 cell line) Expression analysis through WB	E-cadherin ↑, α-SMA ↓, type Col I ↓, and Vimentin ↓	[26]
	1, 5, and 10 μM of mogrol	PLFs cells expression analysis through WB	α-SMA ↓, (LOXL2 ↓), collagen I ↓, and phosphorylation forms of Smad2/3 ↓ and increased p-AMPK ↑	
	10 μM of mogrol	The protein expression was measured by immunofluorescence staining	NOX4 ↓ protein expression in TGF-β1-treated PLFs	
Anti-osteoporosis (inhibition of TRAF6/MAPK/NF-κB signaling pathway)	(0, 5, 10, 20 μM)	BMM stimulated with RANKL	Osteoclasts ↓ (236.67 ± 37.07 to 20.0 ± 6.08) at 20 μM mogrol.	[25]
	0, 5, 10, 20 μM	Bone resorption in bovine bone slices through SEM	Bone resorption ↓	
	20 μM	RNA-Seq	lilrb4a ↑ and Fcgr3 ↑ osteoclastogenesis suppressive genes) MMP9 ↓, OSCAR ↓, and ACP5 ↓ (osteoclastogenesis genes)	
	20 μM	RT–PCR was utilized to confirm osteoclastogenesis marker genes expression	MMP9 ↓, CTSK ↓, ATP6v0d2 ↓, ACP5 ↓, and DCSTAMP ↓	
	20 μM	WB	Phosphorylation of JNK ↓, ERK ↓, P65 ↓, and P38 ↓ Expression of c-FOS ↓, NFATc1 ↓, TRAF6 ↓, and Siglec-15 ↓ Suppressed the degradation of IκBα ↓	
	20 μM	Immunofluorescence staining	Mogrol efficiently blocked P65 nuclear translocation, which is required for NF-κB activation	

AICAR: 5-aminoimidazole-4-carboxamide-1-beta-D-ribofuranoside; AMPK: adenosine monophosphate-activated protein kinase; CCK8: Cell Counting Kit-8 assay; DSS: dextran sulfate sodium; ELISA: enzyme-linked immunosorbent assay; HTRF: homogeneous time-resolved fluorescence assay; LOXL2: lysyl oxidase-like protein 2; LPS: lipopolysaccharide; MTT: 3-(4,5-dimethylthiazole-2-yl)-2,5-diphenyl tetrazolium bromide; PLFs: primary lung fibroblasts; PMA: phorbol myristate acetate; Sirt1: sirtuin 1; SPA: scintillation proximity assay; WB: Western blotting analysis; ↑: significantly up-regulated; ↓: significantly down-regulated.

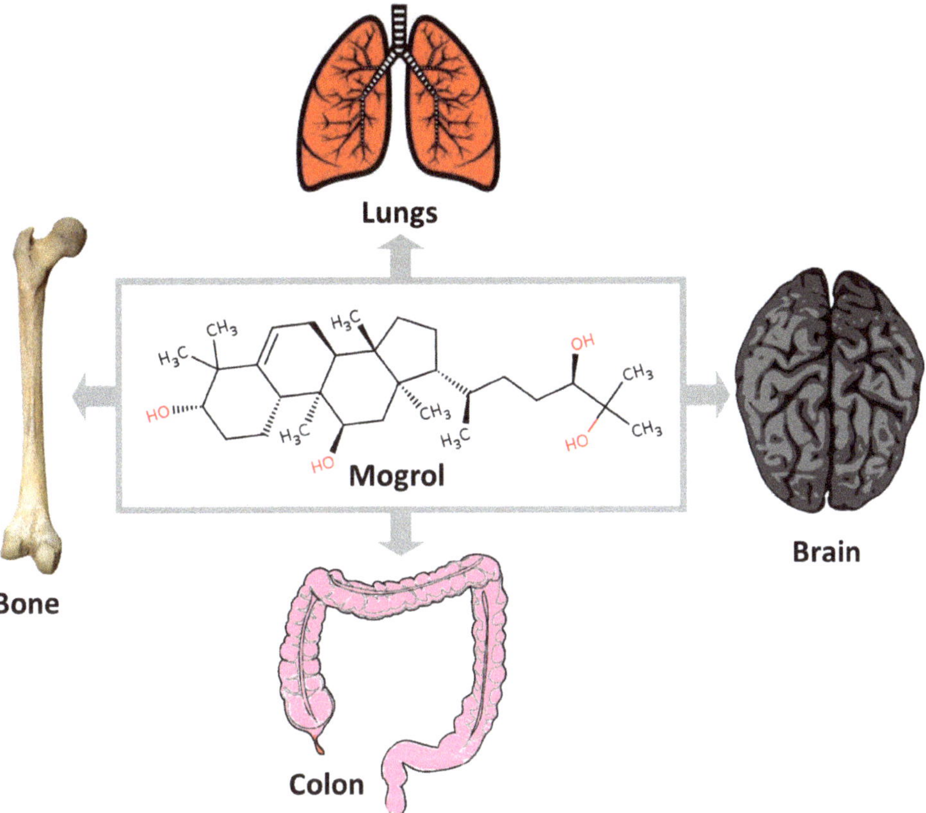

Figure 2. Protective role of mogrol against different organs in animal studies.

4.3. Anticancer Activity of Mogrol

Initially, the anticancer activities of mogrol have been studied in different cancer cell lines in in vitro experiments [32–34]. Later, in vivo studies also supported the anticancer potential of mogrol, and studies were also conducted to decipher the mechanism behind the anticancer activity. Researchers also modified the structure of mogrol to enhance its activity in different studies.

4.3.1. In Vitro Anticancer Activity of Mogrol

In a study, mogrol was derived from *S. grosvenorii* to study its anticancer effect on the human leukemia cell line, K562 cells. The antiproliferative activity of mogrol was studied by the 3-(4,5-dimethylthiazole-2-yl)-2,5-diphenyl tetrazolium bromide (MTT) assay. Mogrol inhibited the growth of K562 cells in a dose- and time-dependent manner (at 0.1, 1, 10, 100, and 250 µM concentration). Further, staining assays reveal the dose-dependent apoptosis of K562 cells through mogrol treatment. Similarly, the population of necrotic/post-apoptotic cells increased significantly (in a dose-dependent manner) to 25.56% from 5.50% with mogrol treatment [32]. Mogrol also caused the growth arrest in the G0/G1 phase of the cell cycle in K562 cells as the percentage of the G0/G1 phase increased from 36.48% to 77.41% with increasing mogrol concentrations. Furthermore, Western blot analysis suggested that mogrol reduces the phosphorylation of ERK1 and ERK2 phosphorylation and also suppresses the level of Bcl-2 in a concentration-dependent manner in K562 cells (Table 1)

Similarly, mogrol also suppressed the level of p-STAT3 and improved the expression of the p21 protein in K562 cells, which supports anticancer activity [38,39].

In a study inspired by the earlier reported anticancer activity of mogrol, the anticancer potential of mogrol and synthesized derivatives of mogrol through modification at the C24 and C25 positions was studied in two cancer cell lines (i.e., A549 and CNE1 cell lines). The mogrol derivative resulting from the addition of a tetrahydro-β-carboline structure was found to have the highest antiproliferative activity against both cell lines in the MTT assay [33]. Increased antitumor activity of mogrol derivative (with tetrahydro-β-carboline) may be the intramolecular synergistic effect as the tetrahydro-β-carboline compounds reported for antitumor activity in the literature [40].

Similarly, the anticancer activity of mogrol and a series of novel synthesized derivatives of mogrol was further studied on human lung cancer cells such as A549 and NCI-H460. The in vitro cytotoxicity of the mogrol and synthesized mogrol derivatives were studied through the CCK8 assay against both human lung cancer cell lines. Like the previous studies, the mogrol has shown a significant antiproliferative effect on the A549 cell line (IC_{50} of 27.78 ± 0.98 μM) [33]. Some derivatives of mogrol, especially the replacement of hydroxyl groups at C11 with ester containing the 1,2,3-triazole ring, exhibited significant antiproliferative activity against both cell lines [34]. Like earlier studies on mogrol, the mogrol derivative was also shown to significantly increase the proportions of cells in G0/G1 and down-regulate the level of p-STAT3 in a dose-dependent manner [32].

Recently, mogrol and synthesized three series of novel mogrol derivatives were used for anticancer activity in non-small cell lung cancer cell lines. Similarly, mogrol treatment significantly reduced the cell proliferation of lung cancer cell lines (A549 and H1975) [7]. In the study, mogrol, such as indole fused derivatives and derivatives containing α, β-unsaturated ketone moiety, had better anti-inflammatory and anticancer activities than mogrol, respectively [7].

Further, considering the insight from the anticancer activity of the synthesized derivative of mogrol in previous and preliminary studies, the C11 and C3 locations in mogrol were selected for the synthesis of two series of ester derivatives of mogrol. The activity of mogrol and its derivatives was studied in different human cancer cell lines (A549, NCI-H460, and CNE1). In comparison with mogrol, most of the derivatives (especially the esterification in the C11 position of mogrol) were found to have higher antiproliferative activity against cancer cell lines. Like mogrol in the previous study, the effective derivatives of mogrol also suppressed the STAT3 expression in the study, which suggests a similar action mechanism as mogrol. The study supports that the inhibition of STAT3 may be an important mechanism for anticancer activity and is expected to help the development of mogrol-based anticancer drugs [35].

Recently, the anticancer effect of mogrol on four cancer cell lines of human lung cancer (A549, H1299, H1975, and SK-MES-1) with normal control bronchial epithelial cells (HBE) was studied before animal experiments. A significant decrease in the survival rate was observed in all cell lines A549, H1299, H1975, and SK-MES-1, in a concentration-dependent manner. Additionally, the surface area (morphology) and the number of lung cancer cells A549, H1299, and SK-MES-1 decreased significantly with the mogrol treatment. Mogrol also inhibits the migration of all studied cell line lung cancer cells but has no significant effect on non-cancerous HBE in the scratch-wound migration assay. The autophagy marker LC3 was found to be significantly increased with mogrol treatment in all four cancer cell lines as compared to respective controls. Further, the adenovirus expressing fluorescent-mRFP-GFP-LC3 was used to visualize autophagic flux; the mogrol significantly increased the autophagosome (yellow dots) as well as autolysosome (green dots) numbers in all four lung cancer cells.

Further, in both A549 and SK-MES-1 cells, the mogrol treatment significantly increased the ratios of p-AMPK/AMPK in a concentration-dependent manner, along with the increased LC3-II level and decreased cell survival rate. Supplementation with compound c (AMPK inhibitor) significantly reversed the level of LC3II and decreased the autophagosomes. Furthermore, mogrol treatment also significantly increased p53 phosphorylation compared to the respective control in A549 and SK-MES-1 cells. Furthermore, the protein expression of cell cycle-dependent proteins such as p53, PUMA, and cell cycle inhibitory proteins p21 and p27 increased significantly with the mogrol treatment. Additionally, mogrol treatment in both A549 and SK-MES-1 cells resulted in a significant decrease in the levels of antiapoptotic protein (Bcl-2) and the ratios of Bcl-2 to Bax. Encouraging results of mogrol in all four lung cancer cells inspired further animal study [36].

4.3.2. In Vivo Anticancer Activity of Mogrol

In the in vivo experiment, male thymus-deficient mice (BALB/C, 20 g) were used for anticancer activity. Lung cancer cells (A549) were injected subcutaneously into the mice, and mogrol treatment started when the tumor size increased to 5×5 mm^3. The weight and volume of the tumor decreased (by 69.18% and 66.22%, respectively) significantly in the mogrol treatment group compared to the control mice group. Additionally, two weeks of mogrol treatment did not show a significant difference in the body weight of tumor-bearing mice with and without mogrol treatment. Similarly, there was no significant effect on the parameters of the cardiac function of mice in echocardiography between mogrol treatment and control mice [36].

4.4. Mogrol Activity against Ulcerative Colitis

Earlier reported anti-inflammatory activity of mogrol encouraged the analysis of the mogrol against ulcerative colitis (UC), a chronic inflammatory bowel disease [37]. Initially, the significant activity of mogrol against UC was observed in the mice model, and the mechanism behind the anti-colitis activity was explored in in vitro cell line experiments.

4.4.1. In Vivo Anti-UC Activity of Mogrol

The dextran sulfate sodium (DSS)-induced mice were used as the model for UC in the study. Mogrol, at a dose of 5 mg/kg body weight (BW), weakly prevented weight loss and colon shortening, and reduced the disease activity index (DAI) caused by DSS in the mice (Table 2). Inhibitive effects of mogrol on the inflammatory infiltration in colonic tissues were also observed in the histopathological examination. Additionally, the mogrol significantly decreased the level of mRNA of IL-17 (anti-inflammatory factor) and increased the level of IL-10 (pro-inflammatory factor). The activation of NLRP3 inflammasome was markedly repressed in the mogrol treatment group, which is known to be activated in UC. Similarly, the levels of mRNA of NLRP3 and IL-1β were also reduced in the colons of mice of mogrol treated group. Furthermore, the mogrol also distinctly inhibited the degradation of IκBα. These results support that mogrol can suppress too great a release of inflammatory factors and also suppress the activation of inflammation-related pathways to support the prevention of colon damage in a similar way to how the Mesalazine drug protects against DSS-induced damage in UC [37]. The study supports the protective role of mogrol through its inflammatory effect, which may be effective for other organs (Figure 2).

Mogrol also protected the integrity of tight junction (TJ) in the intestinal epithelium, as mRNA and protein expression of tight junction proteins (occludin and ZO-1), which were down-regulated due to DSS application in the colon tissues of mice, significantly increased with mogrol treatment. Similarly, mogrol treatment also significantly raised the mRNA and protein expression of SIRT1as compared with DSS administered group. AMPK has been known to enhance intestinal barrier function and can restore the assembly of TJ. In previous studies, the mogrol is known as an AMPK activator, hence, AMPK phosphorylation was also studied [4,6]. The mogrol treatment significantly promoted AMPK activation, which was reduced due to DSS-induced colitis in mice [37].

Table 2. Pharmacological activities of mogrol studied in the in vivo experiments.

Activity (and Probable Mechanism)	Model	Dose	Method	Results	Ref.
Anticancer (activating AMPK signaling and p53 pathway)	Male thymus-deficient mice (BALB/C)	10 mg/Kg three times a week for 2 weeks	Mogrol injected intraperitoneally into the mice	The weight and volume of tumor decreased (by 69.18% and 66.22%, respectively	[36]
Anti-UC (through promoting AMPK activation and inhibition of inflammation)	Female C57BL/6 mice DSS-induced mouse UC model	Mogrol (1 or 5 mg/kg) for 7 days orally	Weight monitoring, DAI, and colon length calculation (metformin as positive control drug used in the study)	Weight loss ↓, DAI ↓, and colon shortening ↓	[37]
			Histopathological examination (metformin as positive control drug used in the study)	Inflammatory infiltration in colonic tissues ↓	
			RT-PCR	IL-10 ↑, IL-17 ↓, NLRP3 ↓, IL-1β ↓, occludin ↑, SIRT1 ↑, and ZO-1 ↑	
			WB	IκBα ↑, occludin ↑, ZO-1 ↑, and p-AMPK/AMPK ↑	
Anti-PF (through activation of AMPK-mediated signaling pathways and inhibition of NF-κB signaling pathways)	Male C57BL/6 mice	1, 5, and 10 mg/kg	The lung was used for measuring pulmonary index	Positive drug nintedanib Pulmonary index ↓ and weight loss ↓	[26]
		1, 5, and 10 mg/kg	Histology of lungs (Nintedanib used as positive drug group)	Alveolar wall thickening ↓, neutrophil infiltration ↓, edema ↓, high collagen production ↓, area of collagen fibrils ↓ and inflammatory degree ↓	
		1, 5, and 10 mg/kg	Expression analysis of lung tissues through RT-PCR and WB	α-SMA ↓, Col I ↓, and TGF-β1 ↓	
		10 μM of mogrol	Expression analysis in lung tissues through WB	LOXL2 ↓, NOX4 ↓, deacetylase Sirt1 ↓, Smad2/3 phosphorylation ↓, and p-AMPK ↑ in lung tissues	
Antiosteoporosis (inhibition of TRAF6/MAPK/NF-κB signaling pathway)	Female C5BL/6J mice ovaries were removed	10 mg/kg mogrol intraperitoneally every second day for 42 days	Micro-CT 3D reconstructions indicated that mogrol decreased bone loss of femurs in OVX mice	BV/TV ↑, Tb. Th ↑, Tb. N ↑ and Cs. Th ↑ of the mogrol group were considerably higher than the vehicle group; BS/BV ↓ was inversely lower	[25]
			Histological assessment	BV/TV ↑ of the mogrol group was dramatically higher than that of the vehicle group	
			Immunohistochemistry TRAP staining of femurs	Oc. S/BS ↓ and N. Oc/B ↓ of the mogrol group were substantially lower than those of the vehicle group	

Table 2. *Cont.*

Activity (and Probable Mechanism)	Model	Dose	Method	Results	Ref.
Neuroprotective (suppression of NF-κB mediated inflammation)	Male ICR mice AB-induced memory impairment	20, 40, 80 mg/kg intragastric administration	Morris water maze test and Y-maze test (2 mg/kg donepezil as positive control drug used in the study)	At all doses, significantly improved memory impairment	[41]
			Immunohistochemical analyses (donepezil as positive control drug used in the study)	Iba1-positive cells↓	
			Hoechst assay	Number of Hoechst-positive cells of the DG ↓	
			WB	NF-κB p65 ↓ IL-1β ↓, IL-6 ↓, TNF-α ↓, cleaved-caspase3/pro-caspase-3 ↓ and the Bcl-2/Bax ↑	
	Male ICR mice LPS-induced memory impairment	20, 40, 80 mg/kg intragastric administration	Morris water maze test, Y-maze test, and novel object recognition test	In all the tests, mogrol improved memory impairment	[42]
			Immunohistochemical analyses	Iba1-positive cells ↓	
			WB of mouse in the mouse hippocampus and frontal cortex	NF-κB p65 ↓ IL-1β ↓, IL-6 ↓, and TNF-α ↓	

AB: amyloid β-peptide; AICAR: 5-aminoimidazole-4-carboxamide-1-beta-D-ribofuranoside; AMPK: adenosine monophosphate-activated protein kinase; BS/BV: bone surface/bone volume; BV/TV: bone volume/tissue volume; CCK8: Cell Counting Kit-8 assay; Cs. Th: cross-sectional thickness; DAI: disease activity index; DG: dentate gyrus; DSS: dextran sulphate sodium; HTRF: homogeneous time-resolved fluorescence assay; LOXL2: lysyl oxidase-like protein 2; LPS; lipopolysaccharide; MTT: 3-(4,5-dimethylthiazole-2-yl)-2,5-diphenyl tetrazolium bromide; N. Oc/B: number of osteoclasts/bone perimeter; Oc. S/BS: osteoclast surface/bone surface area; PF: pulmonary fibrosis; PLFs: primary lung fibroblasts; PMA: phorbol myristate acetate; SEM: scanning electron microscopy; Sirt1: sirtuin 1; SPA: scintillation proximity assay; Tb. N: trabecular number; Tb. Th: trabecular thickness; UC: ulcerative colitis; WB: Western blotting analysis; ↑: significantly up-regulated; ↓: significantly down-regulated.

4.4.2. In Vitro Anti-Colitis Activity of Mogrol

Furthermore, to study the mechanism behind the mogrol intestinal epithelial cell homeostasis with mogrol in colitis, TNF-α-stimulated NCM460 cells (human intestinal epithelial cells) were utilized.

Mogrol treatment at (10 µM) significantly restored the protein expression of occluding and ZO-1, which was significantly reduced by the TNF-α, and there was no effect on the proliferation of normal cells. Additionally, the mogrol treatment reorganized the flat-shaped microtubule network of tubulin stimulated by TNF-α, which may stimulate the assembly of the tight junction (Table 1). Similarly, Mogrol treatment elevated the bcl-2/bax ratio, which was significantly reduced with the application of TNF-α; these results support the role of mogrol in the homeostasis of intestinal epithelial in inflammatory conditions (Table 1). Further, mogrol is the AMPK activator found to increase the phosphorylation of AMPK in the TNF-α-stimulated NCM460 cells, and this may be the main reason for the protective activity of mogrol, as coadministration of mogrol with compound C (an AMPK inhibitor) results in a reversal of the protective activity, such as a reduction in both AMPK phosphorylation and the expression of TJ proteins (ZO-1 and occludin) [37].

Similarly, macrophages (THPM) derived from the THP-1 cells, stimulated with LPS and ATP, were utilized for mogrol treatment to study its role in inflammatory cytokines. Similarly, like NCM460 cells, without affecting normal cell proliferation, the mogrol treatment at 10 µM significantly reduces the caspase-1 activation and the secretions of IL-1β. The effect of mogrol in the study appeared to be similar to that of metformin (a positive drug) [37].

4.5. Activity of Mogrol against the Pulmonary Fibrosis

The activity of mogrol was also studied in different experiments (in vitro and in vivo) against pulmonary fibrosis (PF), a chronic interstitial lung disease that does not have an effective treatment [43]. Furthermore, an increase in air pollution and occupational exposure to metal or wood dust may worsen the risk of PF [44].

4.5.1. In Vitro Anti-PF Activity of Mogrol

The epithelial–mesenchymal transition (EMT) is one of the key mechanisms concomitants with PF associated with the down-regulation of E-cadherin and up-regulation of α-SMA, type Col I, and Vimentin. In the study, Mogrol was found to reverse the expression of E-cadherin, α-SMA, type Col I, and Vimentin in TGF-β1-treated mouse type II alveolar epithelial cells MLE-12 at both 5 and 10 µM concentrations. Similarly, in primary lung fibroblasts (PLFs) cells, the up-regulated protein levels of α-SMA, Lysyl oxidase-like protein 2 (LOXL2), collagen I, and phosphorylation forms of Smad2/3 due to TGF-β1 were restored with the mogrol treatment at 10 µM concentration (Table 1). AMPK activation and NOX4 activity can stimulate lung myofibroblast differentiation, and mogrol also significantly restored the unusually decreased p-AMPK and amplified NOX4 protein expression in TGF-β1-treated PLFs [26].

4.5.2. In Vivo Anti-PF Activity of Mogrol

Furthermore, to study the antifibrotic role of mogrol in animals, a bleomycin-induced lung fibrosis model (Male C57BL/6 mice) was used. In the bleomycin-induced PF mice group, an increase in pulmonary index and a decrease in weight and survival rate were observed. Mogrol treatment significantly reduced weight loss and pulmonary index compared to the bleomycin group at both doses (5 or 10 mg/kg), similar to the antifibrotic effects of the nintedanib (a positive drug at a dose of 40 mg/kg) group [26,45]. The histology of the lungs in the staining study revealed bleomycin-induced alveolar wall thickening, neutrophil infiltration, edema, and higher collagen production in the lung tissues compared to normal. In the mogrol treatment group, these lung damages were distinctly inhibited by decreasing the area of collagen fibrils and inflammatory degree compared to the bleomycin group. Like in the in vitro study, in the mogrol treatment group, the protein and mRNA expression of α-SMA and Col I, which was increased in the bleomycin group, was decreased in lung tissue. Similarly, the expression of LOXL2 in lung tissues was significantly decreased in the mogrol (high dose) and nintedanib groups. Further, to investigate the mechanism behind the antifibrotic mechanism of mogrol through the TGF-β1/Smad2/3 signaling pathway, the expression of mRNA and proteins of TGF-β1 and Smad2/3 phosphorylation was reduced in the mogrol treatment group; their expression was enhanced in the bleomycin-administered group. Importantly, like in vitro experiments, compared to the bleomycin group, the mogrol treatment group also had a significant increase in the activation of AMPK in lung tissues. Mogrol also significantly restored abnormal activation of bleomycin-induced deacetylase sirtuin 1 (Sirt1) compared with the bleomycin group in lung tissue. Sirt1 is important in TGF-β1 signaling and myofibroblast transdifferentiation; hence, both TGF-β1/Smad and the AMPK/Sirt1 signaling pathways may be important for mogrol activity. Moreover, nuclear receptor subfamily 4 group A 1 (NR4A1), which was increased in the bleomycin group, is known to be important in both fibrogenesis and TGF-β1 signaling and was significantly suppressed in the mogrol treatment group. Similarly, mogrol also suppressed bleomycin-induced NOX4 mRNA and protein expression in the lung tissue. Interestingly, coadministration of compound c (AMPK inhibitor) with mogrol reversed the protective effect of mogrol, i.e., the bleomycin-induced lung inflammation and fibrosis process in mice. AMPK activation may be the important mechanism for mogrol activity; hence, molecular docking analysis was also conducted to study the interaction and binding energy of mogrol with AMPK, and mogrol had a slightly better docking score than AMP. Results from the study suggest the anti-fibrosis activity of mogrol results at least partly from the activation of AMPK [26,46].

4.6. Anti-Osteoporosis Activity of Mogrol

Osteoporosis is a common disease around the globe, and elders and especially elderly women, are in the high-risk group [47]. The antiosteoporosis activity of mogrol was also discovered in both in vitro and in vivo studies.

4.6.1. In Vitro Antiosteoporosis Activity of Mogrol

The bone marrow macrophages (BMM) were stimulated with a receptor activator of nuclear factor-kB ligand (RANKL) treated with different concentrations of mogrol. The tartrate-resistant acid phosphatase (TRAP) positive multinucleated osteoclasts were decreased with the treatment of mogrol in a dose-dependent manner (at 5, 10, and 20 μM). The early stages (1–3 days) of osteoclast formation were especially strongly inhibited by the mogrol. The mogrol also shows an inhibitory effect on osteoclast resorption function. The bone resorption in the bovine bone slices implanted with BMM after the formation of osteoclast cells was studied. Scanning electron microscopy examination has shown that the mogrol significantly reduced the resorption area in a dose-dependent manner.

RNA-seq analysis can be used to decipher the gene expression mechanism associated with the treatment [48,49]. To study the gene expression changes behind the mogrol activity on RANKL-induced BMMs, RNA-seq analysis was used, which showed that the osteoclastogenesis suppressive genes (lilrb4a and Fcgr3) were up-regulated and osteoclastogenesis genes (MMP9, OSCAR, and ACP5) were down-regulated with the mogrol treatment. Further, the reduced expression of marker genes of osteoclastogenesis (such as MMP9, CTSK, ATP6v0d2, ACP5, and DCSTAMP) in the mogrol treatment was confirmed through RT-PCR analysis (Table 1).

Mogrol was found to suppress the MAPK/NF-κB signaling pathway by suppressing the phosphorylation of JNK, ERK, P65, and p38. Mogrol also suppressed the degradation of IκBα and nuclear translocation of P65, which is important for the NF-κB activity supporting the antiosteoporosis activity [50].

The expression of downstream transcription factors c-FOS and NFATc1, which was significantly increased in BMM activated with RANKL, was found to be reduced through mogrol treatment [25].

4.6.2. In Vivo Antiosteoporosis Activity of Mogrol

Encouraging osteoporosis activity of mogrol in in vitro experiments paves the path for in vivo studies. In vivo osteoporosis activity of mogrol was conducted through a mouse model after ovariectomy (OVR). After 42 days of mogrol treatment (at the dose of 10 mg/kg), Micro-CT 3D reconstructions showed decreased bone loss of the femurs in OVR mice. Quantitative results showed that the bone volume/tissue volume (BV/TV), trabecular thickness (Tb. Th), trabecular number (Tb. N), and cross-sectional thickness (Cs. th) of the mogrol treatment group were distinctly greater than the vehicle group. Further, in histological assessment, the BV/TV was again found to be higher in the mogrol group (Table 2). The immunohistochemistry TRAP staining of femurs showed a reduction in osteoclast surface/bone surface area (Oc. S/BS) and the number of osteoclasts/bone perimeter (N. Oc/B) of the mogrol treatment group compared to the vehicle group [25]. Both in vitro and in vivo studies suggested the role of mogrol in the suppression of osteoporosis, and it may be developed as a therapeutic against osteoclast-mediated osteolytic disorders such as osteoporosis (Table 2) [25].

4.7. In Vivo Neuroprotective Activity of Mogrol

The anti-inflammatory activity of mogrol was the motivation behind the basis to analyze its neuroprotective activity against neurotoxic conditions due to inflammatory agents such as LPS and amyloid β-peptide (AB) [51,52].

First, the neuroprotective activity of mogrol was studied in mouse models by analyzing memory impairment caused by AB. In both the Morris water maze test and Y-maze test, mogrol has significantly reduced AB-induced memory impairment in ICR mice at all doses used in the study (20, 40, 80 mg/kg). Mogrol also reduced the over-activation of microglia, as Iba1-positive cells in the hippocampus, which were significantly increased due to AB application, were reduced with mogrol treatment. Similarly, mogrol also inhibited the Hoechst-positive cells of the dentate gyrus (DG). The mechanism behind the neuroprotective activity of mogrol may be the reduction in NF-κB mediated neuroinflammation, as AB can induce the expression of NF-κB and associated inflammatory factors. In further experiments, mogrol treatment was found to suppress the NF-κB p65 and associated inflammatory factors such as IL-1β, IL-6, and TNF-α (Table 2) [41]. Similarly, the mogrol inhibited the ratio of cleaved-caspase-3 and pro-caspase-3 found to be increased due to AB, and increased the ratio of Bcl-2 and Bax found to be decreased due to AB toxicity.

Later, in a similar study, the neuroprotective effect of mogrol against lipopolysaccharide (LPS)-induced memory impairment and neuroinflammatory responses was studied in mice. In memory impairment tests such as the Morris water maze, Y-maze, and novel object recognition tests, mogrol treatment significantly reduced the memory impairment in ICR mice caused due to LPS at all doses used in the study (20, 40, 80 mg/kg). Like in the previous study, Mogrol reduced microglial overactivation as it reduced the Iba1-positive cells in the hippocampus that increased significantly increased due to LPS. Similarly, mogrol was found to suppress the NF-κB p65 and associated inflammatory factors such as IL-1β, IL-6, and TNF-α in both the frontal cortex and hippocampus region of the brain (Figure 3) [42]. In both studies, it can also be concluded that the protective effect of mogrol in the brain may be at least partly associated with the anti-inflammatory activity of mogrol (Figure 2).

- **Anti osteoporosis**
  - lilrb4a↑
  - Fcgr3↑
  - MMP9↓
  - OSCAR↓
  - ACP5↓
  - CTSK↓
  - ATP6v0d2↓
  - DCSTAMP↓
  - phosphorylation of p-JNK↓,
  - p-ERK↓
  - p-P65↓
  - p-P38↓
  - c-FOS↓
  - NFATc1↓
  - TRAF6↓
  - Siglec-15↓
  - IκBα↓
  - blocked P65 nuclear translocation

- **Anti-obesity/Anti diabetes**
  - p-AMPK/AMPK↑
  - p-CREB/ CREB↓
  - C/EBPβ↓
  - CRE-mediated transcription↓

- **Anti-ulcerative colitis**
  - Bcl-2/bax↑
  - TNF-α↓
  - Cleaved-caspase1/pro-caspase1↓
  - IL-10↑
  - IL-17↓
  - NLRP3↓
  - IL-1β↓
  - occludin↑
  - SIRT1↑
  - ZO-1↑
  - IκBα↑
  - p-AMPK/AMPK↑

- **Neuroprotective**
  - NF-κB p65↓
  - IL-1β↓
  - IL-6↓
  - TNF-α↓
  - cleaved-caspase3/pro-caspase-3↓
  - Bcl-2/Bax↑

- **Anti-inflammatory**
  - NO production↓(17%)
  - TNF-α↓
  - IL-6↓

- **Anticancer**
  - P21↑
  - p-ERK/ERK↓
  - p-STAT3/STAT3↓
  - Cleaved caspase3↑
  - LC3-II↑
  - p-AMPK/AMPK↑
  - p-P53/P53↑
  - PUMA↑
  - P27↑
  - Bcl-2↓

- **Anti-pulmonary fibrosis**
  - E-cadherin↑
  - Vimentin↓
  - p-Smad2/3↓
  - α-SMA↓
  - Col I↓
  - TGF-β1↓
  - LOXL2↓
  - NOX4↓
  - deacetylase Sirt1↓
  - Smad2/3 phosphorylation↓
  - p-AMPK↑

---

p-AMPK/AMPK↑
- Anti osteoporosis
- Anti-ulcerative colitis
- Anti-obesity/Anti diabetes
- Anticancer
- Neuroprotective
- Anti-pulmonary fibrosis

TNF-α↓
- Anti-ulcerative colitis
- Neuroprotective
- Anti-inflammatory

p-ERK/ERK↓
- Anti osteoporosis
- Anticancer

Bcl-2/bax↑, IL-6↓, and IL-1β↓
- Anti-ulcerative colitis
- Neuroprotective

IκBα↑
- Anti osteoporosis
- Anti-ulcerative colitis

Figure 3. Different targets/markers were found to be associated with the pharmacological activities of mogrol. (Studied genes common in more than one activity are listed with white background and upward (↑) and downward (↓) arrows represent significantly up-regulated and down-regulated genes respectively)

5. Discussion and Future Directions

The positive result of recent studies evaluating the pharmacological activity of mogrol accelerates the efforts for the development of mogrol as a therapeutic agent. The protective effect of mogrol on different organs also supported its therapeutic potential against different diseases (Figure 2). Pharmacokinetic studies also support the potential of mogrol as a drug candidate compared to mogrosides. Still, the development of mogrol as a therapeutic or supplement required considerable effort in the precise direction. The antiobesity and antidiabetic activity of mogrol, which are mainly based on the activation of AMPK, were studied in various studies. However, AMPK is the crucial target of metabolism/energy-related diseases. However, there is a need for in vivo experiments that could establish antiobesity and antidiabetic activities before clinical evaluation. Similarly, the anticancer effect of mogrol was observed in different cancer cell lines, suggesting its anticancer effect against important cancers, including lung cancer, but limited animal studies have been conducted for its support. Hence, more animal studies are required and suggested for the anticancer activity of mogrol before clinical studies [53]. The anti-inflammatory activity of mogrol is also one of the most promising activities, which might be the important factor behind its other pharmacological properties, such as neuroprotective, anticancer, and anti-UC activities [37,41,42,54,55]. Along with important inflammatory mediators such as (IL-6 and TNF-α), it was found that it can suppress NF-κB, which is an important central component in the pathways involved in various important diseases related to chronic inflammation such as different cancer, rheumatoid arthritis, inflammatory bowel diseases, neuro-inflammatory-related diseases, etc. [41,42,56,57]. It would be interesting to see the effect of mogrol on these diseases. Therefore, it is suggested to study the role of mogrol against chronic inflammation-related diseases.

The neuroprotective activity of mogrol observed in both studies was protective against memory impairment in a mouse model by suppressing inflammatory pathways, especially NF-κB signaling. Neuroprotection studies may be further required to establish with more experiments, mainly through different animal models, before clinical evaluation. Similarly, mogrol may also be studied against other neurodegenerative diseases and disorders in which neuro-inflammation is an important component of disease etiology, such as Parkinson's disease [58].

However, the anti-PF activity of mogrol was studied only through two cell-based experiments and one animal experiment. It could be a valuable activity to consider in future studies as PF is a deadly disease with no cure [43]. In other studies, the lung protective effect of mogrol was also observed [4]. However, the development of mogrol as a therapeutic against PF is in the preliminary stage, and more animal models are proposed to be used in the near future to develop mogrol as a therapeutic against it. Importantly, the role of AMPK activation was observed as one of the important molecular mechanisms for PF activity. Hence, the activity of mogrol derivatives that were found to be better AMPK activators may also have been suggested for their anti-PF activity.

A similar suggestion can be followed for anti-UC studies in the near future, such as the PF study on AMPK inhibitors that can reverse the positive effect of mogrol on anti-UC activity. An important aspect is the molecular mechanism behind pharmacological activities, which was also significantly explored for various activities of mogrol (Figure 3).

The molecular mechanism gives an insight into the possible therapeutic targets that may be used to further develop and optimize the effect of therapeutic [59]. Finally, it can be summarized that the activation of AMPK is an important mechanism for most of the pharmacological activities of mogrol, including antiobesity, antidiabetes, anticancer, anti-ulcerative colitis, anti-PF, and antiosteoporosis. Further, anti-inflammation through suppression of the NF-κB pathway and/or inflammatory cytokines is an important mechanism that also supports different pharmacological activities of mogrol, including neuroprotective, anti-UC, anti-PF, and anticancer activities. Inhibition of ERK and STAT3 and activation of the p53 pathway were also found to be important contributors to the anticancer activity of mogrol. Similarly, the CREB activation may also contribute to the antiobesity and diabetes activity of mogrol. Pathways and/or genes found to be targeted in different experiments and associated with more pharmacological activities of mogrol can be considered as important reliable targets to be developed mogrol and its derivatives as therapeutic.

Derivatives of mogrol are synthesized by modifying the chemical group in the selected position in the mogrol. Some of these derivatives have been found to have increased pharmacological activities compared to mogrol, such as anticancer and AMPK activator [7,23]. The binding of mogrol through in silico molecular docking also provided the initial understanding of the interaction of mogrol with its important target AMPK [26]. It is suggested to use similar docking studies to rationally design and synthesize the mogrol derivative for increased activity with AMPK and, subsequently, better biological activities. Overall, mogrol is a potential phytochemical that can be taken for further evaluation in the in vivo pharmacological experiments consulting the currently proposed suggestion, and later clinical studies may be conducted after safety evaluation.

Author Contributions: H.-J.L. and V.J. designed this study. V.J. wrote the manuscript; H.-J.L. reviewed the manuscript intensively; H.-J.L. and V.J. edited the manuscript. All authors have read and agreed to the published version of the manuscript.

Funding: This research was funded by The Cooperative Research Program for Agriculture Science and Technology Development (project No. PJ01701902), Rural Development Administration, Republic of Korea.

Data Availability Statement: Not applicable.

Conflicts of Interest: The authors declare no conflict of interest.

References

1. Atanasov, A.G.; Zotchev, S.B.; Dirsch, V.M.; Supuran, C.T. Natural products in drug discovery: Advances and opportunities. *Nat. Rev. Drug Discov.* **2021**, *20*, 200–216. [CrossRef] [PubMed]
2. Lee, D.; Choi, S.; Yamabe, N.; Kim, K.H.; Kang, K.S. Recent findings on the mechanism of cisplatin-induced renal cytotoxicity and therapeutic potential of natural compounds. *Nat. Prod. Sci.* **2020**, *26*, 28–49.
3. Dzubak, P.; Hajduch, M.; Vydra, D.; Hustova, A.; Kvasnica, M.; Biedermann, D.; Markova, L.; Urban, M.; Sarek, J. Pharmacological activities of natural triterpenoids and their therapeutic implications. *Nat. Prod. Rep.* **2006**, *23*, 394–411. [CrossRef] [PubMed]
4. Harada, N.; Ishihara, M.; Horiuchi, H.; Ito, Y.; Tabata, H.; Suzuki, Y.A.; Nakano, Y.; Yamaji, R.; Inui, H. Mogrol derived from Siraitia grosvenorii mogrosides suppresses 3T3-L1 adipocyte differentiation by reducing cAMP-response element-binding protein phosphorylation and increasing AMP-activated protein kinase phosphorylation. *PLoS ONE* **2016**, *11*, e0162252. [CrossRef]
5. Liu, B.; He, R.; Xiong, R.; Fu, T.; Li, N.; Geng, Q. Effect and mechanism of mogrol on lipopolysaccharide-induced acute lung injury. *Chin. J. Emerg. Med.* **2022**, *31*, 777–782. [CrossRef]
6. Chen, X.-b.; Zhuang, J.-j.; Liu, J.-h.; Lei, M.; Ma, L.; Chen, J.; Shen, X.; Hu, L.-h. Potential AMPK activators of cucurbitane triterpenoids from Siraitia grosvenorii Swingle. *Bioorganic Med. Chem.* **2011**, *19*, 5776–5781. [CrossRef]
7. Song, J.-R.; Li, N.; Wei, Y.-L.; Lu, F.-L.; Li, D.-P. Design and synthesis of mogrol derivatives modified on a ring with anti-inflammatory and anti-proliferative activities. *Bioorg. Med. Chem. Lett.* **2022**, *74*, 128924. [CrossRef]
8. Li, H.; Li, R.; Jiang, W.; Zhou, L. Research progress of pharmacological effects of Siraitia grosvenorii extract. *J. Pharm. Pharmacol.* **2021**, *74*, 953–960. [CrossRef]
9. Wu, J.; Jian, Y.; Wang, H.; Huang, H.; Gong, L.; Liu, G.; Yang, Y.; Wang, W. A Review of the Phytochemistry and Pharmacology of the Fruit of Siraitia grosvenorii (Swingle): A Traditional Chinese Medicinal Food. *Molecules* **2022**, *27*, 6618. [CrossRef]
10. Bhusari, S.; Rodriguez, C.; Tarka Jr, S.M.; Kwok, D.; Pugh, G.; Gujral, J.; Tonucci, D. Comparative In vitro metabolism of purified mogrosides derived from monk fruit extracts. *Regul. Toxicol. Pharmacol.* **2021**, *120*, 104856. [CrossRef]
11. Murata, Y.; Ogawa, T.; Suzuki, Y.A.; Yoshikawa, S.; Inui, H.; Sugiura, M.; Nakano, Y. Digestion and absorption of Siraitia grosvenori triterpenoids in the rat. *Biosci. Biotechnol. Biochem.* **2010**, *74*, 673–676. [CrossRef]
12. Xiao, R.; Liao, W.; Luo, G.; Qin, Z.; Han, S.; Lin, Y. Modulation of Gut Microbiota Composition and Short Chain Fatty Acid Synthesis by Mogroside V in an In Vitro Incubation System. *ACS Omega* **2021**, *6*, 25486–25496. [CrossRef] [PubMed]
13. Wang, S.; Xu, X.; Lv, X.; Liu, Y.; Li, J.; Du, G.; Liu, L. Construction and Optimization of the de novo Biosynthesis Pathway of Mogrol in Saccharomyces cerevisiae. *Front. Bioeng. Biotechnol.* **2022**, *10*, 919526. [CrossRef]
14. Liao, J.; Liu, T.; Xie, L.; Mo, C.; Huang, X.; Cui, S.; Jia, X.; Lan, F.; Luo, Z.; Ma, X. Plant Metabolic Engineering by Multigene Stacking: Synthesis of Diverse Mogrosides. *Int. J. Mol. Sci.* **2022**, *23*, 10422. [CrossRef]
15. Preusche, M.; Ulbrich, A.; Schulz, M. Culturing Important Plants for Sweet Secondary Products under Consideration of Environmentally Friendly Aspects. *Processes* **2022**, *10*, 703. [CrossRef]

16. Li, J.; Mu, S.; Yang, J.; Liu, C.; Zhang, Y.; Chen, P.; Zeng, Y.; Zhu, Y.; Sun, Y. Glycosyltransferase engineering and multi-glycosylation routes development facilitating synthesis of high-intensity sweetener mogrosides. *Iscience* **2022**, *25*, 105222. [CrossRef]

17. Luo, Z.; Qiu, F.; Zhang, K.; Qin, X.; Guo, Y.; Shi, H.; Zhang, L.; Zhang, Z.; Ma, X. In vitro AMPK activating effect and in vivo pharmacokinetics of mogroside V, a cucurbitane-type triterpenoid from Siraitia grosvenorii fruits. *RSC Adv.* **2016**, *6*, 7034–7041. [CrossRef]

18. Luo, Z.; Zhang, K.; Shi, H.; Guo, Y.; Ma, X.; Qiu, F. Development and Validation of a Sensitive LC–MS-MS Method for Quantification of Mogrol in Rat Plasma and Application to Pharmacokinetic Study. *J. Chromatogr. Sci.* **2017**, *55*, 284–290. [CrossRef]

19. LIU, S.-g.; DAI, X.-j.; ZHONG, D.-f.; ZHANG, C.-f.; CHEN, X.-y. Identification of metabolites and pharmacokinetics of mogrol in rat plasma. *Acta Pharm. Sin.* **2017**, *12*, 1452–1457.

20. Subedi, L.; Lee, S.E.; Madiha, S.; Gaire, B.P.; Jin, M.; Yumnam, S.; Kim, S.Y. Phytochemicals against TNFα-Mediated Neuroinflammatory Diseases. *Int. J. Mol. Sci.* **2020**, *21*, 764. [CrossRef] [PubMed]

21. Yu, H.; Lin, L.; Zhang, Z.; Zhang, H.; Hu, H. Targeting NF-κB pathway for the therapy of diseases: Mechanism and clinical study. *Signal Transduct. Target. Ther.* **2020**, *5*, 209. [CrossRef] [PubMed]

22. Kim, J.; Kim, M.-G.; Jeong, S.H.; Kim, H.J.; Son, S.W. STAT3 maintains skin barrier integrity by modulating SPINK5 and KLK5 expression in keratinocytes. *Exp. Dermatol.* **2022**, *31*, 223–232. [CrossRef] [PubMed]

23. Wang, J.; Liu, J.; Xie, Z.; Li, J.; Li, J.; Hu, L. Design, synthesis and biological evaluation of mogrol derivatives as a novel class of AMPKα2β1γ1 activators. *Bio. Med. Chem. Lett.* **2020**, *30*, 126790. [CrossRef]

24. Additives, E.P.o.F.; Flavourings; Younes, M.; Aquilina, G.; Engel, K.H.; Fowler, P.; Frutos Fernandez, M.J.; Fürst, P.; Gürtler, R.; Gundert-Remy, U.; et al. Safety of use of Monk fruit extract as a food additive in different food categories. *EFSA J.* **2019**, *17*, e05921.

25. Chen, Y.; Zhang, L.; Li, Z.; Wu, Z.; Lin, X.; Li, N.; Shen, R.; Wei, G.; Yu, N.; Gong, F. Mogrol Attenuates Osteoclast Formation and Bone Resorption by Inhibiting the TRAF6/MAPK/NF-κB Signaling Pathway In vitro and Protects Against Osteoporosis in Postmenopausal Mice. *Front. Pharmacol.* **2022**, *13*, 803880. [CrossRef] [PubMed]

26. Liu, B.; Yang, J.; Hao, J.; Xie, H.; Shimizu, K.; Li, R.; Zhang, C. Natural product mogrol attenuates bleomycin-induced pulmonary fibrosis development through promoting AMPK activation. *J. Funct. Foods* **2021**, *77*, 104280. [CrossRef]

27. Qin, X.; Xiaojian, S.; Ronggan, L.; Yuxian, W.; Zhunian, T.; Shouji, G.; Heimbach, J. Subchronic 90-day oral (Gavage) toxicity study of a Luo Han Guo mogroside extract in dogs. *Food Chem. Toxicol.* **2006**, *44*, 2106–2109. [CrossRef]

28. Liu, C.; Dai, L.; Liu, Y.; Dou, D.; Sun, Y.; Ma, L. Pharmacological activities of mogrosides. *Future Med. Chem.* **2018**, *10*, 845–850. [CrossRef]

29. Suzuki, Y.A.; Tomoda, M.; Murata, Y.; Inui, H.; Sugiura, M.; Nakano, Y. Antidiabetic effect of long-term supplementation with Siraitia grosvenori on the spontaneously diabetic Goto–Kakizaki rat. *Br. J. Nutr.* **2007**, *97*, 770–775. [CrossRef]

30. Xiangyang, Q.; Weijun, C.; Liegang, L.; Ping, Y.; Bijun, X. Effect of a Siraitia grosvenori extract containing mogrosides on the cellular immune system of type 1 diabetes mellitus mice. *Mol. Nutr. Food Res.* **2006**, *50*, 732–738. [CrossRef]

31. Desjardins, E.M.; Steinberg, G.R. Emerging role of AMPK in brown and beige adipose tissue (BAT): Implications for obesity, insulin resistance, and type 2 diabetes. *Curr. Diabetes Rep.* **2018**, *18*, 1–9. [CrossRef] [PubMed]

32. Liu, C.; Zeng, Y.; Dai, L.-H.; Cai, T.-Y.; Zhu, Y.-M.; Dou, D.-Q.; Ma, L.-Q.; Sun, Y.-X. Mogrol represents a novel leukemia therapeutic, via ERK and STAT3 inhibition. *Am. J. Cancer Res.* **2015**, *5*, 1308. [PubMed]

33. Wang, L.; Li, L.; Fu, Y.; Li, B.; Chen, B.; Xu, F.; Li, D. Separation, synthesis, and cytotoxicity of a series of mogrol derivatives. *J. Asian Nat. Prod. Res.* **2019**, *22*, 663–677. [CrossRef]

34. Song, J.-R.; Li, N.; Li, D.-P. Synthesis and anti-proliferation activity of mogrol derivatives bearing quinoline and triazole moieties. *Bioorg. Med. Chem. Lett.* **2021**, *42*, 128090. [CrossRef] [PubMed]

35. Li, N.; Song, J.; Li, D. Synthesis and Antiproliferative Activity of Ester Derivatives of Mogrol through JAK2/STAT3 Pathway. *Chem. Biodivers.* **2022**, *19*, e202100742. [CrossRef]

36. Li, H.; Liu, L.; Chen, H.-y.; Yan, X.; Li, R.-l.; Lan, J.; Xue, K.-y.; Li, X.; Zhuo, C.-l.; Lin, L. Mogrol suppresses lung cancer cell growth by activating AMPK-dependent autophagic death and inducing p53-dependent cell cycle arrest and apoptosis. *Toxicol. Appl. Pharmacol.* **2022**, *444*, 116037. [CrossRef]

37. Cheng, R.; Wang, J.; Xie, H.; Li, R.; Shimizu, K.; Zhang, C. Mogrol, an aglycone of mogrosides, attenuates ulcerative colitis by promoting AMPK activation. *Phytomedicine* **2021**, *81*, 153427.

38. Jin, W. Role of JAK/STAT3 signaling in the regulation of metastasis, the transition of cancer stem cells, and chemoresistance of cancer by epithelial–mesenchymal transition. *Cells* **2020**, *9*, 217. [CrossRef]

39. Xiao, B.-D.; Zhao, Y.-J.; Jia, X.-Y.; Wu, J.; Wang, Y.-G.; Huang, F. Multifaceted p21 in carcinogenesis, stemness of tumor and tumor therapy. *World J. Stem Cells* **2020**, *12*, 481. [CrossRef]

40. Spindler, A.; Stefan, K.; Wiese, M. Synthesis and investigation of tetrahydro-β-carboline derivatives as inhibitors of the breast cancer resistance protein (ABCG2). *J. Med. Chem.* **2016**, *59*, 6121–6135. [CrossRef]

41. Chen, G.; Liu, C.; Meng, G.; Zhang, C.; Chen, F.; Tang, S.; Hong, H.; Zhang, C. Neuroprotective effect of mogrol against Aβ1–42-induced memory impairment neuroinflammation and apoptosis in mice. *J. Pharm. Pharmacol.* **2019**, *71*, 869–877. [CrossRef] [PubMed]

42. Wang, H.; Meng, G.-L.; Zhang, C.-T.; Wang, H.; Hu, M.; Long, Y.; Hong, H.; Tang, S.-S. Mogrol attenuates lipopolysaccharide (LPS)-induced memory impairment and neuroinflammatory responses in mice. *J. Asian Nat. Prod. Res.* **2020**, *22*, 864–878. [CrossRef] [PubMed]

43. Somogyi, V.; Chaudhuri, N.; Torrisi, S.E.; Kahn, N.; Müller, V.; Kreuter, M. The therapy of idiopathic pulmonary fibrosis: What is next? *Eur. Respir. Rev.* **2019**, *28*, 190021. [CrossRef] [PubMed]

44. Jegal, Y.; Park, J.S.; Kim, S.Y.; Yoo, H.; Jeong, S.H.; Song, J.W.; Lee, J.H.; Lee, H.L.; Choi, S.M.; Kim, Y.W.; et al. Clinical Features, Diagnosis, Management, and Outcomes of Idiopathic Pulmonary Fibrosis in Korea: Analysis of the Korea IPF Cohort (KICO) Registry. *Tuberc. Respir. Dis.* **2022**, *85*, 185–194. [CrossRef]

45. Richeldi, L.; Du Bois, R.M.; Raghu, G.; Azuma, A.; Brown, K.K.; Costabel, U.; Cottin, V.; Flaherty, K.R.; Hansell, D.M.; Inoue, Y. Efficacy and safety of nintedanib in idiopathic pulmonary fibrosis. *N. Engl. J. Med.* **2014**, *370*, 2071–2082. [CrossRef]

46. Ding, Y.; Wang, L.; Liu, B.; Ren, G.; Okubo, R.; Yu, J.; Zhang, C. Bryodulcosigenin attenuates bleomycin-induced pulmonary fibrosis via inhibiting AMPK-mediated mesenchymal epithelial transition and oxidative stress. *Phytother. Res.* **2022**, *36*, 3911–3923. [CrossRef]

47. Lee, S.-K.; Jun, D.-S.; Lee, D.-K.; Baik, J.-M. Clinical Characteristics of Elderly People with Osteoporotic Vertebral Compression Fracture Based on a 12-Year Single-Center Experience in Korea. *Geriatrics* **2022**, *7*, 123. [CrossRef]

48. Jaiswal, V.; Park, M.; Lee, H.-J. Comparative Transcriptome Analysis of the Expression of Antioxidant and Immunity Genes in the Spleen of a Cyanidin 3-O-Glucoside-Treated Alzheimer's Mouse Model. *Antioxidants* **2021**, *10*, 1435. [CrossRef]

49. Park, M.; Kim, K.H.; Jaiswal, V.; Choi, J.; Chun, J.L.; Seo, K.M.; Lee, M.-J.; Lee, H.-J. Effect of black ginseng and silkworm supplementation on obesity, the transcriptome, and the gut microbiome of diet-induced overweight dogs. *Sci. Rep.* **2021**, *11*, 1–14. [CrossRef]

50. Chun, K.-H.; Jin, H.C.; Kang, K.S.; Chang, T.-S.; Hwang, G.S. Poncirin inhibits osteoclast differentiation and bone loss through down-regulation of NFATc1 in vitro and in vivo. *Biomol. Ther.* **2020**, *28*, 337. [CrossRef]

51. Dang, T.K.; Hong, S.-M.; Dao, V.T.; Tran, P.T.T.; Tran, H.T.; Do, G.H.; Hai, T.N.; Nguyet Pham, H.T.; Kim, S.Y. Anti-neuroinflammatory effects of alkaloid-enriched extract from Huperzia serrata on lipopolysaccharide-stimulated BV-2 microglial cells. *Pharm. Biol.* **2023**, *61*, 135–143. [CrossRef] [PubMed]

52. Sanjay; Kim, J.Y. Anti-inflammatory effects of 9-cis-retinoic acid on β-amyloid treated human microglial cells. *Eur. J. Inflamm.* **2022**, *20*, 1721727×221143651. [CrossRef]

53. Mak, I.W.; Evaniew, N.; Ghert, M. Lost in translation: Animal models and clinical trials in cancer treatment. *Am. J. Transl. Res.* **2014**, *6*, 114. [PubMed]

54. Subedi, L.; Lee, J.H.; Yumnam, S.; Ji, E.; Kim, S.Y. Anti-Inflammatory Effect of Sulforaphane on LPS-Activated Microglia Potentially through JNK/AP-1/NF-κB Inhibition and Nrf2/HO-1 Activation. *Cells* **2019**, *8*, 194. [CrossRef] [PubMed]

55. Alshehri, A.; Ahmad, A.; Tiwari, R.K.; Ahmad, I.; Alkhathami, A.G.; Alshahrani, M.Y.; Asiri, M.A.; Almeleebia, T.M.; Saeed, M.; Yadav, D.K.; et al. In Vitro Evaluation of Antioxidant, Anticancer, and Anti-Inflammatory Activities of Ethanolic Leaf Extract of Adenium obesum. *Front. Pharmacol.* **2022**, *13*, 847534. [CrossRef]

56. Kim, S.Y.; An, S.; Park, D.K.; Kwon, K.A.; Kim, K.O.; Chung, J.-W.; Kim, J.H.; Kim, Y.J. Efficacy of iron supplementation in patients with inflammatory bowel disease treated with anti-tumor necrosis factor-alpha agents. *Ther. Adv. Gastroenterol.* **2020**, *13*, 1756284820961302. [CrossRef]

57. Ji, M.; Ryu, H.J.; Baek, H.-M.; Shin, D.M.; Hong, J.H. Dynamic synovial fibroblasts are modulated by NBCn1 as a potential target in rheumatoid arthritis. *Exp. Mol. Med.* **2022**, *54*, 503–517. [CrossRef]

58. Subedi, L.; Gaire, B.P.; Kim, S.-Y.; Parveen, A. Nitric Oxide as a Target for Phytochemicals in Anti-Neuroinflammatory Prevention Therapy. *Int. J. Mol. Sci.* **2021**, *22*, 4771. [CrossRef]

59. Zhang, W.; Huai, Y.; Miao, Z.; Qian, A.; Wang, Y. Systems pharmacology for investigation of the mechanisms of action of traditional Chinese medicine in drug discovery. *Front. Pharmacol.* **2019**, *10*, 743. [CrossRef]

Review

Juglans regia Linn.: A Natural Repository of Vital Phytochemical and Pharmacological Compounds

Aeyaz Ahmad Bhat [1], Adnan Shakeel [2], Sadaf Rafiq [3], Iqra Farooq [4], Azad Quyoom Malik [1], Mohammed E. Alghuthami [5], Sarah Alharthi [6,7], Husam Qanash [8,9,*] and Saif A. Alharthy [10,11,*]

[1] Department of Chemistry, Lovely Professional University, Phagwara 144411, India
[2] Department of Botany, Aligarh Muslim University, Aligarh 202002, India
[3] Division of Floriculture and Landscape Architecture, Sher-e-Kashmir University of Agricultural Sciences and Technology of Kashmir, Srinagar 190025, India
[4] CSIR—Indian Institute of Integrative Medicine, Jammu 180001, India
[5] GCC Accreditation Centre GAC, AL Safarat, Riyadh 12511, Saudi Arabia
[6] Center of Advanced Research in Science and Technology, Taif University, P.O. Box 11099, Taif 21944, Saudi Arabia
[7] Department of Chemistry, College of Science, Taif University, P.O. Box 11099, Taif 21944, Saudi Arabia
[8] Department of Medical Laboratory Science, College of Applied Medical Sciences, University of Ha'il, Hail 55476, Saudi Arabia
[9] Molecular Diagnostics and Personalized Therapeutics Unit, University of Ha'il, Hail 55476, Saudi Arabia
[10] Department of Medical Laboratory Sciences, King Abdulaziz University, P.O. Box 80216, Jeddah 21589, Saudi Arabia
[11] Toxicology and Forensic Sciences Unit, King Fahd Medical Research Center, King Abdulaziz University, P.O. Box 80216, Jeddah 21589, Saudi Arabia
* Correspondence: h.qanash@uoh.edu.sa (H.Q.); saalharthy@kau.edu.sa (S.A.A.); Tel.: +966-165351752 (H.Q.); +966-555556291 (S.A.A.)

Citation: Bhat, A.A.; Shakeel, A.; Rafiq, S.; Farooq, I.; Malik, A.Q.; Alghuthami, M.E.; Alharthi, S.; Qanash, H.; Alharthy, S.A. *Juglans regia* Linn.: A Natural Repository of Vital Phytochemical and Pharmacological Compounds. *Life* **2023**, *13*, 380. https://doi.org/10.3390/life13020380

Academic Editors: Efstathia Papada, Charalampia Amerikanou and Marisa Colone

Received: 10 December 2022
Revised: 26 January 2023
Accepted: 26 January 2023
Published: 30 January 2023

Abstract: *Juglans regia* Linn. is a valuable medicinal plant that possesses the therapeutic potential to treat a wide range of diseases in humans. It has been known to have significant nutritional and curative properties since ancient times, and almost all parts of this plant have been utilized to cure numerous fungal and bacterial disorders. The separation and identification of the active ingredients in *J. regia* as well as the testing of those active compounds for pharmacological properties are currently of great interest. Recently, the naphthoquinones extracted from walnut have been observed to inhibit the enzymes essential for viral protein synthesis in the SARS-CoV-2. Anticancer characteristics have been observed in the synthetic triazole analogue derivatives of juglone, and the unique modifications in the parent derivative of juglone have paved the way for further synthetic research in this area. Though there are some research articles available on the pharmacological importance of *J. regia*, a comprehensive review article to summarize these findings is still required. The current review, therefore, abridges the most recent scientific findings about antimicrobial, antioxidant, anti-fungal, and anticancer properties of various discovered and separated chemical compounds from different solvents and different parts of *J. regia*.

Keywords: antibacterial; antifungal; antioxidant; anticancer; juglone; *Juglans regia*

1. Introduction

Juglans regia Linn. belongs to the family Juglandaceae and is an aromatic transient tree that grows in abundance in the North-Western Himalayas of Kashmir that produces most of the world's walnuts, accounting for around 88% of total walnut production [1]. The bark of the tree is gray, and the tree has longitudinally fissured leaves that are alternating, imparipinnate, sessile, elliptic, or oblong-lanceolate. Flowers are unisexual and there are around two genera and fifteen species in the family. The leaves are evenly organized, measuring 25 40 cm in length, unevenly pinnate with five to nine leaflets and placed in a regular pattern. The medicinal plant *Juglans regia* Linn. has been used extensively in traditional medicine for a variety of illnesses, including helminthiasis, diarrhea, sinusitis, stomach aches, arthritis, asthma, eczema, scrofula, skin disorders, and various endocrine diseases such as diabetes mellitus, anorexia, thyroid dysfunctions, cancer, and infectious diseases [2,3]. The green fruit ripens in the autumn, when the entire fruit, including the husk, drops from the tree. The huge seed contains a rich flavor and a thin edible shell.

The phytochemistry of the tree has extensively been studied and many important phytochemical activities have been exploited [3]. However, the number of elements may vary from one species to another in a different place based on several parameters, such as geographical location, time, temperature, genetic makeup, and other factors. Several studies on the phytochemical examination of the tree's various components with numerous health benefits have been performed (Figure 1 and Table 1). According to these studies, the chemical components found in walnuts fluctuate depending on the climate. The oil is rich in polyunsaturated fatty acids, tocopherols and phytosterols. The walnut fruits are valuable, and tasty walnut leaves contain many chemicals, the most prominent of which are Aesculin, Taxifolin-pantocid, Quercetin-glucronide, Kaempferol-rhamnoside, Syringetin-O-Hexoside, Myricetin-3-O-glucoside, Myricetin-3-O-pantocid, and Epicatechin [3,4].

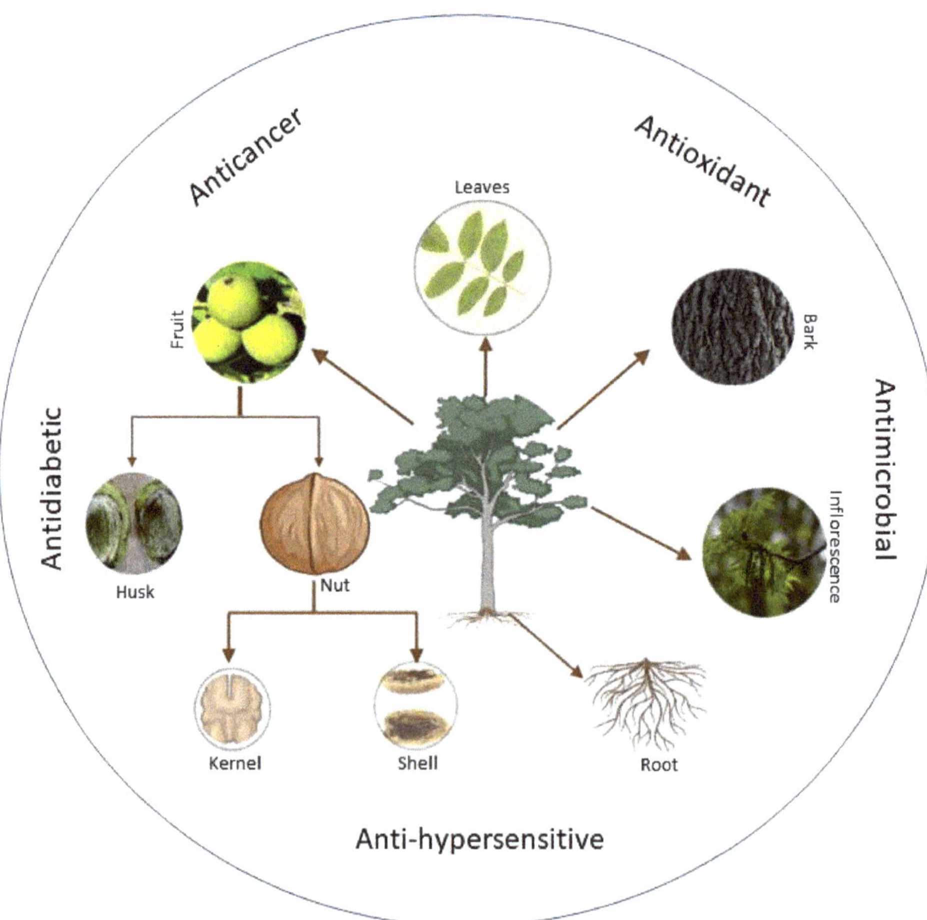

Figure 1. Different parts of *Juglans regia* with medicinal properties.

Table 1. List of compounds in *Juglans regia* and their pharmacological or biological activities.

Compound	Class	Structure	Biological Activity	References
Gallic Acid	Phenolic		Anti-inflammatory	[5,6]
Protocatechuic acid	Phenolic		Anti-inflammatory and antiapoptotic activities.	[7]
Ferulic acid	Phenolic		Anticancerous	[8]
Sinapate	Phenolic		Antioxidant, Antimicrobial	[9]
Protocatechuic acid derivative	Phenolic		Anti-inflammatory	[10]
Ellagic acid	Phenolic		Cytotoxic, Anti-proliferative	[11]
p-hydroxybenzoic acid	Phenolic		Antimicrobial	[12]
p-coumaric acid	Phenolic		Antioxidant	[13]

Table 1. *Cont.*

Compound	Class	Structure	Biological Activity	References
Quercetin 3-galactosid	Phenolic	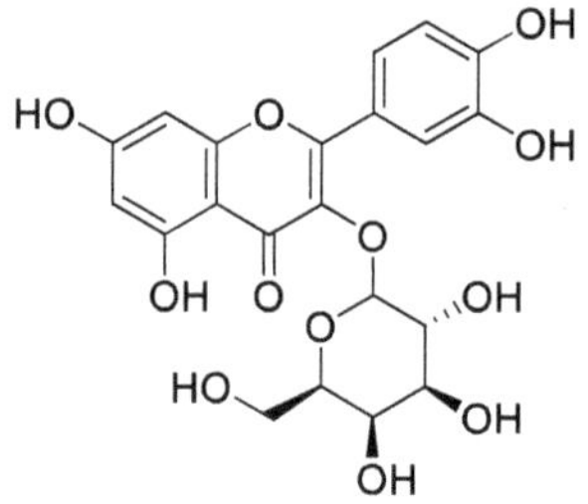	Antimicrobial, Antioxidant	[14]
Galactose	Phenolic	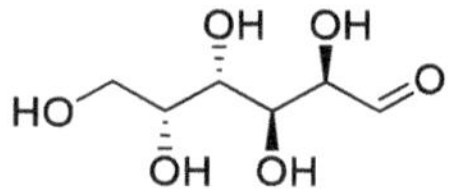	Anticancer, Antidiabetic	[15]
Pentose	Phenolic	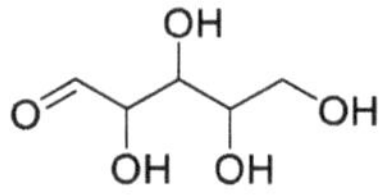	Amino acid synthesis	[16,17]
Arabinose	Phenolic	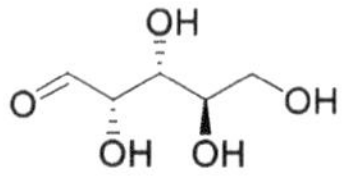	Antimicrobial	[18]
Xylose	Phenolic	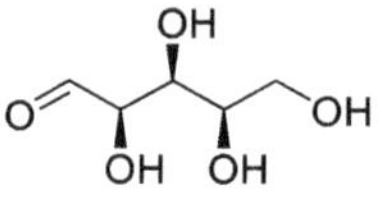	Antimicrobial	[19,20]
Rhamnose	Phenolic	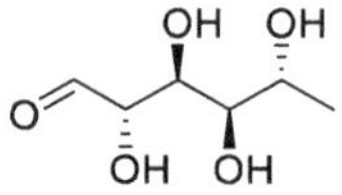	Antioxidant, Antimicrobial	[21]
Juglone	Phenolic	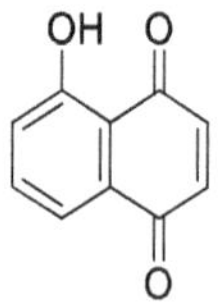	Antibacterial, Anticancer, Antioxidant	[22]
Caffeic acid	Phenolic	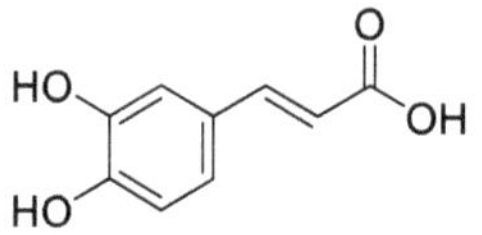	Antioxidant, Anti-inflammatory and Anticarcinogenic activity	[23]
Vannilic acid	Phenolic	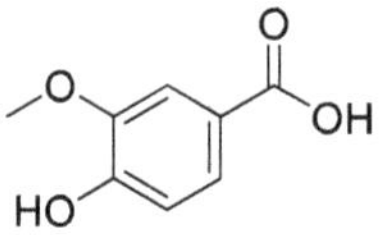	Antibacterial, Antioxidant	[24]

Table 1. *Cont.*

Compound	Class	Structure	Biological Activity	References
Quercetin	Flavonoid		Analgesic, Antibacterial, Antiviral	[25]
Myricetin	Flavonoid		Anticancer, Antidiabetic, Antibacterial, Antiviral	[26,27]
Kaempferol	Flavonoid		Acute and chronic inflammation	[28]
Apigenin	Flavone		Anti-inflammatory, Antioxidant, Neuroprotective	[29]
Luteolin	Flavone		Anticancer, Anti-inflammatory	[30]
Daidzein	Isoflavonoid		Neurobiological activities	[31]
Naringenin	Flavanone		Antiviral, Antibacterial, Antioxidant	[32]

Table 1. *Cont.*

Compound	Class	Structure	Biological Activity	References
Hesperetin	Flavonone	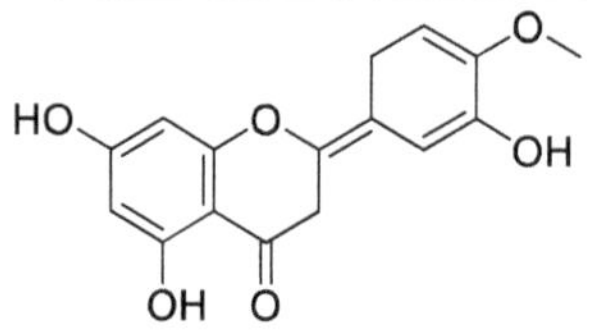	Scavenging activity and Cardioprotective activity	[33,34]
Gallocatechin	Flavanol	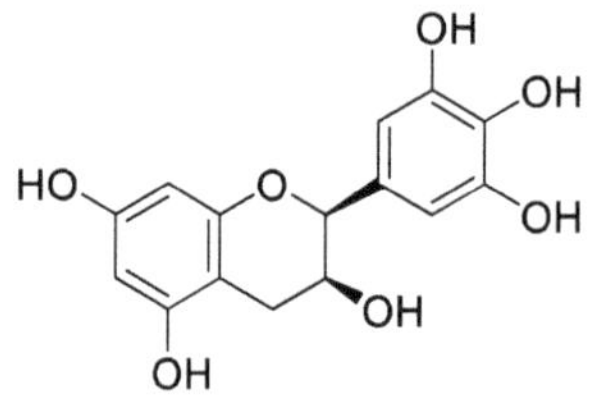	Antioxidant, Antitumor	[35]
Sitosterol	Steroid	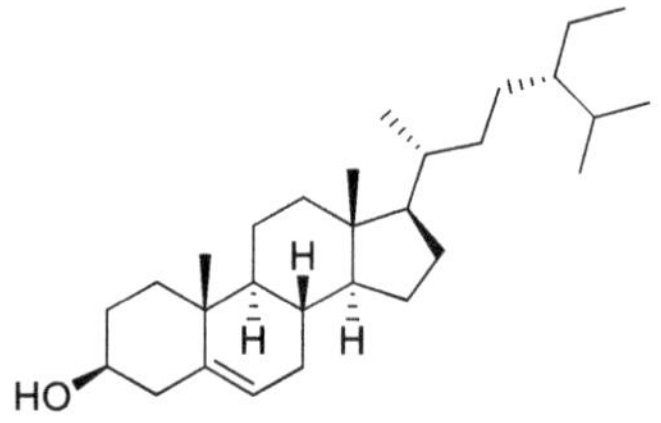	Anticancer, Antidiabetic, Antimicrobial	[36]
Stigmast-5-en-3β,7α-diol	Steroid	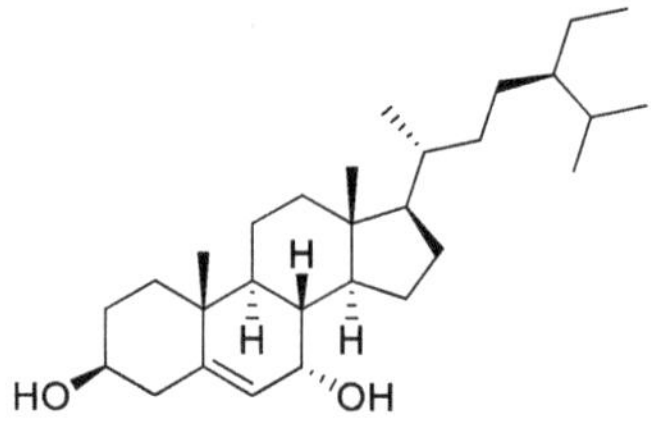	Antimicrobial	[37]
Campesterol	Steroid	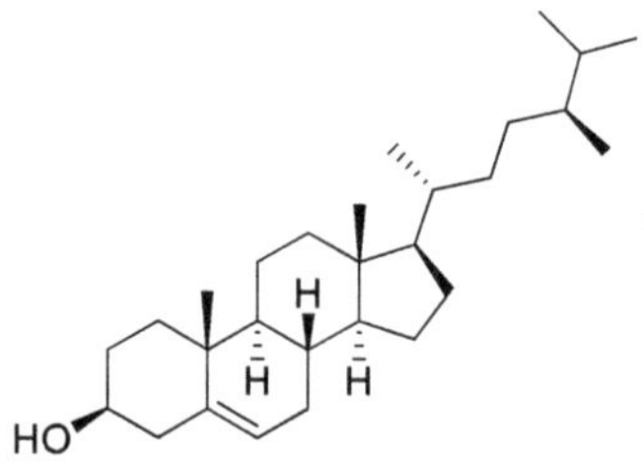	Anti-inflammatory, Anticancer, Antidiabetic	[38]
Stigmasterol	Steroid	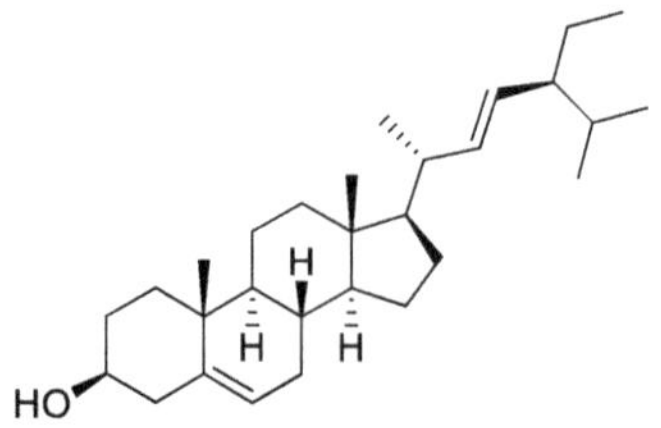	Anticancerous Dyslipidemia, diabetes and metabolic syndrome	[39]

Table 1. *Cont.*

Compound	Class	Structure	Biological Activity	References
Oleanic acid	Tereponoid	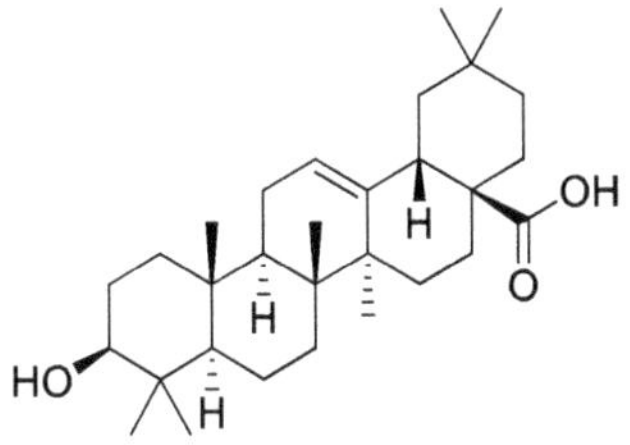	Dyslipidemia, diabetes, and metabolic syndrome	[40]
3-alpha-Corosolic acid	Tereponoid	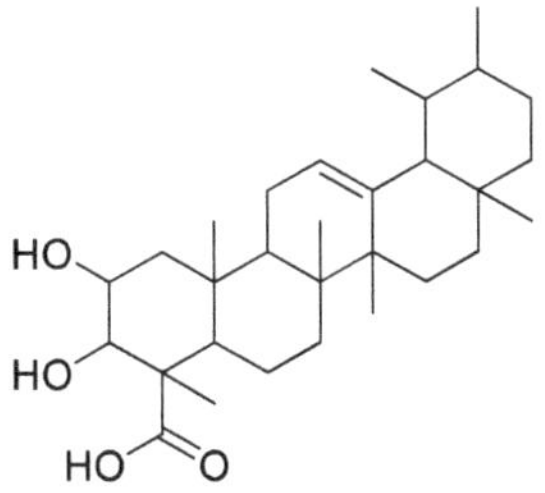	Antidiabetic, Anti-obesity, Anti-inflammatory	[41]
Urosolic acid	Tereponoid	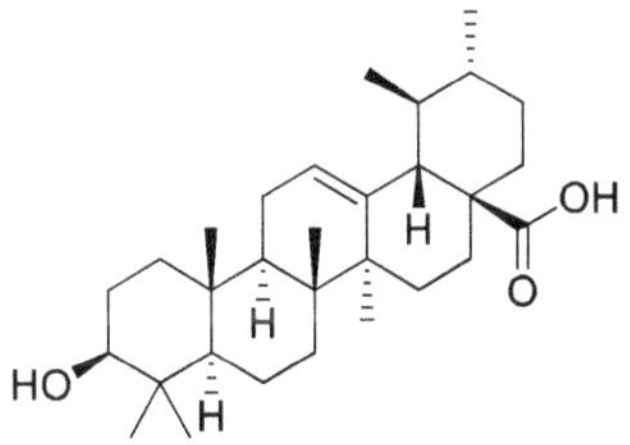	Anti-inflammatory, Anticancer, Antidiabetic, Antioxidant	[42]
3-Epikatonic acid	Tereponoid	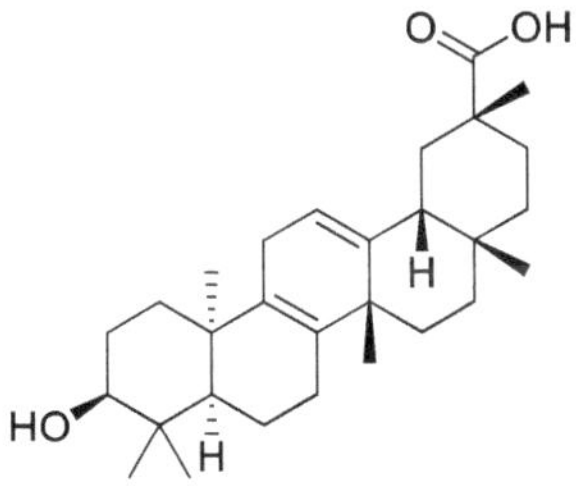	Cytotoxic	[43]

2. Traditional and Ethnobotanical Uses

For ages, the plant has been utilized in tropical medicine to treat cutaneous irritation and excessive perspiration in the hands and feet [44]. The leaves are traditionally used to treat sinusitis and stomach aches and are also used throughout the world as an antibacterial, anthelmintic, antidiarrheal, hypoglycemic, tonic, and depurative medicine [45,46]. In Turkish traditional medicine, fresh leaves are applied to the naked body or the forehead to reduce fever or to swollen joints to treat rheumatic agony [47]. The wood is durable and perfect for furniture and contains essential oils for aromatic essence. As a supplementary ointment, the leaves of this plant are applied directly to cure sunburn, superficial burns, scalp itching, dandruff, and other skin disorders [48]. The plant has been used to treat inflammatory diseases, diabetes, and cardiac illness in Palestine [49] as well as to help older men's vascular and prostate health [50]. In addition to using bark and unripe fruit for piscicidal action, the Lotha, Angami, and Sumi tribes of Kohima (Nagaland) also employed leaves of *J. regia* as astringents, anthelmintics, and treatments for dermatitis [51].

3. Pharmacological Applications

The use of medicinal herbs has increased substantially in recent years, resulting in the utilization of both traditional remedies and medicinal plants for the novel drug discoveries, and the development of more effective pharmaceuticals and nutraceuticals with fewer side effects. The main focus of this review is to explore the pharmacological applications, phytochemistry, and the therapeutic benefits of the plant extracts on antimicrobial, antioxidant, anticancer, and anti-inflammatory activities and many others (Table 2).

Table 2. Phytochemical analysis of different extracts, their effect, dosage, and model organism or experimental reference screened.

Pharmacological Effect	Extract	Model/Experimental-Reference	Dosage-Range	References
Antibacterial	Hull extract of ethanol, ethyl-acetate and water	*E. coli, B. subtilis, K. aerogenes, S. Aureus*	5 mg/mL	[52,53]
Antibacterial	Aqueous leaf extract	Gram-positive and Gram-negative bacteria	0.1 mg/mL (MIC)	[54,55]
Antibacterial	Essential oil and components	Gram-positive and Gram-negative bacteria	Wide range	[56]
Antioxidant	Ethyl acetate, butanol, ether, and aq. Extracts of (kernels, husk, and leaves)	NR	125 µL	[57]
Antioxidant	Leaf extract (ethanol and water)	NR	34.5 and 56.4 µg/mL	[58,59]
Analgesic and anti-inflammatory	Aqueous and ethanolic extract of leaf	Wistar rats, human arthritis	250, 500, 1000, 1500 mg/kg	[60]
Antiviral	Methanolic extract	HSC, polio virus and SINV	2 mg/mL	[61]
Antidiabetic	Leaf extract	Human type-ii diabetic patients	Wide range	[62]
Anticancer	Juglone	Intestinal carcinogenesis	Wide range	[63]
Anticancer	Methanol and aqueous extract of leaf	Mice melanoma (B16F10) and human melanoma (A375) cell lines	Wide range	[64]
Anticancer	Juglone and modified juglone	(NCl-H322 and A549) Lung cancers	Wide range	[65]
Anticancer	Chloroform leaf extract	Human oral and breast cancer cell lines	Wide range	[66]

3.1. Antibacterial Activity

The disc diffusion method (zone of inhibition) was used to identify and examine the antibacterial activity of *Juglans regia* hull extracts and it was revealed that these extracts possessed an appreciable antibacterial activity against a variety of bacterial species, including *E. coli, B. subtilis, K. aerogene,* and *S. aureus* [67]. The findings also implied that *J. regia* green hull extract could be useful in the treatment of acne due to the virtue of having anti-inflammatory effects. Some leaf extracts of *J. regia* showed the ability to prevent the growth of *K. pneumonia* at minimum inhibitory concentrations (MIC) of 100 mg/mL [68]; other extracts from the hull prepared in different solvents were able to stop the development of *P. aeruginosa* and *E. Coli* at MIC of 50 and 100 mg/mL, respectively [69]. Several antibacterial chemicals identified from *Juglans regia* pellicle are represented (Figure 2).

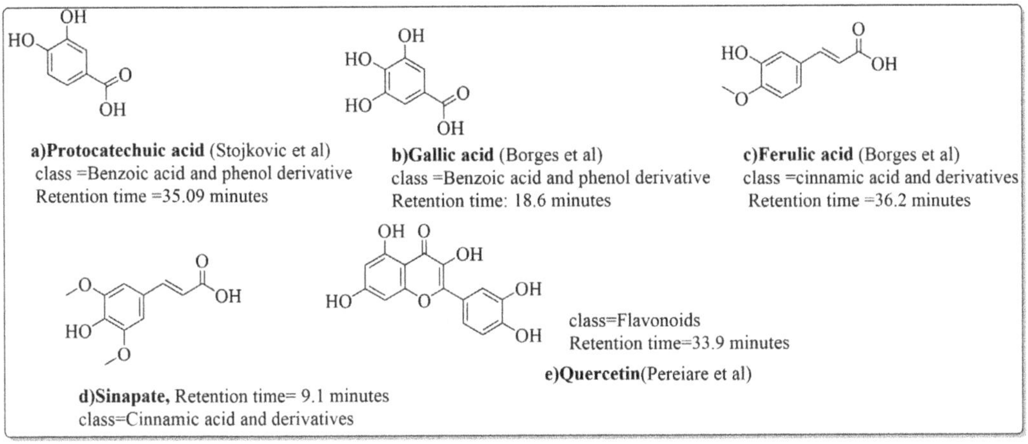

a)**Protocatechuic acid** (Stojkovic et al)
class =Benzoic acid and phenol derivative
Retention time =35.09 minutes

b)**Gallic acid** (Borges et al)
class =Benzoic acid and phenol derivative
Retention time: 18.6 minutes

c)**Ferulic acid** (Borges et al)
class =cinnamic acid and derivatives
Retention time =36.2 minutes

class=Flavonoids
Retention time=33.9 minutes
e)**Quercetin**(Pereiare et al)

d)**Sinapate**, Retention time= 9.1 minutes
class=Cinnamic acid and derivatives

Figure 2. Phenolic compounds having antibacterial activities from pellicle extract of *J. regia*, Nowicki et al. [69].

Pereira et al. [70] investigated the antibacterial activity of aqueous extracts from *Juglans regia* leaves against Gram-positive and Gram-negative microorganisms. The aqueous extracts mostly contained phenolic compounds which were obtained by shade drying the leaves, followed by the extraction of the compounds with various solvents, such as ether, alcohol, and water. By using HPLC-DAD, phenolic substances were identified and quantified. The chromatograms obtained at 320 and 350 nm revealed the presence of the different compounds (Figure 3, compounds 1-6). The obtained extracts were screened against Gram-positive (*B. subtilis*, *Staphylococcus aureus*, *Bacillus cereus*) and Gram-negative (*Escherichia coli*, *Pseudomonas aeruginosa*, *Klebsiella pneumoniae*) bacteria which revealed that Gram-positive bacteria development was inhibited to a greater extent than the inhibition in the case of Gram-negative bacteria, clearly indicating the fact that the ether, alcohol, and water extracts turned out to be favorable for the inhibition activity against Gram-positive bacteria. It was also observed that *B. cereus* was the most suspectable microorganism at MIC of 0.1 mg/mL. The order of inhibition of the above-made extracts from different solvents was found to be *B. cereus* > *S. aureus* > *B. subtilis*.

The antibacterial characteristics of the essential oil from the leaves and its constituents have been examined by Rather et al. [71]. In this study, the essential oil was extracted from the leaves of *J. regia* and analyzed using GC-MS, yielding a total of 38 compounds, accounting for 92.7% of the oil substance, including major components such as β-pinene (30.5%), α-pinene (15.1%), germacrene D (14.4%), β-caryophyllene (15.5%), and Limonene (3.6%). Disc diffusion and microdilution methods were used to test all essential oils and individual constituents for antibacterial activity against a group of Gram-positive (*B. subtilis* MTCC-441, *S. epidermidis* MTCC-43, *S. aureus*) and Gram-negative (*B. subtilis* MTCC-441, *S. epidermidis* MTCC-435, *P. vulgaris* MTCC-321, *Salmonella typhi*, *Pseudomonas aeruginosa* MTCC-1688, *Klebsiella pneumonia*, *Shigella dysenteriae*, and *Escherichia coli*) bacteria. It should be noted that the essential oil and its components were effective against all bacterial strains screened for antibacterial activity with a wide range of activity. However, it was discovered that Gram positive bacteria were found to be more inhibited compared to Gram-negative bacteria in the development and cell division processes. The potent compounds at the maximum inhibition category include α-pinene, β pinene, β-caryophyllene, and Germacrene D. For the most inhibited Gram-positive bacteria (*S. epidermidis*, *B. subtilis and S. aureus*), the MIC recorded were 15.62, 15.62, and 15.62 g/mL for *J. regia* essential oil, 48.31, 47.21, and 45.62 g/mL for α-pinene, 46.55, 46.55, and 41.33 g/mL for β-pinene, respectively, clearly indicating the fact that the essential oil present in the leaves of the plant played a key role in the inhibition of different bacteria. Gentamicin had the lowest potency in this category, with MIC values of 5.95, 5.90, and 3.90 g/mL, respectively, for the above said Gram-positive bacteria. The findings indicate that these three bacterial species are the most susceptible to essential oils and components. The changes in the activity against different bacteria might be ascribed to structural differences between the microbes and the different developmental stages in different bacteria [71,72].

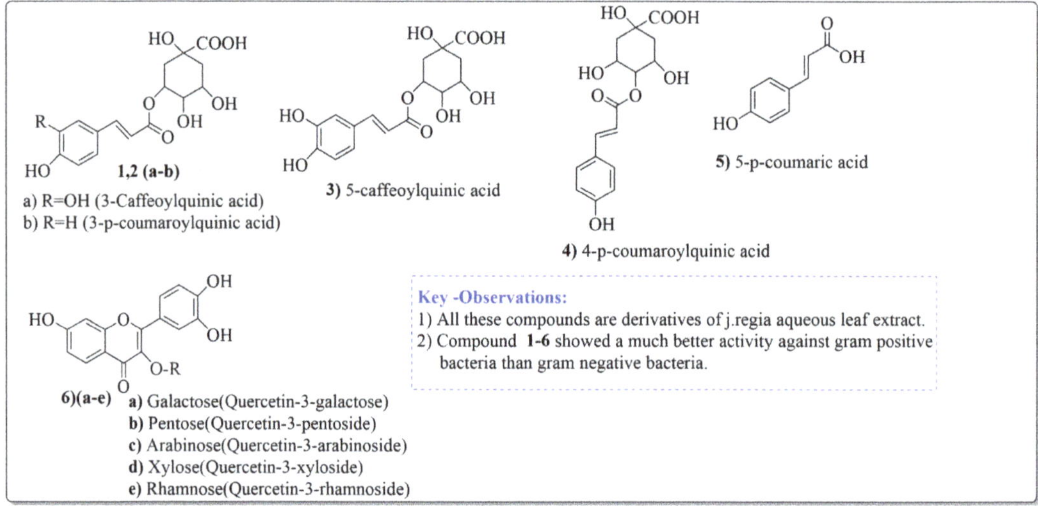

Figure 3. Extracted phenolic compounds obtained from *J. regia*, Rahman et al. [70].

3.2. Antioxidant Activity

Juglans regia leaves are high in flavonoid content, which has been linked to regulating immunological function and boosting anticancer activity, which is a global threat nowadays [73]. The antioxidant potential of walnut kernels, husks, and leaf extracts formed in different solvents, such as ethyl acetate, butanol, methanol, ether, and aqueous methanol, have been evaluated using a variety of methods. Some common and highly placed methods include the reducing power method, lipid oxidation inhibition method, and DPPH radical scavenging activity. Moreover, various studies have validated the antioxidant activity of the flavonoids present in leaves and fruit of *J. regia* and it was established that almost all the extracts prepared in different solvents possessed this property [74,75]. Antioxidant phenolic components from *J. regia* walnut kernels, which were extracted and fractionalized based on the principle of polarity and interactions with the solvent system used (Petroleum ether, Ethyl-acetate, n-butanol, and aqueous solvent) and on the basis of which they were designated as petroleum ether fraction (PEF), ethyl-acetate fraction, n-butanol fraction, and others [76]. Column chromatography over silica gel eluted with increasing polarity was used to separate the more active fractions (EEF and BUF) (Figure 4).

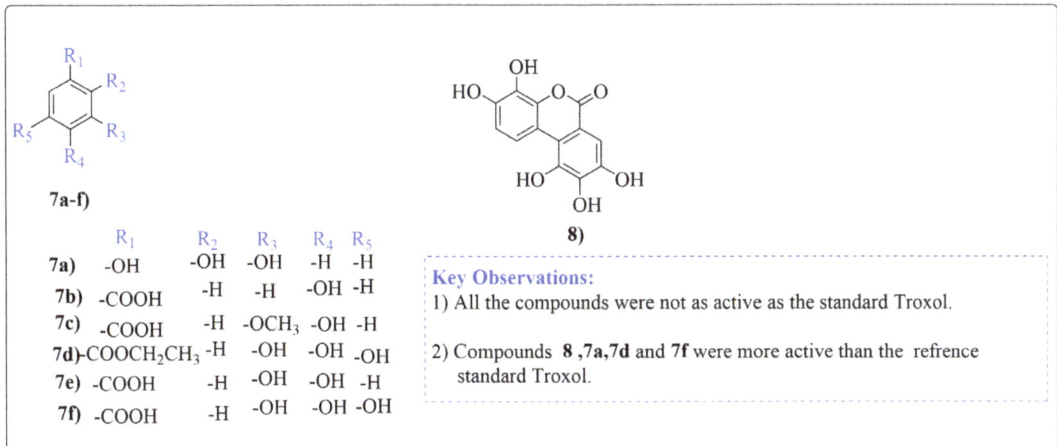

Figure 4. Phenolic antioxidant compounds from walnut kernels, Zhai et al. [76].

All the derivative compounds (7a–f and 8) were screened for α, α-diphenyl-β-picrylhydrazyl (DPPH) free radical scavenging activity using Trolox IC_{50} = 0.026 mM as a reference standard. The IC_{50} values revealed that the EEF and BUF fractions with (IC_{50} = 0.83 and 0.88 μM) were the most active in terms of DPPH free radical scavenging among all the used fractions (PEF, EEF, BUF, and AF), as described earlier. Compounds 8, 7f, 7d, and 7a, however, demonstrated considerable DPPH scavenging capabilities as depicted by IC_{50} values of (0.007, 0.011, 0.013, and 0.015 μM), respectively, indicating that they were more active than the reference standard Trolox (IC_{50} = 0.026 μM). For the first time, spectroscopic methods were used to isolate and identify seven phenolic compounds with significant antioxidant activities in *J. regia*: pyrogallol (7a), p-hydroxybenzoic acid (7b), vanillic acid (7c), ethyl gallate (7d), protocatechuic acid (7e), gallic acid (7f), and 3,4,8,9,10-pentahydroxydibenzo [b, d] pyran-6-one (8). The DPPH scavenging capacity of these compounds was ranked in the following order: 7 > 6 ≥ 4 ≥1 > Trolox ≥ 5 > 3 > 2. The findings of this study suggested that the number of hydroxyls in these phenolic compounds' aromatic rings may influence their antioxidant activities. Some important structure-activity relationships of the derivative compound (7f) indicate the essence of the COOH group as R_1 with other groups as $R_3 = R_4 = R_5$ = OH. This activity can be enhanced in the near future research by introducing some new acid electron withdrawing groups (EWG) as the replacement of R_1 in the said derivative (7f), e.g., introduction of carbonyl, thiol, and aldehyde. The antioxidant activity, which also showed a sharp increase in case of the derivative (7d), can be reasonably attributed to the presence of $R_1 = COOCH_2CH_3$ group with the aid of $R_3 = R_4 = R_5$ = OH in the moiety.

J. regia leaf extracts are effective scavengers of pro-oxidant reactive species and can be used as a readily available source of natural antioxidants. The growing interest in replacing synthetic food antioxidants with natural ones has fueled research on vegetable sources and raw material screening in the search for new antioxidants. An ethanol:water (4:6) extract of Juglans regia leaves was evaluated in vitro for its putative scavenging effects on reactive oxygen species (ROS) [hydroxyl radical (HO.), superoxide radical ($O_2{}^-$, peroxyl radical (ROO·), hydrogen peroxide (H_2O_2.)] and reactive nitrogen species (RNS) [nitric oxide (NO·), and peroxynitrite anion ($ONOO^-$)].The extract demonstrated strong scavenging activity against all reactive species tested, with all IC_{50}s found at the lg/mL level. The IC_{50}s (mean ± SE) for ROS $O_2{}^-$ and H_2O_2 radical were found to be 47.6, 4.6 and 383, 17 lg/mL, respectively. In addition, the antioxidant activity discovered in this study can be used to validate the use of *J. regia* leaf extracts in the treatment of inflammatory diseases. The current study found that the *J. regia* leaf can be used as a readily available source of natural antioxidants. The extract under investigation could aid in the prevention of lipid peroxidation as well as the protection of food, excipient bases, and medicines from oxidative damage. Furthermore, the observed antioxidant activity can at least partially justify the use of *J. regia* leaves as a therapeutic agent in inflammatory diseases. Nonetheless, its potential toxicity should be addressed before any practical application. [77].

By using DPPH and hydroxyl radical scavenging assays, Zurek et al. [78] investigated the antioxidant activity of the leaf's essential oil and some of its constituents. The antioxidant activity was measured using the 2,2-diphenyl-1-picrylhydrazyl (DPPH) radical scavenging activity assay, the copper ion reduction assay (CUPRAC), metal chelating ability (ChA), and the ability to scavenge superoxide ($O2^-$) and hydroxyl (OH^-) radicals. These methods are regarded as important indicators of plant samples' antioxidant potential. The free radical scavenging DPPH had the lowest IC_{50} value in this study, corresponding to the highest antioxidant activity. The calculated half maximum inhibitory concentration (IC_{50}) value was 22.34, 2.70 g/mL, which was higher than the positive control ascorbic acid (5.00, 0.01 g/mL). The chelating capacity test (71.69, 0.02 g/mL), superoxide scavenging (147.06, 0.27 g/mL), and hydroxyl radicals test (41.85, 0.09 g/mL) all had low IC_{50} values, indicating high antioxidant activity. The findings revealed that *J. regia* flowers were found to have a high ability to scavenge free radicals, which could be attributed to their high content of bioactive compounds. Furthermore, the results revealed a strong relationship between the phenolic compounds studied and their antioxidant activities.

3.3. Analgesic and Anti-Inflammatory Properties

J. regia extracts from aqueous (2.87 and 1.64 g/kg) and ethanolic (2.044 and 1.17 g/kg) solutions have demonstrated antinociceptive effectiveness in a hot plate test, according to Hosseinzadeh et al. [79]. The hot-plate test was performed on eight male and female mice groups. The metal surface temperature was kept constant at 55 °C. Before and after drug administration, the latency to a discomfort reaction (licking paws or jumping) was measured. 30 min after the extracts were given to groups of eight male and female mice, they were given an intraperitoneal injection of 0.7% *v/v* acetic acid solution (volume of injection 0.1 mL/10 g body wt.). The number of writhing produced in these animals was counted for 30 min after acid injection. The anti-inflammatory activity against acute inflammation was evaluated in mice using the xylene-induced ear edema method. The mice were divided into eight groups. 30 min after the different doses of extract were injected intravenously, diclofenac and 0.03 mL of xylene were applied to the anterior and posterior surfaces of the right ear. The left ear was used as a control. The mice were sacrificed two hours after xylene application, and both ears were removed. Circular sections were excised and weighed using a cork borer with a 9 mm diameter. The weight gain caused by the irritant was calculated by subtracting the weight of the untreated left ear section from the weight of the treated right ear section. In mice, the anti-inflammatory activity against chronic inflammation was assessed using the cotton pellet granuloma method. The 30 mg dental cotton pellets were sterilized in an air oven at 121 °C for 20 min before being impregnated with 0.4 mL of an ampicillin aqueous solution. Under ketamine (65 mg/kg body weight) and xylazine (6.5 mg/kg body weight) anesthesia, two cotton pellets, one on each side, were implanted subcutaneously in the shoulder region of mice. The extract and diclofenac were administered once daily for seven days. The rats were killed on day 8, and the pellets and surrounding granulation tissue were dried at 60 °C for 24 h. In mice, the intraperitoneal LD_{50} values of *J. regia* aqueous and ethanolic leaves extract were 5.5 g/kg (4.1–6.5) and 3.3 g/kg (3.1–3.5), respectively, with maximum nonfatal doses of 4.1 g/kg and 2.93 g/kg. However, it should be noted that both extracts displayed an anti-inflammatory effect in xylene at lower dosages. The extracts demonstrated antinociceptive action in the Writhing test that was not inhibited by naloxone (lifesaving medication). The extracts were found to have anti-inflammatory properties in the case of chronic inflammation. *J. regia* leaves have been shown to have an antinociceptive action via non-opioid receptors as well as an anti-inflammatory impact against acute and chronic inflammation, making them an effective drug with analgesic and anti-inflammatory properties against rheumatoid arthritis [80].

3.4. Antidepressant Activity

Depression has been classified as a mood disorder and has been defined as a feeling of sadness, loss, or anger [81]. Treatments derived from the *Juglans regia* L. flower and its leaf extracts have been found to be highly successful in treating depression. The effects of the forced swimming test (FST) and tail suspension test (TST) in mice were quite evident [82].

3.4.1. Forced Swimming Test

The immobility time was calculated by observing the muscle movements of rats that were placed in a pool of water. A glass cylinder with a diameter of 25 cm and a height of 23 cm was filled with water to a height of 12 cm. The water temperature was 23 °C. Each rat was given a single injection of the extract. The animals were tested thirty minutes later. For the first two minutes, each animal was given time to get used to the new circumstances before having their immobility time recorded. For the following six minutes, conditions of increased muscle control interrupted with periods of immobility. For the next four minutes, immobility was timed using a stopwatch.

3.4.2. Tail Suspension Test

The tail-suspension test was the second method used to evaluate the extract's antidepressant effect. Rats were tested thirty minutes after a single drug or vehicle injection. A cord which was about 50 cm long was stretched between two metal tripods at a height of 70 cm, to which the rats were taped by the tail. The rats became immobile after a period of vigorous motor activity, and the time was recorded with a stopwatch for a total of 4 min. When rats hung passively and completely motionless, they were considered immobile.

The mean duration of immobility in the control group was 188.33, 2.16 s, whereas it was 151.16, 2.56 s, in the fluoxetine group. The reduction in immobility was found to be statistically significant ($p < 0.05$). The decrease in immobility in the extract groups was also significant ($p < 0.05$) for both doses. For 100 mg/kg and 150 mg/kg body weight, the total duration of immobility was found to be 160.66, 3.76 s and 154.83, 4.32 s, respectively. Regarding the effect on immobility in the tail suspension test, in the control group, the mean duration of immobility was 193.33, 1.96 s, whereas in the fluoxetine group, it was 147.16, 2.48 s. The reduction in immobility was found to be statistically significant ($p < 0.05$). The total duration of immobility for 100 mg/kg body weight was found to be 168.39 s and 148.66 1.75 s for 150 mg/kg body weight. The decrease in immobility in the extract groups was also significant ($p < 0.05$) for both doses [83]. This remarkable antidepressant activity was attributed to the phenol and flavonoid contents, especially quercetin. In addition, the presence of omega-3 fatty acids in *Juglans regia* fruit extracts may have depressive properties [84]. More research is needed to fully understand the mechanism of antidepressant activity.

3.5. Antiviral Activities

Mouhajir et al. [85] studied the activity of *Juglans regia* methanol extracts. At noncytotoxic concentrations, the 2 mg/mL concentration was tested against Sindbis virus (SINV), herpes simplex virus (HSV) and poliovirus. Vardhini [86] also used a computational method to investigate juglone's antiviral activities and the results were well supported by molecular docking studies. In this study, the ligand which had the highest binding affinity had a dock score of 114.967 against ASP 29, ASP 30, and ASP 30, which are the hydrogen bonds that existed between the ligand and the protein molecule. Using phytochemical and chromatographic methods to separate components from *J. regia*, other researchers [85] investigated the impact of anti-HIV activity in vitro. The impact of anti-HIV activity in vitro was assessed using MT4 cells and HIV-III B virus. The target research was conducted using BIACORE 3000 molecules linked equipment. Angeli et al. [87] studied the walnut pellicle extracts and derived some handful number of antiviral compounds which possessed the inhibitory activity against HSV-1 and HSV-2 replication. The ID_{50} (con. which inhibited 50% virus formation) were found to be 10 and 8 µg/mL for HSV-1 and HSV-2, respectively. However, the walnut pellicle extract was ineffective against Echovirus9 (ECHO-9), Poliovirus1 (polio1), Coxsackievirus B1 (coxsackie B1) and Adenovirus (Adenoid). The compounds extracted from walnut pellicle extract were found to be active against viral diseases (Figure 5).

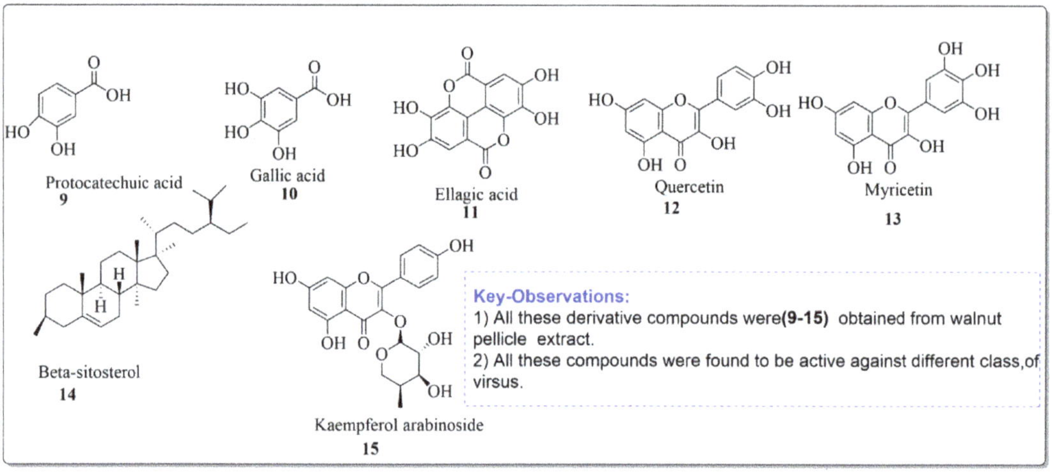

Figure 5. Compound derivatives from walnut pellicle extract, Alkhawajah et al. [87].

The global outbreak of COVID-19 was caused by SARS-CoV-2, a positive-sense single-stranded RNA coronavirus. The virus's main protease (Mpro), the major enzyme processing viral polyproteins, contributed to SARS-CoV-2 replication and transcription in host cells and has been identified as an appealing target in drug discovery. A series of 1,4-naphthoquinones with juglone (compound isolated from leaves of walnut) skeletons were synthesized and tested for inhibitory efficacy against SARS-CoV-2 Mpro. At a concentration of 10 µM, more than half of the tested naphthoquinones effectively inhibited the target enzyme, with an inhibition rate of more than 90%. The characteristics of substituents and their position on the juglone core scaffold were identified as key ingredients for enzyme inhibitory activity in the structure-activity relationships (SARs) analysis. The most active compound, 2-acetyl-8-methoxy-1,4-naphthoquinone, had an IC_{50} value of 72.07–4.84 nM against Mpro-mediated hydrolysis of the fluorescently labelled peptide, which was much higher than shikonin as the positive control. In molecular docking studies, it fit well into the active site cavity of the enzyme by forming hydrogen bonds with adjacent amino acid residues. The results of in vitro antiviral activity testing revealed that the most potent Mpro inhibitor could significantly suppress SARS-CoV-2 replication in Vero E6 cells at low micromolar concentrations, with an EC_{50} value of about 4.55 µM. Under the conditions tested, it was non-toxic to the host Vero E6 cells. The current study suggested that the juglone skeleton could serve as a primary template for the development of potent Mpro inhibitors [88].

Enzyme Inhibition Mechanism

The inhibitory activity of the prepared quinones against Mpro of SARS-CoV-2 was evaluated using a fluorescently labelled short peptide containing a Q-S scissile bond. In the first library of compounds, we tested the enzymatic inhibition rate of several naturally occurring naphthoquinones (juglone 2, 7-methyl juglone, lawsone, plumbagin and shikonin), 9,10-anthraquinones (emodin, rhein and aloe emodin), and synthetic vitamin

K3 against SARS-CoV-2 Mpro at 10 μM. The primary screening results showed that the majority of the natural quinones were ineffective, with inhibition rates of less than 10% at 10 μM. Vitamin K3 was also inactive, with a 12.7% inhibition rate. The positive control was the natural naphthoquinone shikonin, which had previously been identified as a strong Mpro inhibitor (IC_{50} = 15.75 8.22 μM). At a concentration of 10 M, it had a moderate inhibitory effect on the target enzyme. In the first library of naphthoquinones, juglone and 7-methyl juglone inhibited Mpro completely, resulting in the loss of its hydrolytic efficacy. The lead compounds for further structural modifications were the two natural naphthoquinones. In the second library, juglone and 7-methyl juglone derivatives were synthesized by adding a few groups to their naphthoquinone scaffold and modifying the phenolic hydroxyl group. The results indicated that almost all of the juglone derivatives in the second library retained their high inhibitory potency at concentrations of 10 M and 1 M. A few analogues exhibited significantly higher potency than the parent compounds juglone and 7-methyl juglone at 0.1 μM. The compounds with an enzymatic inhibition rate of more than 25% at 0.1 μM concentration were then subjected to IC_{50} value screening.

3.6. Antidiabetic Activity

J. regia leaf extract has been thought to be advantageous for the treatment of diabetes mellitus since ancient times, and this notion has been scientifically proven to be effective. Blood glucose, glycosylated hemoglobin, LDL, lipid, and cholesterol levels were found to be reduced by alcohol extracts from the plant's leaves. Streptozotocin-treated rats were administered by treating with 200 and 400 mg/kg leaf extracts of *J. regia* leaf extract for 28 days, and it alleviated hyperglycemia through reduced glycosylated hemoglobin and enhanced insulin sensitivity [89]. They looked at how *J. regia* leaf extract affected hyperglycemia in type 2 diabetic patients. The leaves of *J. regia* were initially freshly harvested, then washed and dried at 25 °C temperature in the shade before being powdered, and the powder was extracted at room temperature using the percolation technique and 70% aqueous ethanol. The solvent was eliminated using Whatman paper filters and the crude extract was standardized by counting the total phenol concentration after evaporating at a temperature of no more than 40 °C under decreased pressure. Later, capsules of *Juglans regia* and a placebo with the same appearance were created. The *J. regia* capsules included 100 mg of leaf extract powder. In the manufacture of the leaf extract powder, toast powder was chosen as the placebo and as the excipient. The research was conducted on a group of individuals who were divided into two groups: those who had received *J. regia* and those who had received a placebo. The levels of the constituents fasting blood glucose (FBG), HbA1c, total cholesterol, and triglyceride were measured in *J. regia*-treated patients and they were found to be on a lower side than those in the placebo group, with no adverse effects. Finally, for type 2 diabetes, three months of treatment with 100 mg of *Juglans regia* leaf extract twice daily improved glucose control with no noticeable side effects. The results showed that FBG, HbA1c, total cholesterol, and triglyceride levels in Juglans regia-treated patients were significantly lower than in the baseline and placebo groups. When compared to the placebo group, patients in the Juglans regia group were significantly more satisfied with their treatment. Except for more GI events (especially mild diarrhea) associated with extract treatment at the start of the study, no liver, kidney, or other side effects were observed in the groups.

Testicular dysfunction is a complication of diabetes, and *Juglans regia L.* leaf extract contains phenolic compounds with hypoglycemic and antioxidative properties. Nasiry et al. investigated whether *J. regia* leaf extract could protect against the negative effects of diabetes on oxidative stress, testis histology and testosterone hormone production [90]. In their research, four groups of male rats were used: a control (nondiabetic) group given saline, a diabetic group, a diabetic + *J. regia* group that received *J. regia* leaf extract and a *J. regia* group (nondiabetic) that received *J. regia* leaf extract only. They looked at histopathological and histomorphometric changes, serum testosterone, malondialdehyde (MDA), glutathione (GSH), superoxide dismutase (SOD), and catalase (CAT) levels to see how *J. regia L.* leaf extract affected testicular functions in diabetic animals. In the testis of diabetic rats, there was a reduction in MDA as well as an improvement in antioxidant status; *J. regia* leaf extract attenuated these abnormalities. In diabetic rats, it was found that there was a significantly decrease in the levels of testosterone, GSH, SOD, and other antioxidant biomarkers; these levels were restored after the *J. regia* leaf extract was administrated. After administering the *J. regia* leaf extract, the MDA level and improved antioxidant status in the testis of diabetic rats were found. Because of its antioxidant, anti-inflammatory and anti-apoptotic properties, *J. regia* leaf extract may have protective effects against diabetic dysfunction in the testis, according to the findings.

3.7. Anticancer Activity

Cancer treatment has long been a difficulty for medical science, and research into a variety of medicinal plants has resulted in a potential treatment for the disease's early stages [91,92]. Even while cancer has been identified as a burden on human civilization, no comprehensive cure has yet been established [93]. Juglone has been demonstrated to inhibit intestinal carcinogenesis in mice, suggesting that it might be a potential chemo preventive medication for neoplasia in human intestines [94]. The human carcinoma cells lines HCT-15 cells, HL-60 cells, and doxorubicin-resistant HL-60R cells have all demonstrated the potency of juglone as a strong cytotoxin [95]. Human cancer renal cell lines A-498 and 769-P, as well as the colon tumor cell lines Caco-2, were inhibited in a concentration-dependent way by walnut methanolic extract out from the seeds, greenish husks, and leaves of *Juglans regia*. All extracts inhibited the growth in 769-P renal and Caco-2 colon cancer cells (IC$_{50}$ values of 0.352 and 0.229 mg/mL, respectively; range, 0.226 to 0.29 mg/mL), but the walnut extract of leaves was more effective at suppressing cell proliferation than green husks and extracts of seeds (IC$_{50}$ values of 0.352 and 0.229 mg/mL, respectively).

Constituents of *J. regia* chloroform leaf extract was tested for anti-proliferative and apoptotic effects on human breast (MCF-7) and oral tumor (BHY) cell lines. The components were extracted from shrub leaves which were air dried, pulverized, and then percolated in n-hexane for 24 h. After three days of extraction, all extracts were filtered and dried with a rotatory evaporator. At a lower pressure, the solvent, n-hexane, was extracted. The remaining powder was then suspended and extracted to obtain the chloroform fraction, which was further purified using chromatography (Figure 6). The proliferative and apoptotic activities of the compounds mentioned above (16–22) were investigated. Derivatives (20 and 22) were found to be significantly cytotoxic to MCF-7 cell lines, whereas compounds (16, 21, 22) significantly inhibited BHY cell proliferation. According to the IC$_{50}$ values, MCF-7 cell lines were also the most sensitive to virtually all chemicals. Compounds (21) (IC$_{50}$ = 50.98 μM) and (22) (IC$_{50}$ = 21.30 μM) were selectively active against both cancer cell lines, MCF-7 and BHY, but were significantly less effective against normal cells. The compounds (16–22) suppressed cell population development in human tumor cell lines MCF-7 and BHY, as well as mouse fibroblast cell lines, using the MIT test at 24, 48, and 72 h (NIH-3T3). However, the best proliferation activities were obtained after 72 h. It is worth mentioning that compounds 21 and 22, which are plant flavonoids, and a plant naphthoquinone can be further studied well to explore their new biological activities such as antidiabetic and antibacterial activities. Moreover, these compounds induced apoptosis in MCF-7 cell lines by the well-known mechanism caspase-3 independent pathway [95].

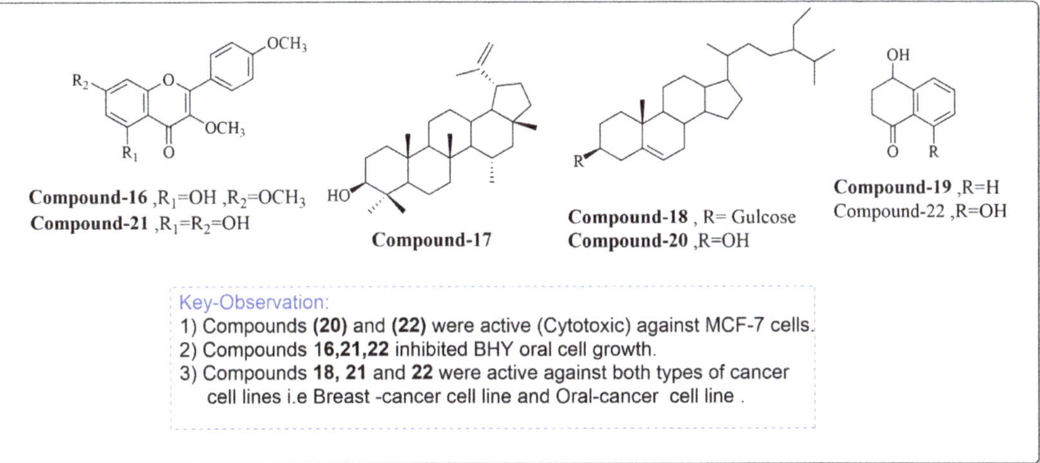

Figure 6. Compounds having antiproliferative and apoptotic activity against MCF-7 and BHY cancer cell lines, Noumi et al. [95].

One of the important synthetic breakthroughs was given by Zhang et al. [96] as they developed modified Juglone from *J. regia* as a strong cytotoxic drug against carcinoma cell lines of lungs by isolating and modifying the compound. The synthesis started with a solution of Juglone (23) in acetonitrile (isolated from *J. regia* roots). At 20 °C, the solution was portioned with NaH and agitated continuously for 10 min. Propargyl bromide was added drop by drop, and the suspension was again agitated for two hours. Ethyl acetate was used to extract the reaction mixture, which was then purified on a silica gel column to get the derivative (24). The derivative (24) was then treated with organic azide, sodium ascorbate, water:butanol (1:1), and copper) (ii) pentahydrate to yield the final target derivatives (25a–30a). Juglone derivates having cytotoxic effects are depicted in Figure 7. All the synthesized target derivatives (25a–30a) were evaluated through cytotoxic assays against lung cancer cell lines (NCl-H322 and A549) using BEZ-235 (IC$_{50}$ = 9.80 μM) as a positive control. Among all the synthesized derivatives of Juglone, 25a and 26a bearing o-nitro and o-cyano phenyl moieties, respectively, displayed the most potency (IC$_{50}$ = 4.72, 8.90 and 4.67, 7.94 μM) against both the lung cancer cell lines (NCl-H322 and A549), respectively. Both these derivatives even displayed better activity results than the standard positive control BEZ-235 (IC$_{50}$ = 9.80 μM). Derivatives 27a, 27b, and 27c with ortho, meta, and para-methoxy R moieties, respectively, were very less potent towards the cancer cell lines. Similarly, derivatives (28 a, b, c) bearing o, m, and p-bromophenol moieties were lesser active than juglone (IC$_{50}$ = 19.32 μM) and standard BEZ-235 as displayed by the IC$_{50}$ values (24.34, 21.82, 24.55 μM), respectively. Moreover, derivatives 29a and 30a bearing a phenyl and o-chlorophenyl displayed slight better activity than juglone; however, they were less potent than the standard BEZ-235 against the lung cancer cell lines (NCl-H322 and A549). From the above results, it can be clearly stated that the introduction of EWG at ortho position of the phenyl ring in the said derivative (25a, 26a) turned out to be favorable for the net potency of the derivatives, which can be enhanced in the new research by introducing some other EWG's at ortho position of the phenyl ring, such as the EWG group carbonyl, carboxyl, CF$_3$, or fluorine at ortho position. Moreover, from structure activity relationship (SAR) studies, it is clear that changing the position of groups from ortho to meta in (25b and 26b) or from ortho to para in (25c and 26c) in the above structures showed a dramatic decrease in the activity as depicted by the IC$_{50}$ values of 25b, 26b, 25c, 26c (IC$_{50}$ = 15.6, 17.22, 10.96, 13.30 μM), respectively, providing clear proof that the respective position of the groups to influence the activity is necessary to impart activity to the new derivatives.

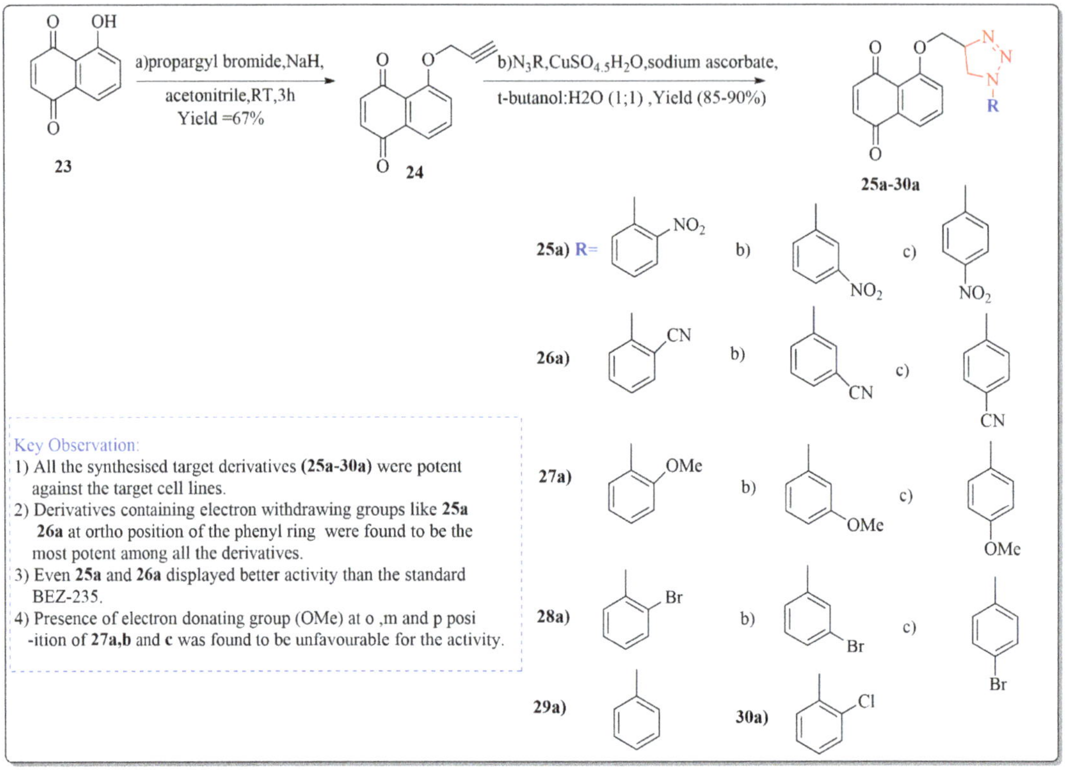

Figure 7. Synthesis of modified *juglone* derivatives (triazole analogs) as cytotoxic agents, Haque et al. [96].

Shah et al. [97] also studied the anti-proliferative and cytotoxic effects of *J. regia* leaf extracts (methanol and aqueous extracts) at different concentrations on growth inhibitions of cell lines of mice melanoma (B16F10) and human melanoma (A375). The extract concentrations prepared in this experiment included (0.1, 0.15, 0.2, 0.25, 0.3, 0.35, 0.40, 0.45, and 0.5 mg/mL of extract/mL). The normal lymphocyte cell lines were observed to show negligible sensitivity towards the extracts. The cytotoxic activity was screened for 72 h of treatment to the (B16F10) and (A375) cell lines, and it was revealed that methanolic extracts at different concentrations showed potent activity (cell anti-proliferation) against mice melanoma with IC_{50} = 0.234 mg/mL compared to IC_{50} = 0.304 mg/mL on human melanoma cell lines. Likewise, aqueous extracts of the above-mentioned concentrations also showed good activity with (IC_{50} = 0.298 and 0.350 mg/mL) against mice and human cell melanoma, respectively. Even though both methanolic and aqueous extracts had significant anti-proliferative action against mice and human cancer cell lines, significant activity was detected for methanol extracts generated at the same concentrations in this investigation.

3.8. Antifungal Activity

The antifungal activity of aqueous and solvent extracts from leaf and bark helps in medicinal use because these extracts showed a wide range of activity against fungus using various methods such as agar dilution, disc diffusion, agar streak dilution, and the Radish method [98]. The antifungal activity was demonstrated by various research, which includes the major breakthrough contributed by D. Wianowska et al. by comparison of the antifungal activity of different extracts of *Juglans regia* cultivars and the major compound juglone. This study compares the antifungal effects of juglone and extracts from walnut green husks of the cultivars Lake, Koszycki, UO1, UO2, and non-grafted against plant pathogenic fungi such as *Alternaria alternata*, *Rhizoctonia solani*, *Botrytis cinerea*, *Fusarium culmorum*, *Phytophthora infestans*, as well as *Ascosphaera apis*. The data obtained demonstrate that the antifungal activities of the extracts can be modulated by their other constituents and are not always dependent on the antifungal activity of juglone. This enables us to draw the conclusion that juglone is not the only element in walnut green husk extracts that inhibits mycelial growth. It was discovered that phenolic compounds were in charge of the extracts' activity and that they could change juglone's antifungal properties [99]. Similarly, the activity was demonstrated by Hubert Sytykiewicz et al. [100] by conducting a study to assess the antifungal efficacy of four extract fractions (methanolic, ethyl acetate, alkaloid, and hydrolyzed methanolic) derived from *Juglans regia* (L.) leaves against pathogenic *Candida albicans* strains. One reference strain (*C. albicans* ATCC 900et 29) and 140 isolates from various biological samples, including skin lesions, sputum, urine, and feces, were used to test the yeasts. The highest anticandidal activity was found in the methanolic extract from walnut leaves, followed by the alkaloid fraction, which had a slightly lower antifungal efficacy. Etyl acetate and hydrolyzed methanolic preparate had the lowest levels of growth rate inhibition for the examined fungal pathogens [101].

Similar antifungal activity was assessed using the disc diffusion method with extract concentrations of 100, 200, and 300 g/mL/disc, using the standard ketoconazole (40 g/mL/disc). Selective fungistatic activity was demonstrated by the extracts against some species. Different levels of inhibitory activity were present in all extracts against all types of fungi. The study revealed that acetone and chloroform extracts significantly inhibited the growth of *Alternaria alternata* and *Trichoderma virens*, respectively. Moreover, methanolic extract demonstrated significant activity against *Aspergillus niger* [102]. The other studies which have been carried out by analyzing the different extract of *Juglans regia* and the antifungal activity have demonstrated that, no matter the type of extract evaluated for the antifungal activity, juglone is the primary component of walnut green husk extracts which demonstrated the maximum antifungal activity. However, the activity varied depending on the composition of other vital extract constituents and the kind of fungal infection that was treated [100,103].

3.9. Cardiovascular Activity

Walnuts have been found to have high quantities of omega-3 and omega-6 polyunsaturated fatty acids (PUFA). While some studies have connected omega-6 PUFA to an enhanced proinflammatory vascular response, the bulk of investigations have indicated that these components have no deleterious consequences on human cardiovascular health. It has been also discovered that in non-hyperlipidemic persons, eating walnuts regularly (30–100 g/day) lessened the cardiovascular risk factors [104]. Consuming nuts on a regular basis has been linked to a lower risk of both fatal and non-fatal myocardial infarction. According to epidemiological studies, compared to people who never consumed nuts, those who consumed nuts five or more times per week had a 50% lower risk of coronary heart disease [105,106]. Green walnut hull extract suppressed protein secretion and thrombin-induced platelet aggregation by 50% in an in vitro investigation, with no deleterious effects on platelets at a dosage of 50 mg/mL. Walnut green hull extract's antiplatelet activity is most likely due to its polyphenolic components and antioxidant capabilities. As a result, it can be regarded a thrombotic disease candidate as well [107].

3.10. Brain Enhancing Activity

A healthy functioning brain requires sufficient amounts of water, vitamins (such as folate, thiamine, vitamins B6, and B12), lipoic acid, lutein, and n-3 fatty acids for proper and improved functioning. Walnuts are high in n-3-linolenic acid (ALA), as well as a variety of other potentially neuro-regenerative substances such as phenolic acid (ellagic acid), gamma tocopherol (vitamin E), folate, melatonin, flavonoids, and a plant-based omega-3 fatty acid. It is worth noting that walnuts placed in second place among 1113 foods that were tested for antioxidant levels [108]. Memory is the retaining process of a learning experience across time. By delivering the right stimuli, a single memory can be retrieved. Polyphenols have been demonstrated to exhibit properties which influence the key neuronal signaling pathways in memory and learning. Based on obesity, hypercholesterolemia, and oxidative stress, walnut polyphenolic extracts improved learning by 42% and memory in hypercholesterolemic rats [109]. A walnut diet containing 6% walnut oil protected male rats from neurotoxicity caused by the chemotherapeutic drug cisplatin, according to another study. The results demonstrated that walnut administration enhanced cognitive and motor skills, implying that including walnut in one's diet may be beneficial in combating chemotherapy-induced motor and cognitive dysfunction [108]. It was also revealed that walnuts enhanced learning skills, locomotor activity, memory, anxiety, and motor coordination in the transgenic mice model against Alzheimer's disease by 6–9 percent [110]. The above studies which were carried in vivo have clearly opened a new path of the research to demonstrate the same effect on humans while making sure the earlier conclusion and the ample number of dosages are delivered hand in hand alongside with the safety of the subject.

4. Toxicity Activity on Plants and Animals

After conducting toxicology research on humans and animals, it was discovered that in the acute oral toxicity trial, Wistar female rats were given several doses of methanolic extracts from 10 to 5000 mg/kg of *J. regia* septum for 14 days. In a sub chronic study, the Wistar rats were fed the extract orally at a rate of 1000 mg kg^{-1} for 28 days. No harmful effects or deaths were observed for the extract even at a concentration of 5000 mg kg^{-1}. Further in the study, there were no notable morphological or histological alterations in the studied tissues [111]. According to the Yang et al. [112], walnut polyphenols increased immune function by lowering oxidative stress. Tropical walnut use has been also linked to skin irritation and discoloration in many cases reported in the past, clearly indicating the fact that taking an excess diet of walnut fruit is also harmful to human health [113]. This fact was also well supported by developing enormous blisters and skin discoloration in a 65-year-old woman after she consumed 15 kg of walnuts in three days [114]. It was also revealed that walnut aqueous extract decreased the cyclophosphamide toxicity and protected metabolizing and antioxidant enzymes at the time of treatment on the woman. The goal of the current study was to determine whether walnut extract can reduce the toxicity of the anticancer drug cyclophosphamide (CP) while also protecting against the disruption of enzymes that help the body break down drugs and fight free radicals [113]. Human erythrocyte forward scatter was significantly reduced after 24 h of exposure to juglone (5 µM). Juglone (1–5 µM) markedly raised the proportion of annexin V-binding cells. Juglone (5 µM) markedly increased the amount of ceramide at the erythrocyte surface and decreased the amount of erythrocyte adenosine triphosphate (ATP). Juglone stimulates suicidal erythrocyte death or eryptosis, at least in part, by increasing the abundance of ceramides (lipid molecule), depleting energy, and activating protein kinase (PKC) [114]. Walnut's detrimental effects on animals, notably horses, have also been reported. Moreover, walnut heartwood can induce laminitis, an inflammatory disease in horses. As a result, the black walnut extract model has been created to study the many parameters connected to horse laminitis [115].

To investigate and validate the much promising study in relation to toxicity caused by the walnut excess diet, Iwamoto et al. investigated the effects of walnut consumption on serum lipids and blood pressure in Japanese subjects to determine whether it would also be beneficial as a component of the Japanese diet. In a crossover design, they randomly assigned 20 men and 20 women to one of two mixed natural diets for four weeks each. Both diets adhered to the typical Japanese diet (reference diet) and included the same foods and macronutrients, with the exception that the walnut diet provided 12.5% of its energy from walnuts (43–57 g/d) (offset by lesser amounts of fatty foods, meat, and visible fat). When comparing the walnut diet to the reference diet, total cholesterol levels were 0.21 mmol/L lower for women ($p < 0.01$) and 0.16 mmol/L lower for men ($p = 0.05$). When they followed the walnut diet, the LDL cholesterol levels were 0.18 mmol/L lower for the men ($p = 0.13$) and 0.22 mmol/L lower for the women ($p < 0.01$). The walnut diet also decreased the concentration of apolipoprotein B and the ratio of LDL to HDL cholesterol ($p < 0.05$) [116].

5. Conclusions

The current review addresses the most prominent published research on *J. regia* L. in medical sciences, the plant's traditional and modern scientific applications, as well as scientific validation of the stated biological activity in vivo and in vitro. Synthetic triazole derivatives suggested in the manuscript can be modified to create new potent molecules that can be tested for a variety of biological activities such as anticancer, antidiabetic, antibacterial, and many others with promising results that will stand in the research gap. Furthermore, various extracts (aqueous or methanolic) from walnut leaves, flowers, or fruits might be utilized in concentration-dependent ways in a new study to make improvements in the relevant field. To the best of our knowledge, we have not reported any deceptive activity after working on our own green fruit and leaf extract of the plant, as well as manipulation of other work in the same field. However, more human trials are needed to determine all *J. regia* Linn extracts' much anticipated and promising capabilities and activities.

Author Contributions: All authors equally contributed through initiating a draft of the review, performing the artwork, and editing the manuscript. All authors have read and agreed to the published version of the manuscript.

Funding: This research received no external funding.

Institutional Review Board Statement: Not applicable.

Informed Consent Statement: Not applicable.

Data Availability Statement: Not applicable.

Acknowledgments: All authors are grateful to King Abdulaziz University (KAU), University of Ha'il (UOH), Taif University (TU), Lovely Professional University (LPU), and Aligarh Muslim University (AMU) for providing the basic infrastructure to write this review article and especially KAU, UOH, and TU for giving access to the Saudi Digital Library (SDL).

Conflicts of Interest: The authors declare no conflict of interest.

References

1. Acarsoy Bilgin, N. Morphological Characterization of Pollen in Some Varieties of Walnut (*Juglans regia*). *Int. J. Fruit Sci.* **2022**, *22*, 471–480. [CrossRef]
2. Hayes, D.; Angove, M.J.; Tucci, J.; Dennis, C. Walnuts (Juglans regia) chemical composition and research in human health. *Crit. Rev. Food Sci. Nutr.* **2016**, *56*, 1231–1241. [CrossRef] [PubMed]
3. Gupta, A.; Behl, T.; Panichayupakaranan, P. A review of phytochemistry and pharmacology profile of *Juglans regia*. *Obes. Med.* **2019**, *16*, 100142. [CrossRef]
4. Dawid-Pać, R. Medicinal plants used in treatment of inflammatory skin diseases. *Postep. Dermatol. Alergol.* **2013**, *30*, 170–177. [CrossRef] [PubMed]
5. Bourais, I.; Elmarrkechy, S.; Taha, D.; Mourabit, Y.; Bouyahya, A.; El Yadini, M.; Machich, O.; El Hajjaji, S.; El Boury, H.; Dakka, N.; et al. A Review on Medicinal Uses, Nutritional Value, and Antimicrobial, Antioxidant, Anti-Inflammatory, Antidiabetic, and Anticancer Potential Related to Bioactive Compounds of *J. regia*. *Food Rev. Int.* **2022**, 1–51. [CrossRef]
6. Chand, S.P.; Arif, H. Depression. In *StatPearls*; StatPearls Publishing: Treasure Island, FL, USA, 2022.
7. Nabavi, S.; Ebrahimzadeh, M.; Nabavi, S.; Mahmoudi, M.; Rad, S. Biological activities of *Juglans regia* flowers. *Rev. Bras. Farmacogn.* **2011**, *21*, 465–470. [CrossRef]
8. Mahmoudi, M.; Ebrahimzadeh, M.; Ansaroudi, F.; Nabavi, S.F.; Nabavi, S.M. Antidepressant and antioxidant activities of Artemisia absinthium L. at flowering stage. *Afr. J. Biotechnol.* **2009**, *8*, 7170–7175. [CrossRef]
9. Rath, B.P.; Pradhan, D. Antidepressant Activity of *Juglans regia* L. Fruit Extract. *Int. J. Toxicol. Pharmacol. Res.* **2009**, *1*, 24–29.
10. Mouhajir, F.; Hudson, J.B.; Rejdali, M.; Towers, G.H.N. Multiple Antiviral Activities of Endemic Medicinal Plants Used by Berber Peoples of Morocco. *Pharm. Biol.* **2001**, *39*, 364–374. [CrossRef]
11. Vardhini, S.R. Exploring the antiviral activity of juglone by computational method. *J. Recept. Signal Transduct.* **2014**, *34*, 456–457. [CrossRef]

12. Liu, Z.M.; Wen, R.X.; Ma, H.T.; Yang, Y.S.; Wang, X.L.; Lv, X.H.; Li, Z.L. Investigation on inhibition of HIV III B virus with extractions of *Juglans regia*. *Zhongguo Zhong Yao Za Zhi* **2008**, *33*, 2535–2538. [PubMed]

13. D'Angeli, F.; Malfa, G.A.; Garozzo, A.; Li Volti, G.; Genovese, C.; Stivala, A.; Nicolosi, D.; Attanasio, F.; Bellia, F.; Ronsisvalle, S.; et al. Antimicrobial, Antioxidant, and Cytotoxic Activities of *Juglans regia* L. Pellicle Extract. *Antibiotics* **2021**, *10*, 159. [CrossRef] [PubMed]

14. Cui, J.; Jia, J. Discovery of juglone and its derivatives as potent SARS-CoV-2 main proteinase inhibitors. *Eur. J. Med. Chem.* **2021**, *225*, 113789. [CrossRef] [PubMed]

15. Pan, A.; Sun, Q.; Manson, J.E.; Willett, W.C.; Hu, F.B. Walnut Consumption Is Associated with Lower Risk of Type 2 Diabetes in Women. *J. Nutr.* **2013**, *143*, 512–518. [CrossRef] [PubMed]

16. Nasiry, D.; Khalatbary, A.R.; Ahmadvand, H.; Talebpour Amiri, F.B. Effects of Juglans regia L. leaf extract supplementation on testicular functions in diabetic rats. *Biotech. Histochem.* **2021**, *96*, 41–47. [CrossRef]

17. Qanash, H.; Yahya, R.; Bakri, M.M.; Bazaid, A.S.; Qanash, S.; Shater, A.F.; TM, A. Anticancer, antioxidant, antiviral and antimicrobial activities of Kei Apple (*Dovyalis caffra*) fruit. *Sci. Rep.* **2022**, *12*, 5914. [CrossRef]

18. Wani, A.K.; Akhtar, N.; Sharma, A.; El-Zahaby, S.A. Fighting Carcinogenesis with Plant Metabolites by Weakening Proliferative Signaling and Disabling Replicative Immortality Networks of Rapidly Dividing and Invading Cancerous Cells. *Curr. Drug Deliv.* **2023**, *202*, 2. [CrossRef]

19. Al-Rajhi, A.M.H.; Qanash, H.; Almuhayawi, M.S.; Al Jaouni, S.K.; Bakri, M.M.; Ganash, M.; Salama, H.M.; Selim, S.; Abdelghany, T.M. Molecular Interaction Studies and Phytochemical Characterization of Mentha pulegium L. Constituents with Multiple Biological Utilities as Antioxidant, Antimicrobial, Anticancer and Anti-Hemolytic Agents. *Molecules* **2022**, *27*, 4824. [CrossRef]

20. Sugie, S.; Okamoto, K.; Rahman, K.M.; Tanaka, T.; Kawai, K.; Yamahara, J.; Mori, H. Inhibitory effects of plumbagin and juglone on azoxymethane-induced intestinal carcinogenesis in rats. *Cancer Lett.* **1998**, *127*, 177–183. [CrossRef]

21. Segura-Aguilar, J.; Jönsson, K.; Tidefelt, U.; Paul, C. The cytotoxic effects of 5-OH-1,4-naphthoquinone and 5,8-diOH-1,4-naphthoquinone on doxorubicin-resistant human leukemia cells (HL-60). *Leuk. Res.* **1992**, *16*, 631–637. [CrossRef]

22. Salimi, M.; Ardestaniyan, M.H.; Mostafapour Kandelous, H.; Saeidnia, S.; Gohari, A.R.; Amanzadeh, A.; Sanati, H.; Sepahdar, Z.; Ghorbani, S.; Salimi, M. Anti-proliferative and apoptotic activities of constituents of chloroform extract of *Juglans regia* leaves. *Cell Prolif.* **2014**, *47*, 172–179. [CrossRef] [PubMed]

23. Zhang, X.B.; Zou, C.L.; Duan, Y.X.; Wu, F.; Li, G. Activity guided isolation and modification of juglone from Juglans regia as potent cytotoxic agent against lung cancer cell lines. *BMC Complement. Altern. Med.* **2015**, *15*, 396. [CrossRef]

24. Shah, T.I.; Sharma, E.; Shah, G.A. Anti-proliferative, Cytotoxicity and Anti-oxidant Activity of uglans regia Extract. *Am. J. Cancer Prev.* **2015**, *3*, 45–50.

25. Bazaid, A.S.; Aldarhami, A.; Patel, M.; Adnan, M.; Hamdi, A.; Snoussi, M.; Qanash, H.; Imam, M.; Monjed, M.K.; Khateb, A.M. The Antimicrobial Effects of Saudi Sumra Honey against Drug Resistant Pathogens: Phytochemical Analysis, Antibiofilm, Anti-Quorum Sensing, and Antioxidant Activities. *Pharmaceuticals* **2022**, *15*, 1212. [CrossRef]

26. Wianowska, S.; Garbaczewska, A.; Cieniecka-Roslonkiewicz, A.L.; Dawidowicz, A.; Jankowska, A. Comparison of antifungal activity of extracts from different *Juglans regia* cultivars and juglone. *Microb. Pathog.* **2016**, *100*, 263–267. [CrossRef] [PubMed]

27. Choub, V.; Ajuna, H.B.; Won, S.-J.; Moon, J.-H.; Choi, S.-I.; Maung, C.E.H.; Kim, C.-W.; Ahn, Y.S. Antifungal Activity of Bacillus velezensis CE 100 against Anthracnose Disease (Colletotrichum gloeosporioides) and Growth Promotion of Walnut (Juglans regia L.) Trees. *Int. J. Mol. Sci.* **2021**, *22*, 10438. [CrossRef]

28. Upadhyay, V.; Kambhoja, S.; Harshaleena, K. Antifungal activity and preliminary phytochemical analysis of stem bark extracts of Juglans regia linn. *IJPBA* **2010**, *1*, 442.

29. Sytykiewicz, H.; Chrzanowski, G.; Czerniewicz, P.; Leszczyński, B.; Sprawka, I.; Krzyżanowski, R.; Matok, H. Antifungal Activity of Juglans regia (L.) Leaf Extracts Against Candida albicans Isolates. *Pol. J. Environ. Stud.* **2015**, *24*, 1339–1348. [CrossRef]

30. Oliveira, I.; Sousa, A.; Ferreira, I.C.; Bento, A.; Estevinho, L.; Pereira, J.A. Total phenols, antioxidant potential and antimicrobial activity of walnut (*Juglans regia* L.) green husks. *Food Chem. Toxicol.* **2008**, *46*, 2326–2331. [CrossRef]

31. Hamazaki, T.; Okuyama, H. The Japan Society for Lipid Nutrition recommends to reduce the intake of linoleic acid. A review and critique of the scientific evidence. *World Rev. Nutr. Diet* **2003**, *92*, 109–132.

32. Czernichow, S.; Thomas, D.; Bruckert, E. n-6 Fatty acids and cardiovascular health: A review of the evidence for dietary intake recommendations. *Br. J. Nutr.* **2010**, *104*, 788–796. [CrossRef] [PubMed]

33. Simopoulos, A.P. The importance of the ratio of omega-6/omega-3 essential fatty acids. *Biomed. Pharmacother.* **2002**, *56*, 365–379. [CrossRef]

34. El Haouari, M.; Rosado, J.A. Medicinal Plants with Antiplatelet Activity. *Phytother. Res.* **2016**, *30*, 1059–1071. [CrossRef] [PubMed]

35. Haider, S.; Batool, Z.; Tabassum, S.; Perveen, T.; Saleem, S.; Naqvi, F.; Javed, H.; Haleem, D.J. Effects of walnuts (Juglans regia) on learning and memory functions. *Plant Foods Hum. Nutr.* **2011**, *66*, 335–340. [CrossRef]

36. Shi, D.; Chen, C.; Zhao, S.; Ge, F.; Liu, D.; Song, H. Effects of Walnut Polyphenol on Learning and Memory Functions in Hypercholesterolemia Mice. *J. Food Nutr. Res.* **2014**, *2*, 450–456. [CrossRef]

37. Shabani, M.; Nazeri, M.; Parsania, S.; Razavinasab, M.; Zangiabadi, N.; Esmaeilpour, K.; Abareghi, F. Walnut consumption protects rats against cisplatin-induced neurotoxicity. *Neurotoxicology* **2012**, *33*, 1314–1321. [CrossRef] [PubMed]

38. Muthaiyah, B.; Essa, M.M.; Lee, M.; Chauhan, V.; Kaur, K.; Chauhan, A. Dietary supplementation of walnuts improves memory deficits and learning skills in transgenic mouse model of Alzheimer's disease. *J. Alzheimers Dis.* **2014**, *42*, 1397–1405. [CrossRef]

39. Ravanbakhsh, A.; Mahdavi, M.; Jalilzade-Amin, G.; Javadi, S.; Maham, M.; Mohammadnejad, D.; Rashidi, M.R. Acute and Subchronic Toxicity Study of the Median Septum of Juglans regia in Wistar Rats. *Adv. Pharm. Bull.* **2016**, *6*, 541–549. [CrossRef]

40. Yang, L.; Ma, S.; Han, Y.; Wang, Y.; Guo, Y.; Weng, Q.; Xu, M. Walnut Polyphenol Extract Attenuates Immunotoxicity Induced by 4-Pentylphenol and 3-methyl-4-nitrophenol in Murine Splenic Lymphocyte. *Nutrients* **2016**, *8*, 287. [CrossRef]

41. Hausen, B.M. *Woods Injurious to Human Health*; De Gruyter: Berlin, Germany; Boston, MA, USA, 2016.

42. Bonamonte, D.; Foti, C.; Angelini, G. Hyperpigmentation and contact dermatitis due to Juglans regia. *Contact Dermat.* **2001**, *44*, 101–102. [CrossRef]

43. Utiérrez Ortiz, A.L.; Berti, F.; Navarini, L.; Crisafulli, P.; Colomban, S.; Forzato, C. Aqueous extracts of walnut (*Juglans regia* L.) leaves: Quantitative analyses of hydroxycinnamic and chlorogenic acids. *J. Chromatogr. Sci.* **2018**, *56*, 753–760. [CrossRef] [PubMed]

44. Calabrò, S.; Alzoubi, K.; Bissinger, R.; Jilani, K.; Faggio, C.; Lang, F. Enhanced eryptosis following juglone exposure. *Basic Clin. Pharmacol. Toxicol.* **2015**, *116*, 460–467. [CrossRef] [PubMed]

45. Belknap, J.K. Black walnut extract: An inflammatory model. *Vet. Clin. N. A. Equine Pract.* **2010**, *26*, 95–101. [CrossRef]

46. Carvalho, M.; Ferreira, P.J.; Mendes, V.S.; Silva, R.; Pereira, J.A.; Jerónimo, C.; Silva, B.M. Human cancer cell antiproliferative and antioxidant activities of *Juglans regia* L. *Food Chem. Toxicol.* **2010**, *48*, 441–447. [CrossRef]

47. Arslan, H.; Ondul Koc, E.; Ozay, Y.; Canli, O.; Ozdemir, S.; Tollu, G.; Dizge, N. Antimicrobial and antioxidant activity of phenolic extracts from walnut (*Juglans regia* L.) green husk by using pressure-driven membrane process. *J. Food Sci. Technol.* **2022**, *60*, 73–83. [CrossRef]

48. Almeida, I.; Fernandes, E.; Lima, J.; Costa, P.; Bahia, M. Walnut (*Juglans regia*) leaf extracts are strong scavengers of pro-oxidant reactive species. *Food Chem.* **2008**, *106*, 1014–1020. [CrossRef]

49. Żurek, N.; Pawłowska, A.; Pycia, K.; Grabek-Lejko, D.; Kapusta, I.T. Phenolic Profile and Antioxidant, Antibacterial, and Antiproliferative Activity of Juglans regia L. Male Flowers. *Molecules* **2022**, *27*, 2762. [CrossRef]

50. Hosseinzadeh, H.; Behravan, E.; Soleimani, M.M. Antinociceptive and Anti-inflammatory Effects of Pistacia vera LeafExtract in Mice. *Iran. J. Pharm. Res.* **2011**, *10*, 821–828.

51. Bhat, A.A.; Tandon, N.; Tandon, R. Pyrrolidine derivatives as antibacterial agents, current status and future prospects: A patent review. *Pharm. Patent Analyst.* **2022**. [CrossRef]

52. Chiavaccini, L.; Hassel, D.M.; Shoemaker, M.L.; Charles, J.B.; Belknap, J.K.; Ehrhart, E.J. Detection of calprotectin and apoptotic activity within the equine colon from horses with black walnut extract-induced laminitis. *Vet. Immunol. Immunopathol.* **2011**, *144*, 366–373. [CrossRef]

53. Stacey Lockyer, Anne E de la Hunty, Simon Steenson, Ayela Spiro, Sara A Stanner, Walnut consumption and health outcomes with public health relevance—A systematic review of cohort studies and randomized controlled trials published from 2017 to present. *Nutr. Rev.* **2023**, *81*, 26–54. [CrossRef]

54. Amaral, J.S.; Casal, S.; Pereira, J.A.; Seabra, R.M.; Oliveira, B.P. Determination of sterol and fatty acid compositions, oxidative stability, and nutritional value of six walnut (*Juglans regia* L.) cultivars grown in Portugal. *J. Agric. Food Chem.* **2003**, *51*, 7698–7702. [CrossRef] [PubMed]

55. Colaric, M.; Veberic, R.; Solar, A.; Hudina, M.; Stampar, F. Phenolic acids, syringaldehyde, and juglone in fruits of different cultivars of *Juglans regia* L. *J. Agric. Food Chem.* **2005**, *53*, 6390–6396. [CrossRef] [PubMed]

56. Trandafir, I.; Cosmulescu, S.; Botu, M.; Nour, V. Antioxidant activity, and phenolic and mineral contents of the walnut kernel (*Juglans regia* L.) as a function of the pellicle color. *Fruits* **2016**, *71*, 173–184. [CrossRef]

57. Akbari, V.; Jamei, R.; Heidari, R.; Jahanban Esfahlan, A. Antiradical activity of different parts of Walnut (Juglans regia L.) fruit as a function of genotype. *Food Chem.* **2012**, *135*, 2404–2410. [CrossRef]

58. Owens, N.; Lee, D. The use of micro bubble flotation technology in secondary and tertiary produced water treatment—A technical comparison with other separation technologies. In Proceedings of the the the TUV NEL, 5th Produced Water Workshop, Aberdeen, Scotland, 30–31 May 2007.

59. Altun, T.; Pehlivan, E. Removal of Cr (VI) from aqueous solutions by modified walnut shells. *Food Chem.* **2012**, *132*, 693–700. [CrossRef]

60. Ding, D.; Zhao, Y.; Yang, S.; Shi, W.; Zhang, Z.; Lei, Z.; Yang, Y. Adsorption of cesium from aqueous solution using agricultural residue–Walnut shell: Equilibrium, kinetic and thermodynamic modeling studies. *Water Res.* **2013**, *47*, 2563–2571. [CrossRef]

61. Kazemipour, M.; Ansari, M.; Tajrobehkar, S.; Majdzadeh, M.; Kermani, H.R. Removal of lead, cadmium, zinc, and copper from industrial wastewater by carbon developed from walnut, hazelnut, almond, pistachio shell, and apricot stone. *J. Hazard. Mater.* **2008**, *150*, 322–327. [CrossRef]

62. Köhler, S.J.; Cubillas, P.; Rodríguez-Blanco, J.D.; Bauer, C.; Prieto, M. Removal of cadmium from wastewaters by aragonite shells and the influence of other divalent cations. *Environ. Sci. Technol.* **2007**, *41*, 112–118. [CrossRef]

63. Pehlivan, E.; Altun, T. Biosorption of chromium (VI) ion from aqueous solutions using walnut, hazelnut and almond shell. *J. Hazard. Mater.* **2008**, *155*, 378–384. [CrossRef]

64. Saadat, S.; Karimi-Jashni, A. Optimization of Pb (II) adsorption onto modified walnut shells using factorial design and simplex methodologies. *Chem. Eng. J.* **2011**, *173*, 743–749. [CrossRef]

65. Saifuddin, M.; Kumaran, P. Removal of heavy metal from industrial wastewater using chitosan coated oil palm shell charcoal. *Electron. J. Biotechnol.* **2005**, *8*, 43–53.

66. Saqib, A.N.S.; Waseem, A.; Khan, A.F.; Mahmood, Q.; Khan, A.; Habib, A.; Khan, A.R. Arsenic bioremediation by low cost materials derived from Blue Pine (Pinus wallichiana) and Walnut (*Juglans regia* L.). *Ecol. Eng.* **2013**, *51*, 88–94. [CrossRef]
67. Shah, J.; Jan, M.R.; Haq, A.; Khan, Y. Removal of Rhodamine B from aqueous solutions and wastewater by walnut shells: Kinetics, equilibrium and thermodynamics studies. *Front. Chem. Sci. Eng.* **2013**, *7*, 428–436. [CrossRef]
68. Soleimani, M.; Kaghazchi, T. Low-Cost Adsorbents from Agricultural By-Products Impregnated with Phosphoric Acid. *Adv. Chem. Eng. Res.* **2014**, *3*, 34–41.
69. Nowicki, P.; Pietrzak, R.; Wachowska, H. Sorption properties of active carbons obtained from walnut shells by chemical and physical activation. *Catal. Today* **2010**, *150*, 107–114. [CrossRef]
70. Rahman, S. Evaluation of filtering efficiency of walnut granules as deep-bed filter media. *J. Petrol. Sci. Eng.* **1992**, *7*, 239–246. [CrossRef]
71. Zare, H.; Najafpour, G.; Rahimnejad, M.; Tardast, A.; Gilani, S. Biofiltration of ethyl acetate by Pseudomonas putida immobilized on walnut shell. *Bioresour. Technol.* **2012**, *123*, 419–423. [CrossRef]
72. Blumenschein, C.D.; Severing, K.; Boyle, E. Walnut shell filtration for oil and solids removal from steel mill recycle systems. *AISE Steel Technol.* **2001**, *78*, 33–37.
73. Srinivasan, A.; Viraraghavan, T. Removal of oil by walnut shell media. *Bioresour. Technol.* **2008**, *99*, 8217–8220. [CrossRef]
74. Srinivasan, A.; Viraraghavan, T. Oil removal from water using biomaterials. *Bioresour. Technol.* **2010**, *101*, 6594–6600. [CrossRef] [PubMed]
75. Wei, Q.; Ma, X.; Zhao, Z.; Zhang, S.; Liu, S. Antioxidant activities and chemical profiles of pyroligneous acids from walnut shell. *J. Anal. Appl. Pyrolysis* **2010**, *88*, 149–154. [CrossRef]
76. Zhai, M.; Shi, G.; Wang, Y.; Mao, G.; Wang, D.; Wang, Z. Chemical compositions and biological activites of pyroligneous acids from walnut shell. *Bioresources* **2015**, *10*, 1715–1729. [CrossRef]
77. Meshkini, A.; Tahmasbi, M. Anti-platelet aggregation activity of walnut hull extract via suppression of ROS generation and caspase activation. *J. Acupunct. Meridian Stud.* **2017**, *10*, 193–203. [CrossRef]
78. Fernández-Agulló, A.; Pereira, E.; Freire, M.S.; Valentão, P.; Andrade, P.B.; González-Álvarez, J.; Pereira, J.A. Influence of solvent on the antioxidant and antimicrobial properties of walnut (*Juglans regia* L.) green husk extracts. *Ind. Crops Prod.* **2013**, *42*, 126–132. [CrossRef]
79. Van Hellemont, J. *Compendium de Phytotherapie*; Association Pharmaceutique: Bruxelles, Belgium, 1986.
80. Bruneton, J. *Pharmacognosy, Phytochemistry, Medicinal Plants*; Intercept: Hampshire, UK, 1999.
81. Girzu, M.; Carnat, A.; Privat, A.-M.; Fialip, J.; Carnat, A.-P.; Lamaison, J.-L. Sedative effect of walnut leaf extract and juglone, an isolated constituent. *Pharm. Biol.* **1998**, *36*, 280–286. [CrossRef]
82. Hadis, E.; Mostafa, H.-Z.S.; Saeedeh, N.; Mehdi, M. Study of the effects of walnut leaf on the levels of a number of Blood Biochemical Factors in normal male rats fed with high cholesterol diet. *Clin. Biochem.* **2011**, *44*, S331. [CrossRef]
83. Hosseini, S.; Fallah Huseini, H.; Larijani, B.; Mohammad, K.; Najmizadeh, A.; Nourijelyani, K.; Jamshidi, L. The hypoglycemic effect of Juglans regia leaves aqueous extract in diabetic patients: A first human trial. *DARU J. Pharm. Sci.* **2014**, *22*, 19. [CrossRef]
84. Hosseini, S.; Mehrzadi, S.; Najmizadeh, A.R.; Kazem, M.; Alimoradi, H.; Fallah Huseini, H. Effects of Juglans regia L. leaf extract on hyperglycemia and lipid profiles in type two diabetic patients: A randomized double-blind, placebo-controlled clinical trial. *J. Ethnopharmacol.* **2014**, *152*, 451–456. [CrossRef]
85. Mahmoodi, M.; Eghbali, H.; Hosseini Zijoud, S.M.; Pourrashidi, A.; Mohamadi, A.; Borhani, M.; Hassanshahi, G.; Rezaeian, M. Study of the effects of walnut leaf on some blood biochemical parameters in hypercholesterolemic rats. *Biochem. AnalBiochem.* **2011**, *1*, 1–2. [CrossRef]
86. Pitschmann, A.; Zehl, M.; Atanasov, A.G.; Dirsch, V.M.; Heiss, E.; Glasl, S. Walnut leaf extract inhibits PTP1B and enhances glucose-uptake in vitro. *J. Ethnopharmacol.* **2014**, *152*, 599–602. [CrossRef] [PubMed]
87. Alkhawajah, A.M. Studies on the antimicrobial activity of Juglans regia. *Am. J. Chin. Med.* **1997**, *25*, 175–180. [CrossRef] [PubMed]
88. Qa'dan, F.; Thewaini, A.-J.; Ali, D.A.; Afifi, R.; Elkhawad, A.; Matalka, K.Z. The antimicrobial activities of Psidium guajava and Juglans regia leaf extracts to acne-developing organisms. *Am. J. Chin. Med.* **2005**, *33*, 197–204. [CrossRef] [PubMed]
89. Valdebenito, D.; Farías, D.; Oyanedel, E.; Castro, M.; Lampinen, B.; Tixier, A.; Saa, S. The morphology of a Walnut (*Juglans regia* L.) shoot is affected by its position in the canopy and correlated to the number and size of its fruits. *Sci. Hortic.* **2017**, *220*, 303–309. [CrossRef]
90. Claudot, A.-C.; Ernst, D.; Sandermann, H.; Drouet, A. Chalcone synthase activity and polyphenolic compounds of shoot tissues from adult and rejuvenated walnut trees. *Planta* **1997**, *203*, 275–282. [CrossRef]
91. Radix, P.; Seigle-Murandi, F.; Charlot, G. Walnut blight: Development of fruit infection in two orchards. *Crop. Prot.* **1994**, *13*, 629–631. [CrossRef]
92. Mahoney, N.; Molyneux, R.J.; Campbell, B.C. Regulation of aflatoxin production by naphthoquinones of walnut (*Juglans regia*). *J. Agric. Food Chem.* **2000**, *48*, 4418–4421. [CrossRef]
93. Taha, N.A.; Al-wadaan, M.A. Utility and importance of walnut, Juglans regia Linn: A review. *Afr. J. Microbiol. Res.* **2011**, *5*, 5796–5805.
94. Kale, A.; Shah, S.; Gaikwad, S.; Mundhe, K.; Deshpande, N.; Salvekar, J. Elements from Stem Bark of Orchard Tree-Juglans regia. *Int. J. Chemtech. Res.* **2010**, *2*, 548–550.

95. Noumi, E.; Snoussi, M.; Hajlaoui, H.; Valentin, E.; Bakhrouf, A. Antifungal properties of Salvadora persica and *Juglans regia* L. extracts against oral Candida strains. *Eur. J. Clin. Microbiol. Infect. Dis.* **2010**, *29*, 81–88. [CrossRef]
96. Haque, R.; Bin-Hafeez, B.; Parvez, S.; Pandey, S.; Sayeed, I.; Ali, M.; Raisuddin, S. Aqueous extract of walnut (*Juglans regia* L.) protects mice against cyclophosphamide-induced biochemical toxicity. *Hum. Exp. Toxicol.* **2003**, *22*, 473–480. [CrossRef] [PubMed]
97. Bhatia, K.; Rahman, S.; Ali, M.; Raisuddin, S. In vitro antioxidant activity of Juglans regia L. bark extract and its protective effect on cyclophosphamide-induced urotoxicity in mice. *Redox Rep.* **2006**, *11*, 273–279. [CrossRef] [PubMed]
98. Nirmla Devi, T.; Apraj, V.; Bhagwat, A.; Mallya, R.; Sawant, L.; Pandita, N. Pharmacognostic and Phytochemical Investigation of *Juglans regia* Linn. bark. *Pharm. J.* **2011**, *3*, 39–43.
99. Bhat, A.A.; Iqubal, S.; Nitin, T.; Runjhun, T. Structure activity relationship (SAR) and anticancer activity of pyrrolidine derivatives: Recent developments and future prospects (A review). *Eur. J. Med. Chem.* **2022**, *246*, 114954. [CrossRef] [PubMed]
100. Abdelghany, T.M.; Yahya, R.; Bakri, M.M.; Ganash, M.; Amin, B.H.; Qanash, H. Effect of Thevetia peruviana seeds extract for microbial pathogens and cancer control. *Int. J. Pharmacol.* **2021**, *17*, 643–655. [CrossRef]
101. Freeman, P. Tyler's Herbs of Choice-The Therapeutic Use of Phytomedicinals. J. E. Robbers and V. E. Tyler. New York, NY: Haworth Herbal Press. 2000. ISBN 0-7890- 0160-8. *Br. J. Nutr.* **2000**, *84*, 583. [CrossRef]
102. Fujita, T.; Sezik, E.; Tabata, M.; Yesilada, E.; Honda, G.; Takeda, Y.; Tanaka, T.; Takaishi, Y. Traditional medicine in Turkey VII. Folk medicine in middle and west Black Sea regions. *Econ. Bot.* **1995**, *49*, 406–422. [CrossRef]
103. Kaileh, M.; Vanden Berghe, W.; Boone, E.; Essawi, T.; Haegeman, G. Screening of indigenous Palestinian medicinal plants for potential anti-inflammatory and cytotoxic activity. *J. Ethnopharmacol.* **2007**, *113*, 510–516. [CrossRef]
104. Spaccarotella, K.J.; Kris-Etherton, P.M.; Stone, W.L.; Bagshaw, D.M.; Fishell, V.K.; West, S.G.; Lawrence, F.R.; Hartman, T.J. The effect of walnut intake on factors related to prostate and vascular health in older men. *Nutr. J.* **2008**, *7*, 13. [CrossRef]
105. Dominic, R.; Ramanujam, S.N. Traditional knowledge and ethnobotanical uses of piscicidal plants of Nagaland, north east India. *Indian J. Nat. Prod. Resour.* **2012**, *3*, 582–588.
106. Bazaid, A.S.; Alamri, A.; Almashjary, M.N.; Qanash, H.; Almishaal, A.A.; Amin, J.; Binsaleh, N.K.; Kraiem, J.; Aldarhami, A.; Alafnan, A. Antioxidant, Anticancer, Antibacterial, Antibiofilm Properties and Gas Chromatography and Mass Spectrometry Analysis of Manuka Honey: A Nature's Bioactive Honey. *Appl. Sci.* **2022**, *12*, 9928. [CrossRef]
107. Sharma, P.; Verma, P.K.; Sood, S.; Pankaj, N.K.; Agarwal, S.; Raina, R. Neuroprotective potential of hydroethanolic hull extract of Juglans regia L. on isoprenaline induced oxidative damage in brain of Wistar rats. *Toxicol. Rep.* **2021**, *8*, 223–229. [CrossRef] [PubMed]
108. Pereira, J.; Oliveira, I.; Sousa, A.; Valentão, P.; Andrade, P.; Ferreira, I.; Ferreres, F.; Bento, A.; Seabra, R.; Estevinho, L. Walnut (*Juglans regia* L.) leaves: Phenolic compounds, antibacterial activity and antioxidant potential of different cultivars. *Food Chem. Toxicol.* **2007**, *45*, 2287–2295. [CrossRef] [PubMed]
109. Bennacer, A.; Sahir-Halouane, F.; Aitslimane-Aitkaki, S.; Oukali, Z.; Oliveira, I.V.; Rahmouni, N.; Aissaoui, M. Structural characterization of phytochemical content, antibacterial, and antifungal activities of Juglans regia L. leaves cultivated in Algeria. *Biocatal. Agric. Biotechnol.* **2022**, *40*, 102304. [CrossRef]
110. Pereira, J.A.; Oliveira, I.; Sousa, A.; Ferreira, I.C.; Bento, A.; Estevinho, L. Bioactive properties and chemical composition of six walnut (*Juglans regia* L.) cultivars. *Food Chem. Toxicol.* **2008**, *46*, 2103–2111. [CrossRef]
111. Rather, M.A.; Dar, B.A.; Dar, M.Y.; Wani, B.A.; Shah, W.A.; Bhat, B.A.; Ganai, B.A.; Bhat, K.A.; Anand, R.; Qurishi, M.A. Chemical composition, antioxidant and antibacterial activities of the leaf essential oil of Juglans regia L. and its constituents. *Phytomedicine* **2012**, *19*, 1185–1190. [CrossRef]
112. Fang, Y.Z.; Yang, S.; Wu, G. Free radicals, antioxidants, and nutrition. *Nutrition* **2002**, *18*, 872–879. [CrossRef]
113. Gaunt, M.W.; Turner, S.L.; Rigottier-Gois, L.; Lloyd-Macgilp, S.A.; Young, J.P. Phylogenies of atpD and recA support the small subunit rRNA-based classification of rhizobia. *Int. J. Syst. Evol. Microbiol.* **2001**, *51*, 2037–2048. [CrossRef]
114. Okatan, V.; Bulduk, I.; Kaki, B.; Gundesli, M.A.; Usanmaz, S.; Alas, T.; Hajizadeh, H.S. Identification and quantification of biochemical composition and antioxidant activity of walnut pollens. *Pak. J. Bot.* **2021**, *53*, 1–10. [CrossRef]
115. Wani, A.K.; Hashem, N.M.; Akhtar, N.; Singh, R.; Madkour, M.; Prakash, A. Understanding microbial networks of farm animals through genomics, metagenomics and other meta-omic approaches for livestock wellness and sustainability—A Review. *Ann. Anim. Sci.* **2022**, *22*, 839–853. [CrossRef]
116. Iwamoto, M.; Sato, M.; Kono, M.; Hirooka, Y.; Sakai, K.; Takeshita, A.; Imaizumi, K. Walnuts lower serum cholesterol in Japanese men and women. *J. Nutr.* **2000**, *130*, 171–176. [CrossRef] [PubMed]

Review

The Role of Plant-Derived Natural Products in the Management of Inflammatory Bowel Disease—What Is the Clinical Evidence So Far?

Mariela Martinez Davila and Efstathia Papada *

Division of Medicine, University College London, London WC1E 6JF, UK
* Correspondence: e.papada@ucl.ac.uk

Abstract: Inflammatory bowel diseases (IBD), including Crohn's disease (CD) and ulcerative colitis (UC), are a major healthcare challenge worldwide. Disturbances in the immune system and gut microbiota followed by environmental triggers are thought to be part of the aetiological factors. Current treatment for IBD includes corticosteroids, immunosuppressants, and other biologic agents; however, some patients are still unresponsive, and these are also linked to high financial load and severe side effects. Plant-derived natural products are rich in phytochemicals and have been used as healing agents in several diseases since antiquity due to their antioxidant, anti-inflammatory, and immunomodulatory properties, as well as gut microbiota modulation. Numerous in vitro and in vivo studies have shown that phytochemicals act in key pathways that are associated with the pathogenesis of IBD. It is also reported that the use of plant-derived natural products as complementary treatments is increasing amongst patients with IBD to avoid the side effects accompanying standard medical treatment. This review summarises the relevant evidence around the use of plant-derived natural products in the management of IBD, with specific focus on the clinical evidence so far for Curcumin, Mastiha, *Boswellia serrata,* and *Artemisia absinthium.*

Keywords: phytochemicals; inflammatory bowel disease; gut microbiota; Mastiha; Curcumin; *Boswellia serrata*; *Artemisia absinthium*

Citation: Davila, M.M.; Papada, E. The Role of Plant-Derived Natural Products in the Management of Inflammatory Bowel Disease—What Is the Clinical Evidence So Far? *Life* **2023**, *13*, 1703. https://doi.org/10.3390/life13081703

Academic Editor: Othmane Merah

Received: 7 July 2023
Revised: 31 July 2023
Accepted: 1 August 2023
Published: 8 August 2023

1. Introduction

Inflammatory bowel disease (IBD) is a gastrointestinal illness that is characterised by chronic inflammation of the gastrointestinal (GI) tract, with the main two types including ulcerative colitis (UC) and Crohn's disease (CD) [1]. UC and CD are characterised by periods of relapse and remission, and despite sharing similar characteristics and symptoms such as abdominal pain, fatigue, reduced appetite, and diarrhoea, [2], they differ in several aspects. UC is characterised by a continuous pattern of inflammation that can affect the colon and rectum at the level of the mucosa [3], while CD can attack any part of the GI tract (mouth to anus) with transmural inflammation in an intermittent pattern [4].

Over the last few decades, a global rise in the incidence of IBD has been observed, and although IBD has been traditionally considered a disease of the Western world, there is an increasing incidence in newly industrialised countries in Asia, including China and India, as emerging data have shown. However, the reasons for this swift increase in the occurrence of IBD are not fully understood [5].

The pathogenesis of IBD has not been fully elucidated yet, although it seems to be a result of an abnormal response to normal antigens of the gastrointestinal tract. More specifically, a combination of genetic, immunological, and environmental factors initially contributes to destruction of the intestinal epithelial barrier and an increase in intestinal permeability, leading to an influx of immune cells in the intestinal lumen. This dysregulation of the immunomodulation of intestinal mucosa leads to abnormal function of activated T cells, mononuclear cells, and macrophages causing lesions in epithelial cells and chronic inflammation [6]. Perturbations in gut microbiota characterised by depleted diversity, reduced abundance of short chain fatty acid (SCFA) producers, and enriched proinflammatory microbes such as adherent/invasive *Escherichia coli (E. coli)* and H2S producers [7], may also contribute to the inflammatory state in IBD through affecting either the immune system or the metabolic pathways.

The medical management of CD and UC includes a wide range of pharmacological agents such as aminosalicylates, corticosteroids, immunomodulators, biological therapy, and antibiotics. Surgical management is considered when patients do not respond to the medications, and it is still necessary in 30–40% of patients with CD and 20–30% of patients with UC [8]. The main goal of the conventional treatment is to induce and maintain disease remission by achieving clinical and endoscopic healing. Unfortunately, some patients do not respond to the treatment, and usually, several medications are associated with serious side effects. Also considering the chronic, progressive nature of IBD and the increased economic burden of the medical treatment, dietary interventions and plant-derived natural products are gaining attention as alternative or complementary therapies as some patients seek a more natural approach in long-term IBD management [9].

Plant-derived natural products have been well known for their favourable effects on human health since antiquity for pain management, wound healing, gut health, and several other indications. As such, they have attracted scientific interest over the last decades with several studies on identification, isolation, and quantification of their compounds and their biomedical properties. Most of the evidence regarding the effects on plant-derived natural products on IBD comes from in vitro studies and animal models, but there are also numerous clinical trials which have been conducted in humans.

Within this narrative review, we will recap the main mechanisms of action of phytochemicals on IBD, as well as the current evidence from clinical trials investigating the effects of selected plant-derived bioactive compounds rich in phytochemicals on the clinical course of IBD with a specific focus on their anti-inflammatory, antioxidant, and immunomodulatory effects, as well as their impact on gut microbiota modulation. There are several plant-derived natural products evaluated in vitro and in animal models of IBD, but within this review, we focused on reviewing the evidence on some of the most well studied products in human trials in IBD, including curcumin, Mastiha, *Boswellia serrata (BS),* and *Artemisia absinthium (AA).* The primary search tools in this literature review were Pubmed and Google Scholar. We used keywords and search terms such as "phytochemicals and IBD", "phytochemicals and UC", "phytochemicals and CD", "natural products and IBD", "natural products and UC", "natural products and CD". Human trials that evaluated the therapeutic effects of natural products in IBD were included. Additional searches were performed using "Curcumin", "Mastiha", "Mastic gum", "Chios mastic", *"Pistacia lentiscus",* *"Boswellia serrata",* and *"Artemisia absinthium".* The reference lists from the retrieved articles were also checked for other relevant studies.

2. Plant-Derived Natural Products in IBD

Plant-derived natural products have been traditionally used to provide benefits in terms of health and disease. Their favourable effects have been attributed mainly to their content in phytochemicals of bioactive elements that are naturally present in fruits, vegetables, grains, and other plants [10], and that exhibit antioxidant, anti-inflammatory, and anti-proliferative activity [11]. There are around 10,000 different phytochemicals, and they are categorised mainly into polyphenols, glycosinolates, carotenoids, alkaloids, and terpenes [12]. In vitro and animal models have shown that these are implicated in several biological processes, including scavenging of free radicals, induction of anti-inflammatory responses, maintenance of the homeostatic regulation of the gut microbiota, activation of the intestinal T regulatory cells, maintenance of the mucosal barrier integrity, and inflammatory pathway control [13].

Phytochemicals have shown to be effective in key pathways that are associated with the pathogenesis of IBD, as shown in Figure 1, especially through cytokine regulation and reduction in oxidative stress [14,15]. Some of the mechanisms implicated in the pathogenesis of IBD are the release of tumour necrosis factor-alpha (TNF-a) from infiltrating immune cells partnered with the overexpression of nuclear factor kappa B (NF-kB); increased cytokine concentrations, such as interleukin (IL)-6, IL-12 in CD, and IL-13 in UC [16]; and increased production of reactive oxygen species (ROS) exacerbating oxidative stress [17]. These factors create a vicious cycle that eventually leads to the disruption of gut homeostasis and altered intestinal function [18].

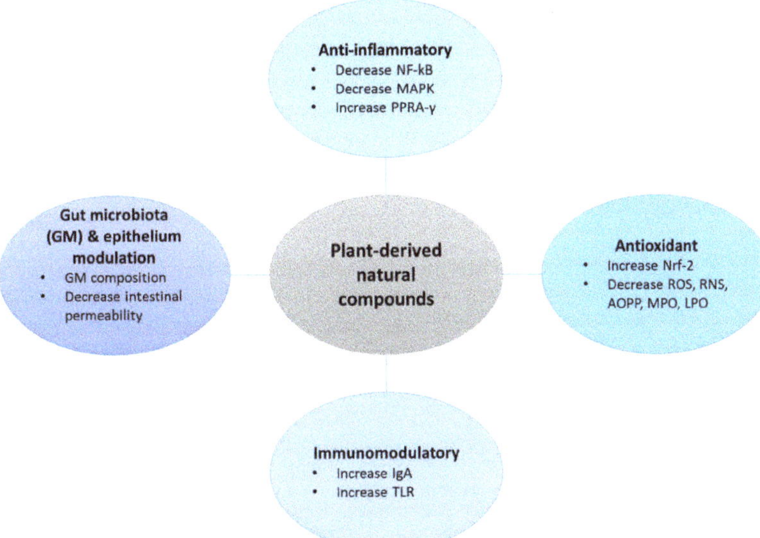

Figure 1. Effects of plant-derived natural compounds on IBD. AOPP = advanced oxidation protein products, IgA = immunoglobulin A, LOP = lipid-oxidised products, NF-kB = nuclear factor kappa B, MAPK = mitogen-activated protein kinase, MPO = myeloperoxidase, PPAR-γ = peroxisome proliferator-activated receptor, ROS = reactive oxygen species, RNS = reactive nitrogen species, TNF-a = tumour necrosis factor alpha, TLR = Toll-like receptors, Nrf-2 = nuclear factor erythroid 2–related factor 2.

Additionally, there is strong evidence that the pathogenesis of IBD is also related to toll-like receptors (TLRs), important mediators of inflammatory pathways in the gut, which play a key role in mediating immune responses. TLR4, one class of TLRs, is thought to play a role in intestinal inflammatory diseases, as it is implicated in the maturation of dendritic cells and differentiation of T helper cells (Th1 and Th2). Furthermore, it can induce the differentiation of macrophages to an M1 phenotype, therefore producing proinflammatory cytokines [19]. TLR4 can also trigger the activation of NF-kB and mitogen-activated protein kinases (MAPKs), as well as the induction of cytokines and inflammation-related enzymes [20]. Studies suggest that certain phytochemicals, such as polyphenols, exert their anti-inflammatory effects through the inhibition of TLR4/NF-kB-mediated signalling pathways and by downregulating the expression of pro-inflammatory mediators [19].

Phytochemicals' health benefits are also linked with the regulation of different microRNAs, which are implicated in the regulation of the intestinal epithelial barrier, T-cell differentiation, and the Th17 signalling pathway. They also interfere in some of the inflammatory signalling pathways (NF-kB) and signal transducer and activator of transcription STAT/IL-6 pathways [21].

3. Modulation of Gut Microbiota

The gut microbiota (GM) can be defined as the community of microorganisms in the gastrointestinal tract and seems to play an important role in human health, including in terms of immune, metabolic and neurobehavioural traits [22]. IBD has been associated with dysbiosis, which is defined as a decrease in gut microbial diversity due to an imbalance between commensal and pathogenic microorganisms [23]. The GM of patients with IBD is characterised by low microbial diversity and decreased abundance of *Bifidobacterium* spp., *Lactobacillus* spp., *and F. prausnitzii,* while having an abundance of pathologic bacteria such as *E. coli* and *Clostridium difficile (C. difficile)* [7].

Phytochemicals in plant-derived natural products are poorly absorbed in the small intestine; therefore, they reach the large intestine, where they can exert a prebiotic and selective antimicrobial effect [24]. The proposed mechanism behind these health benefits is through the metabolites produced by the GM, rather than the bioactive compounds per se [25]. These metabolites can effectively modulate the GM mainly by inducing the growth of some beneficial bacterial populations, such as *Lactobacillus* spp. and *Bifidobacterium* spp., and by triggering the generation of SCFAs [25]. One example is resveratrol, a polyphenol found in berries, grapes, and groundnuts, which showed an increase in the abundance of *Lactobacillus* spp. and *Bifidobacterium* spp. while inhibiting the production of harmful bacteria such as *E. coli* in mice models [26,27]. Additionally, another trial showed that mice fed with resveratrol had an increased amount of SCFA producers like *Butyricicoccus* spp., *Ruminococcus* spp., and *Roseburia* spp. [28]. Furthermore, epigallocatechin (EGCG), a polyphenol found in green tea, showed a prebiotic effect by significantly enriching SCFA-producing bacteria, such as *Akkermansia* spp., while attenuating colitis symptoms in a mouse model of UC when supplemented as prophylaxis [29].

Other mechanisms by which phytochemicals and GM are related include the interaction with the immune system, especially by immunoglobulin A (IgA), regulating the degree of colonisation and preventing dysbiosis [30], as it has been reported with some carotenoids (b-carotene and astaxanthin) [31]. They also include influence on the mucin layer and tight junction integrity [32], as well as synergistic anti-inflammatory effects with other compounds like omega-3 fatty acids [33].

4. Evidence from Human Studies

As mentioned above, plenty of in vitro and animal studies have evaluated the effects of plant-derived natural products in IBD. Data from human studies are also available, but to a significantly lesser extent compared to animal and in vitro studies. However, as it is estimated that around 21–60% of patients with IBD use complementary or alternative therapies (herbal remedies, dietary supplements) due to a perception of less toxicity or harm, and up to 75% do not discuss it with their physicians [34], it is essential to explore the current evidence regarding the benefits and risks. In this part we will discuss the clinical evidence on the effects of some plant-derived natural products, including curcumin, Mastiha, *Boswellia serrata,* and *Artemisia absinthium* on patients living with IBD.

4.1. Curcumin

Curcumin is the biologically active, phenolic component of turmeric (*Curcuma longa*) [35]. Turmeric consists of several significant constituents isolated from the rhizome—structurally-related curcuminoids, including curcumin as the most important and the main active compound (Figure 2a). It has been characterised by antimicrobial, antioxidant, and immunomodulatory properties in several in vitro and in vivo studies, and it is one of the most evaluated compounds in humans. In the intestinal mucosa, curcumin can reduce the levels of ROS, superoxide anions, and malondialdehyde (MDA) [36]. Additionally, curcumin strongly inhibits the expression of NF-kB and downstream signalling of proinflammatory cytokines such as TNF-a [37].

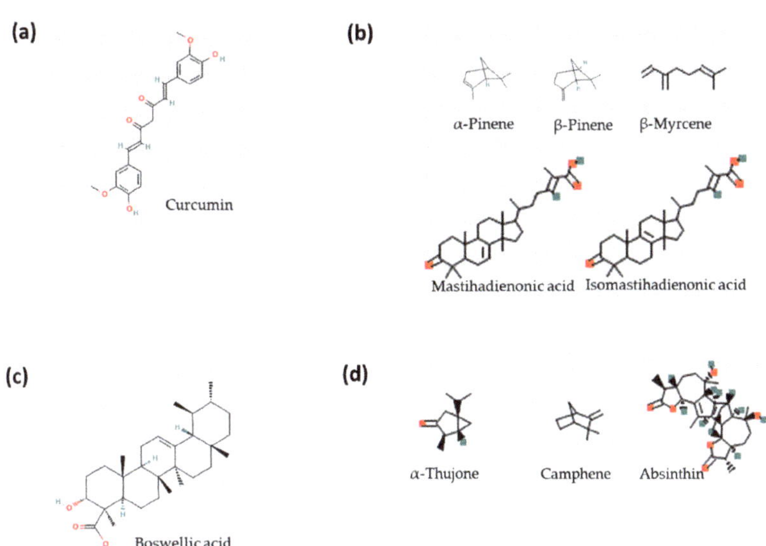

Figure 2. Main bioactive compounds of (**a**) *Curcuma longa* (curcumin [38]); (**b**) Mastiha (α-pinene [39], β-pinene [40], β-myrcene [41], mastihadienonic acid [42], isomastihadienonic acid [43]); (**c**) Boswellia serrata (boswellic acid [44]); (**d**) Artemisia absinthium (α-thujone [45], camphene [46], absinthin [47]). Source of chemical structures: PubChem [48].

Several trials in patients with UC showed positive results in terms of inducing remission with no adverse effects, as well as improving endoscopic and clinical activity (Table 1). A pilot study in five patients with UC reported a significant improvement in the number and quality of stools ($p < 0.02$) with the supplementation of 550 mg of curcumin twice a day for one month and then 550 mg three times a day for another month, as assessed by a global score. In the same study, curcumin's efficacy as an add-on therapy to existing treatments of CD was also tested using 360 mg of curcumin three times a day for one month, and then four times a day for two months, in five patients with CD. The Crohn's Disease Activity Index (CDAI) score and CRP levels for all subjects fell [49]. Another trial compared curcumin supplementation combined with either sulfasalazine or mesalamine versus placebo in eighty-nine patients with quiescent UC. Patients received curcumin capsules (2 g/day) for six months. The results showed an improvement both in the clinical activity index (CAI) and the endoscopic index (EI), therefore suppressing the morbidity associated with UC [50].

Given the high use of complementary alternatives, a forced-dose titration study in patients with CD and UC tested curcumin's tolerability and safety. Patients initially received 500 mg twice per day for three weeks, and the dose was increased up to 1 g twice per day at week 3 for a total of 3 weeks, and then titrated again to 2 g twice per day at week 6 for three weeks. Validated measures of disease activity were used, such as the Paediatric Ulcerative Colitis Activity Index (PUCAI), Paediatric Crohn's Disease Activity Index (PCDAI), and Monitoring of Side Effect System Score, and all patients tolerated curcumin well with no serious side effects reported [51]. It is essential to mention that this study was performed in a paediatric population; therefore, further research is required in order to fully assess the safety and efficacy in larger studies including patients of different ages.

Forty-five patients with distal UC were randomised to either NCB-02 enema (140 mg of NCB-02 (curcumin) preparation dissolved in 20 mL of water) or placebo, both complemented with oral 5-ASA (800 mg twice daily), for eight weeks. At week 8, clinical remission and improvements in endoscopy were higher in the NCB-02 group compared to the placebo, but the difference did not reach statistical significance [52].

Curcumin's efficacy was studied as an add-on therapy with mesalamine treatment (3 g/day for 1 month with continued mesalamine) to induce remission in fifty patients with active UC. The combination of curcumin and mesalamine was significantly superior to the combination of placebo and mesalamine in clinical (reduction > 3 points in Simple Clinical Colitis Activity Index (SCCAI)) and endoscopic remission (partial Mayo score < 1) [53]. Disease activity improvement was reported in seventy patients with mild-to-moderate UC when supplemented with curcumin 1500 mg/day for 8 weeks compared to placebo [54]. Clinical outcomes and quality of life were significantly improved in curcumin-treated patients. Additionally, curcumin supplementation reduced serum high-sensitivity C-reactive protein (hs-CRP) concentration and the erythrocyte sedimentation rate (ESR) significantly, biomarkers which are usually elevated in IBD.

The increased interest in using supplements as a co-adjuvant treatment for IBD has led to the development of products with enhanced properties; new forms of curcumin have been synthesised with higher absorption rates than conventional curcumin powder [55]. A nanomicellar curcuminoid formula with higher bioavailability than conventional formulas was used to test the effectiveness of the treatment of UC. Fifty-six patients with mild-to-moderate UC were randomly assigned to receive the formula (80 mg, 3 times/day orally) plus 3 g/day mesalamine or placebo plus mesalamine for 4 weeks [56]. There was a significant improvement in the score for urgency of defecation in the formula supplemented group, while an increase in self-reported well-being and reduced clinical activity were also reported.

Another study in thirty patients with CD tested a newly synthesised curcumin derivative (Theracurmin) with a higher absorption rate for 12 weeks, showing a significant improvement in clinical remission rates ($p = 0.020$) and endoscopic measures ($p = 0.032$) compared with the placebo with no adverse effects [57]. Lastly, another bioenhanced form of curcumin with higher bioavailability (BEC) was studied versus a placebo in sixty-nine patients with mild-to-moderate UC on standard doses of mesalamine for up to 12 months [58]. Clinical response, clinical remission, and endoscopic remission were evaluated at 6 weeks and 3 months, then followed up at 6 and 12 months to assess the maintenance of remission. At 6 weeks, these were significantly higher in the BEC group compared with the placebo. A total of 95% of BEC responders maintained clinical remission compared to none in placebo group at 6 months [58].

Although the aforementioned studies have reported some promising results regarding the use of curcumin as an adjunct to the standard medical treatment for patients with IBD, a prospective randomised double-blind placebo-controlled trial comparing the remission-inducing effect of oral curcumin and mesalamine using a placebo or 2.4 g of mesalamine in patients with ulcerative colitis of mild-to-moderate severity concluded that a dose of 450 mg/day for 8 weeks was not effective in terms of inducing remission [59]. Another study in patients with CD undergoing bowel resection was performed to assess whether curcumin was efficient for the prevention of post-operative recurrence of CD [60]. Patients received 3 g/day oral curcumin plus azathioprine 2.5 mg/day or an identical placebo for 6 months. The results suggested that curcumin was no more effective than the placebo in preventing CD recurrence, as no significant differences between groups were reported. Thus, there is still need for further research with increased sample sizes to be able to reach safe conclusions on the use of curcumin in the management of IBD.

Table 1. Clinical evidence on the effects of curcumin on IBD.

Aspect Evaluated	Sample	Duration	Dose	MAIN Outcomes	Reference
Effect on IBD	5 patients with UC	2 months	550 mg of curcumin twice daily for 1 month and then 3 times/d for another month for UC	Improved global score (number and quality of stools) ($p < 0.02$), serologic indexes, sedimentation rate, and CRP decrease within normal limits	[49]
	5 patients with CD	3 months	360 mg of curcumin 3 times/day for 1 month and then 360 mg 4 times/day for 2 months	Decrease in CDAI (mean reduction of 55 points); decrease in sedimentation rate (mean reduction of 10 mm/h); CRP was reduced by a mean of 0.1 mg/dL	
Efficacy as a maintenance therapy in quiescent UC	89 patients with quiescent UC	6 months	Curcumin capsules 2 g/day plus sulfasalazine or mesalamine or placebo plus sulfasalazine or mesalamine	Improved clinical activity index (CAI) ($p = 0.038$) and endoscopic index (EI) ($p = 0.0001$)	[50]
Tolerability in children with IBD	11 patients with IBD(6 with CD and 5 UC)	9 weeks	500 mg curcumin twice daily plus 1 g increase twice daily at week 3 plus 2 g twice daily at week 6	Curcumin was well tolerated in doses up to 2 g twice per day	[51]
Efficacy and safety in distal UC	45 patients with distal UC	8 weeks	Curcumin enema plus oral 5-ASA or placebo enema plus 5-ASA	Higher remission in patients with curcumin treatment ($p = 0.14$); improvement in endoscopic parameters ($p = 0.29$)	[52]
Add-on therapy with optimised mesalamine in UC	50 patients with active mild-to-moderate UC	1 month	3 g/day curcumin capsules with continued mesalamine or placebo capsules plus mesalamine	Higher clinical improvement and remission in patients on curcumin vs. placebo ($p < 0.01$) Higher endoscopic remission in the curcumin group vs. placebo ($p = 0.043$).	[53]
Effect on inducing remission in UC	62 patients with active mild-to-moderate UC	8 weeks	450 mg/day curcumin plus mesalamine OR placebo plus mesalamine	No significant differences between groups	[59]
Effect of a nano formulation of curcuminoids in UC	56 patients with mild-to-moderate UC	4 weeks	80 mg curcuminoid formula, 3 times/day orally plus mesalamine or placebo plus mesalamine	Score for urgency of defecation reduced; mean SCCAI score lower in curcuminoid group ($p = 0.050$)	[56]
Theracurmin efficacy and safety in CD.	30 patients with active-to-moderate CD	12 weeks	360 mg/day Theracurmin or placebo	CDAI score significantly decreased ($p = 0.005$) in Theracurcumin group Reduction in endoscopic CD severity ($p = 0.032$) compared to baseline	[57]

Table 1. *Cont.*

Aspect Evaluated	Sample	Duration	Dose	MAIN Outcomes	Reference
Effect on improvement of disease activity in UC.	70 patients with mild-to-moderate UC	8 weeks	1500 mg/day curcumin or placebo	Improvement in clinical outcomes Changes in SCCAI ($p < 0.001$) and IBDQ ($p = 0.006$) higher in curcumin group; serum hs-CRP and ESR decreased ($p = 0.01$ and $p = 0.02$)	[54]
Effect in preventing post-operative recurrence of CD	62 patients with CD undergoing bowel resection	6 months	Azathioprine (2.5 mg/kg) plus oral curcumin (3 g/day) or placebo	Curcumin not more effective than placebo to prevent CD recurrence.	[60]
Efficacy of a novel bio enhanced curcumin as add on therapy in UC.	69 patients with mild-to-moderate UC	Up to 12 months (6-week, 3-month, 6-month, and 12-month follow-ups)	50 mg of bio-enhanced curcumin (BEC) twice a day or placebo plus standard dose of mesalamine	Higher clinical and endoscopic remission compared to placebo ($p < 0.01$) at 6 weeks Maintenance of remission in 95% (6-mo) and 84% (12-mo) of BEC responders compared to none in the placebo.	[58]

IBD = inflammatory bowel disease, UC = ulcerative colitis, CRP = C-reactive protein, CDAI = Crohn's Disease Activity Index, CAI = Clinical Activity Index, EI = Endoscopic Index, CD = Crohn's Disease, SCCAI = Simple Clinical Colitis Activity Index, IBDQ = Inflammatory Bowel Disease Questionnaire, ESR = erythrocyte sedimentation rate, hs-CRP = high-sensitivity C-Reactive Protein, BEC = bio-enhanced curcumin.

4.2. Mastiha

Mastiha, is a natural product of the Mediterranean basin consisting of a plethora of bioactive constituents, including phenolic compounds, phytosterols, arabino-galactanes proteins, and 30% of a natural polymer (poly-β-myrcene). However, it is a particularly concentrated source of terpenes, such as monoterpenes (i.e., α-pinene, β-pinene, and β-myrcene) and triterpenes (i.e., mastihadienonic acid and isomastihadienonic acid) (Figure 2b) [61]. It is obtained as a dried resinous exudate from stems and branches of the tree *Pistacia lentiscus (Pistacia lentiscus L. var latifolius Coss or Pistacia lentiscus var. Chia)* and has been used since antiquity for its anti-inflammatory and antioxidant properties [62]. The European Medicines Agency has recognised Mastiha as a herbal medicinal product for the following indications: (a) mild dyspeptic disorders and (b) symptomatic treatment of minor inflammations of the skin and as an aid in the healing of minor wounds [63]. Previous data on Mastiha has suggested that it could have favourable effects on the clinical course and inflammatory biomarkers of patients with IBD (Table 2). A pilot study in patients with active CD evaluated the effectiveness of Mastiha administration on the clinical course and blood inflammatory markers. The study included a healthy control group, and participants received 4 weeks of treatment with Mastiha capsules (6 capsules/day, 0.37 g/cap). The results suggested that Mastiha significantly decreased the CDAI and plasma levels of IL-6 and CRP [64]. Peripheral blood mononuclear cells (PBMC) were also evaluated before and after treatment, showing a reduction in TNF-alpha secretion and an increase in macrophage migration inhibitory factor (MIF) release. These findings pointed towards an inhibition of random migration and chemotaxis of monocytes/macrophages [65]. A randomised, double-blind, placebo-controlled trial in patients with IBD assessed the effects of a Mastiha supplement on oxidative stress biomarkers and the plasma-free amino acid (AA) profiles of patients with active IBD (CD and UC). Participants were allocated to receive Mastiha (2.8 g/day) or a placebo for three months, being either under no treatment or under stable medical therapy. A favourable effect on oxidative stress biomarkers was documented, as oxidised low-density lipoprotein (OxLDL), OxLDL/HDL, and OxLDL/LDL decreased significantly in the Mastiha group. Mastiha also ameliorated a reduction in plasma free AAs in patients with UC taking the placebo [66]. It has been proposed that Mastiha can also exert a prebiotic effect, as it showed a regulatory role in faecal lysozyme. In addition to a significant improvement in the IBDQ score, results from the same trial showed a significant decrease in faecal lysozyme in the Mastiha group [67]. Lysozyme is an antimicrobial protein that regulates innate immune response, and its increased expression is correlated with dysbiosis and inflammation [68], suggesting that Mastiha's effect could be prebiotic as well.

A randomised, double-blind, placebo-controlled clinical trial in 68 patients with IBD in remission examined the effects of Mastiha on the clinical course and AA profiles of patients. The results showed that AAs such as valine, alanine, proline, glutamine, and tyrosine significantly increased only in the placebo group compared with the baseline, and the change between the two groups was significantly different. As AAs are considered an early prognostic marker of disease activity, these results may suggest a potential role of Mastiha in remission maintenance, although Mastiha was not proven superior to the placebo in terms of the remission rate [69].

Table 2. Clinical evidence on the effects of Mastiha on IBD.

Aspect Evaluated	Sample	Duration	Dose	Main Outcomes	Reference
Effects on the clinical course of active CD	10 patients with active CD and 8 healthy controls	4 weeks	6 Mastiha caps/day (2.2 g/day)	Significant reduction in CDAI ($p = 0.05$), CRP ($p = 0.028$), and plasma IL-6 (0.027). Total antioxidant potential (TAP) significantly increased in the Mastiha group ($p = 0.036$)	[64]
Effects on cytokine production of circulating mononuclear cells in active CD				Reduction in TNF-a secretion from PBMC in the Mastiha group ($p = 0.028$) and an increase in MIF ($p = 0.026$). No significant changes in IL-6, MCP-1, or GSH.	[65]
Effects on oxidative stress and plasma-free amino acid profile in active IBD	60 patients with active IBD	3 months	2.8 g/day Mastiha (4 tabs 700 mg) plus conventional medical treatment or Placebo plus conventional medical treatment	Significant decrease in oxLDL ($p = 0.031$), oxLDL/HDL ($p = 0.020$) and oxLDL/LDL ($p = 0.005$) in the Mastiha group and amelioration of the decrease in plasma-free AAs in patients with UC.	[66]
Effects on QoL, inflammatory biomarkers, and clinical course in IBD				IBDQ significantly improved ($p = 0.004$); significant decrease in faecal lysozymes ($p = 0.018$) and fibrinogen ($p = 0.006$) in the Mastiha group. Significant increase in faecal lactoferrin ($p = 0.001$) and calprotectin ($p = 0.029$) in the placebo group	[67]
Effect on the clinical course and amino acid profile in inactive IBD	68 patients with inactive IBD	6 months	2.8 g/day Mastiha plus conventional medical treatment or Placebo plus conventional medical treatment	Attenuation of the increase in free AA levels in the Mastiha group; significant decrease in oxidative stress biomarkers	[69]
Regulatory effect on IL-17A serum levels in IBD	43 patients with UC and 86 patients with CD	3 months in active and 6 months in inactive IBD	2.8 g/day Mastiha plus conventional medical treatment or Placebo plus conventional medical treatment	Increase in serum IL-17A in the Mastiha group ($p = 0.006$) and significant difference between Mastiha and placebo in the mean change in inactive IBD.	[70]
Anti-inflammatory activity through regulation of miRNA in IBD	60 patients with IBD (endoscopy-proven CD or UC) with usual medical treatment	3–6 months	2.8 g/day Mastiha or placebo	miR-155 increased in the placebo group in active UC ($p = 0.054$), while it was prevented by Mastiha.	[21]

CD = Crohn's disease, CDAI = Crohn's Disease Activity Index, CRP = C-reactive protein, IL-6 = interleukin-6, TAP = total antioxidant capacity, TNF-a = tumour necrosis Factor-alpha, PBMC = peripheral blood mononuclear cells, MIF = macrophage migration inhibitory factor, MCP-1 = monocyte chemoattractant protein-1, GSH = glutathione, IBD = inflammatory bowel disease, UC = ulcerative colitis, oxLDL = oxidised LDL, AAs = amino acids, QoL = quality of life, IBDQ = Inflammatory Bowel Disease Questionnaire, IL-17A = interleukin 17A, miRNA = microRNA, miR-155 = microRNA 155.

In recent years, biological therapies that target different molecular pathways have been developed, such as TNF blockade, IL-6, IL-12/IL-23, and IL-17 pathways. IL-17 blocking agents have been applied in several anti-inflammatory diseases, but unfortunately, in IBD, clinical benefits have not been established without adverse effects. Mastiha has been shown to act in the IL-17 pathway, and its use as a safe adjunct therapy has been proposed in IBD. A double-blind, placebo-controlled, parallel group study explored Mastiha's immunomodulatory effect on IL-17A serum levels in patients with active and inactive IBD. IL-17A has been studied for its anti-inflammatory nature and a potential protective role in intestinal pathology. The participants received natural Mastiha at a dose of 2.8 g/day or an identical placebo (6 months for patients in remission and 3 months for patients in relapse). The results showed a significant increase in serum IL-17A in patients with inactive UC, while decreasing significantly only with the placebo treatment. No significant differences were reported in active disease. Additionally, Mastiha seemed to influence the stool metabolic profile of patients in remission, as there were increased levels of glycine and tryptophan, both related to therapeutic effects and immunoregulatory mechanisms, such as Th17 cell differentiation. These findings suggest that Mastiha has a potential immunomodulatory role in quiescent IBD [70].

MicroRNAs (miRNAs), which are one of the most studied epigenetic mechanisms, are implicated in the regulation of the intestinal barrier and cell membrane trafficking. They also interfere with inflammatory pathways, such as NF-kB and the signal transducer and activator of the transcription (STAT)/IL-6 pathways. In IBD, miRNAs are usually up-regulated and are also involved in T-cell differentiation and Th17 signalling pathways (especially microRNA-155). Circulating miRNAs are considered a valuable tool to indicate the physiological state of the tissue from which they come, and their modulation by certain phytochemicals has been studied over the years. A study in subsets of patients with inflammatory conditions, including IBD, evaluated whether a common route exists in the anti-inflammatory activity of Mastiha, specifically through the regulation of miRNA levels. Participants received Mastiha in tablets at doses of 2.8 g/day or identical placebo tablets adjunct to conventional medical treatment (6 months for patients in remission and 3 months for patients in relapse). The results showed that, particularly in patients with UC in relapse, miRNA-155 increased in the placebo group significantly, while this increase was prevented by mMastiha supplementation. These findings proposed a regulatory role for Mastiha in circulating levels of miR-155, a critical player in Th17 differentiation and function [21]. However, this was the first study linking Mastiha to epigenetic mechanisms; therefore, it should be confirmed by more studies in larger cohorts.

4.3. Boswellia serrata

Boswellia serrata (BS) is a gum resin rich in terpenes, such as boswellic acid (Figure 2c). It is obtained from the *Boswellia serrata* tree, but evidence regarding its effects on human IBD has been contradictory so far (Table 3). There are several mechanisms of action that contribute to its anti-inflammatory and immunomodulatory activities, such as the ability to inhibit the formation of leukotrienes, which act as potent mediators of inflammatory disorders [71]. A randomised, double-blind, verum-controlled, parallel group study in 102 patients with active CD showed that a BS extract could be as effective as mesalamine, but not superior, in reducing the CDAI score [72]. However, a double-blind, placebo-controlled, randomised, parallel study in 108 outpatients with CD in clinical remission randomised to Boswelan, a BS extract, (3 × 2 capsules/day; 400 mg each) or placebo for 52 weeks was unable to demonstrate superiority in maintaining remission when compared to the placebo, although it showed good tolerability [73]. As bioavailability has been a topic of concern, a novel delivery form of *Boswellia serrata* extract (BSE) with enhanced bioavailability was tested [74]. The study was conducted in 43 patients with UC supplemented with either 250 mg/day of BSE or no supplement for 4 weeks. The results showed a significant positive effect in the supplemented group for all the evaluated parameters, such as intestinal pain, evident and occult blood in stools, bowel movements and cramps, watery stools, etc. The faecal concentration of calprotectin was also significantly decreased in the supplemented group [74]. The present results highlight the importance of future research that will assess the effects of BS, both in maintaining remission and in the management of active disease.

4.4. Artemisia absinthium

Also known as wormwood, *Artemisia absinthium (AA)* is a herbaceous plant considered a very important species in the history of medicine [75]. It contains several groups of phytochemicals (Figure 2d), such as terpenes (e.g., α-thujone, camphene), lactones (e.g., absinthin), flavonoids, phenolic acids, coumarins, and tannins; therefore, it has been widely used as a therapeutic aid in digestive disorders and other clinical conditions [75]. Some in vitro studies have reported that Artemisia species and isolated compounds mainly act by supressing TNF-a and other interleukins [76,77].

A double-blind placebo study in CD patients assessed whether *AA* could reduce the patients' dependence on corticosteroids (Table 4). Participants receiving a daily dose of steroids at an equivalent of 40 mg or less of prednisone for at least three weeks were administered an *AA* containing herbal blend capsule (3 × 500 mg/day) or placebo for ten weeks. The steroid dose was maintained stable until week two; after that, tapering off started and was completed at week 10. Already, after six weeks, there was a significantly higher number of patients who showed clinical improvement (CDAI score of 70 or more) in the *AA* group as compared to placebo, and this continued beyond week 10. Improvements were also reported in mood and quality of life, as measured with the 12-item Hamilton Depression Scale (HAMD). These findings suggest that *AA* might have a steroid-sparing effect on CD patients. However, it is important to mention that five patients from this group showed little response to the *AA* treatment, suggesting that there could be a group of patients who are resistant to the treatment [78].

Table 3. Clinical evidence on the effects of *Boswellia serrata* on IBD.

Aspect Evaluated	Sample	Duration	Dose	Main Outcomes	Reference
Safety and efficacy of BS extract H15 on active CD	102 participants	8 weeks	3.6 g/day BS extract H15 OR mesalamine	Reduction in CDAI score, but no significant superiority compared to mesalamine ($p = 0.061$)	[72]
Effect and safety of long-term therapy in CD	108 patients with CD in remission	52 weeks	Boswelan 3 × 2 capsules/day (400 mg each) OR placebo	Good tolerability and safety; no superiority versus placebo as maintenance therapy ($p = 0.85$)	[73]
Effect of BS extract (BSE) in UC	43 participants with UC in remission for at least 1 year	4 weeks	250 mg/day BSE in a novel lecithin-based delivery form (Casperome®) OR no supplementation	Significant improvement in diffuse intestinal pain, blood in stools, bowel movements and cramps, and reduction in calprotectin levels ($p < 0.05$)	[74]

CD = Crohn's disease, CDAI = Crohn's Disease Activity Index, BS = Boswellia Serrata, BSE = Boswellia Serrata extract, UC = ulcerative colitis.

Table 4. Clinical evidence on the effects of *Artemisia absinthium* on IBD.

Aspect Evaluated	Sample	Duration	Dose	Main Outcomes	Reference
Steroid-sparing effect on CD	40 participants with CD	20 weeks	*AA* containing herbal blend (3 × 500 mg/day) (SedaCrohn®) plus steroids or placebo	Significantly higher clinical improvement using CDAI in SedaCrohn® group ($p < 0.01$)	[78]
TNF-α suppressing effect on CD	20 participants with active CD	6 weeks	3 capsules SedaCrohn® 3 times/day (250 mg of powdered *AA* each one) plus conventional medical treatment or placebo	Significant reduction in CDAI and TNF-a levels compared with placebo	[79]

CD = Crohn's disease, CDAI = Crohn's Disease Activity Index, TNF-a = tumour necrosis factor-alpha, *AA*= Artemisia absinthium

As mentioned previously, TNF-a appears to play a central role in the pathogenesis of CD. A clinical trial studied whether the *AA* TNF-a suppression effect was present in 20 patients with CD. Patients were given, in addition to their basic CD therapy, a dried powdered *AA* treatment (3 × 750 mg) for 6 weeks. A control group was included. The results showed a significant decrease in TNF-a serum levels in the *AA* group as compared to the control group. Additionally, *AA*-treated patients showed significant clinical improvements in their CDAI, IBDQ, and HAMD scores. These findings suggest that *AA* can be a potential adjuvant for the TNF-a-mediated diseases, although further research is needed [79].

5. Bioavailability and Safety Considerations

Bioavailability can be defined as the fraction of the active form of a substance that reaches systemic circulation unaltered and is absorbed and used by the body [80]. Some phytochemicals show low levels of stability, as they are highly metabolised or rapidly eliminated [81]; this has been a challenge when using them as therapeutic agents [82], and the research in this field is of great importance for their clinical use.

Many plant-derived products show remarkable potential in vitro, but due to their poor absorption, the effect in vivo has not been completely demonstrated [83]. Several factors, such as the transformation during metabolic pathways or during processing [84], solubility [83], gender differences, and dosage seem to affect bioavailability [85,86]. Previous research on phytochemical bioavailability showed a peak in plasma antioxidant capacity 1–2 h after intake [87]; however, research on the absorption, metabolism, and distribution of these compounds in the human body remains limited. Different strategies can be implemented to increase bioavailability, such as extraction of the active ingredient or combination with other compounds that enhance absorption rates and could be of great therapeutic use. For example, evidence shows that the absorption of curcumin is increased by 20 times in humans and 1.56 times in rats when co-administered with 20 mg of piperine [88]. However, a recent study reported that liver injury due to turmeric appears to be increasing in the United States, perhaps reflecting usage patterns or increased combination with black pepper [89], indicating that further research regarding safety and education of the public on the use of plant-derived natural compounds is necessary. Data on safety are very limited for the majority of these compounds, since most studies are short-term and use different formulations and dosages, making it difficult to reach safe conclusions.

6. Future Implications and Conclusive Remarks

IBD incidence has been rising exponentially worldwide, and an attempt to improve treatment has been a priority over the years. The recurrent pattern of IBD demands constant treatment, representing a significant financial load to individual patients and healthcare systems, especially in developing countries. Additionally, there are still many challenges, as many adverse effects and/or patients who are non-responsive to treatments are still reported. Emerging evidence shows that a combined treatment using standard medications and plant-derived natural products might provide higher remission rates and decrease adverse effects in IBD. The use of plant-derived natural products in patients with IBD has demonstrated improvement in several aspects assessed by tools such as CDAI, IBDQ, and SCAAI scores, which have been validated and implemented in clinical practice. However, it is worth mentioning that these improvements might not always correlate with endoscopic scores or faecal biomarkers of inflammation, and the indices currently in use are limited to predicting long-term outcomes like surgery, relapse, and disability. Even though evidence on this topic has been increasing, human trials are very limited and sample sizes are quite small. Therefore, it is essential to prioritise clinical trials in order to facilitate the development of official recommendations or guidelines on how to use them safely. Further research focusing on phytochemical bioavailability, optimal doses, and safety is needed for the development of phytochemical-rich products with enhanced therapeutic properties for chronic inflammatory conditions, including IBD.

Author Contributions: M.M.D. wrote the first draft of the manuscript; E.P. conceptualised the content of the manuscript and reviewed and edited the manuscript. All authors have read and agreed to the published version of the manuscript.

Funding: This research received no external funding.

Institutional Review Board Statement: Not applicable.

Informed Consent Statement: Not applicable.

Data Availability Statement: Not applicable.

Conflicts of Interest: The authors declare no conflict of interest.

References

1. Bischoff, S.C.; Bernal, W.; Dasarathy, S.; Merli, M.; Plank, L.D.; Schütz, T.; Plauth, M. ESPEN practical guideline: Clinical nutrition in liver disease. *Clin. Nutr.* **2020**, *39*, 3533–3562. [CrossRef] [PubMed]
2. Wędrychowicz, A.; Zając, A.; Tomasik, P. Advances in nutritional therapy in inflammatory bowel diseases: Review. *World J. Gastroenterol.* **2016**, *22*, 1045–1066. [CrossRef] [PubMed]
3. Kayal, M.; Shah, S. Ulcerative Colitis: Current and Emerging Treatment Strategies. *J. Clin. Med.* **2019**, *9*, 94. [CrossRef]
4. Roda, G.; Chien Ng, S.; Kotze, P.G.; Argollo, M.; Panaccione, R.; Spinelli, A.; Kaser, A.; Peyrin-Biroulet, L.; Danese, S. Crohn's disease. *Nat. Rev. Dis. Primers* **2020**, *6*, 22. [CrossRef] [PubMed]
5. Mak, W.Y.; Zhao, M.; Ng, S.C.; Burisch, J. The epidemiology of inflammatory bowel disease: East meets west. *J. Gastroenterol. Hepatol.* **2020**, *35*, 380–389. [CrossRef] [PubMed]
6. Novak, E.A.; Mollen, K.P. Mitochondrial dysfunction in inflammatory bowel disease. *Front. Cell Dev. Biol.* **2015**, *3*, 62. [CrossRef] [PubMed]
7. Sultan, S.; El-Mowafy, M.; Elgaml, A.; Ahmed, T.a.E.; Hassan, H.; Mottawea, W. Metabolic Influences of Gut Microbiota Dysbiosis on Inflammatory Bowel Disease. *Front. Physiol.* **2021**, *12*, 715506. [CrossRef]
8. Ferrari, L.; Krane, M.K.; Fichera, A. Inflammatory bowel disease surgery in the biologic era. *World J. Gastrointest. Surg.* **2016**, *8*, 363–370. [CrossRef]
9. De Conno, B.; Pesce, M.; Chiurazzi, M.; Andreozzi, M.; Rurgo, S.; Corpetti, C.; Seguella, L.; Del Re, A.; Palenca, I.; Esposito, G.; et al. Nutraceuticals and Diet Supplements in Crohn's Disease: A General Overview of the Most Promising Approaches in the Clinic. *Foods* **2022**, *11*, 1044. [CrossRef]
10. Zhang, Y.-J.; Gan, R.-Y.; Li, S.; Zhou, Y.; Li, A.-N.; Xu, D.-P.; Li, H.-B. Antioxidant Phytochemicals for the Prevention and Treatment of Chronic Diseases. *Molecules* **2015**, *20*, 21138–21156. [CrossRef]
11. Thakur, M.; Singh, K.; Khedkar, R. 11—Phytochemicals: Extraction process, safety assessment, toxicological evaluations, and regulatory issues. In *Functional and Preservative Properties of Phytochemicals*; Prakash, B., Ed.; Academic Press: Cambridge, MA, USA, 2020.
12. Campos-Vega, R.; Oomah, B.D. Chemistry and classification of phytochemicals. In *Handbook of Plant Food Phytochemicals: Sources, Stability and Extraction*; Tiwari, B.K., Brunton, N.P., Brennan, C., Eds.; Wiley-Blackwell: Chichester, UK, 2013; pp. 7–8.
13. Larussa, T.; Imeneo, M.; Luzza, F. Potential role of nutraceutical compounds in inflammatory bowel disease. *World J. Gastroenterol.* **2017**, *23*, 2483–2492. [CrossRef] [PubMed]
14. Kaulmann, A.; Bohn, T. Bioactivity of Polyphenols: Preventive and Adjuvant Strategies toward Reducing Inflammatory Bowel Diseases-Promises, Perspectives, and Pitfalls. *Oxid. Med. Cell. Longev.* **2016**, *2016*, 9346470. [CrossRef] [PubMed]
15. Hossen, I.; Hua, W.; Ting, L.; Mehmood, A.; Jingyi, S.; Duoxia, X.; Yanping, C.; Hongqing, W.; Zhipeng, G.; Kaiqi, Z.; et al. Phytochemicals and inflammatory bowel disease: A review. *Crit. Rev. Food Sci. Nutr.* **2020**, *60*, 1321–1345. [CrossRef] [PubMed]
16. Holleran, G.; Lopetuso, L.; Petito, V.; Graziani, C.; Ianiro, G.; Mcnamara, D.; Gasbarrini, A.; Scaldaferri, F. The Innate and Adaptive Immune System as Targets for Biologic Therapies in Inflammatory Bowel Disease. *Int. J. Mol. Sci.* **2017**, *18*, 2020. [CrossRef]
17. Cao, S.; Wang, C.; Yan, J.; Li, X.; Wen, J.; Hu, C. Curcumin ameliorates oxidative stress-induced intestinal barrier injury and mitochondrial damage by promoting Parkin dependent mitophagy through AMPK-TFEB signal pathway. *Free Radic. Biol. Med.* **2020**, *147*, 8–22. [CrossRef] [PubMed]
18. Sodagari, H.; Aryan, Z.; Abdolghaffari, A.H.; Rezaei, N.; Sahebkar, A. Immunomodulatory and anti-inflammatory phytochemicals for the treatment of inflammatory bowel disease (IBD). *J. Pharmacopunct.* **2018**, *21*, 294–295. [CrossRef]
19. Sameer, A.S.; Nissar, S. Toll-Like Receptors (TLRs): Structure, Functions, Signaling, and Role of Their Polymorphisms in Colorectal Cancer Susceptibility. *BioMed. Res. Int.* **2021**, *2021*, 1157023. [CrossRef]
20. Gohda, J.; Matsumura, T.; Inoue, J.-I. Cutting Edge: TNFR-Associated Factor (TRAF) 6 Is Essential for MyD88-Dependent Pathway but Not Toll/IL-1 Receptor Domain-Containing Adaptor-Inducing IFN-β (TRIF)-Dependent Pathway in TLR Signaling. *J. Immunol.* **2004**, *173*, 2913–2917. [CrossRef]
21. Amerikanou, C.; Papada, E.; Gioxari, A.; Smyrnioudis, I.; Kleftaki, S.A.; Valsamidou, E.; Bruns, V.; Banerjee, R.; Trivella, M.G.; Milic, N.; et al. Mastiha has efficacy in immune-mediated inflammatory diseases through a microRNA-155 Th17 dependent action. *Pharmacol. Res.* **2021**, *171*, 105753. [CrossRef]

22. Valdes, A.M.; Walter, J.; Segal, E.; Spector, T.D. Role of the gut microbiota in nutrition and health. *BMJ* **2018**, *361*, k2179. [CrossRef]
23. Sartor, R.B.; Mazmanian, S.K. Intestinal microbes in inflammatory bowel diseases. *Am. J. Gastroenterol. Suppl.* **2012**, *1*, 15. [CrossRef]
24. Rechner, A.R.; Smith, M.A.; Kuhnle, G.; Gibson, G.R.; Debnam, E.S.; Srai, S.K.S.; Moore, K.P.; Rice-Evans, C.A. Colonic metabolism of dietary polyphenols: Influence of structure on microbial fermentation products. *Free Rad. Biol. Med.* **2004**, *36*, 212–225. [CrossRef]
25. Santhiravel, S.; Bekhit, A.E.-D.A.; Mendis, E.; Jacobs, J.L.; Dunshea, F.R.; Rajapakse, N.; Ponnampalam, E.N. The Impact of Plant Phytochemicals on the Gut Microbiota of Humans for a Balanced Life. *Int. J. Mol. Sci.* **2022**, *23*, 8124. [CrossRef]
26. Campbell, C.L.; Yu, R.; Li, F.; Zhou, Q.; Chen, D.; Qi, C.; Yin, Y.; Sun, J. Modulation of fat metabolism and gut microbiota by resveratrol on high-fat diet-induced obese mice. *Diabetes Metab. Syndr. Obes.* **2019**, *12*, 97–107. [CrossRef]
27. Larrosa, M.; Yañéz-Gascón, M.J.; Selma, M.V.; González-Sarrías, A.; Toti, S.; Cerón, J.J.; Tomás-Barberán, F.; Dolara, P.; Espín, J.C. Effect of a low dose of dietary resveratrol on colon microbiota, inflammation and tissue damage in a DSS-induced colitis rat model. *J. Agric. Food Chem.* **2009**, *57*, 2211–2220. [CrossRef]
28. Zhuang, Y.; Huang, H.; Liu, S.; Liu, F.; Tu, Q.; Yin, Y.; He, S. Resveratrol Improves Growth Performance, Intestinal Morphology, and Microbiota Composition and Metabolism in Mice. *Front Microbiol.* **2021**, *12*, 726878. [CrossRef]
29. Wu, Z.; Huang, S.; Li, T.; Li, N.; Han, D.; Zhang, B.; Xu, Z.Z.; Zhang, S.; Pang, J.; Wang, S.; et al. Gut microbiota from green tea polyphenol-dosed mice improves intestinal epithelial homeostasis and ameliorates experimental colitis. *Microbiome* **2021**, *9*, 184. [CrossRef] [PubMed]
30. Palm, N.W.; De Zoete, M.R.; Cullen, T.W.; Barry, N.A.; Stefanowski, J.; Hao, L.; Degnan, P.H.; Hu, J.; Peter, I.; Zhang, W.; et al. Immunoglobulin A coating identifies colitogenic bacteria in inflammatory bowel disease. *Cell* **2014**, *158*, 1000–1010. [CrossRef] [PubMed]
31. Lyu, Y.; Wu, L.; Wang, F.; Shen, X.; Lin, D. Carotenoid supplementation and retinoic acid in immunoglobulin A regulation of the gut microbiota dysbiosis. *Exp. Biol. Med.* **2018**, *243*, 613–620. [CrossRef] [PubMed]
32. Pozuelo, M.; Panda, S.; Santiago, A.; Mendez, S.; Accarino, A.; Santos, J.; Guarner, F.; Azpiroz, F.; Manichanh, C. Reduction of butyrate- and methane-producing microorganisms in patients with Irritable Bowel Syndrome. *Sci. Rep.* **2015**, *5*, 12693. [CrossRef] [PubMed]
33. Peluso, I.; Romanelli, L.; Palmery, M. Interactions between prebiotics, probiotics, polyunsaturated fatty acids and polyphenols: Diet or supplementation for metabolic syndrome prevention? *Int. J. Food. Sci. Nutr.* **2014**, *65*, 259–267. [CrossRef]
34. Lin, S.C.; Cheifetz, A.S. The use of complementary and alternative medicine in patients with Inflammatory Bowel Disease. *Gastroenterol. Hepatol.* **2018**, *14*, 415–425.
35. Burge, K.; Gunasekaran, A.; Eckert, J.; Chaaban, H. Curcumin and Intestinal Inflammatory Diseases: Molecular Mechanisms of Protection. *Int. J. Mol. Sci.* **2019**, *20*, 1912. [CrossRef] [PubMed]
36. Ukil, A.; Maity, S.; Karmakar, S.; Datta, N.; Vedasiromoni, J.R.; Das, P.K. Curcumin, the major component of food flavour turmeric, reduces mucosal injury in trinitrobenzene sulphonic acid-induced colitis. *Br. J. Pharmacol.* **2003**, *139*, 209–218. [CrossRef]
37. Jian, Y.T.; Mai, G.F.; Wang, J.D.; Zhang, Y.L.; Luo, R.C.; Fang, Y.X. Preventive and therapeutic effects of NF-kappaB inhibitor curcumin in rats colitis induced by trinitrobenzene sulfonic acid. *World J. Gastroenterol.* **2005**, *11*, 1747–1752. [CrossRef]
38. National Center for Biotechnology Information. PubChem Compound Summary for CID 969516, Curcumin. Available online: https://pubchem.ncbi.nlm.nih.gov/compound/Curcumin (accessed on 31 July 2023).
39. National Center for Biotechnology Information. PubChem Compound Summary for CID 6654, alpha-PINENE. Available online: https://pubchem.ncbi.nlm.nih.gov/compound/alpha-PINENE (accessed on 31 July 2023).
40. National Center for Biotechnology Information. PubChem Compound Summary for CID 14896, beta-Pinene. Available online: https://pubchem.ncbi.nlm.nih.gov/compound/beta-Pinene (accessed on 31 July 2023).
41. National Center for Biotechnology Information. PubChem Compound Summary for CID 31253, Myrcene. Available online: https://pubchem.ncbi.nlm.nih.gov/compound/Myrcene (accessed on 31 July 2023).
42. National Center for Biotechnology Information. PubChem Compound Summary for CID 5951616, Masticadienonic acid. Available online: https://pubchem.ncbi.nlm.nih.gov/compound/Masticadienonic-acid (accessed on 31 July 2023).
43. National Center for Biotechnology Information. PubChem Compound Summary for CID 15559978, Isomasticadienonic acid. Available online: https://pubchem.ncbi.nlm.nih.gov/compound/Isomasticadienonic-acid (accessed on 31 July 2023).
44. National Center for Biotechnology Information. PubChem Compound Summary for CID 168928, beta-Boswellic acid. Available online: https://pubchem.ncbi.nlm.nih.gov/compound/beta-Boswellic-acid (accessed on 31 July 2023).
45. National Center for Biotechnology Information. PubChem Compound Summary for CID 261491, Thujone. Available online: https://pubchem.ncbi.nlm.nih.gov/compound/Thujone (accessed on 31 July 2023)
46. National Center for Biotechnology Information. PubChem Compound Summary for CID 6616, Camphene. Available online: https://pubchem.ncbi.nlm.nih.gov/compound/Camphene (accessed on 31 July 2023).
47. National Center for Biotechnology Information. PubChem Compound Summary for CID 442138, Absinthin. Available online: https://pubchem.ncbi.nlm.nih.gov/compound/Absinthin (accessed on 31 July 2023).
48. Kim, S.; Chen, J.; Cheng, T.; Gindulyte, A.; He, J.; He, S.; Li, Q.; Shoemaker, B.A.; Thiessen, P.A.; Yu, B.; et al. PubChem 2023 update. *Nucleic. Acids. Res.* **2023**, *51*, D1373–D1380. [CrossRef]

49. Holt, P.R.; Katz, S.; Kirshoff, R. Curcumin therapy in inflammatory bowel disease: A pilot study. *Dig. Dis. Sci.* **2005**, *50*, 2191–2193. [CrossRef] [PubMed]

50. Hanai, H.; Iida, T.; Takeuchi, K.; Watanabe, F.; Maruyama, Y.; Andoh, A.; Tsujikawa, T.; Fujiyama, Y.; Mitsuyama, K.; Sata, M.; et al. Curcumin maintenance therapy for ulcerative colitis: Randomized, multicenter, double-blind, placebo-controlled trial. *Clin. Gastroenterol. Hepatol.* **2006**, *4*, 1502–1506.e1. [CrossRef]

51. Suskind, D.L.; Wahbeh, G.; Burpee, T.; Cohen, M.; Christie, D.; Weber, W. Tolerability of curcumin in pediatric inflammatory bowel disease: A forced-dose titration study. *J. Pediatr. Gastroenterol. Nutr.* **2013**, *56*, 277–279. [CrossRef]

52. Singla, V.; Pratap Mouli, V.; Garg, S.K.; Rai, T.; Choudhury, B.N.; Verma, P.; Deb, R.; Tiwari, V.; Rohatgi, S.; Dhingra, R.; et al. Induction with NCB-02 (curcumin) enema for mild-to-moderate distal ulcerative colitis—A randomized, placebo-controlled, pilot study. *J. Crohns. Colitis.* **2014**, *8*, 208–214. [CrossRef]

53. Lang, A.; Salomon, N.; Wu, J.C.; Kopylov, U.; Lahat, A.; Har-Noy, O.; Ching, J.Y.; Cheong, P.K.; Avidan, B.; Gamus, D.; et al. Curcumin in Combination with Mesalamine Induces Remission in Patients with Mild-to-Moderate Ulcerative Colitis in a Randomized Controlled Trial. *Clin. Gastroenterol. Hepatol.* **2015**, *13*, 1444–1449.e1. [CrossRef] [PubMed]

54. Sadeghi, N.; Mansoori, A.; Shayesteh, A.; Hashemi, S.J. The effect of curcumin supplementation on clinical outcomes and inflammatory markers in patients with ulcerative colitis. *Phytother. Res.* **2020**, *34*, 1123–1133. [CrossRef] [PubMed]

55. Lopresti, A.L. The problem of Curcumin and its bioavailability: Could its gastrointestinal influence contribute to its overall health-enhancing effects? *Adv. Nutr.* **2018**, *9*, 41–50. [CrossRef] [PubMed]

56. Masoodi, M.; Mahdiabadi, M.A.; Mokhtare, M.; Agah, S.; Kashani, A.H.F.; Rezadoost, A.M.; Sabzikarian, M.; Talebi, A.; Sahebkar, A. The efficacy of curcuminoids in improvement of ulcerative colitis symptoms and patients' self-reported well-being: A randomized double-blind controlled trial. *J. Cell. Biochem.* **2018**, *119*, 9552–9559. [CrossRef]

57. Sugimoto, K.; Ikeya, K.; Bamba, S.; Andoh, A.; Yamasaki, H.; Mitsuyama, K.; Nasuno, M.; Tanaka, H.; Matsuura, A.; Kato, M.; et al. Highly Bioavailable Curcumin Derivative Ameliorates Crohn's Disease Symptoms: A Randomized, Double-Blind, Multicenter Study. *J. Crohns. Colitis.* **2020**, *14*, 1693–1701. [CrossRef]

58. Banerjee, R.; Pal, P.; Penmetsa, A.; Kathi, P.; Girish, G.; Goren, I.; Reddy, D.N. Novel Bioenhanced Curcumin with Mesalamine for Induction of Clinical and Endoscopic Remission in Mild-to-Moderate Ulcerative Colitis: A Randomized Double-Blind Placebo-controlled Pilot Study. *J. Clin. Gastroenterol.* **2021**, *55*, 702–708. [CrossRef]

59. Kedia, S.; Bhatia, V.; Thareja, S.; Garg, S.; Mouli, V.P.; Bopanna, S.; Tiwari, V.; Makharia, G.; Ahuja, V. Low dose oral curcumin is not effective in induction of remission in mild to moderate ulcerative colitis: Results from a randomized double blind placebo controlled trial. *World J. Gastrointest. Pharmacol. Ther.* **2017**, *8*, 147–154. [CrossRef]

60. Bommelaer, G.; Laharie, D.; Nancey, S.; Hebuterne, X.; Roblin, X.; Nachury, M.; Peyrin-Biroulet, L.; Fumery, M.; Richard, D.; Pereira, B.; et al. Oral Curcumin No More Effective Than Placebo in Preventing Recurrence of Crohn's Disease After Surgery in a Randomized Controlled Trial. *Clin. Gastroenterol. Hepatol.* **2020**, *18*, 1553–1560.e1. [CrossRef]

61. Papada, E.; Kaliora, A.C. Antioxidant and Anti-Inflammatory Properties of Mastiha: A Review of Preclinical and Clinical Studies. *Antioxidants* **2019**, *8*, 208. [CrossRef]

62. Soulaidopoulos, S.; Tsiogka, A.; Chrysohoou, C.; Lazarou, E.; Aznaouridis, K.; Doundoulakis, I.; Tyrovola, D.; Tousoulis, D.; Tsioufis, K.; Vlachopoulos, C.; et al. Overview of Chios Mastic Gum (*Pistacia lentiscus*) Effects on Human Health. *Nutrients* **2022**, *14*, 590. [CrossRef]

63. European Medicines Agency. 2015. Available online: http://www.ema.europa.eu/docs/en_GB/document_library/Herbal_ _Herbal_monograph/2015/07/WC500190099.pdf (accessed on 1 July 2023).

64. Kaliora, A.C.; Stathopoulou, M.G.; Triantafillidis, J.K.; Dedoussis, G.V.; Andrikopoulos, N.K. Chios mastic treatment of patients with active Crohn's disease. *World J. Gastroenterol.* **2007**, *13*, 748–753. [CrossRef] [PubMed]

65. Kaliora, A.C.; Stathopoulou, M.G.; Triantafillidis, J.K.; Dedoussis, G.V.; Andrikopoulos, N.K. Alterations in the function of circulating mononuclear cells derived from patients with Crohn's disease treated with mastic. *World J. Gastroenterol.* **2007**, *13*, 6031–6036.

66. Papada, E.; Forbes, A.; Amerikanou, C.; Torović, L.; Kalogeropoulos, N.; Tzavara, C.; Triantafillidis, J.K.; Kaliora, A.C. Antioxidative Efficacy of a *Pistacia Lentiscus* Supplement and Its Effect on the Plasma Amino Acid Profile in Inflammatory Bowel Disease: A Randomised, Double-Blind, Placebo-Controlled Trial. *Nutrients* **2018**, *10*, 1779. [CrossRef] [PubMed]

67. Papada, E.; Gioxari, A.; Amerikanou, C.; Forbes, A.; Tzavara, C.; Smyrnioudis, I.; Kaliora, A.C. Regulation of faecal biomarkers in inflammatory bowel disease patients treated with oral mastiha (*Pistacia lentiscus*) supplement: A double-blind and placebo-controlled randomised trial. *Phytother. Res.* **2019**, *33*, 360–369. [CrossRef]

68. Coulombe, G.; Langlois, A.; De Palma, G.; Langlois, M.-J.; Mccarville, J.L.; Gagné-Sanfaçon, J.; Perreault, N.; Feng, G.-S.; Bercik, P.; Boudreau, F.; et al. SHP-2 Phosphatase Prevents Colonic Inflammation by Controlling Secretory Cell Differentiation and Maintaining Host-Microbiota Homeostasis. *J. Cell. Physiol.* **2016**, *231*, 2529–2540. [CrossRef]

69. Papada, E.; Amerikanou, C.; Torović, L.; Kalogeropoulos, N.; Tzavara, C.; Forbes, A.; Kaliora, A.C. Plasma free amino acid profile in quiescent Inflammatory Bowel Disease patients orally administered with Mastiha (*Pistacia lentiscus*); a randomised clinical trial. *Phytomedicine* **2019**, *56*, 40–47. [CrossRef] [PubMed]

70. Amerikanou, C.; Dimitropoulou, E.; Gioxari, A.; Papada, E.; Tanaini, A.; Fotakis, C.; Zoumpoulakis, P.; Kaliora, A.C. Linking the IL-17A immune response with NMR-based faecal metabolic profile in IBD patients treated with Mastiha. *Biomed. Pharmacother.* **2021**, *138*, 111535. [CrossRef]

71. Varma, K.; Haponiuk, J.T.; Gopi, S. 7-Antiinflammatory activity of Boswellia. In *Inflammation and Natural Products*; Gopi, S., Amalraj, A., Kunnumakkara, A., Thomas, S., Eds.; Academic Press: London, UK, 2021; pp. 147–159.

72. Gerhardt, H.; Seifert, F.; Buvari, P.; Vogelsang, H.; Repges, R. Therapy of active Crohn disease with Boswellia serrata extract H 15. *Z. Gastroenterol.* **2001**, *39*, 11–17. [CrossRef]

73. Holtmeier, W.; Zeuzem, S.; Preiβ, J.; Kruis, W.; Böhm, S.; Maaser, C.; Raedler, A.; Schmidt, C.; Schnitker, J.; Schwarz, J.; et al. Randomized, placebo-controlled, double-blind trial of Boswellia serrata in maintaining remission of Crohn's disease: Good safety profile but lack of efficacy. *Inflamm. Bowel. Dis.* **2011**, *17*, 573–582. [CrossRef] [PubMed]

74. Pellegrini, L.; Milano, E.; Franceschi, F.; Belcaro, G.; Gizzi, G.; Feragalli, B.; Dugall, M.; Luzzi, R.; Togni, S.; Eggenhoffner, R.; et al. Managing ulcerative colitis in remission phase: Usefulness of Casperome®, an innovative lecithin-based delivery system of Boswellia serrata extract. *Eur. Rev. Med. Pharmacol. Sci.* **2016**, *20*, 2695–2700.

75. Szopa, A.; Pajor, J.; Klin, P.; Rzepiela, A.; Elansary, H.O.; Al-Mana, F.A.; Mattar, M.A.; Ekiert, H. *Artemisia absinthium* L.-Importance in the History of Medicine, the Latest Advances in Phytochemistry and Therapeutical, Cosmetological and Culinary Uses. *Plants* **2020**, *9*, 1063. [CrossRef] [PubMed]

76. Choi, S.C.; Choi, E.J.; Oh, H.M.; Lee, S.; Lee, J.K.; Lee, M.S.; Shin, Y.I.; Choi, S.J.; Chae, J.R.; Lee, K.M.; et al. DA-9601, a standardized extract of *Artemisia asiatica*, blocks TNF-alpha-induced IL-8 and CCL20 production by inhibiting p38 kinase and NF-kappaB pathways in human gastric epithelial cells. *World J. Gastroenterol.* **2006**, *12*, 4850–4858.

77. Hatziieremia, S.; Gray, A.I.; Ferro, V.A.; Paul, A.; Plevin, R. The effects of cardamonin on lipopolysaccharide-induced inflammatory protein production and MAP kinase and NFkappaB signalling pathways in monocytes/macrophages. *Br. J. Pharmacol.* **2006**, *149*, 188–198. [CrossRef]

78. Omer, B.; Krebs, S.; Omer, H.; Noor, T.O. Steroid-sparing effect of wormwood (*Artemisia absinthium*) in Crohn's disease: A double-blind placebo-controlled study. *Phytomedicine* **2007**, *14*, 87–95. [CrossRef]

79. Krebs, S.; Omer, T.N.; Omer, B. Wormwood (*Artemisia absinthium*) suppresses tumour necrosis factor alpha and accelerates healing in patients with Crohn's disease—A controlled clinical trial. *Phytomedicine* **2010**, *17*, 305–309. [CrossRef] [PubMed]

80. Olivares-Morales, A.; Hatley, O.J.; Turner, D.; Galetin, A.; Aarons, L.; Rostami-Hodjegan, A. The use of ROC analysis for the qualitative prediction of human oral bioavailability from animal data. *Pharm. Res.* **2014**, *31*, 720–730. [CrossRef]

81. Manach, C.; Scalbert, A.; Morand, C.; Rémésy, C.; Jiménez, L. Polyphenols: Food sources and bioavailability. *Am. J. Clin. Nutr.* **2004**, *79*, 727–747. [CrossRef]

82. Siviero, A.; Gallo, E.; Maggini, V.; Gori, L.; Mugelli, A.; Firenzuoli, F.; Vannacci, A. Curcumin, a golden spice with a low bioavailability. *J. Herb. Med.* **2015**, *5*, 57–70. [CrossRef]

83. Kesarwani, K.; Gupta, R.; Mukerjee, A. Bioavailability enhancers of herbal origin: An overview. *Asian Pac. J. Trop Biomed.* **2013**, *3*, 253–266. [CrossRef] [PubMed]

84. Johnson, I.T. Phytochemicals and health. In *Handbook of Plant Food Phytochemicals: Sources, Stability and Extraction*; Tiwari, B.K., Brunton, N.P., Brennan, C., Eds.; Wiley-Blackwell: Chichester, UK, 2013; p. 50.

85. Domínguez-Perles, R.; Auñón, D.; Ferreres, F.; Gil-Izquierdo, A. Gender differences in plasma and urine metabolites from Sprague-Dawley rats after oral administration of normal and high doses of hydroxytyrosol, hydroxytyrosol acetate, and DOPAC. *Eur. J. Nutr.* **2017**, *56*, 215–224. [CrossRef]

86. García-Villalba, R.; Larrosa, M.; Possemiers, S.; Tomás-Barberán, F.A.; Espín, J.C. Bioavailability of phenolics from an oleuropein-rich olive (*Olea europaea*) leaf extract and its acute effect on plasma antioxidant status: Comparison between pre- and post-menopausal women. *Eur. J. Nutr.* **2014**, *53*, 1015–1027. [CrossRef]

87. Papada, E.; Gioxari, A.; Brieudes, V.; Amerikanou, C.; Halabalaki, M.; Skaltsounis, A.L.; Smyrnioudis, I.; Kaliora, A.C. Bioavailability of Terpenes and Postprandial Effect on Human Antioxidant Potential. An Open-Label Study in Healthy Subjects. *Mol. Nutr. Food Res.* **2018**, *62*, 1700751. [CrossRef] [PubMed]

88. Shoba, G.; Joy, D.; Joseph, T.; Majeed, M.; Rajendran, R.; Srinivas, P.S. Influence of piperine on the pharmacokinetics of curcumin in animals and human volunteers. *Planta Med.* **1998**, *64*, 353–356. [CrossRef] [PubMed]

89. Halegoua-Demarzio, D.; Navarro, V.; Ahmad, J.; Avula, B.; Barnhart, H.; Barritt, A.S.; Bonkovsky, H.L.; Fontana, R.J.; Ghabril, M.S.; Hoofnagle, J.H.; et al. Liver Injury Associated with Turmeric-A Growing Problem: Ten Cases from the Drug-Induced Liver Injury Network [DILIN]. *Am. J. Med.* **2023**, *136*, 200–206. [CrossRef] [PubMed]

MDPI

St. Alban-Anlage 66

4052 Basel

Switzerland

www.mdpi.com

Life Editorial Office

E-mail: life@mdpi.com

www.mdpi.com/journal/life